Open Problems in Algebraic Combinatorics

Proceedings of Symposia in
PURE MATHEMATICS

Volume 110

Open Problems in Algebraic Combinatorics

Conference
Open Problems in Algebraic Combinatorics
May 16–22, 2022
University of Minnesota, Twin Cities
Minneapolis, Minnesota

Christine Berkesch
Benjamin Brubaker
Gregg Musiker
Pavlo Pylyavskyy
Victor Reiner
Editors

AMS AMERICAN MATHEMATICAL SOCIETY
Providence, Rhode Island

2020 *Mathematics Subject Classification.* Primary 05E05, 13F60, 52B05, 82B20, 14N15.

Library of Congress Cataloging-in-Publication Data

Names: Berkesch, Christine, 1982– editor. | Brubaker, Ben, 1976– editor. | Musiker, Gregg, editor. | Pylyavskyy, Pavlo, 1982– editor. | Reiner, Victor, 1965– editor.

Title: Open problems in algebraic combinatorics / Christine Berkesch, Benjamin Brubaker, Gregg Musiker, Pavlo Pylyavskyy, Victor Reiner, editors.

Description: Providence, Rhode Island : American Mathematical Society, [2024] | Series: Proceedings of symposia in pure mathematics, 0082-0717 ; volume 110 | Includes bibliographical references.

Identifiers: LCCN 2024011846 | ISBN 9781470473334 (paperback) | ISBN 9781470477974 (ebook)

Subjects: LCSH: Combinatorial analysis. | AMS: Combinatorics – Algebraic combinatorics – Symmetric functions and generalizations. | Commutative algebra – Arithmetic rings and other special rings – Cluster algebras. | Convex and discrete geometry – Polytopes and polyhedra – Combinatorial properties (number of faces, shortest paths, etc.). | Statistical mechanics, structure of matter – Equilibrium statistical mechanics – Lattice systems (Ising, dimer, Potts, etc.) and systems on graphs. | Algebraic geometry – Projective and enumerative geometry – Classical problems, Schubert calculus.

Classification: LCC QA164 .O64 2024 | DDC 511/.6–dc23/eng/20240412

LC record available at https://lccn.loc.gov/2024011846

∞ The paper used in this book is acid-free and falls within the guidelines
established to ensure permanence and durability.
Visit the AMS home page at `https://www.ams.org/`

10 9 8 7 6 5 4 3 2 1   29 28 27 26 25 24

Contents

Preface

What is algebraic combinatorics? This is not an easy question, as modern algebraic combinatorics encompasses a huge swath of topics. Two key features come to mind:

- Concern with exact, as opposed to approximate, results.
- Emphasis on interaction with other areas of mathematics, such as algebra, topology, geometry, and physics.

Just as many fruitful and beautiful ideas in mathematics had their genesis in physics and the sciences, combinatorialists have recognized that many of their most beloved, rich, and well-studied combinatorial notions are avatars of creatures inhabiting these other fields. Progress in algebraic combinatorics is often achieved by extracting the elementary core from these concepts arising elsewhere, then analyzing and systematizing them. They then acquire a life of their own, woven into the fabric of algebraic combinatorics. This has happened to symmetric functions, monomial ideals, Coxeter groups, tropical geometry, crystal graphs, cluster algebras, and many other beautiful topics.

This is both a blessing and a curse. Mathematicians in other areas can convince the world that their work is important by appealing to a plethora of internally motivated conjectures, many open for decades. By contrast, algebraic combinatorialists often find themselves discovering something completely new, perhaps motivated by another area, yet maybe a bit foreign to outsiders in its methods and emphasis. There are notable exceptions of course, as there exists a (rather short) list of famous open problems and conjectures in the area. Some get solved, and some seem to remain open forever, but the overall number appears small.

The idea of the Open Problems in Algebraic Combinatorics conference, its associated blog, and this volume, is to help people working in the area tell each other about the best open problems and conjectures encountered in their research. We believe that the creation of such a forum is useful, especially to young researchers. Algebraic combinatorics remains a vibrant area, with new fascinating discoveries happening all the time—some you will find in this volume. We are very grateful to all the contributors of this volume—as well as to all the speakers of the original conference—for generously sharing their ideas. We hope that this publication will help our beloved area become even more lively.

Christine Berkesch
Benjamin Brubaker
Gregg Musiker
Pavlo Pylyavskyy
Victor Reiner
Editors

Proceedings of Symposia in Pure Mathematics
Volume **110**, 2024
https://doi.org/10.1090/pspum/110/02011

$(\mathrm{GL}_k \times \mathbb{S}_n)$-modules of multivariate diagonal harmonics

François Bergeron

ABSTRACT. This is the first in a series of papers in which we describe explicit structural properties of spaces of diagonal rectangular harmonic polynomials in k sets of n variables, both as GL_k-modules and $\mathbb{S}_n$-modules, as well as some of there relations to areas such as Algebraic Combinatorics, Representation Theory, Algebraic Geometry, Knot Theory, and Theoretical Physics. Our global aim is to develop a unifying point of view for several areas of research of the last two decades having to do with Macdonald Polynomials Operator Theory, Diagonal Coinvariant Spaces, Rectangular-Catalan Combinatorics, the Delta-Conjecture, Hilbert Scheme of Points in the Plane, Khovanov-Rozansky Homology of (m, n)-Torus links, etc.

CONTENTS

1. Introduction

Initiated at the beginning of the 1990s, the work of Garsia and Haiman [**15**] relating to Macdonald polynomials sparked a strong and long-lasting interest in modules of "diagonal harmonic polynomials" in two sets of variables, whose overall dimension is $(n+1)^{n-1}$. Links to parking functions on the combinatorial side, and Hilbert Schemes of points in the plane [**18, 19**] on the Algebraic Geometry one, were soon established. These links have become much better understood in recent years. Since its inception, this line of inquiry has opened many fruitful areas of research, including recent ties with Khovanov-Rozansky homology of (m, n)-torus links (see [**17, 21–23, 29**]), and ties to the study of super-spaces of Bosons-Fermions. A significant part of this story involves a formula for the (bi)graded character of the above-mentioned modules in the form $\nabla(e_n)$, where ∇ is an operator (introduced in [**4**]) having the (combinatorial) Macdonald symmetric functions as joint eigenfunctions, here applied to the degree n elementary symmetric function e_n.

2020 *Mathematics Subject Classification.* Primary 05E05, 05E10; Secondary 20C32.

This operator has many interesting properties on its own, and several important questions about it are still being actively investigated (see [**8**, **9**, **21**, **24**]).

Our aim is to describe several structural properties of the ($\mathrm{GL}_k \times \mathbb{S}_n$)-module, denoted by $\mathcal{E}_n^{\langle k\rangle}$, of k-variate diagonal harmonic polynomials (previously studied in [**7**]), which specialize at $k=2$ to the module of diagonal harmonic polynomials of Garsia and Haiman. As discussed in [**7**], there is a filtration of $\mathbb{S}_n$-module (over the field $\mathbb{Q}$)

$$\text{(1.1)} \qquad \mathbb{Q} = \mathcal{E}_n^{\langle 0\rangle} \hookrightarrow \mathcal{E}_n^{\langle 1\rangle} \hookrightarrow \mathcal{E}_n^{\langle 2\rangle} \hookrightarrow \cdots\cdots \hookrightarrow \mathcal{E}_n^{\langle k\rangle} \hookrightarrow \cdots\cdots \hookrightarrow \mathcal{E}_n := \lim_{k\to\infty} \mathcal{E}_n^{\langle k\rangle},$$

which "stabilizes" when k becomes large enough.

The first non-trivial module, *i.e.* $\mathcal{E}_n^{\langle 1\rangle}$, of this filtration plays a crucial role in several subjects, up to natural isomorphisms. Under different guises, it appears as the cohomology ring of the full-flag manifold, as the coinvariant space of the symmetric group $\mathbb{S}_n$, as well as the space of $\mathbb{S}_n$-harmonic polynomials. These are well-known results going back to the 1950s (see [**26**]). This module carries a graded version of the regular representation of $\mathbb{S}_n$, and it may simply be described as the linear span of all derivatives (to all orders) of the Vandermonde determinant $\boldsymbol{V}(x)$ in the variables $x=(x_1,\ldots,x_n)$. Among several interesting bases, one of the simplest is the set of partial derivatives $\partial x^{\boldsymbol{d}}\boldsymbol{V}(x)$, for $\boldsymbol{d}=(d_1,\ldots,d_n)\in\mathbb{N}^n$ such that $0\le d_k\le k-1$, where $\partial x^{\boldsymbol{d}} := \partial x_1^{d_1}\cdots\partial x_n^{d_n}$. From this, it may be seen that its graded dimension is the q-analogue of $n!$. In fact, the corresponding graded irreducible decomposition as a $\mathbb{S}_n$-module is entirely encoded in terms of the Macdonald polynomial $\widetilde{H}_n(q;\boldsymbol{z})$, in variables $\boldsymbol{z}=(z_i)_{i\in\mathbb{N}}$ (discussed further in the sequel).

The work of Garsia-Haiman, on the case $k=2$, is seminal in this whole line of investigation. The corresponding space, denoted by $\mathcal{E}_n^{\langle 2\rangle}$, may be obtained from $\mathcal{E}_n^{\langle 1\rangle}$ by further closing with respect to **higher polarization** operators: $\sum_{i=1}^n y_i\partial x_i^j$, for any $j\ge 1$ (as well as similar operators obtained by exchanging the role of x and y). Since the early 1990s, many new lines of investigation have been added to their original framework. Noteworthy among recent developments are the successive appearances: of "rational", and "rectangular" Catalan combinatorics (see [**1**–**3**, **8**]); as well as interesting ties between these combinatorial settings and the elliptic Hall algebra, introduced by Burban-Schiffmann (see [**12**]). This last context offers a broad extension to the spectrum of operators first considered in [**5**], particularly those that appear in Theorem 4.4 therein.

In parallel, an explicit study of $\mathcal{E}_n^{\langle 3\rangle}$ (obtained by closing $\mathcal{E}_n^{\langle 2\rangle}$ with respect to polarization operators involving a third set of variables) was started about 10 years ago. Most fundamental questions about it remain open, but its study has suggested new (hard to prove) combinatorial identities linked to the Tamari Lattice as discussed in [**10**, **11**]. The general k-framework is also considered in a previous paper [**7**], where some broad properties of the modules of k-variate diagonal coinvariant (for any finite complex reflection groups), considering the inductive limit $\mathcal{E}_n := \lim_{k\to\infty}\mathcal{E}_n^{\langle k\rangle}$ as a $\mathrm{GL}_\infty \times\mathbb{S}_n$-module (with commuting actions), and its decomposition:

$$\text{(1.2)} \qquad \mathcal{E}_n \simeq \bigoplus_{\mu\vdash n}\bigoplus_{\lambda} (\mathcal{W}_\lambda\otimes\mathcal{V}_\mu)^{\oplus c_{\lambda,\mu}},$$

into GL_∞-polynomial-irreducibles (the $\mathcal{W}_\lambda$'s), and $\mathbb{S}_n$-irreducibles ($\mathcal{V}_\mu$'s).

A crucial missing part in our previous work on the general case was an explicit description of this irreducible decomposition of $\mathcal{E}_n$ (the values of the $c_{\lambda,\mu}$). We are currently going to describe the structure of this decomposition, as see that it is characterized by a strikingly small set of data. This involves a precise and explicit link between various $\mathbb{S}_n$-isotypic components of the modules under study, established in part via Vertex Operators (see [**30**]). A surprising corollary of our approach is that we can reconstruct the alternating component of $\mathcal{E}_n$ from the only knowledge of (hook-shape components) of the module $\mathcal{E}_n^{\langle k\rangle}$, with $k = \lfloor\frac{n-1}{2}\rfloor$ (rather than having to calculate up to $k = n-1$).

Many seemingly independent aspects of the theory for $k = 2$ are nicely explained and tied together via general properties of the $\mathcal{E}_n$, and their decomposition into irreducible components. As we will see most of these ties cannot be explained if one stays in the restricted context of $k = 2$. In other terms, it is only by going to the general stable framework that the simplicity of the underlying structure is revealed. In particular, we establish a surprising connection with the "Delta operators" Δ_{e_a} (see [**5**]), which generalize the ∇ operator, shedding new light on an open conjecture of Haglund-Remmel-Wilson (see [**13, 16**]). Indeed, one of our main conjecture states that

CONJECTURE 1 (Delta-by-Skewing). *For $n = a + b + 1$, we have*

$$\Delta'_{e_a}\, e_n(\boldsymbol{z}) = ((e_b^{\perp} \otimes \mathrm{Id})\, \mathcal{E}_n)(q, t; \boldsymbol{z}). \tag{1.3}$$

Let us underline that, in parsing the right-hand side of Equation 1.3 (see subsection 2.2 for details), one must be careful to consider that $\mathcal{E}_n$ stands for an expression of the form given by Equation 2.3, encoding a $\mathrm{GL}_\infty \times \mathbb{S}_n$-Frobenius. It expands as a linear combination of formal tensors $s_\lambda \otimes s_\mu$, to which it makes sense to apply an operator such as $(e_b^{\perp} \otimes \mathrm{Id})$. Hence, it is only after computing the formal expression $((e_b^{\perp} \otimes \mathrm{Id})\, \mathcal{E}_n)$, that one may evaluate it in the explicit parameters (q, t) and the variables $\boldsymbol{z}$.

It is further noteworthy that Conjecture 1 establishes a connection with the "Superspace" of Bosons-Fermions, the study of which is ongoing. In a nutshell, the generic Frobenius characteristic of the superspace is conjectured to be given by the plethysm $\mathcal{E}_n[\boldsymbol{q} - \varepsilon \boldsymbol{u}; \boldsymbol{z}]$. See the appendix for plethysms involving "ε", and in particular Equation 6.4.

2. Spaces of multivariate diagonal harmonic polynomials

Let us set up our context and recall some classical notions (see also the appendix). Let $\boldsymbol{x}$ be a $\infty \times n$ matrix of variables:

$$\boldsymbol{x} = \begin{pmatrix} x_{11} & x_{12} & \cdots & x_{1n} \\ x_{21} & x_{22} & \cdots & x_{2n} \\ \vdots & \vdots & \ddots & \vdots \end{pmatrix}.$$

We think of each row $(x_{i1}, x_{i2}, \ldots, x_{in})$ of $\boldsymbol{x}$ as a "set of n variables". We sometimes restrict $\boldsymbol{x}$ to its first k rows, thus getting a $k \times n$ matrix denoted by $\boldsymbol{x}_{\langle k\rangle}$; and then consider the ring $\mathcal{R}_n^{\langle k\rangle} := \mathbb{Q}[\boldsymbol{x}_{\langle k\rangle}]$, of polynomials in the variables $\boldsymbol{x}_{\langle k\rangle}$. Letting k go to infinity, we get the limit ring $\mathcal{R}_n := \mathbb{Q}[\boldsymbol{x}]$ of polynomials in the variables $\boldsymbol{x}$. To each $\infty \times n$ matrix of positive integers $\alpha = (a_{ij})$, of finite support, we associate the **monomial**

$$\boldsymbol{x}^\alpha := \prod_{ij} x_{ij}^{a_{ij}},$$

in $\mathcal{R}_n$. The **ith-degree** of $\boldsymbol{x}^\alpha$ is the sum of the entries lying on row i of α:

$$\deg_i(\boldsymbol{x}^\alpha) := a_{i1} + a_{i2} + \ldots + a_{in}.$$

It follows from the constraint on α, that the **degree** of $\boldsymbol{x}^\alpha$, defined as

$$\deg(\boldsymbol{x}^\alpha) := (\deg_1(\boldsymbol{x}^\alpha), \deg_2(\boldsymbol{x}^\alpha), \ldots, \deg_i(\boldsymbol{x}^\alpha), \ldots),$$

lies in the semi-group $\mathbb{N}^\infty$ of **finite support sequences of integers**. In this way, one gets two gradings, one by degree and the other by **total degree**, on the ring $\mathcal{R} = \mathcal{R}_n$

$$\mathcal{R} = \bigoplus_{\boldsymbol{d}\in\mathbb{N}^\infty} \mathcal{R}_{\boldsymbol{d}}, \qquad \text{and} \qquad \mathcal{R} = \bigoplus_{j\in\mathbb{N}} \mathcal{R}^{(j)},$$

with **homogeneous components** $\mathcal{R}_{\boldsymbol{d}}$ and $\mathcal{R}^{(j)}$, respectively spanned by the monomials $\boldsymbol{x}^\alpha$ having **row sum sequence** equal to $\boldsymbol{d} = (d_1, d_2, \ldots)$, and by monomials of total degree $j = d_1 + d_2 + \cdots = |\boldsymbol{d}|$. In other terms,

$$\mathcal{R}^{(j)} := \bigoplus_{|\boldsymbol{d}|=j} \mathcal{R}_{\boldsymbol{d}}.$$

We denote by $\rho(\alpha)$ the row sum sequence $(d_1, d_2, \ldots)$, where $d_i := a_{i1}+a_{i2}+\ldots+a_{in}$. The $\mathcal{R}_{\boldsymbol{d}}$' are clearly such that $\mathcal{R}_{\boldsymbol{d}}\mathcal{R}_{\boldsymbol{d}'} \subseteq \mathcal{R}_{\boldsymbol{d}+\boldsymbol{d}'}$. For a given $\boldsymbol{d} \in \mathbb{N}^\infty$, a polynomial $F(\boldsymbol{x})$ lies in $\mathcal{R}_{(\boldsymbol{d})}$ if and only if $F(\boldsymbol{q}\cdot\boldsymbol{x}) = \boldsymbol{q}^{\boldsymbol{d}} F(\boldsymbol{x})$, with $\boldsymbol{q}$ standing for a diagonal matrix $\mathrm{diag}(q_1, q_2, \ldots, \ldots)$ of formal "parameters" q_i; writing $\boldsymbol{q}^{\boldsymbol{d}} := \prod_i q_i^{d_1}$, for any $\boldsymbol{d}$ in $\mathbb{N}^\infty$.

2.1. $(\mathrm{GL}_k \times \mathbb{S}_n)$**-Modules.** The ring $\mathcal{R} = \mathcal{R}_n$ (or $\mathcal{R} = \mathcal{R}_n^{\langle k\rangle}$) is equipped with two commuting actions, one of $\mathbb{S}_n$ and one of the group[1] GL_k, for $k \in \mathbb{N}\cup\{\infty\}$, which may be jointly defined by

$$F(\boldsymbol{x}) \longmapsto F(g\cdot\boldsymbol{x}\cdot\sigma), \quad \text{for} \quad g \in \mathrm{GL}_k \quad \text{and} \quad \sigma\in\mathbb{S}_n.$$

Here, elements of $\mathbb{S}_n$ are considered as $n\times n$ permutation matrices. In particular, the above turns $\mathcal{R}$ into a polynomial representations of GL_k. The $\mathbb{S}_n$-action preserves the grading, so that each homogenous component $\mathcal{R}_{\boldsymbol{d}}$ inherits the S_n action. The GL_k-action preserves total degree, so that $\mathcal{R}^{(j)}$ inherits the whole $(\mathrm{GL}_k \times \mathbb{S}_n)$-action.

All the modules $\mathcal{M}$ that we are going to consider in the sequel are homogeneous $(\mathrm{GL}_k \times \mathbb{S}_n)$-submodules of $\mathcal{R}$, and polynomial representations of GL_k (with k in $\mathbb{N}\cup\{\infty\}$). They thus afford a decomposition, graded by degree, on the form

$$\mathcal{M} = \bigoplus_{\boldsymbol{d}\in\mathbb{N}^k} \mathcal{M}_{\boldsymbol{d}}, \qquad \text{with} \qquad \mathcal{M}_{\boldsymbol{d}} := \mathcal{M}\cap\mathcal{R}_{\boldsymbol{d}}.$$

Each homogenous component $\mathcal{M}_{\boldsymbol{d}}$ affords some basis $\mathcal{B}_{\boldsymbol{d}}$, of homogeneous polynomials. The GL_k**-character** of $\mathcal{M}$ (it also said to be its **Hilbert series**) may then simply calculated as

$$\mathcal{M}(\boldsymbol{q}) := \sum_{\boldsymbol{d}\in\mathbb{N}^k} \dim(\mathcal{M}_{\boldsymbol{d}})\, \boldsymbol{q}^{\boldsymbol{d}}.$$

[1] Recall that, for $k=\infty$, this is the group of $\mathbb{N}\times n$ matrices that coincide with the identity matrix in all rows rows but a finite number.

Classical results the representation theory of GL_∞ ensure that these characters are symmetric function of the variables $\boldsymbol{q} = (q_1, q_2, \ldots)$, which expand as positive integer linear combinations of Schur functions $s_\lambda(\boldsymbol{q}) = s_\lambda(q_1, q_2, \ldots)$, indexed by partition λ. Recall that positive integral linear combination of Schur functions are said to be **Schur positive**. Also recall that it is usual to encode characters of representations of $\mathbb{S}_n$ as symmetric functions[2] via their **Frobenius transform**. In this encoding, irreducible $\mathbb{S}_n$-modules correspond to Schur functions $s_\mu(\boldsymbol{z})$, for μ partitions of n. Summing up, the $(\mathrm{GL}_\infty \times \mathbb{S}_n)$-modules $\mathcal{M}$ we consider, have decompositions that can be presented in the format

$$\mathcal{M}(\boldsymbol{q};\boldsymbol{z}) = \sum_{\mu\vdash n}\sum_{\lambda} a_{\lambda\mu} s_\lambda(\boldsymbol{q}) s_\mu(\boldsymbol{z}), \tag{2.1}$$

with the $a_{\lambda\mu}$ standing for multiplicities of $(\mathrm{GL}_\infty \times \mathbb{S}_n)$-irreducibles. Observe that the "restriction" to variables in $\boldsymbol{X}_k$ can simply be obtained by setting $q_i = 0$ for all $i > k$.

2.2. Formal tensors. We will use the following "Schur-$\otimes$-Schur" format to express our formulas for characters

$$\mathcal{M} = \sum_{\mu\vdash n}\sum_{\lambda} a_{\lambda\mu} s_\lambda \otimes s_\mu, \tag{2.2}$$

with a "formal" tensor product only used here to distinguish between Schur functions that are characters of GL_∞ (those on left-hand side), and Schur functions that encode $\mathbb{S}_n$-irreducibles (those on right-hand side). To go back to Equation 2.1, one specializes–evaluates– each of the tensor products $s_\lambda \otimes s_\mu$ in Equation 2.2 to $s_\lambda(\boldsymbol{q})s_\mu(\boldsymbol{z})$ with $\boldsymbol{q} = (q_1, q_2, \ldots, q_k)$ and $\boldsymbol{z} = (z_1, z_2, \ldots)$. It seems entirely natural to denote by $\mathcal{M}(\boldsymbol{q};\boldsymbol{z})$ this specialization process. In the particular case when $k = 2$, we typically set $\boldsymbol{q} = (q, t)$, and rather write $\mathcal{M}(q, t; \boldsymbol{z})$. In many instances, a formal tensor product expression is first calculated for $\mathcal{M}$, which is afterward evaluated in (q, t). Recall that this makes it so that some Schur functions will vanish, if they are indexed by partitions with more than two parts.

2.3. Diagonal harmonic polynomials. As alluded to in the introduction, the modules that we consider are obtained as follows. Starting with the classical Vandermonde determinant $\boldsymbol{V}_n = \boldsymbol{V}_n(\boldsymbol{x})$, in the variables $\boldsymbol{x} = (x_1, \ldots, x_n)$ (the first row of $\boldsymbol{X}$), we consider the smallest graded $\mathbb{S}_n$-submodule $\mathcal{E}_n^{\langle k\rangle}$ of $\mathcal{R}_n^{\langle k\rangle}$ which contains $\boldsymbol{V}_n$, and is closed under:

- partial derivatives with respect to any variables in $\boldsymbol{X}_k$ (the first k rows of $\boldsymbol{X}$); as well as
- **higher polarization** operators $\sum_{i=1}^n u_i \partial v_i^j$, for any pair or rows $(u_i)_i$ and $(v_i)_i$ of $\boldsymbol{X}_k$, and any $j \geq 1$.

The fact that $\mathcal{E}_n^{\langle k\rangle}$ is closed under polarization ensures that it is also a GL_k-submodule of $\mathcal{R}_n^{\langle k\rangle}$. The filtration in Equation 1.1 is compatible with the action $\mathbb{S}_n$; and with the restriction from GL_k to GL_{k-1}, which corresponds to restriction to polynomials in the variables $\boldsymbol{X}_{k-1}$ (clearly contained in $\boldsymbol{X}_k$). Hence, we get an action of $\mathrm{GL}_\infty \times \mathbb{S}_n$ on $\mathcal{E}_n$, by passing to the associated (stable) limit. It has

[2]This is just to say that we have polynomials in a denumerable set of "abstract" variables $\boldsymbol{z} = (z_i)_{i\in\mathbb{N}}$.

been shown in [7] that one reaches stability at $k = n - 1$. This implies that the multiplicities $c_{\lambda\mu}$, in the "Schur-⊗-Schur" character (in the format of Equation 2.2)

$$\mathcal{E}_n = \sum_{\mu \vdash n} \sum_{\lambda} c_{\lambda\mu} s_\lambda \otimes s_\mu, \tag{2.3}$$

are non-vanishing only for partitions λ having at most $n-1$ parts, and having size (sum of parts) at most $\binom{n}{2}$. Each $\mathbb{S}_n$-isotypic components of type μ affords the structure of a GL_∞-module. We denote by $\boldsymbol{c}_\mu := \sum_\lambda c_{\lambda\mu}\, s_\lambda$ the corresponding GL_∞-character, and say that this is the **coefficient** of $s_\mu(\boldsymbol{z})$ in $\mathcal{E}_n(\boldsymbol{q};\boldsymbol{z})$. The character of $\mathcal{E}_n^{\langle k\rangle}$ is readily obtained from $\mathcal{E}_n$ as the evaluation:

$$\mathcal{E}_n(\boldsymbol{q};\boldsymbol{z}) = \sum_{\mu \vdash n} \boldsymbol{c}_\mu(\boldsymbol{q}) s_\mu(\boldsymbol{z}), \tag{2.4}$$

in k parameters $\boldsymbol{q} = q_1, \dots, q_k$ (thus k is specified), and formal variables $\boldsymbol{z} = (z_i)_{i\in\mathbb{N}}$. In particular,

$$\mathcal{E}_n(q;\boldsymbol{z}) = \frac{h_n[\boldsymbol{z}/(1-q)]}{h_n[1/(1-q)]}, \qquad \text{and} \qquad \mathcal{E}_n(q,t;\boldsymbol{z}) = \nabla(e_n)(q,t;\boldsymbol{z}). \tag{2.5}$$

For $n \le 5$, we have the following explicit values for $\mathcal{E}_n$:

$$\begin{aligned}
\mathcal{E}_0 &= 1 \otimes 1,\\
\mathcal{E}_1 &= 1 \otimes s_1,\\
\mathcal{E}_2 &= 1 \otimes s_2 + s_1 \otimes s_{11},\\
\mathcal{E}_3 &= 1 \otimes s_3 + (s_1 + s_2) \otimes s_{21} + (s_{11} + s_3) \otimes s_{111},\\
\mathcal{E}_4 &= 1 \otimes s_4 + (s_1 + s_2 + s_3) \otimes s_{31} + (s_{21} + s_2 + s_4) \otimes s_{22}\\
&\qquad + (s_{11} + s_{21} + s_{31} + s_3 + s_4 + s_5) \otimes s_{211} + (s_{111} + s_{31} + s_{41} + s_6) \otimes s_{1111},\\
\mathcal{E}_5 &= 1 \otimes s_5 + (s_1 + s_2 + s_3 + s_4) \otimes s_{41}\\
&\qquad\quad + (s_{22} + s_{21} + s_{31} + s_{41} + s_2 + s_3 + s_4 + s_5 + s_6) \otimes s_{32}\\
&\qquad + (s_{32} + s_{11} + s_{21} + 2s_{31} + s_{41} + s_{51} + s_3 + s_4 + 2s_5 + s_6 + s_7) \otimes s_{311}\\
&\qquad + (s_{211} + s_{311} + s_{22} + s_{32} + s_{42}\\
&\qquad\qquad + s_{21} + s_{31} + 2s_{41} + 2s_{51} + s_{61} + s_4 + s_5 + s_6 + s_7 + s_8) \otimes s_{221}\\
&\qquad + (s_{111} + s_{211} + s_{311} + s_{411} + s_{33} + s_{32} + s_{42} + s_{52}\\
&\qquad\qquad + s_{31} + 2s_{41} + 2s_{51} + 2s_{61} + s_6 + s_7 + s_{71} + s_8 + s_9) \otimes s_{2111}\\
&\qquad + (s_{1111} + s_{311} + s_{411} + s_{511}\\
&\qquad\qquad + s_{43} + s_{42} + s_{62} + s_{61} + s_{71} + s_{81} + s_{(10)}) \otimes s_{11111}.
\end{aligned}$$

We present in a similar format the expansions of $\nabla(e_n)$ (which corresponds to the restriction of $\mathcal{E}_n$ to the λ's that have at most 2 parts). For sure, these expressions for $\nabla(e_n)$ may be directly calculated from their "usual" expansion as linear combination of Schur function with coefficients in $\mathbb{N}[q,t]$, simply by rewriting these coefficients (which are symmetric polynomials in q and t) in terms of Schur functions. For instance, on rewrites

$$\nabla(e_3)(q,t;\boldsymbol{z}) = s_3(\boldsymbol{z}) + (q + t + q^2 + qt + t^2) s_{21}(\boldsymbol{z}) + (qt + q^3 + q^2t + qt^2 + t^3) s_{111}(\boldsymbol{z})$$

as $\nabla(e_3) = 1 \otimes s_3 + (s_1 + s_2) \otimes s_{21} + (s_{11} + s_3) \otimes s_{111}$. This allows us to express $\mathcal{E}_6$ in the following reasonably compact manner, assuming that the Schur-⊗-Schur expansion of $\nabla(e_6)$ is known.

$$
\begin{aligned}
\mathcal{E}_6 = \nabla(e_6) &+ (s_{221} + s_{411}) \otimes s_{33} \\
&+ (s_{221} + 2s_{321} + s_{421} + s_{211} + 2s_{311} + 2s_{411} + 2s_{511} + s_{611}) \otimes s_{321} \\
&+ (s_{331} + s_{321} + s_{421} + s_{521} + s_{111} + s_{211} + 2s_{311} + 2s_{411} + 2s_{511} + s_{611} + s_{711}) \otimes s_{3111} \\
&+ (s_{3111} + s_{331} + s_{221} + s_{321} + s_{421} + s_{521} + s_{311} + s_{411} + 2s_{511} + s_{611} + s_{711}) \otimes s_{222} \\
&+ (s_{2111} + s_{3111} + s_{4111} + s_{331} + s_{431} + s_{221} + 2s_{321} + 3s_{421} + 2s_{521} + s_{621} \\
&\qquad + s_{211} + s_{311} + 3s_{411} + 3s_{511} + 4s_{611} + 2s_{711} + s_{811}) \otimes s_{2211} \\
&+ (s_{1111} + s_{2111} + s_{3111} + s_{4111} + s_{5111} + s_{331} + 2s_{431} + s_{531} + s_{321} + 2s_{421} + 2s_{521} \\
&\qquad + 2s_{621} + s_{721} + s_{311} + 2s_{411} + 3s_{511} + 3s_{611} + 3s_{711} + 2s_{811} + s_{911}) \otimes s_{21111} \\
&+ (s_{11111} + s_{3111} + s_{4111} + s_{5111} + s_{6111} + s_{421} + s_{431} + s_{441} + s_{521} + s_{531} \\
&\qquad + s_{611} + s_{621} + s_{631} + s_{711} + s_{721} + 2s_{811} + s_{821} + s_{911} + s_{(10,1,1)}) \otimes s_{111111}.
\end{aligned}
$$

2.4. Numerical Specializations. We may evaluate s_μ at $k \in \mathbb{N}$, using the well-known formula

$$s_\mu(k) = s_\mu(\underbrace{1, 1, \ldots, 1}_{k \text{ copies}}, 0, \ldots) = \textstyle\prod_{(i,j)\in\mu} \frac{k+(j-i)}{h(i,j)}, \tag{2.6}$$

where (i, j) runs over the set of cells of μ, and $h(i, j)$ stands for the associated **hook length**. The stableness of Equation 1.1 (discussed in the sequel) implies that the (ungraded) Frobenius characteristic of the $\mathbb{S}_n$-module $\mathcal{E}_n^{\langle k\rangle}$, obtained as $\mathcal{E}_n(k; \boldsymbol{z})$, is polynomial in the parameter k for any given n. For instance, $\mathcal{E}_2(k; \boldsymbol{z}) = k\, s_{11}(\boldsymbol{z}) + s_2(\boldsymbol{z})$, and

$$
\begin{aligned}
\mathcal{E}_3(k; \boldsymbol{z}) &= \left(\tbinom{k+2}{3} + \tbinom{k}{2}\right) s_{111}(\boldsymbol{z}) + \left(\tbinom{k+1}{2} + \tbinom{k}{1}\right) s_{21}(\boldsymbol{z}) + s_3(\boldsymbol{z}) \\
&= \left(\tbinom{k+1}{3} + \tbinom{k-1}{2}\right) e_3(\boldsymbol{z}) + \left(\tbinom{k}{2} + 2\tbinom{k-1}{1}\right) e_{21}(\boldsymbol{z}) + e_{111}(\boldsymbol{z}); \\
\mathcal{E}_4(k; \boldsymbol{z}) &= \left(\tbinom{k+2}{3} + \tbinom{k}{2}\right) s_{111}(\boldsymbol{z}) + \left(\tbinom{k+1}{2} + \tbinom{k}{1}\right) s_{21}(\boldsymbol{z}) + s_3(\boldsymbol{z})
\end{aligned}
$$

Using the fact that $\dim \mathcal{E}_n^{\langle k\rangle} := \langle \mathcal{E}_n(k; \boldsymbol{z}), p_1(\boldsymbol{z})^n\rangle$ and $\dim \mathcal{A}_n^{\langle k\rangle} := \langle \mathcal{E}_n(k; \boldsymbol{z}), e_n(\boldsymbol{z})\rangle$, we deduce from the above that the dimensions of $\mathcal{E}_n^{\langle k\rangle}$ and $\mathcal{A}_n^{\langle k\rangle}$ are also polynomial in k, for any given n. For instance, we have the following positive integer linear combinations of binomial coefficient polynomials

$$
\begin{aligned}
&\dim \mathcal{E}_2^{\langle k\rangle} = k + 1, && \dim \mathcal{A}_2^{\langle k\rangle} = k; \\
&\dim \mathcal{E}_3^{\langle k\rangle} = \tbinom{k}{3} + 5\tbinom{k}{2} + 5\,k + 1, && \dim \mathcal{A}_3^{\langle k\rangle} = \tbinom{k}{3} + 3\tbinom{k}{2} + k; \\
&\dim \mathcal{E}_4^{\langle k\rangle} = \tbinom{k}{6} + 12\tbinom{k}{5} + 51\tbinom{k}{4} + 96\tbinom{k}{3} + 78\tbinom{k}{2} + 23\,k + 1, \\
&\dim \mathcal{A}_4^{\langle k\rangle} = \tbinom{k}{6} + 9\tbinom{k}{5} + 25\tbinom{k}{4} + 29\tbinom{k}{3} + 12\tbinom{k}{2} + k.
\end{aligned}
$$

It would be nice to have an explicit combinatorial understanding of these expressions in general. As alluded to in the introduction, we have explicit (or conjectured) expressions for values of these various polynomials for all n, when $k \leq 3$. Namely.

- for $k = 0$, since $\mathcal{E}_n^{\langle 0\rangle} = \mathbb{Q}$, we have $\mathcal{E}_n(0; \boldsymbol{z}) = s_n(\boldsymbol{z})$;
- for $k = 1$, the module $\mathcal{E}_n^{\langle 1\rangle}$ is the regular representation of $\mathbb{S}_n$, of dimension $n!$, and we have the well-known formula

$$\mathcal{E}_n(1; \boldsymbol{z}) = \sum_{\mu\vdash n} f^\mu\, s_\mu(\boldsymbol{z}) \tag{2.7}$$

$$= e_1(\boldsymbol{z})^n, \tag{2.8}$$

with f^μ denoting the number of standard Young tableaux of shape μ (given by the hook-length formula);

- For $k = 2$, the module $\mathcal{E}_n^{\langle 2\rangle}$ is the module of "parking functions", and we have (see [**27**])

$$\mathcal{E}_n(2;\boldsymbol{z}) = \sum_{\mu\vdash n}(-1)^{n-\ell(\mu)}(n+1)^{\ell(\mu)-1}z_\mu^{-1}\,p_\mu(\boldsymbol{z}) \tag{2.9}$$

$$= \frac{1}{n+1}e_n[(n+1)\,\boldsymbol{z}]. \tag{2.10}$$

From this one easily derives

$$\dim \mathcal{E}_n^{\langle 2\rangle} = (n+1)^{n-1}, \qquad \text{and} \qquad \dim \mathcal{A}_n^{\langle 2\rangle} = \frac{1}{n+1}\binom{2\,n}{n}; \tag{2.11}$$

- For $k = 3$ the following formula is conjectured to hold (see [**10**])

(2.12)

$$\mathcal{E}_n(3;\boldsymbol{z}) = \sum_{\mu\vdash n}(-1)^{n-\ell(\mu)}(n+1)^{\ell(\mu)-2}\,z_\mu^{-1}\,p_\mu(\boldsymbol{z})\,\textstyle\prod_{k\in\mu}\binom{2\,k}{k},$$

$$= \frac{1}{(n+1)^2}\Phi_n[2(n+1)\,\boldsymbol{z}], \quad \text{with} \quad \Phi_n(\boldsymbol{z}) := \textstyle\sum_{\mu\vdash n} f_\mu(\boldsymbol{z})\prod_{k\in\mu}\mathrm{Cat}_k, \tag{2.13}$$

implying that

$$\dim \mathcal{E}_n^{\langle 3\rangle} = 2^n(n+1)^{n-2}, \qquad \text{and} \qquad \dim \mathcal{A}_n^{\langle 3\rangle} = \frac{2}{n(n+1)}\binom{4\,n+1}{n-1}. \tag{2.14}$$

No such simple formulas are known for larger positive values of k. It may be seen that, at $k = -1$, the k-variate polynomial expression $\mathcal{E}_n(k;\boldsymbol{z})$ evaluates to $p_n(\boldsymbol{z})$.

2.5. An intriguing negative evaluation. We have observed that, at $k = -2$, the k-variate polynomial $\dim(\mathcal{E}_n^{\langle k\rangle})$ appears to evaluate to the signed Catalan numbers $(-1)^{n-1}\mathrm{Cat}_{n-1}$. Moreover, this intriguing signed Catalan property seems to afford the refinement:

$$\mathcal{E}_n(-2;\boldsymbol{z}) = (-1)^{n-1}\sum_{\mu\vdash n}\pi(\mu)\,\mathrm{Cat}_{\ell(\mu)-1}\,f_\mu(\boldsymbol{z}), \tag{2.15}$$

where $\pi(\mu)$ is the product of the parts of μ. In turn, this is the specialization at $q = 1$ of the formula

$$(-q)^{n-1}\mathcal{E}_n[-q-1/q;\boldsymbol{z}] = \sum_{\mu\vdash n}\textstyle\prod_{k\in\mu}([k]_{q^2})\,C_{\ell(\mu)-1}(-q)\,f_\mu(\boldsymbol{z}) \tag{2.16}$$

where

$$C_n(q) := \sum_{k=1}^{2n+1}\tbinom{n}{\lfloor (k-1)/2\rfloor}\tbinom{n}{\lfloor k/2\rfloor}q^{k-1};$$

since we may check that $C_n(-1) = \mathrm{Cat}_n$.

3. Main results-conjectures

Considering the "scalar product" such that $\langle f\otimes s_\mu, s_\mu\rangle_\otimes = f$, so that $\langle \mathcal{E}_n, s_\mu\rangle_\otimes$ is the coefficient of s_μ in $\mathcal{E}_n$, we may express our first main "fact" as follows:

CONJECTURE 2 (Hook-Components). *For all n and all $0 \le k \le n-1$, if μ is the hook shape $(k+1, 1^{n-k-1})$, then we have the identity*

$$e_k^\perp \mathcal{A}_n = \langle \mathcal{E}_n, s_\mu \rangle_\otimes. \tag{3.1}$$

In particular, $e_{n-1}^\perp \mathcal{A}_n = 1$.

One of the interesting implication of Equation 3.1, together with Conjecture 3 below, is that we can reconstruct $\mathcal{A}_n$ from (very) partial knowledge of the values of the $\langle \mathcal{E}_n, s_\mu \rangle_\otimes$. To see how this goes, let us first state the following conjecture, defining the **length** $\ell(f)$ of a symmetric function f, to be the maximum number of parts $\ell(\lambda)$ in a partition λ that index a Schur function s_λ occurring with non-zero coefficient a_λ in its Schur expansion $f = \sum_\lambda a_\lambda f_\lambda$. In formula:

$$\ell(f) = \max_{a_\lambda \neq 0} \ell(\lambda).$$

CONJECTURE 3 (Coefficient-Length). *For all partitions μ of n, we have $\ell(\langle \mathcal{E}_n, s_\mu \rangle_\otimes) = n - \mu_1$. In particular, $(n-k-1)$ is the length of the coefficient of s_μ, for the hook-shape $\mu = (k+1, 1^{n-k-1})$.*

For $k = n-1$, this is compatible with Theorem 2 which implies that

$$\mathcal{A}_n = s_{1^{n-1}} + (\text{lower length terms}),$$

so that $\mathcal{A}_n$ is indeed of length $n-1$. As another example, the length of $\langle \mathcal{E}_n, s_{(n-1,1)} \rangle_\otimes$ is conjectured to be equal to 1, so that we get

$$\langle \mathcal{E}_n, s_{(n-1,1)} \rangle_\otimes = \langle \nabla(e_n), s_{(n-1,1)} \rangle_\otimes = s_1 + s_2 + \ldots + s_{n-1}, \tag{3.2}$$

since the second equality is well known.

3.1. $\mathcal{A}_n$-reconstruction. Assuming Conjecture 3, let us illustrate how to reconstruct $\mathcal{A}_n$. We already know that in general $\mathcal{A}_n = s_{1^{n-1}} + (\text{lower length terms})$, so that $e_{n-2}^\perp \mathcal{A}_n = s_1 + (\text{other length 1 terms})$. In fact, from Equation 3.2, we also get for all n that

$$e_{n-2}^\perp \mathcal{A}_n = s_1 + s_2 + \ldots + s_{n-1},$$

which forces

$$\mathcal{A}_n = s_{1^{n-1}} + \sum_{k=3}^{n} s_{(k,1^{n-3})} + (\text{terms of length} < n-2). \tag{3.3}$$

Likewise, all terms of length $n-3$ of $\mathcal{A}_n$ are imposed by the identity

$$e_{n-3}^\perp \mathcal{A}_n = \langle \mathcal{E}_n, s_{(n-2,1,1)} \rangle_\otimes = \langle \nabla(e_n), s_{(n-2,1,1)} \rangle_\otimes, \tag{3.4}$$

which results from the assumption that $\langle \mathcal{E}_n, s_{(n-2,1,1)} \rangle_\otimes$ is of length 2, hence it value is entirely determined by $\nabla(e_n)$. For instance, with $n = 6$, this gives

$$\begin{aligned}\mathcal{A}_6 = s_{11111} &+ s_{3111} + s_{4111} + s_{5111} + s_{6111} \\ &+ s_{441} + s_{431} + s_{531} + s_{631} + s_{421} + s_{521} + s_{621} + s_{721} + s_{821} \\ &+ s_{611} + s_{711} + 2s_{811} + s_{911} + s_{10.11} + (\text{terms of length} \le 2).\end{aligned}$$

The remaining missing terms are thus readily calculated since they correspond exactly to the Schur expansion of $\langle\nabla(e_6), e_6\rangle\rangle_\otimes$, which we give below for completeness sake.

$$\begin{aligned}\langle\nabla(e_6), e_6\rangle_\otimes &= s_{44} + s_{64} + s_{74} + s_{63} + s_{73} + s_{83} + s_{93} + s_{72} + s_{82} + s_{92}\\ &\quad + s_{10.2} + s_{11.2} + s_{10.1} + s_{11.1} + s_{12.1} + s_{13.1} + s_{15}.\end{aligned}$$

Once again, let us underline that for all n, the value of $\nabla(e_n)$ fixes all the components of length at most 2 in $\mathcal{E}_n$.

3.2. Hook-components reconstruction. Now, from this explicit knowledge of $\mathcal{A}_n$, we may calculate all the hook-indexed components of $\mathcal{E}_n$. In particular, we may check that we indeed get back the expression already mentioned for $\langle\mathcal{E}_6, s_{21111}\rangle_\otimes$, namely

$$\begin{aligned}\langle\mathcal{E}_6, s_{21111}\rangle_\otimes &= e_1^\perp \mathcal{A}_n\\ &= \langle\nabla(e_6), s_{21111}\rangle_\otimes + (s_{1111} + s_{2111} + s_{3111} + s_{4111} + s_{5111} + s_{331}\\ &\quad + 2s_{431} + s_{531} + s_{321} + 2s_{421} + 2s_{521} + 2s_{621}\\ &\quad + s_{721} + s_{311} + 2s_{411} + 3s_{511} + 3s_{611}\\ &\quad + 3s_{711} + 2s_{811} + s_{911});\end{aligned}$$

as well as that for $\langle\mathcal{E}_6, s_{3111}\rangle_\otimes$:

$$\begin{aligned}\langle\mathcal{E}_6, s_{3111}\rangle_\otimes &= e_2^\perp \mathcal{A}_n\\ &= \langle\nabla(e_6), s_{3111}\rangle_\otimes + (s_{331} + s_{321} + s_{421} + s_{521} + s_{111} + s_{211} + 2s_{311}\\ &\quad + 2s_{411} + 2s_{511} + s_{611} + s_{711}).\end{aligned}$$

3.3. Partial reconstruction of other components. Similarly, using both the Conjecture 3 and Conjecture 1, we may partially reconstruct other coefficients of $\mathcal{E}_n$, considering that the expansion of $\Delta'_{e_k}(e_n)$ is known for all k and n. Observe that the Conjecture 3 directly implies that, for all n,

$$\langle\mathcal{E}_n, s_{(n-2,2)}\rangle_\otimes = \langle\nabla(e_n), s_{(n-2,2)}\rangle_\otimes \tag{3.5}$$

so that we already have the coefficient of $s_{(n-2,2)}$ fully characterized. Since Conjecture 1 states that the length at most 2 components of $\langle e_k^\perp\mathcal{E}_n, s_\mu\rangle_\otimes$ coincide with those of $\langle\Delta'_{e_{n-1-k}} e_n, s_\mu\rangle$ for all μ, it may be used to infer components of the corresponding coefficients. We may also deduce from Conjecture 1 part of Conjecture 3. For instance, since $\Delta'_{e_1} e_n = \Delta_{e_1} e_n - 1\otimes e_n$ and we have (see [**16**, Prop. 6.1])

$$\Delta_{e_1}(e_n) = \textstyle\sum_{k=1}^n s_{k-1}\otimes e_{n-k}e_k, \tag{3.6}$$

we deduce that $\langle\Delta'_{e_1} e_n, s_\mu\rangle_\otimes = 0$ for all partition μ having first part larger than 2. Hence, Conjecture 1 implies that $\langle e_{n-3}^\perp\mathcal{E}_n, s_\mu\rangle_\otimes = 0$ when $\mu_1 > 2$, implying that $\langle\mathcal{E}_n, s_\mu\rangle_\otimes = 0$ has length at most 2 in those cases. However, for μ such that $\mu_1 = 2$, Equation 3.6 implies that $\langle e_{n-2}^\perp\mathcal{E}_n, s_{(2^k,1^{n-2k})}\rangle_\otimes$ does not vanish and is of length 1. Thus we conclude that

LEMMA 3.1. *Conjecture 1 implies Conjecture 3, for any partition μ such that $\mu_1 = 2$.*

Moreover, using (3.6), Conjecture 1 states that

$$\langle e_{n-2}^{\perp}\mathcal{E}_n, s_{(2^k,1^{n-2k})}\rangle_{\otimes} = \sum_{i=k-1}^{n-k-1} s_i.$$

Since we already know that $\langle e_{n-2}^{\perp}\mathcal{E}_n, s_{(2^k,1^{n-2k})}\rangle_{\otimes} = 0$ if $k \geq 1$, the above identity forces

$$\langle \mathcal{E}_n, s_{(2^k,1^{n-2k})}\rangle_{\otimes} = \sum_{i=k-1}^{n-k-1} s_{i+1,1^{n-3}} + (\text{terms of length} < n-2). \tag{3.7}$$

4. Structure properties of $\mathcal{E}_n$

4.1. Hook restriction property. Multiplicities of hook components in $\mathcal{A}_n$ (the alternating part of $\mathcal{E}_n$), are explicitly known. Indeed, considering in this section that a and b are such that $n = a+b+1$, using Equation 6.4 we may calculate that

$$\frac{1}{q+u}\sum_{b\geq 0} u^b\,(e_b^{\perp} s_{(a\,|\,b)})(q) = \frac{s_{(a\,|\,b)}[q-\varepsilon u]}{q+u} = q^a u^b; \tag{4.1}$$

and that $s_\mu[q-\varepsilon u] = 0$ whenever μ is not a hook. Hence, for any symmetric function f in Λ_n,

$$\frac{f[q-\varepsilon u]}{q+u} = \sum_{a,b} \langle f, s_{(a\,|\,b)}\rangle\, q^a u^b \tag{4.2}$$

may be considered as a generating function for the multiplicities of hook components in f. In the case of $f = \mathcal{A}_n$, the multiplicities are thus readily calculated in view of the following identity, which follows from Equation 4.1.

PROPOSITION 4.1. *For all n, we have*

$$\mathcal{A}_n[q-\varepsilon u] = (q+u)(q^2+u)\cdots(q^{n-1}+u). \tag{4.3}$$

PROOF. We first recall identity Equation 6.7, which may be formulated as

$$\langle \mathcal{E}_n(q;\boldsymbol{z}), s_{(a\,|\,b)}(\boldsymbol{z})\rangle = q^{\binom{b+1}{2}} \begin{bmatrix} n-1 \\ b \end{bmatrix}_q.$$

Observe that Equation 4.1 implies that $(e_a^{\perp}\mathcal{A}_n)(q) = \langle \mathcal{E}_n(q;\boldsymbol{z}), s_{(a\,|\,b)}(\boldsymbol{z})\rangle$, hence using Equation 6.4 we calculate that

$$\begin{aligned}\mathcal{A}_n[q-\varepsilon u] &= \sum_{a+b=n-1} \langle \mathcal{E}_n(q;\boldsymbol{z}), s_{(a\,|\,b)}(\boldsymbol{z})\rangle\, u^a \\ &= \sum_{a+b=n-1} q^{\binom{b+1}{2}} \begin{bmatrix} n-1 \\ b \end{bmatrix}_q u^a \\ &= (q+u)(q^2+u)\cdots(q^{n-1}+u).\end{aligned}$$

□

As another indication of the appropriateness of Equation 4.3, we may directly check that its specialization at $u = 0$ does indeed correspond to the known expression for the hook components of $\mathcal{E}_n^{\langle 1\rangle}$. Indeed, the expression

$$\begin{aligned}\sum_{k=0}^{n-1} \boldsymbol{c}_{(n-k,1^k)}[q]\, z^k &= \sum_{k=0}^{n-1} q^{\binom{k+1}{2}} \begin{bmatrix} n-1 \\ k \end{bmatrix}_q z^k,\\ &= \textstyle\prod_{i=0}^{n-1}(1+q^i\, z),\end{aligned}$$

may easily be seen to follow from Equation 6.7, since $\mathcal{E}_n(q;\boldsymbol{z}) = \widetilde{H}_n(q;\boldsymbol{z})$.

4.2. Conjectured formula for the hook part multiplicities in other hook components. Our calculations suggest that there is a simple elegant expression for the multiplicity $\langle s_\lambda, \boldsymbol{c}_\mu\rangle$ of $s_\lambda \otimes s_\mu$ in $\mathcal{E}_n$, when both $\lambda = (i\,|\,j)$ and $\mu = (a\,|\,b)$ are hook shapes. Indeed, we conjecture the following formula for the corresponding generating function, using the encoding of Equation 4.2.

CONJECTURE 4. *For all $a + b + 1 = n$, we have*

$$\frac{\boldsymbol{c}_{(a\,|\,b)}[q-\varepsilon u]}{q+u} = \begin{bmatrix} n-1 \\ b \end{bmatrix}_q (q^2+u)\cdots(q^b+u), \tag{4.4}$$

using the Gaussian q-binomial notation.

For example, the size collected restriction to hook shapes of $\langle \mathcal{E}_5, s_{(1\,|\,3)}\rangle_\otimes$ is

$$\begin{aligned}\boldsymbol{c}_{(1\,|\,3)}\big|_{\text{hooks}} &= s_{111} + (s_{31}+s_{211}) + (2s_{41}+s_{311}) + (s_6 + 2s_{51} + s_{411})\\ &\quad + (s_7 + 2s_{61}) + (s_8 + s_{71}) + s_9,\end{aligned}$$

and the corresponding polynomial is

$$\begin{aligned}\frac{\boldsymbol{c}_{(1\,|\,3)}[q-\varepsilon u]}{q+u} &= \begin{bmatrix} 4 \\ 3 \end{bmatrix}_q (q^2+u)(q^3+u)\\ &= (q+1)(q^2+1)(q^2+u)(q^3+u)\\ &= u^2 + (q^2u + qu^2) + (2\,q^3u + q^3u^2) + (q^5 + 2\,q^4u + q^3u^2)\\ &\quad + (q^6 + 2\,q^5u) + (q^7 + q^6u) + q^8.\end{aligned}$$

4.3. Length components. Define the **degree** of

$$\sum_{\lambda,\mu} c_{\lambda,\mu} s_\lambda \otimes s_\mu$$

to be the maximum of the values $|\lambda|$, for which $c_{\lambda,\mu} \neq 0$. The **length d component** of $\mathcal{E}_n$ is set to be

$$\mathcal{E}_n^{(d)} := \sum_{\mu\vdash n} \boldsymbol{c}_\mu^{(d)} \otimes s_\mu, \qquad \text{with} \qquad \boldsymbol{c}_\mu^{(d)} = \sum_{\ell(\lambda)=d} c_{\lambda,\mu} s_\lambda. \tag{4.5}$$

We clearly have

$$\mathcal{E}_n := \mathcal{E}_n^{(0)} + \mathcal{E}_n^{(1)} + \ldots + \mathcal{E}_n^{(l)}, \tag{4.6}$$

where $l = \ell(\mathcal{E}_n)$ is the maximal length occurring in terms of $\mathcal{E}_n$. With this notation, Conjecture 3 states that $\ell(\boldsymbol{c}_\mu) = |\mu| - \mu_1$, and it may be shown that

$$\deg(\boldsymbol{c}_\mu^{(j)}) = \binom{n}{2} - \binom{j}{2} - \sum_{i\in\mu}\binom{i}{2}. \tag{4.7}$$

Let $\mathcal{E}_n^{\langle k\rangle}$ (similarly for $\boldsymbol{c}_\mu^{\langle k\rangle}$) stand for the restriction of $\mathcal{E}_n$ to its components of length at most k. In other terms,

$$\mathcal{E}_n^{\langle k\rangle} := \mathcal{E}_n^{(0)} + \mathcal{E}_n^{(1)} + \ldots \mathcal{E}_n^{(k)}. \tag{4.8}$$

For example,

$$\begin{aligned}
\mathcal{E}_4^{(0)} &= 1\otimes s_4,\\
\mathcal{E}_4^{(1)} &= (s_1+s_2+s_3)\otimes s_{31} + (s_2+s_4)\otimes s_{22} + (s_3+s_4+s_5)\otimes s_{211} + s_6\otimes s_{1111},\\
\mathcal{E}_4^{(2)} &= s_{21}\otimes s_{22} + (s_{11}+s_{21}+s_{31})\otimes s_{211} + (s_{31}+s_{41})\otimes s_{1111},\\
\mathcal{E}_4^{(3)} &= s_{111}\otimes s_{1111}.
\end{aligned}$$

Simple general values for length components are:

$$\mathcal{E}_n^{(0)} = 1\otimes s_n, \qquad \text{and} \qquad \mathcal{E}_n^{(n-1)} = e_{n-1}\otimes e_n. \tag{4.9}$$

The following **reduced-length components** can be efficiently used to reconstruct (part of) $\mathcal{E}_n$. We set

$$\varepsilon_n^{(k)} := (e_k^\perp\otimes \mathrm{Id})\,\mathcal{E}_n^{(k)}, \qquad (\text{and} \qquad \alpha_n^{(k)} := e_k^\perp \mathcal{A}_n^{(k)}). \tag{4.10}$$

For example,

$$\begin{aligned}
\varepsilon_4^{(0)} &= 1\otimes s_4,\\
\varepsilon_4^{(1)} &= (1+s_1+s_2)\otimes s_{31} + (s_1+s_3)\otimes s_{22} + (s_2+s_3+s_4)\otimes s_{211} + s_5\otimes s_{1111},\\
\varepsilon_4^{(2)} &= (1+s_1+s_2)\otimes s_{211} + s_1\otimes s_{22} + (s_2+s_3)\otimes s_{1111},\\
\varepsilon_4^{(3)} &= 1\otimes s_{1111}.
\end{aligned}$$

4.4. Description of hook components in terms of $\mathcal{A}_n$. From Equation 3.1 we may calculate calculate all the hook components $\boldsymbol{c}_\mu = \langle\mathcal{E}_n, s_\mu\rangle_\otimes$ directly from the alternant component $\mathcal{A}_n = \langle\mathcal{E}_n, s_{1^n}\rangle_\otimes$, both considered as GL_∞-character of isotypic components of $\mathcal{E}_n$. We will see that this implies that we can reconstruct $\mathcal{A}_n$ from much less information than is apparently needed at *prima facie*. From now on, let us use **Frobenius's notation** for hook shape partitions, writing $(a\,|\,b)$ for the partition $(a+1, 1^b)$ of $n = a+b+1$ (see Figure 1). It is often said that a stands for the **arm** of the hook, while b stands for its **leg**. We use the Cartesian (aka French) convention to draw diagrams, so that the leg goes up.

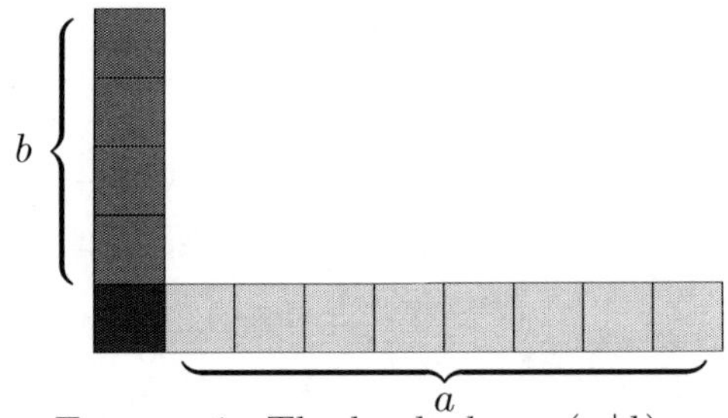

FIGURE 1. The hook shape $(a\,|\,b)$.

5. Link to Delta operators, and the Delta-conjecture

5.1. Skewing versus Δ-operators. The "Delta-Conjecture" of [**16**] proposes an explicit combinatorial description for the evaluation at e_n of the Macdonald "eigenoperator" Δ'_{e_k}. The operators Δ'_{e_k} were originally introduced in [**5**], up to a slight shift in eigenvalues. Our discussion below leads to the conclusion that, for each μ, the coefficient $\langle \Delta'_{e_k}(e_n), s_\mu\rangle_\otimes$ should be Schur positive. Hence the same clearly holds true for $\Delta_{e_k}(e_n)$. This is a slightly stronger positivity statement than that of the original Delta-Conjecture. The special case $\Delta_{e_2}(e_n)$ is considered in [**25**], not referring to a Schur expansion but proving an equivalent result.

Conjecture 1 states that $\Delta'_{e_b}(e_n)$ may be obtained simply by skewing first components in $\mathcal{E}_n$ by the elementary symmetric function e_a, for $a+b+1=n$. Hence, we may consider $(e_a^\perp \otimes \mathrm{Id})\,\mathcal{E}_n$ as a multivariate analogue of $\Delta'_{e_b}(e_n)$. Once again we emphasize that the above conjecture could not have been stated in the original restricted context of two sets of variables, namely inside $\mathcal{R}_n^{\langle 2\rangle}$, in which the Δ_{e_k} operators are usually considered. It has recently been shown (see [**28**]) that Conjecture 1 implies Conjecture 4.

It also follows, together with Equation 3.1, that we have the equality

$$\langle \Delta'_{e_b} e_n, s_{(c|\,d)}\rangle_\otimes = \langle \Delta'_{e_d} e_n, s_{(a\,|\,b)}\rangle_\otimes, \tag{5.1}$$

for all $a+b=c+d=n-1$. This follows directly using known identities involving Macdonald polynomials, thus giving indirect support to Equation 1.3. Keeping the same convention for j and k, we have

$$((e_j^\perp \otimes \mathrm{Id})\,\mathcal{E}_n)(q,1/q;\mathbf{x}) = \sum_{j=0}^{k}(-1)^{k-j}\,\frac{e_j[[n]_q]}{q^{j\,(n-1)}}\,\frac{e_n[[j+1]_q\,\mathbf{x}]}{[j+1]_q} \tag{5.2}$$

5.2. Inferred components of $\mathcal{E}_n$. Just as before, we consider the decomposition of $\mathcal{E}_n$ into its length components $\mathcal{E}_n^{(k)}$, and the associate reduced-length components $\varepsilon_n^{(j)}$ (see Equation 4.10). One may see that the length of $\varepsilon_n^{(j)}$ (that is the maximal length of one of its GL_∞-coefficients) is equal to $\min(j, n-1-j)$.

Together with this length bound, Conjecture 1 implies that we may calculate the various $\varepsilon_n^{(j)}$, for all $0 \le j < n \le 6$, just from the knowledge of the length 2 expressions $\Delta'_{e_k} e_n$. As a matter of fact, for all n, we have the general formulas:

(5.3) $\varepsilon_n^{(0)} = 1 \otimes s_n,$

(5.4) $\varepsilon_n^{(1)} = (e_1^\perp \otimes \mathrm{Id}) \sum_{\mu \vdash n} \sum_{\tau \in \mathrm{SYT}(\mu)} s_{\mathrm{charge}(\tau)} \otimes s_\mu,$

(5.5) $\varepsilon_n^{(2)} = (e_2^\perp \otimes \mathrm{Id}) \nabla(e_n),$

(5.6) $\varepsilon_n^{(n-3)} = \Delta'_{e_2}(e_n) + (s_1 + s_2) \otimes e_n + (1 \otimes e_{n-1} e_1) - \sum_{k=1}^n s_{k-1} \otimes e_k e_{n-k}.$

(5.7) $\varepsilon_n^{(n-2)} = \sum_{k=1}^n s_{k-1} \otimes e_k e_{n-k} - (1 + s_1) \otimes e_n$

(5.8) $\varepsilon_n^{(n-1)} = 1 \otimes e_n.$

5.3. Partial reconstruction of other components. Using both Conjecture 3 and Conjecture 1, we may partially reconstruct coefficients $\boldsymbol{c}_\mu$ (of $\mathcal{E}_n$) for μ's that are not hooks, considering that the expansion of $\Delta'_{e_k}(e_n)$ is known for all k and n. Observe that Conjecture 3 directly implies that, for all n,

$$\langle \mathcal{E}_n, s_{(n-2,2)} \rangle_\otimes = \langle \nabla(e_n), s_{(n-2,2)} \rangle_\otimes, \tag{5.9}$$

so that we already have the coefficient of $s_{(n-2,2)}$ fully characterized, on top of those for all hook-shapes. Since Conjecture 1 states that the length at most 2 components of $\langle e_k^\perp \mathcal{E}_n, s_\mu \rangle_\otimes$ coincide with those of $\langle \Delta'_{e_{n-1-k}} e_n, s_\mu \rangle_\otimes$ for all μ, it may be used to infer components of the corresponding coefficients. We may also deduce from Conjecture 1 part of Conjecture 3. For instance, since $\Delta'_{e_1} e_n = \Delta_{e_1} e_n - 1 \otimes e_n$ and we have (see [**16**, Prop. 6.1])

$$\Delta_{e_1}(e_n) = \sum_{k=1}^n s_{k-1} \otimes e_{n-k} e_k, \tag{5.10}$$

we deduce that $\langle \Delta'_{e_1} e_n, s_\mu \rangle_\otimes = 0$ for all partition μ having first part larger than 2. Hence, Conjecture 1 implies that $\langle e_{n-3}^\perp \mathcal{E}_n, s_\mu \rangle_\otimes = 0$ when $\mu_1 > 2$, implying that $\langle \mathcal{E}_n, s_\mu \rangle_\otimes = 0$ has length at most 2 in those cases. However, for μ such that $\mu_1 = 2$, Equation 3.6 implies that $\langle e_{n-2}^\perp \mathcal{E}_n, s_{(2^k,1^{n-2k})} \rangle_\otimes$ does not vanish and is of length 1. Thus we conclude that

LEMMA 5.1. *Conjecture 1 implies Conjecture 3, for any partition μ such that $\mu_1 = 2$.*

Moreover, using Equation 3.6, Conjecture 1 states that

$$\langle e_{n-2}^\perp \mathcal{E}_n, s_{(2^k,1^{n-2k})} \rangle_\otimes = \sum_{i=k-1}^{n-k-1} s_i.$$

Since we already know that $\langle e_{n-2}^\perp \mathcal{E}_n, s_{(2^k,1^{n-2k})} \rangle_\otimes = 0$ if $k \geq 1$, the above identity forces

$$\langle \mathcal{E}_n, s_{(2^k,1^{n-2k})} \rangle_\otimes = \sum_{i=k-1}^{n-k-1} s_{i+1,1^{n-3}} + (\text{terms of length} < n-2). \tag{5.11}$$

6. The e-positivity phenomenon

As discussed in [**8**], most of the symmetric functions constructed via the elliptic Hall algebra approach, including $\nabla(e_n)$, seem to become e-positive when specialized at $t = 1$. We discuss here an analogous situation for $\mathcal{E}_n$, which corresponds to a multi-parameter "lift" of $\nabla(e_n)$. For this we consider the specialization of any one of the parameters q_i to the value 1. In our context of infinitely many parameters, this is handily realized by the plethystic evaluation at $1+\boldsymbol{q}$ of the GL_∞-coefficients $\boldsymbol{c}_\mu$ of $\mathcal{E}_n$. It is noteworthy that this evaluation is an invertible operation.

For the sake of discussion, let us set $\mathcal{F}_n := \mathcal{E}_n[1+\boldsymbol{q};\boldsymbol{z}]$, and write

$$\mathcal{F}_n = \sum_{\mu\vdash n} \boldsymbol{c}_\mu[1+\boldsymbol{q}]\otimes s_\mu(\boldsymbol{z}) = \sum_{\nu\vdash n} \boldsymbol{d}_\nu \otimes e_\nu(\boldsymbol{z}), \tag{6.1}$$

so that $\boldsymbol{d}_\nu$ stands for the coefficients of $e_\nu(\boldsymbol{z})$ in $\mathcal{F}_n$. Using our scalar product convention, we may also write $\boldsymbol{d}_\nu = \langle \mathcal{F}_n, f_\nu\rangle_\otimes$ (since the forgotten symmetric functions f_ν are dual to the e_ν). Then, as far as we can check experimentally, all of the $\boldsymbol{d}_\nu$ are **Schur positive**. For instance, we have

$$\begin{aligned}\mathcal{F}_4 = 1\otimes e_{1111} &+ (3s_1+2s_2+s_3)\otimes e_{211} + (s_{11}+s_2+s_{21}+s_4)\otimes e_{22}\\ &+ (2s_{11}+s_{21}+2s_3+s_{31}+s_4+s_5)\otimes e_{31} + (s_{111}+s_{31}+s_{41}+s_6)\otimes e_4.\end{aligned}$$

The $\boldsymbol{c}_\nu$ are related to the $\boldsymbol{d}_\mu$ by the formula $\boldsymbol{c}_\mu[1+\boldsymbol{q}] = \sum_\nu K_{\mu'\lambda}\boldsymbol{d}_\lambda$, where $K_{\mu\lambda}$ stands for the usual Kostka number. There is a close tie between this e-positivity phenomenon, and related conjectures. To see this, we recall that the coefficient of e_n in the e-expansion of s_μ vanishes for all μ except hooks. For hooks, we have $\langle s_{(a\,|\,b)}, f_{(n)}\rangle = (-1)^a$. We may then calculate that

$$\begin{aligned}\boldsymbol{d}_{(n)} = \langle \mathcal{E}_n[1+\boldsymbol{q};\boldsymbol{z}], f_{(n)}\rangle_\otimes &= \sum_{\mu\vdash n} \boldsymbol{c}_\mu[1+\boldsymbol{q}]\,\langle s_\mu, f_{(n)}\rangle\\ &= \Big(\sum_{a=0}^{n-1}(-1)^a\,\boldsymbol{c}_{(a\,|\,b)}\Big)[1+\boldsymbol{q}] = \Big(\sum_{k\geq 0}(-1)^a\,e_a^\perp\mathcal{A}_n\Big)[1+\boldsymbol{q}].\end{aligned}$$

If we recall that $f[\boldsymbol{q}-1] = \sum_{a\geq 0}(-1)^a\,e_a^\perp f(\boldsymbol{q})$, we may conclude the above calculation to get

$$\boldsymbol{d}_{(n)} = (\mathcal{A}_n[\boldsymbol{q}-1])[1+\boldsymbol{q}] = \mathcal{A}_n, \tag{6.2}$$

which is Schur positive. To get other similar results, let μ be any partition of n which is largest in dominance order among those such that $\boldsymbol{c}_\mu \neq 0$. Then, it is easy to see that $\boldsymbol{d}_{(n)} = \boldsymbol{c}_{1^n}$, and $\boldsymbol{d}_{\mu'} = \boldsymbol{c}_\mu[1+\boldsymbol{q}]$. In particular, $\boldsymbol{d}_{(1^n)} = 1$. We thus automatically have Schur positivity in the these cases. Experiments suggest that whenever $a\geq 1$, we have

$$\boldsymbol{d}_{(a\,|\,b)} = \sum_{j=0}^{b} \boldsymbol{c}_{(j\,|\,a)}, \tag{6.3}$$

implying Schur-positivity of $\boldsymbol{d}_{(a\,|\,b)}$.

Appendix

Symmetric functions and plethysm. We mainly use Macdonald's notations (see [**20**]). Thus p_μ, e_μ, f_μ, and h_μ respectfully stand for the **power sum**, **elementary**, **forgotten**, and **complete homogeneous** symmetric functions, with indices integer partitions $\mu = \mu_1\mu_2\cdots\mu_k$ of n. Recall that these are symmetric functions that are all homogeneous of degree $n = |\mu| := \mu_1+\mu_2+\ldots+\mu_k$. Assuming that they are evaluated in enough variables, they respectively form bases of the homogenous degree n component Λ_n of the graded ring of symmetric function Λ. It is often useful to avoid writing variables, implicitly making the above assumption.

The **length** $\ell(\mu)$ of a partition μ is the number of non-zero parts (the μ_i) of μ. Recall that partitions are often described in terms of their **Ferrers diagram** (herein in French notation). This diagram is the set of **cells** (i,j) in $\mathbb{N}\times\mathbb{N}$ (considered as points in the usual Cartesian coordinates), with $1 \leq i-1 \leq \mu_{j+1}$, for $0 \leq j < \ell(\mu)$. We often write $(i,j) \in \mu$ when the context makes this convention clear. The row lengths of this diagram are the parts of μ. One writes $\mu \subseteq \lambda$, when the diagram of μ is contained in the diagram of λ. The **skew shape** λ/μ has diagram equal to the set-difference of the respective diagrams of λ and μ. The **conjugate** μ' of μ is the partition with diagram equal to the set $\{(j,i) \mid (i,j) \in \mu\}$. The **hook length** of a cell (i,j) of μ is defined as

$$h(i,j) = h_{ij} = \mu_{j+1} + \mu'_{i+1} - i - j - 1.$$

The **descent set**, $\operatorname{des}(\mu)$, of a partition μ, is the set of i such that $\mu_i > \mu_{i+1}$. For a permutation σ of $\{1,\ldots,n\}$, its **cycle structure** is the partition μ of n whose parts are the lengths of its cycles. One denotes by z_μ the integer such that $n!/z_\mu$ is the number of permutations with cycle structure equal to μ. For partitions μ and ν, the **sum** $\mu+\nu$ is the partition whose parts are $\mu_i+\nu_i$.

A **tableau** of shape μ (or λ/μ) is simply a function $\tau : \mu \to \mathbb{N}$. Thus some value $\tau(i,j) \in \mathbb{N}$ is associated to each cell of μ. If these values are increasing along rows of μ, and strictly increasing along columns, one says that the tableau is a **semi-standard Young** tableau. If moreover, τ is bijective with values going from 1 to $n = |\mu|$, one says that the tableau is **standard**. The **skew-Schur functions** $s_{\lambda/\mu}$ (with $s_\lambda = s_{\lambda/\emptyset}$, that is for $\mu = \emptyset$ the empty partition) may be considered as enumerators of semi-standard tableaux of shape λ/μ. In formula, this says that

$$s_{\lambda/\mu}(\boldsymbol{z}) = \sum_{\tau:\lambda/\mu\to\mathbb{N}} \boldsymbol{z}_\tau, \qquad \text{with} \qquad \boldsymbol{z}_\tau := \textstyle\prod_{c\in\lambda/\mu} z_{\tau(c)},$$

with τ running over the set of semi-standard tableaux of shape λ/μ. Observe that $s_\lambda(z_1,\ldots,z_k) = 0$, whenever $k < \ell(\lambda)$. The Schur functions are orthonormal for the usual **Hall scalar product** on Λ, which may be defined by setting $\langle p_\lambda, p_\mu\rangle := z_\mu \delta_{\lambda,\mu}$. As usual ω stands for the linear multiplicative self adjoint operator such that $\omega\, s_\lambda = s_{\lambda'}$. For this scalar product, the forgotten functions are **dual** to the elementary, *i.e.* $\langle e_\mu, f_\lambda\rangle = \delta_{\lambda\mu}$; and the Schur functions are self-dual. For a given symmetric function f, the linear operator $f^\perp$ is the **adjoint** to the linear operation of multiplication by f. In formula, $\langle f\cdot g_1, g_2\rangle = \langle g_1, f^\perp g_2\rangle$, for any symmetric functions g_1 and g_2. The classical **(dual) Pieri rule** (see [**20**]), states that $e_k^\perp s_\mu = \sum_{\lambda\subset\mu} s_\lambda$, with the indices of the sum running over partitions λ that can be obtained from μ by removing k cells, no two of which lying on the same row. In particular, $e_k^\perp s_\mu$ vanishes if μ has less than k rows. A symmetric function f is said to be **Schur positive** when, for all partition μ, the coefficients of its Schur expansion are polynomials (in some variables) with positive integer coefficients. Sums, products, and "plethysms" (see below) of Schur-positive functions are also Schur-positive. Similarly, one has the notion of ***e*-positivity**. Since all the e_λ's are Schur-positive, we have that e-positivity implies Schur positivity. Moreover, e-positivity is clearly closed under sums and product.

For symmetric function f and g, the **plethysm** $f\circ g = f[g]$ is a special case of λ-ring calculations $f[\boldsymbol{a}]$, in which symmetric function are considered as operators.

The following evaluation rules entirely characterize these, assuming that α and β are scalars, and that $\boldsymbol{a}$ and $\boldsymbol{b}$ lie in some suitable ring:

$$
\begin{array}{rlrl}
\text{(i)} & (\alpha f+\beta g)[\boldsymbol{a}] = \alpha\, f[\boldsymbol{a}]+\beta\, g[\boldsymbol{a}], & \text{(ii)} & (f\cdot g)[\boldsymbol{a}] = f[\boldsymbol{a}]\cdot g[\boldsymbol{a}],\\
\text{(iii)} & p_k[\boldsymbol{a}\pm \boldsymbol{b}] = p_k[\boldsymbol{a}]\pm p_k[\boldsymbol{b}], & \text{(iv)} & p_k[\boldsymbol{a}\cdot \boldsymbol{b}] = p_k[\boldsymbol{a}]\cdot p_k[\boldsymbol{b}],\\
\text{(v)} & p_k[\boldsymbol{a}/\boldsymbol{b}] = p_k[\boldsymbol{a}]/p_k[\boldsymbol{b}], & \text{(vi)} & p_k[p_j] = p_{kj},\\
\text{(vii)} & p_k[p_j\otimes p_\ell] = p_{kj}\otimes p_{k\ell}, & \text{(viii)} & p_k[\varepsilon] = (-1)^k,\\
\text{(ix)} & p_k[x] = x^k,\ \text{if } x \text{ a “variable”}, & \text{(x)} & p_k[c] = c,\ \text{if } x \text{ a “constant”}.
\end{array}
$$

Hence we specify what are variables and what are constants. The first two properties make it clear that any evaluation of the form $f[\boldsymbol{a}]$ may be reduced to instances of the form $p_k[\boldsymbol{a}]$. We also assume that property (iii) extends to denumerable sums. Property (viii) implies, that $f[\varepsilon \boldsymbol{z}] = \omega\, f(\boldsymbol{z})$, for all symmetric function f. It may readily be shown that, for all symmetric function f,

$$
\sum_{k\geq 0} u^k (e_b^\perp f)(q) = f[q-\varepsilon u]. \tag{6.4}
$$

See [**6**, **16**] for more on plethysm.

Macdonald polynomials, and operators. Recall that the set of **combinatorial Macdonald polynomials** $\{\widetilde{H}_\mu(q,t;\boldsymbol{z})\}_{\mu\vdash n}$ forms a linear basis of the ring $\Lambda(q,t)$, of symmetric functions in the variables $\boldsymbol{z}=(z_i)_{i\in\mathbb{N}}$ over the field $\mathbb{Q}(q,t)$. They are uniquely characterized by the equations

$$
\begin{array}{lll}
\text{(i) } \langle s_\lambda(\boldsymbol{z}), \widetilde{H}_\mu[q,t;(1-q)\,\boldsymbol{z}]\rangle = 0, & \text{if} & \lambda \not\geq \mu,\\
\text{(ii) } \langle s_\lambda(\boldsymbol{z}), \widetilde{H}_\mu[q,t;(1-t)\,\boldsymbol{z}]\rangle = 0, & \text{if} & \lambda \not\geq \mu', \text{ and}\\
\text{(iii) } \langle s_n(\boldsymbol{z}), \widetilde{H}_\mu(q,t;\boldsymbol{z})\rangle = 1, & &
\end{array}
$$

involving plethystic notation. See [**19**, Section 3.5] for more details on these dominance order triangularities. It follows from the $n!$-theorem (see [**18**]) that the $\widetilde{H}_\mu$ are Schur positive. The operators ∇ and Δ'_{e_k}, introduced in [**5**], are special instances of **Macdonald eigenoperators**. This is to say that they have the Macdonald polynomials $\widetilde{H}_\mu$ as joint eigenfunctions. Their respective eigenvalues are

$$
\Delta_{e_k}(\widetilde{H}_\mu) := e_k[B_\mu]\,\widetilde{H}_\mu, \qquad \Delta'_{e_k}(\widetilde{H}_\mu) := e_k[B_\mu - 1]\,\widetilde{H}_\mu,
$$

with $B_\mu := \sum_{(i,j)\in\mu} q^i t^j$. On homogeneous symmetric functions of degree n (and only for those), the operator ∇ coincides with $\Delta_{e_n} = \Delta'_{e_{n-1}}$. In other terms, the associated eigenvalues are

$$
T_\mu := q^{\eta(\mu')} t^{\eta(\mu)} = \prod\nolimits_{(i,j)\in\mu} q^i t^j, \qquad \text{with} \qquad \eta(\mu) := \sum\nolimits_k (k-1)\mu_k, \tag{6.5}
$$

for $\mu = \mu_1\cdots\mu_\ell$. Among the many interesting formulas pertaining to the $\widetilde{H}_\mu$ (equivalent to formulas that may be found in [**20**, Exer. 2, page 362]), we have

$$
\langle \widetilde{H}_\mu, s_{(a\,|\,b)}\rangle = e_b[B_\mu - 1], \quad \text{for all} \quad a+b+1=n. \tag{6.6}
$$

In particular,

$$\langle \widetilde{H}_\mu, s_{1^n}\rangle = T_\mu, \qquad \text{and} \qquad \langle \widetilde{H}_n, s_{(a\,|\,b)}\rangle = q^{\binom{b+1}{2}} \begin{bmatrix} n-1 \\ b \end{bmatrix}_q. \tag{6.7}$$

We also have the symmetries

$$\widetilde{H}_\mu(q,t;\boldsymbol{x}) = T_\mu\, \omega \widetilde{H}_\mu(1/q,1/t;\boldsymbol{x}), \qquad \text{and} \qquad \widetilde{H}_\mu(t,q;\boldsymbol{x}) = \widetilde{H}_{\mu'}(q,t;\boldsymbol{x}). \tag{6.8}$$

The Δ-conjecture. In [**16**], one finds an explicit combinatorial formula conjectured to be equal to $\Delta'_{e_k} e_n$. With our particular point of view, it takes the form

$$\Delta'_{e_k} e_n = \sum_{\mu \subseteq \delta_{nn}} \Big(\sum_{\substack{J \supseteq \mathrm{des}(\mu) \\ \# J = k}} q^{(J,\boldsymbol{a})} \Big) \mathbb{L}_\mu(t;\boldsymbol{z}) \tag{6.9}$$

where the indices J (in the inner sum) run over all k-subsets of $\{1,2,\ldots,n-1\}$, and $(J,\boldsymbol{a})$ is shorthand for $\sum_{i\in J} a_i$. We here denote by $\mathbb{L}_\mu(t;\boldsymbol{z})$ the vertical strip LLT-polynomial associated to the path μ. See [**14**] for more on these, in particular for a proof that $\mathbb{L}(1+t;\boldsymbol{z})$ is e-positive. For instance, when $k=0$, the only non-zero term of the outer sum corresponds to $s_{(\emptyset+1^n)/\emptyset} = e_n$, thus agreeing with $\Delta'_{e_0} e_n$. At the opposite end of the spectrum, when $k=n-1$, there is but one term in the inner sum (since J must be equal to $[n]$) which is clearly equal to $q^{\mathrm{area}(\mu)}$. Observe that, at $t=1$, the above expression simplifies to

$$\Delta'_{e_k} e_n \big|_{t=1} = \sum_{\mu \subseteq \delta_{nn}} \Big(\sum_{\substack{J \supseteq \mathrm{des}(\mu) \\ \# J = k}} q^{(J,\boldsymbol{a})} \Big) s_{(\mu+1^n)/\mu}(\boldsymbol{z}). \tag{6.10}$$

The e-expansions of $\mathcal{F}_n$. Partially expressed in terms of the $\mathcal{A}_n$, we have the following values:

$$\begin{aligned}
\mathcal{F}_1 &= 1 \otimes e_1,\\
\mathcal{F}_2 &= \mathcal{A}_2 \otimes e_2 + 1 \otimes e_{11},\\
\mathcal{F}_3 &= \mathcal{A}_3 \otimes e_3 + (2s_1 + s_2) \otimes e_{21} + 1 \otimes e_{111},\\
\mathcal{F}_4 &= \mathcal{A}_4 \otimes e_4 + (e_1^\perp \mathcal{A}_4 + \mathcal{A}_3) \otimes e_{31} + (s_{11} + s_{21} + s_2 + s_4) \otimes e_{22}\\
&\qquad + (3s_1 + 2s_2 + s_3) \otimes e_{211} + 1 \otimes e_{1111},\\
\mathcal{F}_5 &= \mathcal{A}_5 \otimes e_5 + (e_1^\perp \mathcal{A}_5 + \mathcal{A}_4) \otimes e_{41} + (e_2^\perp \mathcal{A}_5 + e_1^\perp \mathcal{A}_4 + \mathcal{A}_3) \otimes e_{311}\\
&\quad + (2s_{111} + 2s_{211} + s_{311} + s_{22} + s_{32} + s_{42} + 2s_{21} + 2s_{31}\\
&\qquad + 3s_{41} + 2s_{51} + s_{61} + 2s_4 + s_6 + s_7 + s_8) \otimes e_{32}\\
&\quad + (s_{22} + 3s_{11} + 4s_{21} + 2s_{31} + s_{41} + 3s_2 + 2s_3 + 2s_4 + 2s_5 + s_6) \otimes e_{221}\\
&\quad + (4s_1 + 3s_2 + 2s_3 + s_4) \otimes e_{2111} + 1 \otimes e_{11111},
\end{aligned}$$

$$
\begin{aligned}
\mathcal{F}_6 = {} & \mathcal{A}_6 \otimes e_6 + (e_1^\perp \mathcal{A}_6 + \mathcal{A}_5) \otimes e_{51} + (e_2^\perp \mathcal{A}_6 + e_1^\perp \mathcal{A}_5 + \mathcal{A}_4) \otimes e_{411} \\
& + (e_3^\perp \mathcal{A}_6 + e_2^\perp \mathcal{A}_5 + e_1^\perp \mathcal{A}_4 + \mathcal{A}_3) \otimes e_{3111} \\
& + (2s_{1111} + 2s_{2111} + s_{3111} + s_{4111} + s_{431} + s_{221} + 2s_{321} + 3s_{421} \\
& \quad + s_{521} + s_{621} + 2s_{211} + 2s_{311} + 5s_{411} + 3s_{511} + 4s_{611} + s_{711} \\
& \quad + s_{811} + s_{44} + s_{54} + 2s_{43} + 2s_{53} + s_{63} + s_{73} + 2s_{32} + 3s_{42} \\
& \quad + 3s_{52} + 3s_{62} + 3s_{72} + s_{82} + s_{92} + 2s_{41} + 2s_{51} + 3s_{61} + 3s_{71} \\
& \quad + 3s_{81} + 3s_{91} + s_{(10,1)} + s_{(11,1)} + 2s_7 + s_9 + 2s_{(11)} + s_{(13)}) \otimes e_{42} \\
& + (s_{1111} + s_{2111} + s_{3111} + s_{331} + s_{221} + s_{321} + s_{421} + s_{521} + s_{211} + 2s_{311} \\
& \quad + 2s_{411} + 2s_{511} + s_{611} + s_{711} + s_{44} + s_{33} + s_{43} + s_{53} + s_{63} + s_{22} \\
& \quad + 2s_{42} + 2s_{52} + s_{62} + s_{72} + s_{82} + 2s_{41} + s_{51} + s_{61} + 2s_{71} + s_{81} \\
& \quad + s_{91} + s_{(10,1)} + s_6 + s_9 + s_{(12)}) \otimes e_{33} \\
& + (2s_{221} + 3s_{321} + s_{421} + 6s_{111} + 8s_{211} + 8s_{311} + 4s_{411} + 3s_{511} + s_{611} \\
& \quad + 2s_{33} + 3s_{43} + s_{53} + 4s_{22} + 8s_{32} + 8s_{42} + 5s_{52} + 4s_{62} + s_{72} \\
& \quad + 6s_{21} + 10s_{31} + 12s_{41} + 12s_{51} + 10s_{61} + 5s_{71} + 4s_{81} \\
& \quad + 6s_4 + 4s_5 + 4s_6 + 4s_7 + 6s_8 + 2s_9 + s_{91} + 3s_{(10)} + s_{(11)}) \otimes e_{321} \\
& + (s_{221} + s_{111} + 2s_{211} + s_{311} + s_{411} + 2s_{22} + s_{32} + s_{42} + s_{52} \\
& \quad + 2s_{21} + 2s_{31} + 2s_{41} + 2s_{51} + s_{61} + s_{71} + s_3 + 2s_5 + s_7 + s_9) \otimes e_{222} \\
& + (3s_{22} + 2s_{32} + s_{42} + 6s_{11} + 9s_{21} 7s_{31} + 5s_{41} + 2s_{51} + s_{61} \\
& \quad + 6s_2 + 6s_3 + 6s_4 + 4s_5 + 5s_6 + 2s_7 + s_8) \otimes e_{2211} \\
& + (5s_1 + 4s_2 + 3s_3 + 2s_4 + s_5) \otimes e_{21111} + 1 \otimes e_{111111}.
\end{aligned}
$$

Length components of $\mathcal{A}_7$.

$$
\begin{aligned}
\mathcal{A}_7^{(1)} &= s_{(21)}, \qquad \mathcal{A}_7^{(6)} = s_{111111}; \\
\mathcal{A}_7^{(2)} &= s_{77} + (s_{76} + s_{86} + s_{96}) + (s_{75} + s_{85} + s_{95} + s_{(10,5)} + s_{(11,5)}) \\
& + (s_{74} + s_{84} + 2s_{94} + 2s_{(10,4)} + 2s_{(11,4)} + s_{(12,4)} + s_{(13,4)}) \\
& + (s_{93} + s_{(10,3)} + 2s_{(11,3)} + 2s_{(12,3)} + 2s_{(13,3)} + s_{(14,3)} + s_{(15,3)}) \\
& + (s_{(11,2)} + s_{(12,2)} + 2s_{(13,2)} + s_{(14,2)} + 2s_{(15,2)} + s_{(16,2)} + s_{(17,2)}) \\
& + (s_{(15,1)} + s_{(16,1)} + s_{(17,1)} + s_{(18,1)} + s_{(19,1)}); \\
\mathcal{A}_7^{(3)} &= s_{443} + s_{633} + (s_{442} + s_{542} + s_{642} + s_{742}) + (s_{532} + s_{632} + s_{732} + s_{832} + s_{932}) \\
& + (s_{522} + s_{722} + s_{822} + s_{922} + s_{(11,2,2)}) + s_{661} + (s_{651} + s_{751} + s_{851}) \\
& + (s_{441} + s_{541} + 2s_{641} + 3s_{741} + 3s_{841} + 2s_{941} + s_{(10,4,1)}) \\
& + (s_{631} + 2s_{731} + 3s_{831} + 3s_{931} + 3s_{(10,3,1)} + 2s_{(11,3,1)} + s_{(12,3,1)}) \\
& + (s_{721} + 2s_{821} + 2s_{921} + 3s_{(10,2,1)} + 3s_{(11,2,1)} + 2s_{(12,2,1)} + 2s_{(13,2,1)} + s_{(14,2,1)}) \\
& + (s_{(10,1,1)} + s_{(11,1,1)} + 2s_{(12,1,1)} + 2s_{(13,1,1)} + 2s_{(14,1,1)} + s_{(15,1,1)} + s_{(16,1,1)}); \\
\mathcal{A}_7^{(4)} &= (s_{4411} + s_{5411} + s_{6411}) + (s_{4311} + s_{5311} + 2s_{6311} + s_{7311} + s_{8311}) \\
& + (s_{4211} + s_{5211} + 2s_{6211} + s_{7211} + 2s_{8211} + s_{9211} + s_{(10,2,1,1)}) \\
& + (s_{6111} + s_{7111} + 2s_{8111} + 2s_{9111} + 2s_{(10,1,1,1)} + s_{(11,1,1,1)} + s_{(12,1,1,1)}); \\
\mathcal{A}_7^{(5)} &= s_{31111} + s_{41111} + s_{51111} + s_{61111} + s_{71111}.
\end{aligned}
$$

Acknowledgments

Much of this work would not have been achieved without the possibility of perusing sufficiently large expressions resulting from difficult explicit computations. These calculations were very elegantly realized by Nicolas Thiéry using the Sage computer algebra system, together with an inspired used of the right mathematical properties of objects considered, and his special expertise of higher Specht modules. Indeed, direct calculations of the relevant symmetric function expressions rapidly become unfeasible, even with powerful computers. Such explicit calculations must thus rely on an artful combination of high-level computer algebra skills, and a subtle understanding of the mathematical structures involved.

References

[1] Drew Armstrong, Nicholas A. Loehr, and Gregory S. Warrington, *Rational parking functions and Catalan numbers*, Ann. Comb. **20** (2016), no. 1, 21–58, DOI 10.1007/s00026-015-0293-6. MR3461934

[2] Drew Armstrong, Victor Reiner, and Brendon Rhoades, *Parking spaces*, Adv. Math. **269** (2015), 647–706, DOI 10.1016/j.aim.2014.10.012. MR3281144

[3] Drew Armstrong, Brendon Rhoades, and Nathan Williams, *Rational associahedra and noncrossing partitions*, Electron. J. Combin. **20** (2013), no. 3, Paper 54, 27, DOI 10.37236/3432. MR3118962

[4] F. Bergeron and A. M. Garsia, *Science fiction and Macdonald's polynomials*, Algebraic methods and q-special functions (Montréal, QC, 1996), CRM Proc. Lecture Notes, vol. 22, Amer. Math. Soc., Providence, RI, 1999, pp. 1–52, DOI 10.1090/crmp/022/01. MR1726826

[5] F. Bergeron, A. M. Garsia, M. Haiman, and G. Tesler, *Identities and positivity conjectures for some remarkable operators in the theory of symmetric functions. Part III*, Methods Appl. Anal. **6** (1999), no. 3, 363–420, DOI 10.4310/MAA.1999.v6.n3.a7. Dedicated to Richard A. Askey on the occasion of his 65th birthday. MR1803316

[6] François Bergeron, *Algebraic combinatorics and coinvariant spaces*, CMS Treatises in Mathematics, Canadian Mathematical Society, Ottawa, ON; A K Peters, Ltd., Wellesley, MA, 2009, DOI 10.1201/b10583. MR2538310

[7] François Bergeron, *Multivariate diagonal coinvariant spaces for complex reflection groups*, Adv. Math. **239** (2013), 97–108, DOI 10.1016/j.aim.2013.02.013. MR3045143

[8] François Bergeron, *Open questions for operators related to rectangular Catalan combinatorics*, J. Comb. **8** (2017), no. 4, 673–703, DOI 10.4310/JOC.2017.v8.n4.a6. MR3682396

[9] François Bergeron, *$(\mathrm{GL}_k \times \mathrm{Sym}_n)$-modules and nabla of hook-indexed Schur functions*, Algebr. Comb. **5** (2022), no. 5, 1033–1051. MR4511159

[10] François Bergeron and Louis-François Préville-Ratelle, *Higher trivariate diagonal harmonics via generalized Tamari posets*, J. Comb. **3** (2012), no. 3, 317–341, DOI 10.4310/JOC.2012.v3.n3.a4. MR3029440

[11] Nantel Bergeron, Cesar Ceballos, and Vincent Pilaud, *Hopf dreams and diagonal harmonics*, J. Lond. Math. Soc. (2) **105** (2022), no. 3, 1546–1600, DOI 10.1112/jlms.12541. MR4411312

[12] Igor Burban and Olivier Schiffmann, *On the Hall algebra of an elliptic curve, I*, Duke Math. J. **161** (2012), no. 7, 1171–1231, DOI 10.1215/00127094-1593263. MR2922373

[13] Erik Carlsson and Anton Mellit, *A proof of the shuffle conjecture*, J. Amer. Math. Soc. **31** (2018), no. 3, 661–697, DOI 10.1090/jams/893. MR3787405

[14] Michele D'Adderio, *e-positivity of vertical strip LLT polynomials*, J. Combin. Theory Ser. A **172** (2020), 105212, 15, DOI 10.1016/j.jcta.2020.105212. MR4054520

[15] A. M. Garsia and M. Haiman, *A remarkable q,t-Catalan sequence and q-Lagrange inversion*, J. Algebraic Combin. **5** (1996), no. 3, 191–244, DOI 10.1023/A:1022476211638. MR1394305

[16] J. Haglund, J. B. Remmel, and A. T. Wilson, *The delta conjecture*, Trans. Amer. Math. Soc. **370** (2018), no. 6, 4029–4057, DOI 10.1090/tran/7096. MR3811519

[17] Jim Haglund, *The combinatorics of knot invariants arising from the study of Macdonald polynomials*, Recent trends in combinatorics, IMA Vol. Math. Appl., vol. 159, Springer, [Cham], 2016, pp. 579–600, DOI 10.1007/978-3-319-24298-9_23. MR3526424

[18] Mark Haiman, *Vanishing theorems and character formulas for the Hilbert scheme of points in the plane*, Invent. Math. **149** (2002), no. 2, 371–407, DOI 10.1007/s002220200219. MR1918676

[19] Mark Haiman, *Combinatorics, symmetric functions, and Hilbert schemes*, in Current developments in mathematics, 2002, Int. Press, Somerville, MA, 2003, pp. 39–111.

[20] I. G. Macdonald, *Symmetric functions and Hall polynomials*, 2nd ed., Oxford Mathematical Monographs, The Clarendon Press, Oxford University Press, New York, 1995. With contributions by A. Zelevinsky; Oxford Science Publications. MR1354144

[21] Anton Mellit, *Toric braids and (m, n)-parking functions*, Duke Math. J. **170** (2021), no. 18, 4123–4169, DOI 10.1215/00127094-2021-0011. MR4348234

[22] Anton Mellit, *Homology of torus knots*, Geom. Topol. **26** (2022), no. 1, 47–70, DOI 10.2140/gt.2022.26.47. MR4404874

[23] Hugh Morton and Peter Samuelson, *The HOMFLYPT skein algebra of the torus and the elliptic Hall algebra*, Duke Math. J. **166** (2017), no. 5, 801–854, DOI 10.1215/00127094-3718881. MR3626565

[24] Andrei Negut, *The shuffle algebra revisited*, Int. Math. Res. Not. IMRN **22** (2014), 6242–6275, DOI 10.1093/imrn/rnt156. MR3283004

[25] Dun Qiu, Jeffrey B. Remmel, Emily Sergel, and Guoce Xin, *On the Schur positivity of $\Delta_{e_2} e_n[X]$*, Electron. J. Combin. **25** (2018), no. 4, Paper No. 4.20, 27, DOI 10.37236/7494. MR3874286

[26] G. C. Shephard and J. A. Todd, *Finite unitary reflection groups*, Canad. J. Math. **6** (1954), 274–304, DOI 10.4153/cjm-1954-028-3. MR59914

[27] Richard P. Stanley, *Parking functions and noncrossing partitions*, Electron. J. Combin. **4** (1997), no. 2, Research Paper 20, approx. 14, DOI 10.37236/1335. The Wilf Festschrift (Philadelphia, PA, 1996). MR1444167

[28] Nancy Wallace, *Explicit combinatorial formulas for some irreducible characters of the $GL_k \times S_n$-module of multivariate diagonal harmonics* (English, with English and French summaries), Sém. Lothar. Combin. **82B** (2020), Art. 71, 12. MR4098292

[29] A. T. Wilson, *Torus link homology and the nabla operator*, J. Combin. Theory Ser. A **154** (2018), 129–144, DOI 10.1016/j.jcta.2017.08.009. MR3718063

[30] Mike Zabrocki, *Vertex operators for standard bases of the symmetric functions*, J. Algebraic Combin. **13** (2001), no. 1, 83–101, DOI 10.1023/A:1008732003019. MR1817706

Département de Mathématiques, LACIM, Université du Québec à Montréal, Canada
URL: `http://bergeron.math.uqam.ca`
Email address: `bergeron.francois@uqam.ca`

Proceedings of Symposia in Pure Mathematics
Volume **110**, 2024
https://doi.org/10.1090/pspum/110/02019

Some open problems on Coxeter groups and unimodality

Francesco Brenti

ABSTRACT. In this paper I present some open problems on Coxeter groups and unimodality, together with the main partial results, and computational evidence, that are known about them.

1. Coxeter groups

Recall that a *Coxeter system* is a pair (W, S) where $S := \{s_1, \ldots, s_n\}$ is a finite set and W is a group having S as generating set and relations of the form

$$(s_i s_j)^{m_{i,j}} = e$$

for all $i, j \in [n]$, where $m_{i,j} \in \mathbb{P} \cup \{+\infty\}$ (where $\mathbb{P} := \{1, 2, 3, \ldots\}$), $m_{i,i} = 1$, $m_{i,j} = m_{j,i} \geq 2$ if $i \neq j$, and there is no relation if $m_{i,j} = +\infty$. Given $w \in W$ we let

$$\ell(w) := min\{k \in \mathbb{N} : \text{ there are } s_{i_1}, \ldots, s_{i_k} \in S \text{ such that } w = s_{i_1} \cdots s_{i_k}\}$$

(where $\mathbb{N} := \mathbb{P} \cup \{0\}$) and call this the *length* of w. The *right* (resp. *left*) *descent set* of w is then

$$D_R(w) := \{s \in S : \ell(w) > \ell(ws)\}$$

and

$$D_L(w) := \{s \in S : \ell(w) > \ell(sw)\}.$$

We let

$$T := \{wsw^{-1} : s \in S, w \in W\}$$

be the set of *reflections* of W, and

$$N_L(w) := \{t \in T : \ell(w) > \ell(tw)\}.$$

Recall that the *Bruhat graph* of W is the directed graph $B(W)$ having W as set of vertices and where, for all $u, v \in W$, $u \to v$ if and only if $v = u\,t$ for some $t \in T$ and $\ell(v) > \ell(u)$, in this case we also write $u \xrightarrow{t} v$. The *Bruhat order* on W is the partial order on W, denoted by $\leq$, that is the transitive closure of the Bruhat graph. For $u, v \in W$ we let $[u, v] := \{z \in W : u \leq z \leq v\}$ and $\ell(u, v) := \ell(v) - \ell(u)$.

It is well known (see, e.g., [**65**, Prop. 5.12] or [**14**, Cor. 7.1.4]) that if (W, S) is a Coxeter system then $\sum_{u \in W} x^{\ell(u)}$ is a rational generating function. The following problem was proposed by Stembridge in [**104**].

2020 *Mathematics Subject Classification.* Primary 05A20, 20F55; Secondary 05E10, 05E16.

The author was partially supported by the MIUR Excellence Department Projects CUP E83C18000100006 and E83C23000330006.

PROBLEM 1.1. *Let (W,S) be a Coxeter system. Is it true that then*

$$\sum_{t\in T} x^{\ell(t)}$$

is a rational generating function?

No partial results on this problem seem to be known.

The *Kazhdan-Lusztig polynomials* of W are a family of polynomials

$$\{P_{u,v}(q)\}_{u,v\in W} \subseteq \mathbb{Z}[q]$$

in one variable q, indexed by pairs of elements of W, that were first defined by Kazhdan and Lusztig in [**71**]. These polynomials have found important applications in various areas of mathematics (see, e.g., [**72**], [**7**], [**41**], and also [**10**], [**14**], [**66**], and the references cited there) and have been generalized in various ways (see, e.g., [**92**], [**101**, §6], [**30**], [**52**], [**79**], [**60**], [**33**], [**63**], to cite just a few). We find it most convenient here to define the parabolic Kazhdan-Lusztig polynomials. Let $J \subseteq S$. Recall that the (left) quotient of W corresponding to J is

$$W^J := \{u \in W : D_L(u) \subseteq S \setminus J\}.$$

The polynomials can be defined in two "Theorem-Definitions" ([**52**]).

THEOREM 1.2. *Let (W,S) be a Coxeter system, $J \subseteq S$, and $x \in \{-1,q\}$. Then there is a unique family of polynomials $\{R^{J,x}_{u,v}\}_{u,v\in W^J} \subseteq \mathbb{Z}[q]$ such that, for all $u,v \in W^J$:*

(i) $R^{J,x}_{u,v} = 0$ *if* $u \not\leq v$;
(ii) $R^{J,x}_{u,u} = 1$;
(iii) *if* $u < v$ *and* $s \in D_R(v)$ *then*

$$R^{J,x}_{u,v}(q) = \begin{cases} R^{J,x}_{us,vs}(q), & \text{if } us < u, \\ (q-1)R^{J,x}_{u,vs}(q) + qR^{J,x}_{us,vs}(q), & \text{if } u < us \in W^J, \\ (q-1-x)R^{J,x}_{u,vs}(q), & \text{if } u < us \notin W^J. \end{cases}$$

THEOREM 1.3. *Let (W,S) be a Coxeter system, $J \subseteq S$, and $x \in \{-1,q\}$. Then there is a unique family of polynomials $\{P^{J,x}_{u,v}\}_{u,v\in W^J} \subseteq \mathbb{Z}[q]$ such that, for all $u,v \in W^J$:*

(i) $P^{J,x}_{u,v} = 0$ *if* $u \not\leq v$;
(ii) $P^{J,x}_{u,u} = 1$;
(iii) $\deg(P^{J,x}_{u,v}) < \frac{1}{2}\ell(u,v)$ *if* $u < v$;
(iv)

$$q^{\ell(u,v)}\, P^{J,x}_{u,v}\left(\frac{1}{q}\right) = \sum_{a\in[u,v]^J} R^{J,x}_{u,a}(q)\, P^{J,x}_{a,v}(q)$$

if $u \leq v$, *where*

$$[u,v]^J := \{a \in W^J : u \leq a \leq v\}.$$

The polynomials $\{R^{J,x}_{u,v}\}_{u,v\in W^J}$ (resp., $\{P^{J,x}_{u,v}\}_{u,v\in W^J}$) whose existence the previous results show are called the *parabolic R-polynomials* (resp., *parabolic Kazhdan-Lusztig polynomials*) of W^J of type x. The polynomials $R_{u,v} := R^{\emptyset,-1}_{u,v}$ $(= R^{\emptyset,q}_{u,v})$ and $P_{u,v} := P^{\emptyset,-1}_{u,v}$ $(= P^{\emptyset,q}_{u,v})$ for $u,v \in W$, are called the *R-polynomials* (resp., *Kazhdan-Lusztig polynomials*) of W.

The most outstanding open problem about Kazhdan-Lusztig polynomials, particularly from a combinatorial point of view, is the so-called "Combinatorial Invariance Conjecture".

CONJECTURE 1.4. *Let (W_1, S_1), (W_2, S_2) be two Coxeter systems, and $u, v \in W_1$, $w, z \in W_2$ be such that $[u, v] \simeq [w, z]$ (isomorphic as posets). Then*

$$P_{u,v}(q) = P_{w,z}(q). \tag{1.1}$$

This conjecture was made by Lusztig [**13**] (see also [**54**, Rem. 7.31]). It may be noted that Kazhdan never conjectured the CIC. However, during a long mathematical conversation in Jerusalem [**70**] he told me: "I believe the [combinatorial invariance] conjecture to be true". The conjecture is known to be true if $\ell(u, v) \leq 4$ ([**54**, 7.31], [**14**, Chap. 5, Exercises 7 and 8]), if $\ell(u, v) \leq 8$ and W_1 and W_2 are both of type A ([**67**]), if $\ell(u, v) \leq 6$ and W_1 and W_2 are both of type B or D ([**67**]), if W_1 and W_2 are both of type $\tilde{A}_2$ ([**42**]), if $[u, v]$ is a lattice ([**54**, 7.23], [**24**, Thm. 6.3]), if one adds the hypothesis that "$P_{u,v}(q) = 1$" ([**43**, Thm. C], [**56**, Prop. 3.3]), if $uv^{-1} \in T$ and W_1 and W_2 are both of type A ([**27**, Cor. 4.7], [**56**, Prop. 3.3]), if $x = u = e$ ([**32**]), and for the coefficient of q if W_1 and W_2 are both simply laced ([**86**]). Note that an equivalent conjecture is obtained if one replaces (1.1) with "$R_{x,y}(q) = R_{u,v}(q)$". It would be desirable to have computational evidence in favor of the CIC, particularly for the exceptional finite Coxeter groups. I have been asked more than once if I believe the CIC. If I had to bet my money, I would say that it's true. The following result, proved in [**56**, Prop. 3.3], has a flavor similar to the CIC, and plays a role in some of the proofs of the CIC in special cases. For $u, v \in W$ let $B(u, v)$ be the directed graph induced on $[u, v]$ by $B(W)$ (so, the vertex set of $B(u, v)$ is $[u, v]$ and, if $x, y \in [u, v]$, there is a directed edge from x to y in $B(u, v)$ if and only if $x \to y$).

THEOREM 1.5. *Let (W_1, S_1), (W_2, S_2) be two Coxeter systems, and $u, v \in W_1$, $w, z \in W_2$ be such that $[u, v] \simeq [w, z]$ (isomorphic as posets). Then $B(u, v) \simeq B(w, z)$ (isomorphic as directed graphs).*

The most natural generalization of the CIC to the parabolic polynomials is that if (W_1, S_1) and (W_2, S_2) are two Coxeter systems, $J_1 \subseteq S_1$, $J_2 \subseteq S_2$, and $u, v \in W_1^{J_1}$, $w, z \in W_2^{J_2}$ are such that $[u, v]^{J_1} \simeq [w, z]^{J_2}$ (as posets) then $P_{u,v}^{J_1,q}(q) = P_{w,z}^{J_2,q}(q)$ (equivalently, $P_{u,v}^{J_1,-1}(q) = P_{w,z}^{J_2,-1}(q)$, $R_{u,v}^{J_1,q}(q) = R_{w,z}^{J_2,q}(q)$, $R_{u,v}^{J_1,-1}(q) = R_{w,z}^{J_2,-1}(q)$). Although this statement holds for some particularly nice and interesting quotients ([**31**], [**36**]) it's false in general ([**38**]). For example, if W_1 and W_2 are both of type B_5, $J_1 = J_2 = S \setminus \{s_3\}$ (numbering as in [**14**, Appendix A1]), $u = [4, 1, 5, 2, 3]$, $v = [5, -4, 1, 2, 3]$, $w = [1, 4, 2, 5, 3]$, and $z = [-4, 1, 5, 2, 3]$ then $[u, v]^{J_1} \simeq [w, z]^{J_2}$ while $P_{u,v}^{J_1,q}(q) = q \neq 0 = P_{w,z}^{J_2,q}(q)$.

A deeper generalization of the CIC to the parabolic setting has been proposed by Marietti in [**83**].

CONJECTURE 1.6. *Let (W_1, S_1), (W_2, S_2) be two Coxeter systems, $J_1 \subseteq S_1$, $J_2 \subseteq S_2$, $u, v \in W_1^{J_1}$, $w, z \in W_2^{J_2}$ and $f : [u, v] \to [w, z]$ be a poset isomorphism such that $f([u, v]^{J_1}) = [w, z]^{J_2}$. Then*

$$P_{u,v}^{J_1,q}(q) = P_{w,z}^{J_2,q}(q). \tag{1.2}$$

Again, an equivalent statement is obtained by substituting (1.2) with

$$\text{“}P_{u,v}^{J_1,-1}(q) = P_{w,z}^{J_2,-1}(q)\text{”}$$

or the analogous equalities for the parabolic R-polynomials. It is clear that Conjecture 1.6 reduces to the CIC if $J_1 = J_2 = \emptyset$, and that it holds if the naive parabolic CIC does. In addition, Conjecture 1.6 is known to hold if $u = w = e$ ([**84**]).

Just as finding a combinatorial interpretation is more satisfactory than proving nonnegativity so finding an explicit algorithm for computing the Kazhdan-Lusztig (or R) polynomial of a pair of elements $u, v \in W$ starting from the Bruhat interval $[u, v]$ as an abstract poset would be more satisfactory than simply proving the CIC. Recently, using techniques from deep learning ([**49**]), a candidate such algorithm has been proposed, in the case that W is a Weyl group of type A, in [**17**]. Let $u, v \in W$, $u \leq v$. For any $z \in [u, v]$ $z < v$, say that $[u, z]$ is *diamond complete* in $[u, v]$ if whenever $a, b, c, d \in [u, v]$ are such that $a \to b$, $b \to d$, $a \to c$, $c \to d$, and $a, b, c \in [u, z]$ then $d \in [u, z]$. Let $[u, z]$ be such a diamond complete subinterval. For any $w \in [u, z]$ let $U(w) := \{x \in [u, v] \setminus [u, z] : w \to x\}$, and consider the partial order induced on $U(w)$ by the Bruhat order. Say that $U(w)$ *spans a hypercube cluster* if whenever $A \subseteq U(w)$ is an antichain then there is a unique embedding of directed graphs $\theta : \mathcal{P}(A) \to B(u, v)$ such that $\theta(\emptyset) = w$ and $\theta(\{a\}) = a$ for all $a \in A$ (where $\mathcal{P}(A)$ is the directed graph having the subsets of A as vertices and where $B \to C$ if and only if $B \subset C$ and $|C \setminus B| = 1$). Finally, say that $[u, z]$ is a *hypercube decomposition* if ($[u, z]$ is diamond complete and) $U(w)$ spans a hypercube cluster for any $w \in [u, z]$. In [**17**] a procedure is given to compute, starting from a hypercube decomposition $[u, z]$ of a Bruhat interval $[u, v]$, and all the Kazhdan-Lusztig polynomials $P_{x,y}(q)$ for $x, y \in [u, v]$ such that $\ell(x, y) < \ell(u, v)$, a polynomial $\widetilde{P}_{u,v,z}(q)$, and the following conjecture is made ([**17**, Conj. 3.8]).

CONJECTURE 1.7. *Let W be a Weyl group of type A and $u, v \in W$, $u \leq v$. Then*

$$P_{u,v}(q) = \widetilde{P}_{u,v,z}(q) \tag{1.3}$$

for any hypercube decomposition $[u, z]$ of $[u, v]$.

It is shown in [**17**, Thm. 3.7] that if $u, v \in S_n$ then $\{w \in [u, v] : w^{-1}(1) = u^{-1}(1)\}$ is a hypercube decomposition of $[u, v]$ and (1.3) holds for this choice. Conjecture 1.7 is true for all intervals in S_n if $n \leq 7$, and for millions of intervals in S_8 and S_9. Note that a Bruhat interval in a general Coxeter group may not have a hypercube decomposition (for example, a 5-crown does not have a hypercube decomposition).

Although the nonnegativity conjecture for the Kazhdan-Lusztig polynomials ([**71**]) has been proved ([**59**]) there is (at least) one other nonnegativity conjecture that is still open. Recall (see [**57**, §2]) that a *reflection ordering* on (W, S) is a total order $\preceq$ on T such that if W' is a dihedral reflection subgroup of W (so $W' :=< J >$ for some $J \subseteq T$ and $S' := |\{t \in T : N_L(t) \cap W' = \{t\}\}| = 2$) then either

$$a \preceq aba \preceq ababa \preceq \cdots \preceq babab \preceq bab \preceq b$$

or

$$a \succeq aba \succeq ababa \succeq \cdots \succeq babab \succeq bab \succeq b$$

where $\{a, b\} := S'$. Equivalently, using the canonical bijection between T and Φ^+ (see, e.g., [**14**, Prop. 4.4.5]), a reflection ordering is a total order on Φ^+ such that if $\alpha, \beta \in \Phi^+$ and $\lambda, \mu \in \mathbb{R}_{>0}$ are such that $\lambda\alpha + \mu\beta \in \Phi^+$ then either $\alpha \preceq \lambda\alpha + \mu\beta \preceq \beta$ or $\alpha \succeq \lambda\alpha + \mu\beta \succeq \beta$. Reflection orderings always exist, and in fact there are many (we refer the reader to [**14**, §5.2], and [**57**] for further information about reflection orderings). If W is of type A then the lexicographic order $(1,2) \prec (1,3) \prec \cdots \prec (1,n) \prec (2,3) \prec \cdots \prec (n-1,n)$ is a reflection ordering.

Let $\prec$ be a reflection ordering of (W,S) and $\Gamma = x_0 \xrightarrow{t_1} x_1 \xrightarrow{t_2} \cdots \xrightarrow{t_k} x_k$ be a directed path in the Bruhat graph $B(W)$. Let a and b be two noncommuting variables. We associate to Γ a monomial $m_{\prec}(\Gamma) := y_1 \cdots y_{k-1}$ in a and b by letting

$$y_j := \begin{cases} a, & \text{if } t_j \prec t_{j+1}, \\ b, & \text{if } t_j \succ t_{j+1}, \end{cases}$$

for $j = 1, \ldots, k-1$, and $m_{\prec}(\Gamma) := 1$ if $k = 1$, and let

$$\widetilde{\psi}_{u,v} := \sum_{\Gamma} m_{\prec}(\Gamma)$$

where the sum is over all directed paths Γ from u to v in $B(W)$. It can be shown (see [**9**, Prop. 1.5], and also [**28**, Prop. 4.4]) that, although $m_{\prec}(\Gamma)$ depends on the reflection ordering used to define it, $\widetilde{\psi}_{u,v}$ does not. Also, it is known (see, e.g., the proof of Theorem 4 in [**6**]) that there is a polynomial $\widetilde{\Phi}_{u,v} \in \mathbb{Z}\langle c,d \rangle$ in two non-commuting variables c and d such that

$$\widetilde{\psi}_{u,v}(a,b) = \widetilde{\Phi}_{u,v}(a+b, ab+ba).$$

The polynomial $\widetilde{\Phi}_{u,v}(c,d)$ is called the *complete cd-index* of u, v. The reason for this terminology lies in the fact that the homogeneous part of highest possible degree in $\widetilde{\Phi}_{u,v}(c,d)$ (where $deg(a) = deg(b) = deg(c) = 1$ and $deg(d) = 2$) namely $\ell(u,v) - 1$ is (by [**103**, Thm. 3.14.2] and the fact that the assignment $\lambda(w,z) := w^{-1}z$ for $u \leq w \lhd z \leq v$ is an EL-labeling of $[u,v]$, [**57**]) the *cd*-index of $[u,v]$ as an Eulerian poset (see, e.g., [**103**, §3.17]). It is known ([**9**, Thm. 4.1]) that the coefficients of the Kazhdan-Lusztig polynomial $P_{u,v}(q)$ are given by explicit linear combinations of the coefficients of $\widetilde{\Phi}_{u,v}(c,d)$. The following conjecture appears in [**9**, Conj. 6.1].

CONJECTURE 1.8. *Let (W,S) be a Coxeter system, and $u, v \in W$, $u \leq v$. Then*

$$\widetilde{\Phi}_{u,v} \in \mathbb{N}\langle c,d \rangle.$$

Conjecture 1.8 has been verified if W is of type A and $\ell(u,v) \leq 7$, and is known to be true for the coefficients of the monomials of highest possible degree by [**69**, Thm. 1.3] since every Bruhat interval $[u,v]$ is a Gorenstein* poset. Some further evidence is presented in [**9**, §6] and [**16**].

It is known ([**89**], see also [**45**]) that if $P(q) \in \mathbb{N}[q]$ is such that $P(0) = 1$ then there are $n \in \mathbb{P}$ and $u, v \in S_n$ such that $P(q) = P_{u,v}(q)$. The following related problem was posed by Björner ([**13**]).

PROBLEM 1.9. *Let (W,S) be a Coxeter system and $u, v \in W$, $u \leq v$. Is it true that then there are a Coxeter system (W',S') and $w \in W'$ such that*

$$P_{u,v}(q) = P_{e,w}(q)\ ?$$

It is easy to see that the answer to this question is negative if one requires that $W' = W$. No partial results on this problem seem to be known.

The proof of the celebrated nonnegativity conjecture for Kazhdan-Lusztig polynomials ([**59**]) makes the following problem even more compelling.

PROBLEM 1.10. *Find a combinatorial interpretation for Kazhdan-Lusztig polynomials.*

So, given a Coxeter system (W, S) and $u, v \in W$, $u \leq v$, one would like to produce (in some explicit combinatorial way) a set $M(u,v)$ and function $s : M(u,v) \to \mathbb{N}$ such that

$$P_{u,v}(q) = \sum_{a \in M(u,v)} q^{s(a)}. \tag{1.4}$$

Problem 1.10 is open, and interesting, even for $q = 1$. Combinatorial interpretations for the Kazhdan-Lusztig polynomials are known if W is a Weyl group and v is a Deodhar element ([**12**, Thms. 2.3 and 5.12]), if (W, S) is a universal Coxeter system ([**55**, Thm. 3.8]), if W is a Weyl group and v is rationally smooth ([**11**, Thm. 2.5], [**71**, Thm. A2]), if the Coxeter graph of (W, S) is acyclic and u and v are Boolean elements ([**85**, Cor. 4.3]), if the Coxeter graph of (W, S) is a cycle and u and v are Boolean elements ([**81**, Thm.4.4]), if W is a Weyl group and $u, v \in W^J$ where (W, W_J) is a Hermitian symmetric pair ($J \subseteq S$) ([**18**], see also [**61**]), and if $[u, v]$ is isomorphic (as a poset) to a Boolean algebra or to the lattice of faces of an $(\ell(u,v) - 1)$-dimensional cube ([**24**, Cor. 6.8 and 6.9]). In addition to the above cases, if W is of type A then combinatorial interpretations are known for various special families of permutations ([**44**], [**45**], [**46**], [**109**]), including if $u = e$, $v([3]) = [n-2, n]$, $v([n-2, n]) = [3]$ and $v(4) > v(5) > \cdots > v(n-3)$ ([**46**, Cor. 5.5], [**44**, Thm. 4.8]). The combinatorial interpretation given in [**75**] (see also [**14**, Chap.5, Ex.39]) in the case that v is a permutation that avoids 3412 must be considered a conjecture since no complete proof of that statement is known. It should be noted that Deodhar in [**53**] constructs, given a Coxeter system (W, S) for which the Kazhdan-Lusztig polynomials have nonnegative coefficients and $u, v \in W$, $u \leq v$, a set $M(u,v)$ and function $s : M(u,v) \to \mathbb{N}$ such that (1.4) holds. However, the definition of $M(u,v)$ is recursive and for this reason this is not generally considered a "combinatorial interpretation". Still, the combinatorial interpretations for Deodhar elements referred to above, and a few others, have been obtained using Deodhar's general framework. For a more algebraic viewpoint on the combinatorial interpretation problem, and Deodhar's construction, see [**78**]. A hint for a general combinatorial interpretation could come from the following result, which is proved in [**88**, Thm. 5.8].

THEOREM 1.11. *Let (W, S) be a Coxeter system and $u, v, w \in W$, $u \leq v \leq w$. Then*

$$P_{v,w}(q) \leq P_{u,w}(q)$$

(coefficientwise).

Just as for the CIC one may consider the combinatorial interpretation problem also for the parabolic Kazhdan-Lusztig polynomials. The nonnegativity of the parabolic Kazhdan-Lusztig polynomials of type q, $\{P^{J,q}_{u,v}(q)\}_{u,v \in W^J}$, is proved in [**77**, Thm. 1.1]. Combinatorial interpretations of these polynomials are known if W is of type A and W^J is a tight quotient ([**31**], [**37**]), if W is a Weyl group and W^J is a quasi-minuscule quotient ([**31**], [**36**], [**38**], see also [**93**], [**76**], [**48**]), and if u and v are Boolean elements and W is of type A ([**82**, Thm. 5.2]). Also, it is known ([**91**, Thm. 5.8]) that it is enough to find a combinatorial interpretation for the parabolic Kazhdan-Lusztig polynomials $\{P^{J,x}_{u,v}\}_{u,v \in W^J}$ in the case that $|J| = |S| - 1$. A hint for a combinatorial interpretation could come from the following result which was conjectured in [**34**] and proved in [**77**, Cor. 8.4].

THEOREM 1.12. *Let (W,S) be a Coxeter system and $I \subseteq J \subseteq S$. Then*

$$P^{J,q}_{u,v}(q) \leq P^{I,q}_{u,v}(q)$$

(coefficientwise) for all $u, v \in W^J$.

The nonnegativity of the parabolic Kazhdan-Lusztig polynomials of type -1 is open in general, but see remark (2) after Theorem 1.1 in [**77**].

2. Unimodality

Recall that a sequence $(a_0, \ldots, a_n)$ is said to be *unimodal* if there is $0 \leq m \leq n$ such that $a_0 \leq \cdots \leq a_{m-1} \leq a_m \geq a_{m+1} \geq \cdots \geq a_n$, and is said to be *symmetric* if $a_j = a_{n-j}$ for all $j = 0, \ldots, n$. It is said to be *log-concave* if $(a_j)^2 \geq a_{j-1}a_{j+1}$ for all $1 \leq j \leq n-1$ and *ultra log-concave* if the sequence $\{\frac{a_j}{\binom{n}{j}}\}_{j=0,\ldots,n}$ is log-concave. We say that a polynomial $\sum_{j=0}^{n} a_j t^j$ is *unimodal* (resp. *symmetric*, *log-concave*, *ultra log-concave*) if the sequence $(a_0, \ldots, a_n)$ has the corresponding property. If $P(t) = \sum_{j=0}^{n} a_j t^j$ is symmetric then there are unique numbers $\gamma_0, \ldots, \gamma_{\lfloor n/2 \rfloor}$ such that

$$P(t) = \sum_{k=0}^{\lfloor n/2 \rfloor} \gamma_k \, t^k (1+t)^{n-2k}.$$

The vector $(\gamma_0, \ldots, \gamma_{\lfloor n/2 \rfloor})$ is called the γ-*vector* of $P(t)$, and the polynomial $P(t)$ is said to be γ-*nonnegative* if $\gamma_k \geq 0$ for all $0 \leq k \leq \lfloor n/2 \rfloor$. It is clear that a γ-nonnegative polynomial is unimodal, and it is not hard to see (see, e.g., [**20**, Rem. 1.3.1]) that if $P(t)$ is a symmetric polynomial with nonnegative coefficients and only real roots then $P(t)$ is γ-nonnegative. Unimodal, log-concave, and γ-nonnegative polynomials appear often in combinatorics, geometry, and algebra (see, e.g., [**100**], [**23**], [**20**], [**4**], and the references cited there). A real (finite or infinite) matrix $A := (A_{i,j})_{i,j \in \mathbb{N}}$ is said to be *totally positive* (or, *TP*, for short, sometimes more appropriately called *totally nonnegative*, [**62**]) if all the minors of A have nonnegative determinant. There is a deep relationship between polynomials with only real roots and totally positive matrices. The following result was first proved in [**58**] (see also [**87**, Thm. 4.5]).

THEOREM 2.1. *Let $P(t) = \sum_{j=0}^{n} a_j t^j$ be a polynomial with nonnegative coefficients. Then $P(t)$ has only real roots if and only if the matrix*

$$(a_{j-i})_{i,j \in \mathbb{N}}$$

is totally positive (where $a_k := 0$ if $k < 0$ or $k > n$).

Totally positive matrices arise in a number of areas in science, including mathematics, statistics, probability, economics, mechanics, and computer science (see, e.g., [**68**], the Foreword of [**87**], [**62**, Sec. 0.2] and the references cited there).

For $\sigma \in S_n$ let

$$L(\sigma) := |\{(i,j) \in [n]^2 : i < j,\ \sigma(i) > \sigma(j),\ i \not\equiv j \pmod 2\}|.$$

The statistic L is known as the *odd length* or *odd inversion number* of σ. The statistic was introduced in [**73**] in relation to formed spaces and has been further studied in [**39**] and [**40**]. The following conjecture appears in [**40**, Conj. 6.2].

CONJECTURE 2.2. *Let $n \in \mathbb{P}$, $n \geq 5$. Then*

$$L_n(x) := \sum_{\sigma \in S_n} x^{L(\sigma)}$$

is symmetric and unimodal.

The conjecture has been verified for $n \leq 11$. The symmetry statement is clear since $L(\sigma w_0) = L(w_0) - L(\sigma)$ for all $\sigma \in S_n$ where w_0 is the longest permutation $w_0 = n\, n-1 \cdots 3\,2\,1$. Note that, in general, $L_n(x)$ is not log-concave and not γ-nonnegative.

Let (W, S) be a Coxeter system. The following result follows easily from the case $J = \emptyset$ of Theorem 1.2.

PROPOSITION 2.3. *Let (W, S) be a Coxeter system. Then there exists a unique family of polynomials $\{\widetilde{R}_{u,v}(t)\}_{u,v \in W} \subseteq \mathbb{N}[t]$ such that*

$$R_{u,v}(q) = q^{\ell(u,v)/2} \widetilde{R}_{u,v}(q^{1/2} - q^{-1/2}).$$

So, knowledge of the $\widetilde{R}$-polynomials is equivalent to knowledge of the R-polynomials. Combinatorial interpretations of the polynomials $\{\widetilde{R}_{u,v}(t)\}_{u,v \in W}$ have been given in [**51**] and [**57**], see [**14**, Thms 5.3.4 and 5.3.7]. It is easy to see that, if $u \leq v$, then $deg(\widetilde{R}_{u,v}) = \ell(u,v)$ and the powers appearing in $\widetilde{R}_{u,v}(t)$ are all of the same parity. More precisely, there is a polynomial $Q_{u,v}(t) \in \mathbb{N}[t]$ such that $\widetilde{R}_{u,v}(t) = Q_{u,v}(t^2)$ if $\ell(u,v) \equiv 0 \pmod 2$ and $\widetilde{R}_{u,v}(t) = t\, Q_{u,v}(t^2)$ if $\ell(u,v) \equiv 1 \pmod 2$. It is of interest to characterize the R-polynomials. In this respect I propose the following conjecture which is a generalization of [**29**, Conj. 7.1].

CONJECTURE 2.4. *Let (W, S) be a finite Coxeter system and $u, v \in W$, $u \leq v$. Then the polynomial $Q_{u,v}(t)$ is log-concave.*

The conjecture has been verified if W is of type F_4, or H_3, or A_n, or B_n, or D_n and $n \leq 5$, and for dihedral groups. Note that the polynomial $Q_{u,v}(t)$ is not, in general, ultra log-concave. For example, if W is of type A_5, $u = 213465$, and $v = 563412$ then $\widetilde{R}_{u,v} = t^2 + 4t^4 + 6t^6 + 5t^8 + t^{10}$ and neither the sequence $(0, 1, 4, 6, 5, 1)$ nor the sequence $(1, 4, 6, 5, 1)$ are ultra log-concave. It is conceivable that the polynomials $Q_{u,v}(t)$ are log-concave for any Coxeter system.

Despite the settling ([**64**]) of the celebrated log-concavity conjecture for chromatic polynomials ([**90**]) there is (at least) one other unimodality statement about them that is still open. Let $G = (V, E)$ be a (simple, loopless) graph on p vertices and $\chi(G; x)$ be its chromatic polynomial. Write

$$\chi(G; x) = \sum_{i=0}^{p} (-1)^{p-i} c_i \langle x \rangle_i$$

where $\langle x \rangle_i := x(x+1)\cdots(x+i-1)$ for $i \geq 1$ and $\langle x \rangle_0 := 1$. The *τ-polynomial* of G is

$$\tau(G; x) := \sum_{i=0}^{p} c_i x^i.$$

The τ-polynomial was first introduced and studied in [**22**] where the following combinatorial interpretation of it is given. Let $\pi \in \Pi(V)$ where $\Pi(V)$ is the set of all set partitions of V, say $\pi = \{B_1, \ldots, B_k\}$. For $B \subseteq V$ let $G[B]$ be the subgraph induced by B (so $G[B] = (B, \{\{i,j\} \in E : i,j \in B\})$) and $G[\pi] := \biguplus_{i=1}^{k} G[B_i]$. Finally, for a graph H, let $a(H)$ be the number of acyclic orientations of H.

THEOREM 2.5. *Let G be a graph. Then*

$$\tau(G;x) = \sum_{\pi \in \Pi(V)} a(G[\pi])\, x^{|\pi|}.$$

The following problem was raised in [**22**, Prob. 7.1].

PROBLEM 2.6. *Does $\tau(G;x)$ have only real roots for all graphs G?*

The answer is yes for all connected graphs on ≤ 8 vertices, and if the chromatic polynomial of G has only real roots ([**22**, Thm. 6.1]).

Let P be a convex polytope of dimension d. Recall (see, e.g., [**102**, Chap. III, §1]) that P is *simplicial* if every proper face of it is a simplex. The *boundary complex* $\Delta(P)$ of P is then the simplicial complex of all the proper faces of P. Being a simplicial complex, $\Delta(P)$ has an h-vector $(h_0(\Delta(P)), \ldots, h_d(\Delta(P)))$ (we refer the reader to, e.g., [**102**, Chap. II, §2, p. 58] for the definition of the h-vector of a simplicial complex). The following famous result is well known (see, e.g., [**102**, Chap. III, Thm. 1.1]).

THEOREM 2.7. *Let P be a simplicial convex polytope. Then the h-vector of $\Delta(P)$ is symmetric and unimodal.*

For the barycentric subdivision of a simplicial convex polytope more can be said ([**35**, Cor. 3]).

THEOREM 2.8. *Let P be a simplicial convex polytope. Then the (generating polynomial of the) h-vector of the barycentric subdivision of P has only real roots.*

The following problem is natural, and has been circulated informally by the author since 2004.

PROBLEM 2.9. *Let $\sum_{i=0}^{d} h_i t^i \in \mathbb{N}[t]$ be a symmetric, monic polynomial with only real roots. Is there a simplicial convex polytope whose h-vector is $(h_0, \ldots, h_d)$?*

The answer is yes if $d \leq 9$ and $h_i \leq 100$ for all $0 \leq i \leq d$. Note that there are numerical characterizations both of h-vectors of simplicial convex polytopes ([**102**, Chap. III, Thm. 1.1]) as well as of monic polynomials with only real roots and nonnegative coefficients (see Theorem 2.1) thus Problem 2.9 is really asking whether one such set of inequalities implies the other one. The following related problem appears in [**74**, Question 4.4].

PROBLEM 2.10. *Let $\sum_{i=0}^{d} h_i t^i \in \mathbb{N}[t]$ be a polynomial with only real roots such that $h_0 = 1$ and $h_0 < h_1 < \cdots < h_k$ for some $0 \leq k \leq d$. Is there a simplicial complex whose f-vector is $(h_0, h_1 - h_0, \ldots, h_k - h_{k-1})$?*

Note that, by [**102**, Chap. III, Thm. 1.1], a positive answer to Problem 2.10 implies a positive answer to Problem 2.9. Related results and problems also appear in [**8**].

Let (W,S) be a Coxeter system. It is of interest, and difficult, to obtain properties of the rank generating function of Bruhat intervals $[u,v]$, $u,v \in W$ (see, e.g., [**15**]). In this respect, I feel that the following holds.

CONJECTURE 2.11. *Let W be a Weyl group, and $u,v \in W$. Then $[u,v]$ is rank log-concave.*

The conjecture has been verified if W is of type A_n and $n \leq 5$, or D_n and $n \leq 5$, or B_n and $n \leq 4$, or B_5 and $\ell(u,v) \geq 20$, or F_4, and for the dihedral groups. Note that the corresponding statement does not hold for finite Coxeter systems. For example, if (W,S) if of type H_3, $u = s_3$, and $v = s_1s_2s_3s_2s_1s_2s_1s_3$ (where $S = \{s_1, s_2, s_3\}$, $m(s_1,s_2) = 5$, and $m(s_2,s_3) = 3$) then the rank generating function of $[u,v]$ is $1 + 3t + 5t^2 + 7t^3 + 10t^4 + 10t^5 + 5t^6 + t^7$.

Many matrices arising in combinatorics are known to be TP (see, e.g., [**21**], [**25**], and [**26**]). It is therefore surprising that for a fundamental and old combinatorial matrix such as the one consisting of the Eulerian numbers this property has not yet been settled. For $n \in \mathbb{P}$ and $k \in \mathbb{N}$ let $A(n,k)$ be the corresponding *Eulerian number* (so, $A(n,k)$ is the number of permutations in S_n that have k descents). The following conjecture was first put forward in [**25**, Conj. 6.10].

CONJECTURE 2.12. *The matrix*

$$A := (A(n+1,k))_{n,k\in\mathbb{N}}$$

is totally positive.

It has been checked that $(A(n+1,k))_{0\leq n,k\leq 44}$ is TP. A more general conjecture, which includes Conjecture 2.12, has been proposed, and proved in some special cases, in [**47**, Conj. 1.4]. We feel that the following stronger property (which could be called "monotone total positivity") actually holds.

CONJECTURE 2.13. *For all $i,j,r \in \mathbb{N}$ the determinant of the submatrix determined by the rows indexed by $i, i+1, \ldots, i+r$ and columns indexed by $j, j+1, \ldots, j+r$ is a monotonically increasing function of $i \in \mathbb{N}$.*

We have checked that this is true if $j + r \leq 44$ and $i + r \leq 44$. Note that Conjecture 2.13 implies Conjecture 2.12 by [**87**, Thm. 2.8].

Acknowledgments

Some of the computations for the research presented in this paper have been carried out using some Maple packages for computing with Coxeter systems and posets developed by Pietro Mongelli and John Stembridge. I would like to thank Mario Marietti and Volkmar Welker for pointing out some relevant references.

References

[1] Ron M. Adin, *On face numbers of rational simplicial polytopes with symmetry*, Adv. Math. **115** (1995), no. 2, 269–285, DOI 10.1006/aima.1995.1057. MR1354672

[2] Henning Haahr Andersen, *The irreducible characters for semi-simple algebraic groups and for quantum groups*, Proceedings of the International Congress of Mathematicians, Vol. 1, 2 (Zürich, 1994), Birkhäuser, Basel, 1995, pp. 732–743. MR1403973

[3] T. Ando, *Totally positive matrices*, Linear Algebra Appl. **90** (1987), 165–219, DOI 10.1016/0024-3795(87)90313-2. MR884118

[4] Christos A. Athanasiadis, *Gamma-positivity in combinatorics and geometry*, Sém. Lothar. Combin. **77** ([2016), Art. B77i, 64. MR3878174

[5] M. F. Atiyah and I. G. Macdonald, *Introduction to commutative algebra*, Addison-Wesley Publishing Co., Reading, Mass.-London-Don Mills, Ont., 1969. MR242802

[6] Margaret M. Bayer and Andrew Klapper, *A new index for polytopes*, Discrete Comput. Geom. **6** (1991), no. 1, 33–47, DOI 10.1007/BF02574672. MR1073071

[7] Alexandre Beĭlinson and Joseph Bernstein, *Localisation de g-modules* (French, with English summary), C. R. Acad. Sci. Paris Sér. I Math. **292** (1981), no. 1, 15–18. MR610137

[8] Jason Bell and Mark Skandera, *Multicomplexes and polynomials with real zeros*, Discrete Math. **307** (2007), no. 6, 668–682, DOI 10.1016/j.disc.2006.07.020. MR2291443

[9] Louis J. Billera and Francesco Brenti, *Quasisymmetric functions and Kazhdan-Lusztig polynomials*, Israel J. Math. **184** (2011), 317–348, DOI 10.1007/s11856-011-0070-0. MR2823980

[10] Sara Billey and V. Lakshmibai, *Singular loci of Schubert varieties*, Progress in Mathematics, vol. 182, Birkhäuser Boston, Inc., Boston, MA, 2000, DOI 10.1007/978-1-4612-1324-6. MR1782635

[11] Sara Billey and Alexander Postnikov, *Smoothness of Schubert varieties via patterns in root subsystems*, Adv. in Appl. Math. **34** (2005), no. 3, 447–466, DOI 10.1016/j.aam.2004.08.003. MR2123545

[12] Sara C. Billey and Brant C. Jones, *Embedded factor patterns for Deodhar elements in Kazhdan-Lusztig theory*, Ann. Comb. **11** (2007), no. 3-4, 285–333, DOI 10.1007/s00026-007-0321-2. MR2376108

[13] A. Björner, private communication, Djursholm, 1992.

[14] Anders Björner and Francesco Brenti, *Combinatorics of Coxeter groups*, Graduate Texts in Mathematics, vol. 231, Springer, New York, 2005. MR2133266

[15] Anders Björner and Torsten Ekedahl, *On the shape of Bruhat intervals*, Ann. of Math. (2) **170** (2009), no. 2, 799–817, DOI 10.4007/annals.2009.170.799. MR2552108

[16] Saúl A. Blanco, *The complete* **cd***-index of dihedral and universal Coxeter groups*, Electron. J. Combin. **18** (2011), no. 1, Paper 174, 16, DOI 10.37236/661. MR2836809

[17] Charles Blundell, Lars Buesing, Alex Davies, Petar Veličković, and Geordie Williamson, *Towards combinatorial invariance for Kazhdan-Lusztig polynomials*, Represent. Theory **26** (2022), 1145–1191, DOI 10.1090/ert/624. MR4510816

[18] Brian D. Boe, *Kazhdan-Lusztig polynomials for Hermitian symmetric spaces*, Trans. Amer. Math. Soc. **309** (1988), no. 1, 279–294, DOI 10.2307/2001170. MR957071

[19] Petter Brändén, *Actions on permutations and unimodality of descent polynomials*, European J. Combin. **29** (2008), no. 2, 514–531, DOI 10.1016/j.ejc.2006.12.010. MR2388387

[20] Petter Brändén, *Unimodality, log-concavity, real-rootedness and beyond*, Handbook of enumerative combinatorics, Discrete Math. Appl. (Boca Raton), CRC Press, Boca Raton, FL, 2015, pp. 437–483. MR3409348

[21] Francesco Brenti, *Unimodal, log-concave and Pólya frequency sequences in combinatorics*, Mem. Amer. Math. Soc. **81** (1989), no. 413, viii+106, DOI 10.1090/memo/0413. MR963833

[22] Francesco Brenti, *Expansions of chromatic polynomials and log-concavity*, Trans. Amer. Math. Soc. **332** (1992), no. 2, 729–756, DOI 10.2307/2154193. MR1069745

[23] Francesco Brenti, *Log-concave and unimodal sequences in algebra, combinatorics, and geometry: an update*, Jerusalem combinatorics '93, Contemp. Math., vol. 178, Amer. Math. Soc., Providence, RI, 1994, pp. 71–89, DOI 10.1090/conm/178/01893. MR1310575

[24] Francesco Brenti, *A combinatorial formula for Kazhdan-Lusztig polynomials*, Invent. Math. **118** (1994), no. 2, 371–394, DOI 10.1007/BF01231537. MR1292116

[25] Francesco Brenti, *Combinatorics and total positivity*, J. Combin. Theory Ser. A **71** (1995), no. 2, 175–218, DOI 10.1016/0097-3165(95)90000-4. MR1342446

[26] F. Brenti, *The applications of total positivity to combinatorics, and conversely*, Total positivity and its applications (Jaca, 1994), Math. Appl., vol. 359, Kluwer Acad. Publ., Dordrecht, 1996, pp. 451–473, DOI 10.1007/978-94-015-8674-0_22. MR1421615

[27] Francesco Brenti, *Combinatorial properties of the Kazhdan-Lusztig R-polynomials for* S_n, Adv. Math. **126** (1997), no. 1, 21–51, DOI 10.1006/aima.1996.1607. MR1440252

[28] Francesco Brenti, *Combinatorial expansions of Kazhdan-Lusztig polynomials*, J. London Math. Soc. (2) **55** (1997), no. 3, 448–472, DOI 10.1112/S0024610797004948. MR1452258

[29] Francesco Brenti, *Kazhdan-Lusztig and R-polynomials from a combinatorial point of view*, Discrete Math. **193** (1998), no. 1-3, 93–116, DOI 10.1016/S0012-365X(98)00137-X. Selected papers in honor of Adriano Garsia (Taormina, 1994). MR1661365

[30] Francesco Brenti, *Twisted incidence algebras and Kazhdan-Lusztig-Stanley functions*, Adv. Math. **148** (1999), no. 1, 44–74, DOI 10.1006/aima.1999.1839. MR1736641

[31] Francesco Brenti, *Kazhdan-Lusztig and R-polynomials, Young's lattice, and Dyck partitions*, Pacific J. Math. **207** (2002), no. 2, 257–286, DOI 10.2140/pjm.2002.207.257. MR1972246

[32] Francesco Brenti, Fabrizio Caselli, and Mario Marietti, *Special matchings and Kazhdan-Lusztig polynomials*, Adv. Math. **202** (2006), no. 2, 555–601, DOI 10.1016/j.aim.2005.01.011. MR2222360

[33] Francesco Brenti, Fabrizio Caselli, and Mario Marietti, *Diamonds and Hecke algebra representations*, Int. Math. Res. Not., posted on 2006, Art. ID 29407, 34, DOI 10.1155/IMRN/2006/29407. MR2250022

[34] F. Brenti, talk at *Festive Combinatorics in honor of Anders Björner's 60th birthday*, Royal Institute of Technology (KTH), Stockholm, Sweden, 5/28-30, 2008.

[35] Francesco Brenti and Volkmar Welker, *f-vectors of barycentric subdivisions*, Math. Z. **259** (2008), no. 4, 849–865, DOI 10.1007/s00209-007-0251-z. MR2403744

[36] Francesco Brenti, *Parabolic Kazhdan-Lusztig polynomials for Hermitian symmetric pairs*, Trans. Amer. Math. Soc. **361** (2009), no. 4, 1703–1729, DOI 10.1090/S0002-9947-08-04458-9. MR2465813

[37] Francesco Brenti, Federico Incitti, and Mario Marietti, *Kazhdan-Lusztig polynomials, tight quotients and Dyck superpartitions*, Adv. in Appl. Math. **47** (2011), no. 3, 589–614, DOI 10.1016/j.aam.2011.02.002. MR2822204

[38] Francesco Brenti, Pietro Mongelli, and Paolo Sentinelli, *Parabolic Kazhdan-Lusztig polynomials for quasi-minuscule quotients*, Adv. in Appl. Math. **78** (2016), 27–55, DOI 10.1016/j.aam.2016.01.005. MR3497995

[39] Francesco Brenti and Angela Carnevale, *Proof of a conjecture of Klopsch-Voll on Weyl groups of type A*, Trans. Amer. Math. Soc. **369** (2017), no. 10, 7531–7547, DOI 10.1090/tran/7197. MR3683117

[40] Francesco Brenti and Angela Carnevale, *Odd length: odd diagrams and descent classes*, Discrete Math. **344** (2021), no. 5, Paper No. 112308, 17, DOI 10.1016/j.disc.2021.112308. MR4210959

[41] J.-L. Brylinski and M. Kashiwara, *Kazhdan-Lusztig conjecture and holonomic systems*, Invent. Math. **64** (1981), no. 3, 387–410, DOI 10.1007/BF01389272. MR632980

[42] Gaston Burrull, Nicolas Libedinsky, and David Plaza, *Combinatorial invariance conjecture for $\tilde{A}_2$*, Int. Math. Res. Not. IMRN **10** (2023), 8903–8933, DOI 10.1093/imrn/rnac105. MR4589090

[43] James B. Carrell, *The Bruhat graph of a Coxeter group, a conjecture of Deodhar, and rational smoothness of Schubert varieties*, Algebraic groups and their generalizations: classical methods (University Park, PA, 1991), Proc. Sympos. Pure Math., vol. 56, Amer. Math. Soc., Providence, RI, 1994, pp. 53–61, DOI 10.1090/pspum/056.1/1278700. MR1278700

[44] Fabrizio Caselli, *Proof of two conjectures of Brenti and Simion on Kazhdan-Lusztig polynomials*, J. Algebraic Combin. **18** (2003), no. 3, 171–187, DOI 10.1023/B:JACO.0000011936.75388.14. MR2036113

[45] Fabrizio Caselli, *A simple combinatorial proof of a generalization of a result of Polo*, Represent. Theory **8** (2004), 479–486, DOI 10.1090/S1088-4165-04-00203-1. MR2110357

[46] F. Caselli and M. Marietti, *Formulas for multi-parameter classes of Kazhdan-Lusztig polynomials in $\mathfrak{S}(n)$*, Discrete Math. **306** (2006), no. 8-9, 711–725, DOI 10.1016/j.disc.2006.02.006. MR2234979

[47] Xi Chen, Bishal Deb, Alexander Dyachenko, Tomack Gilmore, and Alan D. Sokal, *Coefficientwise total positivity of some matrices defined by linear recurrences*, Sém. Lothar. Combin. **85B** (2021), Art. 30, 12. MR4311911

[48] Anton Cox and Maud De Visscher, *Diagrammatic Kazhdan-Lusztig theory for the (walled) Brauer algebra*, J. Algebra **340** (2011), 151–181, DOI 10.1016/j.jalgebra.2011.05.024. MR2813567

[49] A. Davies, P. Veličković, L. Buesing, S. Blackwell1, D. Zheng, N. Tomašev, R. Tanburn, P. Battaglia, C. Blundell, A. Juhàsz, M. Lackenby, G. Williamson, D. Hassabis, P. Kohli, *Advancing mathematics by guiding human intuition with AI*, Nature, **600** (2021), 70–74.

[50] Ewan Delanoy, *Combinatorial invariance of Kazhdan-Lusztig polynomials on intervals starting from the identity*, J. Algebraic Combin. **24** (2006), no. 4, 437–463, DOI 10.1007/s10801-006-0014-7. MR2266379

[51] Vinay V. Deodhar, *On some geometric aspects of Bruhat orderings. I. A finer decomposition of Bruhat cells*, Invent. Math. **79** (1985), no. 3, 499–511, DOI 10.1007/BF01388520. MR782232

[52] Vinay V. Deodhar, *On some geometric aspects of Bruhat orderings. II. The parabolic analogue of Kazhdan-Lusztig polynomials*, J. Algebra **111** (1987), no. 2, 483–506, DOI 10.1016/0021-8693(87)90232-8. MR916182

[53] Vinay V. Deodhar, *A combinatorial setting for questions in Kazhdan-Lusztig theory*, Geom. Dedicata **36** (1990), no. 1, 95–119, DOI 10.1007/BF00181467. MR1065215

[54] M. J. Dyer, *Hecke algebras and reflections in Coxeter groups*, Ph.D. Thesis, University of Sydney, 1987.

[55] Matthew Dyer, *On some generalisations of the Kazhdan-Lusztig polynomials for "universal" Coxeter systems*, J. Algebra **116** (1988), no. 2, 353–371, DOI 10.1016/0021-8693(88)90223-2. MR953157

[56] Matthew Dyer, *On the "Bruhat graph" of a Coxeter system*, Compositio Math. **78** (1991), no. 2, 185–191. MR1104786

[57] M. J. Dyer, *Hecke algebras and shellings of Bruhat intervals*, Compositio Math. **89** (1993), no. 1, 91–115. MR1248893

[58] Albert Edrei, *Proof of a conjecture of Schoenberg on the generating function of a totally positive sequence*, Canad. J. Math. **5** (1953), 86–94, DOI 10.4153/cjm-1953-010-3. MR53176

[59] Ben Elias and Geordie Williamson, *The Hodge theory of Soergel bimodules*, Ann. of Math. (2) **180** (2014), no. 3, 1089–1136, DOI 10.4007/annals.2014.180.3.6. MR3245013

[60] Ben Elias, Nicholas Proudfoot, and Max Wakefield, *The Kazhdan-Lusztig polynomial of a matroid*, Adv. Math. **299** (2016), 36–70, DOI 10.1016/j.aim.2016.05.005. MR3519463

[61] Thomas J. Enright and Brad Shelton, *Categories of highest weight modules: applications to classical Hermitian symmetric pairs*, Mem. Amer. Math. Soc. **67** (1987), no. 367, iv+94, DOI 10.1090/memo/0367. MR888703

[62] Shaun M. Fallat and Charles R. Johnson, *Totally nonnegative matrices*, Princeton Series in Applied Mathematics, Princeton University Press, Princeton, NJ, 2011, DOI 10.1515/9781400839018. MR2791531

[63] R. M. Green and J. Losonczy, *Canonical bases for Hecke algebra quotients*, Math. Res. Lett. **6** (1999), no. 2, 213–222, DOI 10.4310/MRL.1999.v6.n2.a9. MR1689211

[64] June Huh, *Milnor numbers of projective hypersurfaces and the chromatic polynomial of graphs*, J. Amer. Math. Soc. **25** (2012), no. 3, 907–927, DOI 10.1090/S0894-0347-2012-00731-0. MR2904577

[65] James E. Humphreys, *Reflection groups and Coxeter groups*, Cambridge Studies in Advanced Mathematics, vol. 29, Cambridge University Press, Cambridge, 1990, DOI 10.1017/CBO9780511623646. MR1066460

[66] James E. Humphreys, *Representations of semisimple Lie algebras in the BGG category $\mathcal{O}$*, Graduate Studies in Mathematics, vol. 94, American Mathematical Society, Providence, RI, 2008, DOI 10.1090/gsm/094. MR2428237

[67] Federico Incitti, *More on the combinatorial invariance of Kazhdan-Lusztig polynomials*, J. Combin. Theory Ser. A **114** (2007), no. 3, 461–482, DOI 10.1016/j.jcta.2006.06.009. MR2310745

[68] Samuel Karlin, *Total positivity. Vol. I*, Stanford University Press, Stanford, CA, 1968. MR230102

[69] Kalle Karu, *The cd-index of fans and posets*, Compos. Math. **142** (2006), no. 3, 701–718, DOI 10.1112/S0010437X06001928. MR2231198

[70] D. Kazhdan, private communication, Jerusalem, 2001.

[71] David Kazhdan and George Lusztig, *Representations of Coxeter groups and Hecke algebras*, Invent. Math. **53** (1979), no. 2, 165–184, DOI 10.1007/BF01390031. MR560412

[72] David Kazhdan and George Lusztig, *Schubert varieties and Poincaré duality*, Geometry of the Laplace operator (Proc. Sympos. Pure Math., Univ. Hawaii, Honolulu, Hawaii, 1979), Proc. Sympos. Pure Math., XXXVI, Amer. Math. Soc., Providence, RI, 1980, pp. 185–203. MR573434

[73] Benjamin Klopsch and Christopher Voll, *Igusa-type functions associated to finite formed spaces and their functional equations*, Trans. Amer. Math. Soc. **361** (2009), no. 8, 4405–4436, DOI 10.1090/S0002-9947-09-04671-6. MR2500892

[74] Martina Kubitzke and Volkmar Welker, *Enumerative g-theorems for the Veronese construction for formal power series and graded algebras*, Adv. in Appl. Math. **49** (2012), no. 3-5, 307–325, DOI 10.1016/j.aam.2012.08.002. MR3017962

[75] Alain Lascoux, *Polynômes de Kazhdan-Lusztig pour les variétés de Schubert vexillaires* (French, with English and French summaries), C. R. Acad. Sci. Paris Sér. I Math. **321** (1995), no. 6, 667–670. MR1354702

[76] Tobias Lejczyk and Catharina Stroppel, *A graphical description of (D_n, A_{n-1}) Kazhdan-Lusztig polynomials*, Glasg. Math. J. **55** (2013), no. 2, 313–340, DOI 10.1017/S0017089512000547. MR3040865

[77] Nicolas Libedinsky and Geordie Williamson, *The anti-spherical category*, Adv. Math. **405** (2022), Paper No. 108509, 34, DOI 10.1016/j.aim.2022.108509. MR4437613

[78] Nicolas Libedinsky and Geordie Williamson, *Kazhdan-Lusztig polynomials and subexpressions*, J. Algebra **568** (2021), 181–192, DOI 10.1016/j.jalgebra.2020.10.009. MR4166587

[79] George Lusztig and David A. Vogan Jr., *Singularities of closures of K-orbits on flag manifolds*, Invent. Math. **71** (1983), no. 2, 365–379, DOI 10.1007/BF01389103. MR689649

[80] M. Marietti, *Kazhdan-Lusztig Theory: Boolean elements, special matchings and combinatorial invariance*, Ph.D. Thesis, Università di Roma "La Sapienza", 2003.

[81] Mario Marietti, *Boolean elements in Kazhdan-Lusztig theory*, J. Algebra **295** (2006), no. 1, 1–26, DOI 10.1016/j.jalgebra.2005.09.031. MR2188847

[82] M. Marietti, *Parabolic Kazhdan-Lusztig and R-polynomials for Boolean elements in the symmetric group*, European J. Combin. **31** (2010), no. 3, 908–924, DOI 10.1016/j.ejc.2009.06.001. MR2587039

[83] Mario Marietti, *Special matchings and parabolic Kazhdan-Lusztig polynomials*, Trans. Amer. Math. Soc. **368** (2016), no. 7, 5247–5269, DOI 10.1090/tran6676. MR3456179

[84] Mario Marietti, *The combinatorial invariance conjecture for parabolic Kazhdan-Lusztig polynomials of lower intervals*, Adv. Math. **335** (2018), 180–210, DOI 10.1016/j.aim.2018.07.008. MR3836662

[85] Pietro Mongelli, *Kazhdan-Lusztig polynomials of Boolean elements*, J. Algebraic Combin. **39** (2014), no. 2, 497–525, DOI 10.1007/s10801-013-0455-8. MR3159260

[86] Leonardo Patimo, *A combinatorial formula for the coefficient of q in Kazhdan-Lusztig polynomials*, Int. Math. Res. Not. IMRN **5** (2021), 3203–3223, DOI 10.1093/imrn/rnz255. MR4227568

[87] Allan Pinkus, *Totally positive matrices*, Cambridge Tracts in Mathematics, vol. 181, Cambridge University Press, Cambridge, 2010. MR2584277

[88] David Plaza, *Graded cellularity and the monotonicity conjecture*, J. Algebra **473** (2017), 324–351, DOI 10.1016/j.jalgebra.2016.10.040. MR3591153

[89] Patrick Polo, *Construction of arbitrary Kazhdan-Lusztig polynomials in symmetric groups*, Represent. Theory **3** (1999), 90–104, DOI 10.1090/S1088-4165-99-00074-6. MR1698201

[90] Ronald C. Read, *An introduction to chromatic polynomials*, J. Combinatorial Theory **4** (1968), 52–71. MR224505

[91] Paolo Sentinelli, *Isomorphisms of Hecke modules and parabolic Kazhdan-Lusztig polynomials*, J. Algebra **403** (2014), 1–18, DOI 10.1016/j.jalgebra.2014.01.010. MR3166061

[92] Vera Serganova, *Kazhdan-Lusztig polynomials for Lie superalgebra $\mathfrak{gl}(m|n)$*, I. M. Gel′fand Seminar, Adv. Soviet Math., vol. 16, Amer. Math. Soc., Providence, RI, 1993, pp. 151–165. MR1237837

[93] Keiichi Shigechi and Paul Zinn-Justin, *Path representation of maximal parabolic Kazhdan-Lusztig polynomials*, J. Pure Appl. Algebra **216** (2012), no. 11, 2533–2548, DOI 10.1016/j.jpaa.2012.03.027. MR2927185

[94] Richard P. Stanley, *Hilbert functions of graded algebras*, Advances in Math. **28** (1978), no. 1, 57–83, DOI 10.1016/0001-8708(78)90045-2. MR485835

[95] Richard P. Stanley, *Unimodal sequences arising from Lie algebras*, Combinatorics, representation theory and statistical methods in groups, Lect. Notes Pure Appl. Math., vol. 57, Dekker, New York, 1980, pp. 127–136. MR588199

[96] Richard P. Stanley, *The number of faces of a simplicial convex polytope*, Adv. in Math. **35** (1980), no. 3, 236–238, DOI 10.1016/0001-8708(80)90050-X. MR563925

[97] Richard P. Stanley, *An introduction to combinatorial commutative algebra*, Enumeration and design (Waterloo, Ont., 1982), Academic Press, Toronto, ON, 1984, pp. 3–18. MR782306

[98] Richard P. Stanley, *Unimodality and Lie superalgebras*, Stud. Appl. Math. **72** (1985), no. 3, 263–281, DOI 10.1002/sapm1985723263. MR790132

[99] Richard Stanley, *Generalized H-vectors, intersection cohomology of toric varieties, and related results*, Commutative algebra and combinatorics (Kyoto, 1985), Adv. Stud. Pure Math., vol. 11, North-Holland, Amsterdam, 1987, pp. 187–213, DOI 10.2969/aspm/01110187. MR951205

[100] Richard P. Stanley, *Log-concave and unimodal sequences in algebra, combinatorics, and geometry*, Graph theory and its applications: East and West (Jinan, 1986), Ann. New York Acad. Sci., vol. 576, New York Acad. Sci., New York, 1989, pp. 500–535, DOI 10.1111/j.1749-6632.1989.tb16434.x. MR1110850

[101] Richard P. Stanley, *Subdivisions and local h-vectors*, J. Amer. Math. Soc. **5** (1992), no. 4, 805–851, DOI 10.2307/2152711. MR1157293

[102] Richard P. Stanley, *Combinatorics and commutative algebra*, 2nd ed., Progress in Mathematics, vol. 41, Birkhäuser Boston, Inc., Boston, MA, 1996. MR1453579

[103] Richard P. Stanley, *Enumerative combinatorics. Volume 1*, 2nd ed., Cambridge Studies in Advanced Mathematics, vol. 49, Cambridge University Press, Cambridge, 2012. MR2868112

[104] J. Stembridge, *Open Problems Session*, Mathematical Sciences Research Institute, Berkeley, CA, 1997.

[105] John R. Stembridge, *Counterexamples to the poset conjectures of Neggers, Stanley, and Stembridge*, Trans. Amer. Math. Soc. **359** (2007), no. 3, 1115–1128, DOI 10.1090/S0002-9947-06-04271-1. MR2262844

[106] David G. Wagner, *Total positivity of Hadamard products*, J. Math. Anal. Appl. **163** (1992), no. 2, 459–483, DOI 10.1016/0022-247X(92)90261-B. MR1145841

[107] David G. Wagner, *Logarithmic concavity and* $\mathfrak{sl}_2(\mathbf{C})$, J. Combin. Theory Ser. A **94** (2001), no. 2, 383–386, DOI 10.1006/jcta.2000.3144. MR1825794

[108] David G. Wagner, *Multivariate stable polynomials: theory and applications*, Bull. Amer. Math. Soc. (N.S.) **48** (2011), no. 1, 53–84, DOI 10.1090/S0273-0979-2010-01321-5. MR2738906

[109] Alexander Woo, *Permutations with Kazhdan-Lusztig polynomial* $P_{id,w}(q) = 1+q^h$, Electron. J. Combin. **16** (2009), no. 2, Special volume in honor of Anders Björner, Research Paper 10, 32, DOI 10.37236/76. With an appendix by Sara Billey and Jonathan Weed. MR2515773

Dipartimento di Matematica, Universitá di Roma "Tor Vergata", Via della Ricerca Scientifica, 1, 00133 Roma, Italy

Email address: brenti@mat.uniroma2.it

Proceedings of Symposia in Pure Mathematics
Volume **110**, 2024
https://doi.org/10.1090/pspum/110/02012

Some combinatorial problems arising in the dimer model

R. Kenyon

1. Introduction

We collect here a few diverse problems on the dimer model. For background on dimers one can read for example [**6**].

Let $\mathcal{G} = (V, E)$ be a finite, connected, bipartite graph. A *dimer cover*, or *perfect matching*, of $\mathcal{G}$ is a collection $m \subset E$ of edges that covers each vertex exactly once. We denote by $\mathcal{M} = \mathcal{M}(\mathcal{G})$ the set of dimer covers of $\mathcal{G}$.

Let $c : E \to \mathbb{R}_{>0}$ be a positive weight function edges of $\mathcal{G}$. If $m \in \mathcal{M}$ is a dimer cover, we define $c(m) = \prod_{e\in m} c_e$ to be the *weight* of m. This weight function defines a probability measure $\mu = \mu_c$ on $\mathcal{M}$, giving a dimer cover a probability proportional to its weight, that is, $\Pr(m) = \frac{c(m)}{Z}$, where the constant of proportionality Z is the *partition function*: $Z = \sum_{m\in\mathcal{M}} c(m)$.

Essentially the most fundamental problem in the dimer model is the following:

Problem 0. *For a random dimer cover, compute the edge probabilities, that is, compute the probability* $\Pr(e)$ *that any particular edge* e *is used.*

This is easy if we can compute Z as a function of edge weights, since

$$\Pr(e) = \frac{c_e}{Z}\frac{\partial Z}{\partial c_e} = \frac{\partial \log Z}{\partial \log c_e}. \tag{1.1}$$

However on general graphs computing Z is hard, even for constant edge weights: enumerating dimer covers is in fact #P-complete, by a celebrated result of Valiant [**11**]. Note that computing $\Pr(e)$ is in fact just as hard as computing Z, since one can reconstruct Z from integrating $\Pr(e)$ using (1.1). For planar graphs we can however compute Z and thus solve the above problem, using Kasteleyn's method of enumerating dimer covers with determinants, see [**6**] and Section 2.2 below.

One useful change of perspective is to consider the edge weights as a "$\mathbb{C}^*$-local system" on $\mathcal{G}$, that is, as a connection on a line bundle on the graph. This point of view may seem at first just an added level of abstraction, and to be taking us away from combinatorics. But it turns out to have a number of advantages, such as:

(1) It highlights a non-obvious symmetry of the problem: the gauge invariance (see below).

2020 *Mathematics Subject Classification.* Primary 82B20.

This research was supported by NSF grant DMS-1940932 and the Simons Foundation grant 327929.

(2) It generalizes naturally to other gauge groups such as $\mathrm{SL}_n(\mathbb{R})$, with interesting combinatorial interpretations (see Sections 6 and 7 below).
(3) It connects the problem to geometry.

Our choice of problems in this article is (mostly) motivated by this geometric point of view.

2. Preliminaries

2.1. Vector bundles and connections on graphs. What is a bundle with connection on a graph? Fix a vector space $\mathcal{V}$ and assign an isomorphic copy $\mathcal{V}_v$ of $\mathcal{V}$ to each vertex $v \in V$. For each edge $e = uv$ we associate a linear isomorphism $\phi_{uv} : \mathcal{V}_u \to \mathcal{V}_v$ of these vector spaces, with $\phi_{vu} = \phi_{uv}^{-1}$. Generally we have $\phi_{uv} \in \mathrm{GL}(\mathcal{V})$ but we can restrict if desired to other subgroups of $\mathrm{GL}(\mathcal{V})$, and even noninvertible endomorphisms [**2**]. In this paper we will only use $\mathrm{SL}_n(\mathbb{C})$ connections on $\mathbb{C}^n$ or $\mathbb{C}^*$-connections on $\mathbb{C}$. (Here $\mathbb{C}^* = \mathbb{C} \setminus \{0\}$ is the group of $\mathbb{C}$-linear automorphisms of $\mathbb{C}$.) These are called, respectively, SL_n-local systems and $\mathbb{C}^*$-local systems. Sometimes we specialize the latter from $\mathbb{C}^*$ to $\mathbb{R}^*$.

If we choose a basis for each vector space $\mathcal{V}_v$ then each ϕ_{uv} is a matrix. Changing bases by matrices A_v at vertices v results in new edge matrices $\phi'_{uv} = A_u \phi_{uv} A_v^{-1}$. Such changes are called *gauge transformations.* They form a symmetry of the system.

If γ is an oriented loop in $\mathcal{G}$ based at x_0, the *monodromy* ϕ_γ of the connection around γ is the composition of the isomorphisms ϕ_e along γ starting at x_0. The conjugacy class of ϕ_γ is invariant under gauge transformations. Note also that changing the basepoint along γ conjugates the monodromy.

For one-dimensional bundles, with $\mathcal{V} = \mathbb{R}$, each $\phi_{uv} \in \mathbb{R}^*$ is a scalar. Since we are assuming $\mathcal{G}$ is bipartite, we can canonically orient the edges from black to white. Then ϕ_{bw} is a scalar quantity associated to edge bw which we call the *edge weight.* In probability settings we usually take ϕ_{bw} positive. However it is sometimes useful to consider $\mathcal{V} = \mathbb{C}$ and $\phi_{bw} \in \mathbb{C}^*$. Gauge transformations are functions $A : V \to \mathbb{C}^*$ and transform ϕ_{bw} into $\phi'_{bw} = A_b \phi_{bw} A_w^{-1}$.

For an $\mathbb{R}$-bundle, when edge weights ϕ_{bw} are positive, we can associate a probability measure μ as above. Then positive gauge transformations (i.e. when the $A_v > 0$) change the weights of individual dimer covers but not the probability measure μ: If we multiply the edge weights of all edges at a single vertex by a constant $\lambda > 0$, then the weight of every dimer cover is multiplied by λ. This implies that the probability measure μ only depends on the equivalence class of edge weights under gauge transformations, that is, on the "edge weights modulo gauge".

2.2. Planar bipartite graphs and the Kasteleyn matrix. Let $\mathcal{G}$ be a planar bipartite graph with a $\mathbb{C}^*$-local system Φ (equivalently, a set of edge weights in $\mathbb{C}^*$, see the previous section). We define a matrix K, the *Kasteleyn matrix for* $\mathcal{G}$, as follows. The matrix $K = (K_{wb})$ has rows indexing white vertices and columns indexing black vertices, and

$$K_{wb} = \begin{cases} \pm\phi_{bw} & b \sim w \\ 0 & \text{otherwise} \end{cases}$$

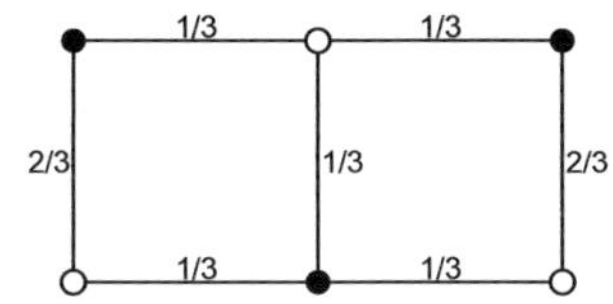

FIGURE 1. A graph and its expected fractional matching (for uniform edge weights). This graph has three dimer covers; two of these use the leftmost vertical edge, so its probability is 2/3.

where the signs are chosen by the "Kasteleyn rule": a face of length ℓ has $\frac{\ell}{2}+1 \bmod 2$ minus signs. Kasteleyn showed in this case that the determinant of K counts weighted dimer covers:

THEOREM 1 ([**4**, **7**]). $|\det K| = \sum_{m\in\mathcal{M}} c(m)$.

As an application, probabilities of one or more edges being in a dimer cover are minors of the inverse of K, see [**5**].

If $\mathcal{G}$ has an SL_n-local system, one can similarly define a Kasteleyn matrix, whose entries are now $n \times n$ matrices themselves; see Sections 6.1,7.2 below.

3. From edge weights to probabilities

We wish to study the mapping from edge weights to probabilities. This requires first defining the correct spaces.

3.1. Fractional matchings. A *fractional matching* of $\mathcal{G} = (V, E)$ is a function $f : E \to [0,1]$ which sums to one at each vertex: $\sum_{u\sim v} f(u) = 1$. A dimer cover m determines a fractional matching, by assigning $f(e) = 1$ for each $e \in m$ and $f(e) = 0$ otherwise. The set of fractional matchings $\Omega(\mathcal{G})$ is a convex polytope in $\mathbb{R}^E$ (in fact a subset of $[0,1]^E$) whose vertices are exactly the dimer covers of $\mathcal{G}$, see [**8**].

A probability measure on dimer covers is a probability measure on the vertices of Ω, and so has a barycenter which is a point in Ω. The coordinate of this barycenter corresponding to an edge is the probability of that edge.

So a more elegant statement of Problem 0 is

PROBLEM 1. *Compute the expected fractional matching.*

See an example in Figure 1.

3.2. Cycles. A *flow* on a graph $\mathcal{G}$ is a function f on oriented edges satisfying $f(-e) = -f(e)$ where $-e$ represents the edge with the reversed orientation. The *cycle space* $H_1(\mathcal{G})$ is the space of divergence-free flows: flows f satisfying $\sum_{u\sim v} f(vu) = 0$ for all v.

Since $\mathcal{G}$ is bipartite, we can orient edges from black to white, and then for any two fractional matchings $f_1, f_2 \in \Omega$, $f_1 - f_2$ represents an element of the cycle space of $\mathcal{G}$, denoted $[f_1 - f_2]$. Fixing a basepoint $f_0 \in \Omega$, we can map Ω linearly and injectively into $H_1(\mathcal{G})$ by $f \mapsto [f - f_0]$.

We say $\mathcal{G}$ is *nondegenerate* if Ω is of dimension $d = |E| - |V| + 1$, the dimension of the cycle space of the graph. A graph is degenerate (that is, not nondegenerate) if there are unused edges (edges which participate in no dimer cover) whose removal does not disconnect the graph [**8**]. Figure 2 shows a degenerate graph.

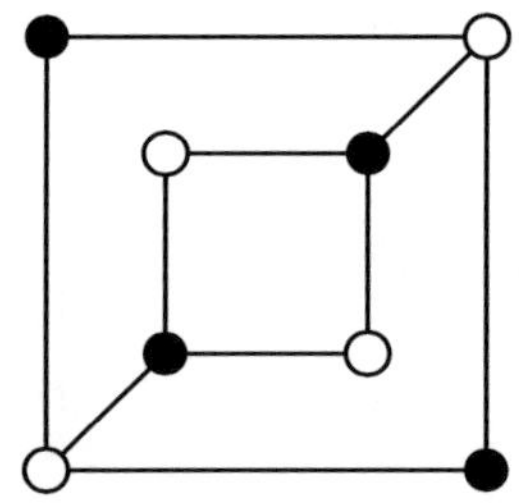

FIGURE 2. A degenerate graph. The dimension of the cycle space is 3 but Ω is of dimension 2. Note that the two diagonal edges are "unused", that is, do not appear in any dimer cover.

3.3. Gauge equivalence. For a bipartite graph, the space of equivalence classes of edge weights modulo gauge transformations, that is, the space of $\mathbb{C}^*$-local systems, is isomorphic to $(\mathbb{C}^*)^d$ where d is the dimension of the cycle space of the graph. Explicitly, for a cycle γ we can define X_γ to be the monodromy of the connection around γ. In terms of edge weights, X_γ is the alternating product of edge weights around γ: assuming we start at a black vertex, X_γ is the first weight, divided by the second, times the third, and so on. Note that X_γ is invariant under gauge transformations. Reversing the orientation inverts the monodromy. If γ ranges over a basis for the cycle space, the set of monodromies $\{X_\gamma\}$ parameterize the space of all edge weights modulo gauge, that is, all $\mathbb{C}^*$-local systems.

For a planar graph, a basis for the cycle space is the set of cycles around the bounded faces. Consequently we can take the "face weights" $\{X_f\}_{f\in F}$ on the bounded faces of the graph to parameterize the choices of edge weights modulo gauge.

Now a perhaps more interesting variant of Problem 0 is:

PROBLEM 2. *For nondegenerate graphs, study the map Ψ from cycle weights $\{X_f\}$ to the expected fractional matching.*

The domain and range of Ψ have the same dimension: the dimension of the cycle space of the graph.

THEOREM 2. *If $\mathcal{G}$ is nondegenerate then Ψ is a diffeomorphism from the space of cycle weights $(\mathbb{R}_{>0})^d$ to $\Omega(\mathcal{G})$.*

PROOF. Fix the gauge by choosing a spanning tree T of $\mathcal{G}$ and setting edge weights 1 on the tree T and weights c_e for $e \notin T$. Let $Z = Z(\{c_e\})$ be the associated partition function. Let p_e be the probability of edge e for $e \notin T$. All the other edge probabilities (and therefore the fractional matching f) are uniquely determined by the probabilities $\{p_e \mid e \notin T\}$: one can determine the value of f on the leaves of the tree, then remove these and continue with the leaves of the subtrees inductively. As in (1.1) we have $p_e = c_e\frac{\partial \log Z}{\partial c_e}$. We can write this as $\vec{p} = \nabla \log Z$ where the gradient is taken with respect to the vector of "edge energies" $(\log c_e)_{e\notin T}$.

Since Z is a positive polynomial (that is, its coefficients are nonnegative), it is, up to scale, the probability generating function of a probability measure on $\mathbb{Z}^d$ of finite support. The Hessian of $\log Z$ with respect to the logs of its variables is the covariance matrix of this probability measure:

$$c_1c_2\frac{\partial^2}{\partial c_1 \partial c_2}\log Z = \frac{c_1c_2}{Z}\frac{\partial^2 Z}{\partial c_1 \partial c_2} - \frac{c_1c_2Z_{c_1}Z_{c_2}}{Z^2} = \mathbb{E}[X_1X_2] - \mathbb{E}[X_1]\mathbb{E}[X_2]$$

where X_1, X_2 are the indicator functions of edges $1, 2$. Any covariance matrix is positive semidefinite, and thus this Hessian is positive semidefinite: $\text{Hess}\log Z \geq 0$. In fact it is positive definite as long as the Newton polytope of Z is of positive volume, that is, if the exponents of the monomials of Z span the cycle space. But this is guaranteed by nondegeneracy. Since $\text{Hess}(\log Z)$ is positive definite, $\log Z$ is strictly convex and so $\nabla \log Z$ is a diffeomorphism onto its image. □

Note that $\det(\nabla\Psi)$ is a rational function of the c_e. What is the degree of Ψ as a rational map on $\mathbb{C}^d$? If we extend Ψ to all of $\mathbb{C}^d$ it appears that the inverse of a point $\vec{p} \in \Omega$ consists in *real* values of c_e (only one preimage consists in positive reals, but there can be multiple non-positive preimages; it is not clear that these are always real, although they are in the small examples we tried). A similar situation, for resistor networks, where reality of the other preimages is proved, can be found in [**1**].

4. Dimer walks

To each dimer cover $m \in \mathcal{M}(\mathcal{G})$ is associated an involution $\pi_m : V \to V$. This is the involution exchanging a vertex with the vertex it is paired with.

PROBLEM 3. *Let $\{m_1, m_2, \dots\}$ be a sequence of i.i.d. dimer covers of $\mathcal{G}$. Study the random walks on S_V, the permutation group on the vertices, defined by the π_{m_i}.*

Another way to state this is to define $g = \frac{1}{Z}\sum_{m\in\Omega_1(\mathcal{G})}\pi(m) \in \mathbb{R}[S_V]$, the element of the group algebra $\mathbb{R}[S_V]$ of S_V defined by a random choice of dimer cover. What can be said about g^n?

As a simple example one can take $\mathcal{G} = K_4$ with vertices $[4] = \{1, 2, 3, 4\}$. The three dimer covers correspond to the three permutations $(12)(34), (13)(24), (14)(23)$ which generate a subgroup of S_4 isomorphic to $(\mathbb{Z}_2)^2$. The dimer random walk just becomes simple random walk on this group, where every element is adjacent to any other element, that is, we have simple random walk on K_4.

For another simple example, let $\mathcal{G}$ be the 3×2 grid. Rather than record the element in S_6, without loss of information we can record simply the permutations of the x-coordinates of the vertices, in S_3. There are 3 dimer covers, and the walk corresponds to multiplication by $g = \frac{1}{3}(1 + e_{(12)} + e_{(23)})$. Its eigenvalues are $1, \frac{2}{3}, \frac{2}{3}, -\frac{1}{3}, 0, 0$.

As a more interesting example, when $\mathcal{G}$ is the $n \times n$ grid graph on a torus, each vertex undergoes a simple random walk on V, since with probability $1/4$ it is matched to any neighbor. These simple random walks are coupled to avoid each other. How quickly does the process mix? What is the expected winding of two walks around each other, as a function of time?

5. Magnetic double dimer model

5.1. Double dimers. Recall that $\mathcal{M} = \mathcal{M}(\mathcal{G})$ is the set of dimer covers of $\mathcal{G}$. A *double dimer cover* is a function $m : E \to \{0, 1, 2\}$ which sums to 2 at each vertex. See Figure 3. A union of two single dimer covers determines a double dimer cover, and (for bipartite graphs) every double dimer cover occurs this way. The map from

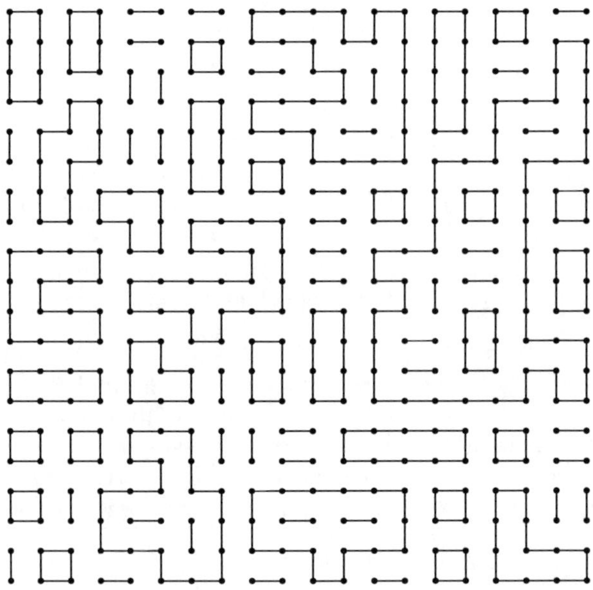

FIGURE 3. A double dimer cover of a grid. Edges of multiplicity 1 or 2 are shown.

pairs of dimer covers to double dimer covers is not, however, injective: each double dimer cover m arises from 2^ℓ ordered pairs of single dimer covers, where ℓ is the number of cycles in m, where the cycles are formed from the edges of multiplicity 1.

Let $\mathcal{M}_2$ be the set of double dimer covers of $\mathcal{G}$. The natural probability measure on $\mathcal{M}_2$ (for us) is *not* the uniform measure. It is, rather, the projection of the uniform measure on $\mathcal{M} \times \mathcal{M}$ under the standard map $\mathcal{M} \times \mathcal{M} \to \mathcal{M}_2$. Thus a double dimer cover m has a probability proportional to 2^ℓ where ℓ is the number of cycles (cycles formed from edges of multiplicity 1; edges of multiplicity 2 do not count as cycles). We call 2^ℓ the *weight* $c(m)$ of m.

PROBLEM 4. *What can be said about the distribution of loops in the double dimer model on* $\mathbb{Z}^2$*?*

Let $\mathcal{G}$ be a bipartite planar graph, and $q \in \mathbb{C}^*$ be a variable. One can define a $\mathbb{C}^*$-connection on $\mathcal{G}$ so that each face has counterclockwise monodromy q, see for example Figure 4. Let $K(q)$ be an associated Kasteleyn matrix. For a double-dimer configuration m_2 define

$$c(m_2) = \prod_{\text{loops } \gamma \text{ of } m_2} (q^{A_\gamma} + q^{-A_\gamma}) \tag{5.1}$$

where A_γ is the area (number of faces) enclosed by γ. We then have

THEOREM 3.

$$\det K(q) \det K(1/q) = \sum_{m_2 \in \mathcal{M}_2} \prod_{\textit{loops } \gamma \textit{ of } m_2} (q^{A_\gamma} + q^{-A_\gamma}). \tag{5.2}$$

PROOF. The quantity $\det K(q) \det K(1/q)$ counts pairs of dimer configurations (m, m'), where m' has the inverted q-weight. In the superposition of m and m',

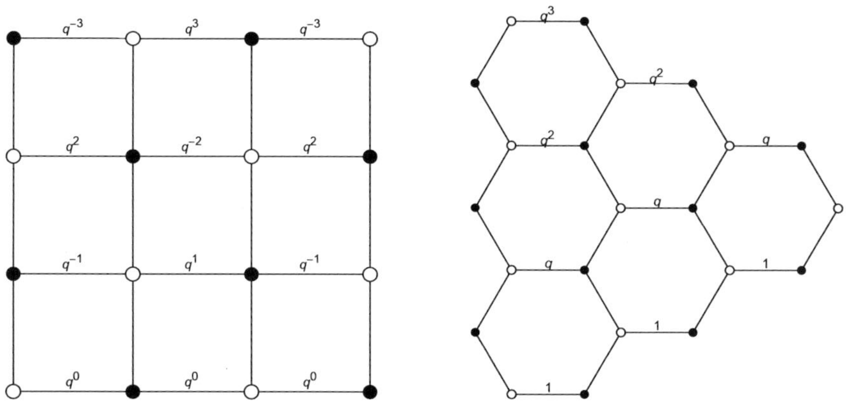

FIGURE 4. Connections with monodromy q per face.

we orient m's dimers from black to white and m''s dimers from white to black, reversing the parallel transport, so that the orientations of loops are consistent. In this way the weight of an oriented loop is given by its monodromy, that is, q^A where A is the area enclosed by the loop. When we sum over both orientations of each loop, we find (5.2). □

It is tempting to use (5.2) to answer Problem 4. For example one can expand $\det K(q) \det K(1/q)$ near $q = 1$ to get various moments. However extracting information about the areas from this does not appear easy.

An alternative approach is to invert the standard Kasteleyn matrix (when $q = 1$) and use it to compute the probability of any given shape of loop. For the square grid $\mathbb{Z}^2$, this can be used to compute the expected density of loops of fixed small area, as follows. The "inverse Kasteleyn matrix" for $\mathbb{Z}^2$ can be obtained as a limit as $n \to \infty$ of $K_{\mathcal{G}_n}^{-1}$ for $\mathcal{G}_n$ the $2n \times 2n$ grid, see [**6**]. There is an explicit formula: when $b - w = (x, y) \in \mathbb{Z}^2$,

$$K^{-1}((0,0),(x,y)) = \frac{1}{(2\pi i)^2} \iint_{|z|=|w|=1} \frac{z^{(-x+y+1)/2} w^{(-x-y+1)/2}}{1 + z + w - zw} \frac{dz}{z} \frac{dw}{w}.$$

Minors of this matrix compute edge probabilities for random dimer covers of $\mathbb{Z}^2$, see [**5**]:

$$Pr((w_1, b_1), \ldots, (w_k, b_k)) = |\det(K^{-1}(b_j, w_i)_{1 \le i,j \le n})|.$$

The expected density of loops of area 1 (the probability that a given face is in such a loop) is the probability that for a pair of single dimer models (m, m') on $\mathbb{Z}^2$, m contains edges $(0,0)(1,0)$ and $(0,1)(1,1)$ and m' contains edges $(0,0)(0,1)$ and $(1,0)(1,1)$, or the reverse. The probability that m contains both edges $(0,0)(1,0)$ and $(0,1)(1,1)$ is, by a short computation, $\frac{1}{8}$. So the expected density of loops of area 1 in the double dimer model is $\frac{2}{8^2} = \frac{1}{32}$. Similarly the density of loops of area 2 can be computed to be $\frac{(\pi-1)^2}{2\pi^4}$. The density for area-3 loops is already more complicated:

$$\frac{3\left(64 - 192\pi^2 + 192\pi^3 - 32\pi^4 - 32\pi^5 + 24\pi^6 - 8\pi^7 + \pi^8\right)}{32\pi^8}.$$

6. SL_2 connections

6.1. Kasteleyn matrix. Consider $\mathcal{G}$ a planar bipartite graph with an SL_2-local system Φ. Let $K = (K_{wb})$ be an associated Kasteleyn matrix for $\mathcal{G}$, that is a matrix with rows indexing white vertices and columns indexing black vertices, and

$$K_{wb} = \begin{cases} \pm\phi_{bw} & b \sim w \\ 0 & \text{otherwise} \end{cases}$$

where the signs are chosen by the same Kasteleyn rule as in the scalar case: a face of length ℓ has $\frac{\ell}{2} + 1 \bmod 2$ minus signs. Here by 0 we mean the zero matrix in $M_2(\mathbb{R})$. Then K is an $N \times N$ matrix with entries in $M_2(\mathbb{R})$. Let $\tilde{K}$ be the $2N \times 2N$ matrix obtained from K by replacing each entry with its 2×2 array of real numbers.

By a theorem of [**2**],

$$\det \tilde{K}(\Phi) = \sum_{m \in \mathcal{M}_2} \prod_{\text{loops in } m} \mathrm{Tr}(\phi_\gamma). \tag{6.1}$$

Here ϕ_γ is the monodromy of the connection Φ around the loop γ, and $\mathrm{Tr}(\phi_\gamma)$ is its trace. Even though ϕ_γ depends on a starting vertex and the orientation of the loop, the trace $\mathrm{Tr}(\phi_\gamma)$ is independent of starting vertex (since the trace of a matrix only depends on its conjugacy class) and orientation (since for matrices $A \in \mathrm{SL}_2$ we have $\mathrm{Tr}(A) = \mathrm{Tr}(A^{-1})$).

6.2. Flat connections and simple laminations. Suppose graph $\mathcal{G}$ is drawn on a multiply-connected planar domain Σ, and Φ is a *flat* SL_2 connection. This means the monodromy of Φ around any cycle in $\mathcal{G}$ which is contractible as a cycle in Σ, is the identity. In this case the trace of the monodromy around a loop γ only depends on its isotopy class as a loop in Σ. In other words two loops in $\mathcal{G}$ which are isotopic as loops in Σ, have monodromies with the same trace.

A *simple lamination* is an isotopy class of finite collections of pairwise disjoint simple closed loops on Σ. Let Λ_2 be the collection of simple laminations.

When Φ is flat we can group the terms in the sum (6.1) according to their isotopy classes:

$$\det \tilde{K}(\Phi) = \sum_{\lambda \in \Lambda_2} C_\lambda \,\mathrm{Tr}(\phi_\lambda) \tag{6.2}$$

with $C_\lambda \in \mathbb{Z}_{\geq 0}$.

By a Theorem of Fock and Goncharov [**3**], the functions $\mathrm{Tr}(\phi_\lambda)$ as λ runs over Λ_2, considered as functions of Φ, are linearly independent and in fact form a linear basis for the regular (i.e. polynomial) functions on the character variety (the space of flat SL_2-local systems Φ modulo gauge). As a consequence C_λ can be in principle determined from $\det \tilde{K}(\Phi)$ as Φ varies over flat connections. Mysteriously, even though for a finite graph $\det \tilde{K}(\Phi)$ is a polynomial function of the matrix entries, and the sum (6.2) is a finite sum, it is not at all clear how to extract the individual C_λ from it.

PROBLEM 5. *How can one extract C_λ from $\det \tilde{K}(\Phi)$ as in (6.2)?*

For an example where we know how to extract these coefficients, consider the simplest nontrivial case when Σ is an annulus. Suppose graph $\mathcal{G}$ is drawn on an annulus with flat connection having monodromy A around the nontrivial isotopy class γ, the generator of $\pi_1(\Sigma)$. Then a simple lamination on $\mathcal{G}$ just consists in some

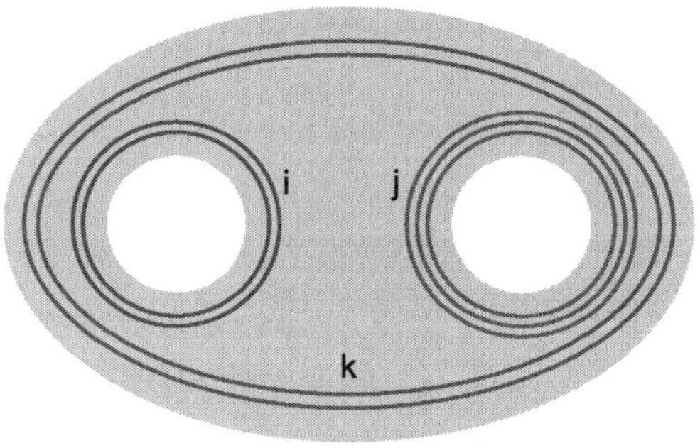

FIGURE 5. Simple laminations on a pair of pants consist in three kinds of loops: those surrounding p_1, p_2 or both.

number n of copies of loops each isotopic to γ (we are ignoring orientation). The isotopy classes Λ_2 are thus in correspondence with the set of nonnegative integers. We can write

$$\det \tilde{K}(A) = \sum_{j=0}^{\infty} C_j \operatorname{Tr}(A)^j.$$

From this expression it is easy to extract C_j: let $\operatorname{Tr}(A) = z$, then the C_j are power series coefficients of $\det \tilde{K}(A)$.

The next simplest planar surface is a disk with two small disks p_1, p_2, removed (also known as a "pair of pants"). In this case the space Λ_2 has a simple parameterization by $(\mathbb{Z}_+)^3$ where (i, j, k) corresponds to the lamination with i, j, and k loops surrounding respectively p_1 only, p_2 only or both p_1 and p_2, see Figure 5. If A, B are the monodromies around p_1, p_2 respectively then given any $x, y, z \in \mathbb{C}^*$ we can choose A, B in $\mathrm{SL}_2(\mathbb{C})$ so that $x = \operatorname{Tr}(A), y = \operatorname{Tr}(B), z = \operatorname{Tr}(AB)$. Then one can extract $C_{i,j,k}$ by a contour integral

$$C_{i,j,k} = \frac{1}{(2\pi i)^3} \int_{(S^1)^3} \frac{\det K(\Phi)}{x^i y^j z^j} \frac{dx}{x} \frac{dy}{y} \frac{dz}{z}.$$

For a general punctured surface Fock and Goncharov [**3**] describe a different (and orthogonal) basis on the space of laminations, the "Peter-Weyl basis", and show that the basis $\{\operatorname{Tr}(\phi_\lambda)\}_{\lambda \in \Lambda_2}$ is obtained from the Peter-Weyl basis by an upper triangular transformation. One route to solving Problem 5 would be to write down this base-change matrix.

7. n-fold dimer model

7.1. n-multiwebs. An *n-multiweb*, or *n-fold dimer cover* in $\mathcal{G}$ is a function $m : E \to \{0, 1, 2, \ldots, n\}$ which sums to n at each vertex. So a 1-multiweb is a dimer cover and a 2-multiweb is a double dimer cover. A union of n single dimer covers is an n-multiweb, and every n-multiweb occurs this way (although not uniquely).

Let $\mathcal{M}_n$ be the set of n-multiwebs of $\mathcal{G}$. The natural probability measure on $\mathcal{M}_n$ is, like in the double dimer case, the projection of the uniform measure on $\mathcal{M}^n$ under the standard map $\mathcal{M}^n \to \mathcal{M}_n$. Unlike the $n = 2$ case, the size of the preimage is difficult to compute in general, in fact #P-complete: if $n = 3$ and a 3-multiweb is a trivalent graph (that is, all edges have multiplicity 1), then the triples of dimer covers in the preimage are in bijection with edge 3-colorings, or

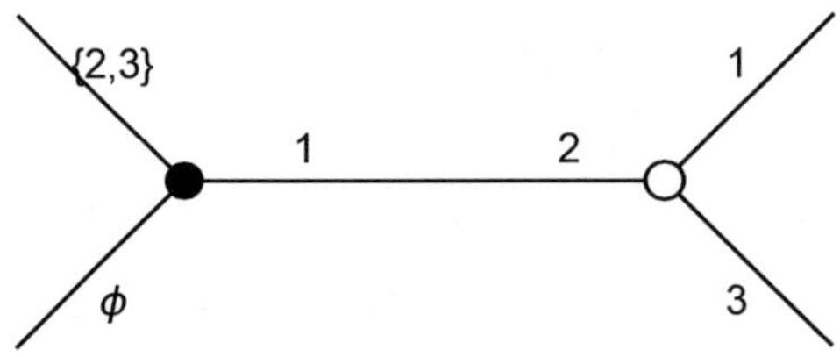

FIGURE 6. 3-coloring of the half-edges of $\mathcal{G}$.

Tait colorings. Tait colorings are colorings of the edges with three colors so that each color appears at each vertex. For planar graphs Tait colorings are dual to proper 4-colorings of the vertices of the dual graph.

7.2. Kasteleyn matrix. Consider $\mathcal{G}$ a planar bipartite graph with an SL_n-local system Φ. Let $K = (K_{wb})$ be a Kasteleyn matrix for $\mathcal{G}$; K is an $N \times N$ matrix with entries in $M_n(\mathbb{R})$. Let $\tilde{K}$ be the $nN \times nN$ matrix obtained from K by replacing each entry with its $n \times n$ array of reals.

By a theorem of [**2**],

$$\det \tilde{K}(\Phi) = \sum_{m \in \mathcal{M}_n} \mathrm{Tr}(\phi_m) \tag{7.1}$$

where we still need to define the trace of an n-multiweb with an SL_n-connection; see the next section.

7.3. Trace of a 3-multiweb. The trace of an n-multiweb m is not simple to define; it is a contraction of n-tensors defined at each vertex. We refer to [**2**] for the general definition and give here a working definition for $n = 3$. We cut each edge of $\mathcal{G}$ into two half-edges, one associated with each of its vertices. We consider colorings c of the half-edges of m with three colors $\{1, 2, 3\}$, so that an edge of multiplicity $j \in \{0, 1, 2, 3\}$ gets a set of j colors, and so that all colors appear at each vertex of $\mathcal{G}$. See Figure 6.

Then on an edge e of multiplicity j we have two sets of colors S, T of size j, with S located near the white vertex and T near the black vertex. We define

$$Tr(c) = \sum_{\text{colorings } c} sgn(c) \prod_e (\phi_{bw})_S^T \tag{7.2}$$

where $(\phi)_S^T$ denotes the S, T minor of ϕ. Here $sgn(c)$ is a sign depending on the cyclic order of colors at each vertex: at a black vertex if the colors are in counterclockwise order we get sign $+$ and otherwise sign $-$, and this convention is reversed at a white vertex; the product of these signs over all vertices is $sgn(c)$. (For edges with multiplicity the colors are assumed in natural order within that set.)

When Φ is the identity connection, and m is planar, $|\mathrm{Tr}(m)|$ is the number of Tait colorings of m (see [**2**]). In other words, all nonzero terms of (7.2) have the same sign. The global sign of $\mathrm{Tr}(m)$ is unimportant since it depends on an artificial choice of vertex ordering.

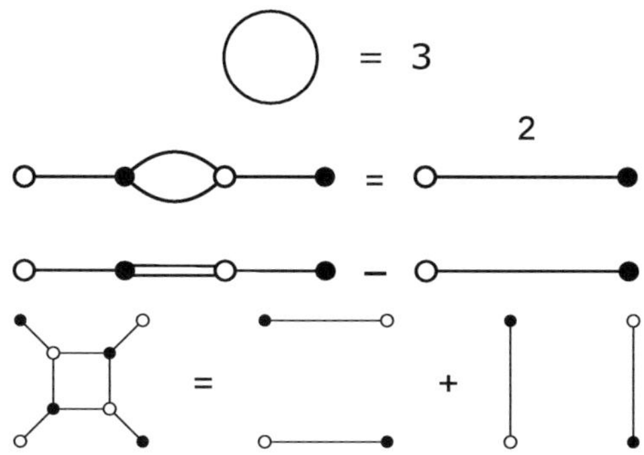

FIGURE 7. Relations between 3-multiwebs: a loop can be removed, increasing the weight of the remaining web by a factor of 3; a bigon can be replaced with a single edge as shown multiplying the weight by 2; a doubled edge can be replaced with a single edge as shown, a square face can be resolved into a linear combination of two webs.

7.4. Flat connections and reduced 3-webs. Suppose $\mathcal{G}$ is drawn on a multiply-connected planar domain Σ, and Φ is a *flat* SL_3 connection.

A vertex in a 3-multiweb is *trivalent* if it has three adjacent edges of multiplicity 1. A 3-multiweb consists in a collection of trivalent vertices connected in pairs by chains of edges with multiplicities alternating between 1 and 2 along the chains.

We say a 3-multiweb is *reduced* or *nonelliptic* if each component, considered as a planar graph by itself, has no contractible faces with 0, 2 or 4 trivalent vertices.

There are certain *skein* relations by which the trace of any 3-multiweb can be written as a linear combination of traces of reduced 3-multiwebs (see Figure 7). The set of isotopy classes of the reduced 3-multiwebs which arise from reducing a given web are well defined.

As a consequence for a flat SL_3-connection we can group the terms in the sum (7.1) according to isotopy classes of reduced 3-multiwebs:

$$\det \tilde{K}(\Phi) = \sum_{\lambda \in \Lambda_3} C_\lambda \operatorname{Tr}(\phi_\lambda). \tag{7.3}$$

By a Theorem of Sikora and Westbury [**10**], the functions $\operatorname{Tr}(\lambda)$ as λ runs over Λ_3 form a linear basis for the regular (i.e. polynomial) functions on the character variety of flat SL_3-local systems modulo gauge. As a consequence C_λ can be in principle determined from $\det \tilde{K}(\Phi)$ as Φ varies over flat connections.

PROBLEM 6. *How can one extract* C_λ *from* $\det \tilde{K}(\Phi)$*?*

Applying the skein relations to a non-reduced 3-multiweb results in a collection of reduced webs which depends on the order in which the skein relations are applied. Even though the isotopy classes of the reduced web are well-defined, the individual webs themselves will depend on the order. Is there a way to make a canonical choice in this reduction process, so that starting from a random 3-multiweb we arrive at a well-defined, canonical probability measure on reduced 3-multiwebs?

References

[1] A. Abrams and R. Kenyon, *Fixed-energy harmonic functions*, Discrete Anal., posted on 2017, Paper No. 18, 21, DOI 10.19086/da.2730. MR3734203

[2] D. C. Douglas, R. Kenyon, and H. Shi, *Dimers, webs, and local systems*, Trans. Amer. Math. Soc. **377** (2024), no. 2, 921–950, DOI 10.1090/tran/8993. MR4688540

[3] V. Fock and A. Goncharov, *Moduli spaces of local systems and higher Teichmüller theory*, Publ. Math. Inst. Hautes Études Sci. **103** (2006), 1–211, DOI 10.1007/s10240-006-0039-4. MR2233852

[4] P. W. Kasteleyn, *Dimer statistics and phase transitions*, J. Mathematical Phys. **4** (1963), 287–293, DOI 10.1063/1.1703953. MR153427

[5] R. Kenyon, *Local statistics of lattice dimers* (English, with English and French summaries), Ann. Inst. H. Poincaré Probab. Statist. **33** (1997), no. 5, 591–618, DOI 10.1016/S0246-0203(97)80106-9. MR1473567

[6] R. Kenyon, *Lectures on dimers*, Statistical mechanics, IAS/Park City Math. Ser., vol. 16, Amer. Math. Soc., Providence, RI, 2009, pp. 191–230, DOI 10.1090/pcms/016/04. MR2523460

[7] G. Kuperberg, *An exploration of the permanent-determinant method*, Electron. J. Combin. **5** (1998), Research Paper 46, 34, DOI 10.37236/1384. MR1663576

[8] L. Lovász and M. D. Plummer, *Matching theory*, North-Holland Mathematics Studies, vol. 121, North-Holland Publishing Co., Amsterdam; North-Holland Publishing Co., Amsterdam, 1986. Annals of Discrete Mathematics, 29. MR859549

[9] A. S. Sikora, SL_n*-character varieties as spaces of graphs*, Trans. Amer. Math. Soc. **353** (2001), no. 7, 2773–2804, DOI 10.1090/S0002-9947-01-02700-3. MR1828473

[10] A. S. Sikora and B. W. Westbury, *Confluence theory for graphs*, Algebr. Geom. Topol. **7** (2007), 439–478, DOI 10.2140/agt.2007.7.439. MR2308953

[11] L. G. Valiant, *The complexity of computing the permanent*, Theoret. Comput. Sci. **8** (1979), no. 2, 189–201, DOI 10.1016/0304-3975(79)90044-6. MR526203

Mathematics Department, Yale University, New Haven, Connecticut
Email address: richard.kenyon@yale.edu

Proceedings of Symposia in Pure Mathematics
Volume **110**, 2024
https://doi.org/10.1090/pspum/110/02005

A positivity conjecture on the structure constants of shifted Jack functions

Per Alexandersson and Valentin Féray

Abstract. We consider Jack polynomials J_λ and their shifted analogue $J^\#_\lambda$. In 1989, Stanley conjectured that $\langle J_\mu J_\nu, J_\lambda\rangle$ is a polynomial with nonnegative coefficients in the parameter α. In this note, we extend this conjecture to the case of *shifted* Jack polynomials.

1. Introduction

In his seminal article on the combinatorics of Jack polynomials [**Sta89**], Stanley made a positivity conjecture on the *Littlewood–Richardson type* coefficients for Jack polynomials, i.e. the coefficients appearing in the Jack basis expansion of the product of two Jack polynomials. Despite intensive work on Jack polynomials on the one side and on generalizations of the Littlewood–Richardson rule on the other side, this problem still remains open.

In this note, we suggest an extension of Stanley's conjecture to *shifted* Jack polynomials, Conjecture 4.1 below. An advantage of the shifted version is that there exist a recursion formula to compute the corresponding Littlewood–Richardson type coefficients (Proposition 6.2), giving another angle of attack for the conjecture. We give numerical data and very partial results to support the conjecture: a general polynomiality result in Section 5 and the analysis of some specific cases in Section 7.

Throughout this note, we use the notation of Macdonald's book [**Mac95**] regarding partitions, symmetric functions, (skew) diagrams, etc. We assume the reader to be acquainted with basic notions in this field.

2. Jack polynomials

A *reverse (semi-standard) tableau* of shape μ is a filling of the boxes of μ with numbers in $\{1,\dots,n\}$ which is weakly decreasing along rows and strictly decreasing along columns. The set of such tableaux is denoted $\mathrm{RSSYT}(\mu,n)$. For T in $\mathrm{RSSYT}(\mu,n)$, we denote $\rho^i(T)$ the skew diagram consisting in boxes of T with entry i. This is always a horizontal strip (no two boxes in the same column).

2020 *Mathematics Subject Classification.* Primary 05E05; Secondary 20C30.
Key words and phrases. Shifted symmetric functions, Jack polynomials.

For a horizontal strip λ/μ, let $R_{\lambda/\mu}$ (resp. $C_{\lambda/\mu}$) denote the set of boxes in a row (resp. column) that intersects the shape λ/μ. We then define

$$\psi_{\lambda/\mu}(\alpha) = \prod_{s \in R_{\lambda/\mu} - C_{\lambda/\mu}} \frac{(\alpha a_\lambda(s) + l_\lambda(s) + \alpha)(\alpha a_\mu(s) + l_\mu(s) + 1)}{(\alpha a_\lambda(s) + l_\lambda(s) + 1)(\alpha a_\mu(s) + l_\mu(s) + \alpha)}. \tag{2.1}$$

The Jack polynomial P_μ in n variables $x_1, \dots, x_n$ may be defined as[1]

$$P_\mu = \sum_{T \in \mathrm{RSSYT}(\mu,n)} \psi_T(\alpha) \prod_{s \in \mu} x_{T(s)}.$$

For $\alpha = 1$, we have $\psi_{\lambda/\mu}(1) = 1$ and P_μ specializes to the Schur polynomial in n variables when $\alpha = 1$.

Let us define the two α-deformations of the hook products, H_λ and H'_λ, as

$$H_\lambda = \prod_{s \in \lambda} (\alpha a_\lambda(s) + l_\lambda(s) + 1), \quad H'_\lambda = \prod_{s \in \lambda} (\alpha a_\lambda(s) + l_\lambda(s) + \alpha).$$

(these are called $c_\lambda(\alpha)$ and $c'_\lambda(\alpha)$ in [**Mac95**, (10.22)]). The integral version of the Jack polynomial J_μ is then given by $J_\mu = H_\mu P_\mu$.

3. Shifted Jack polynomials

We define the *shifted Jack polynomials* $P^\#_\mu$ as

$$P^\#_\mu(x_1, \dots, x_n) = \sum_{T \in \mathrm{RSSYT}(\mu,n)} \psi_T(\alpha) \prod_{s \in \mu} (x_{T(s)} - a'(s) + l'(s)/\alpha), \tag{3.1}$$

where $a'(s) = j - 1$ and $l'(s) = i - 1$ for the box $s = (i, j)$. Thus, the top degree component of $P^\#_\mu$ is given by the ordinary Jack polynomial P_μ. This definition can be found in [**OO97**], see also [**Sah94, KS96**] for older references on these polynomials (with a shift of variables).

Alternatively to the tableaux definition, the Jack shifted polynomials can also be characterized by their values $P^\#_\mu(\lambda)$ on partitions, i.e. when we specialize $x_j = \lambda_j$ (for $j \le n$; appending zeroes to λ if necessary). Let $\Lambda^\alpha(x_1, \dots, x_n)$ be the ring of polynomials which are *shifted symmetric*, that is, symmetric in the variables $x_i - i/\alpha$. The polynomials $P^\#_\mu$ lie in $\Lambda^\alpha(x_1, \dots, x_n)$ and it was shown by F. Knop and S. Sahi in [**KS96**] that $P^\#_\mu$ is the unique polynomial in $\Lambda^\alpha(x_1, \dots, x_n)$ such that

$$P^\#_\mu(\lambda) = \begin{cases} \alpha^{-|\mu|} H'_\mu, & \lambda = \mu \\ 0, & |\lambda| \le |\mu|,\ \lambda \ne \mu. \end{cases} \tag{3.2}$$

Additionally, we have $P^\#_\mu(\lambda) = 0$ whenever $\lambda \not\supseteq \mu$.

There is a J-normalized version of shifted Jack polynomials given as $J^\#_\mu = H_\mu P^\#_\mu$, which is the main topic of study in [**Las08**], where the notation $J^\dagger_\mu$ is used in place of $J^\#_\mu$.

From the definition of $P^\#_\mu$, it is easy to see that

$$P^\#_\mu(x_1, \dots, x_n, 0) = P^\#_\mu(x_1, \dots, x_n)$$

[1] This formula is usually found in the literature with a sum over semi-standard tableaux (and not reverse ones as here). But since the resulting sum is known to be symmetric, this modification is irrelevant. In the next section, using reverse tableaux is however important.

so we can see $P^{\#}_\mu$ as shifted symmetric functions in infinitely many variables and they define a basis in the projective limit Λ^α of the rings $\Lambda^\alpha(x_1,\dots,x_n)$. In fact, for any fixed d, the family $(P^{\#}_\mu)_{|\mu|\le d}$ is a basis of the space of shifted symmetric functions of degree at most d.

4. Shifted Jack structure constants

We define the *Littlewood–Richardson type structure constants for shifted Jack functions*, $c^\lambda_{\mu\nu}$, which are rational functions in α, by the relation

$$P^{\#}_\mu P^{\#}_\nu = \sum_\lambda c^\lambda_{\mu\nu} P^{\#}_\lambda, \text{ or equivalently } J^{\#}_\mu J^{\#}_\nu = \sum_\lambda c^\lambda_{\mu\nu} \frac{H_\mu \cdot H_\nu}{H_\lambda} J^{\#}_\lambda. \tag{4.1}$$

Since $P^{\#}_\mu P^{\#}_\nu$ is a shifted symmetric function of degree at most $|\lambda|$, the above sum runs over partitions λ of size at most $|\mu|+|\nu|$ (since the functions are not homogeneous, we need to take $|\lambda|\le|\mu|+|\nu|$ and we cannot restrict to $|\lambda|=|\mu|+|\nu|$).

For $\alpha=1$, these are Littlewood–Richardson coefficients for shifted Schur functions. A combinatorial interpretation, implying their nonegativity has been found (in the more general context of factorial Schur functions) by Molev [**Mol09**]; see also [**MS99**].

On the other hand, taking the homogeneous component of degree $|\mu|+|\nu|$ in (4.1), we get

$$J_\mu J_\nu = \sum_{\lambda:|\lambda|=|\mu|+|\nu|} c^\lambda_{\mu\nu} \frac{H_\mu \cdot H_\nu}{H_\lambda} J_\lambda.$$

Recall that Jack polynomials are orthogonal for the standard α-deformation of Hall scalar product, namely we have $\langle J_\rho, J_\lambda\rangle = H_\lambda H'_\lambda \delta_{\rho,\lambda}$ (see, e.g. [**Sta89**, Thm. 5.8]). This implies that for $|\lambda|=|\mu|+|\nu|$,

$$\langle J_\mu J_\nu, J_\lambda\rangle = H_\mu H_\nu H'_\lambda c^\lambda_{\mu\nu}.$$

In his 1989 paper, Stanley conjectures that the above quantity $\langle J_\mu J_\nu, J_\lambda\rangle$ is a polynomial with nonnegative coefficients in α, for all $|\lambda|=|\mu|+|\nu|$. Unlike the left-hand side, the right-hand side is also defined for $|\lambda|<|\mu|+|\nu|$. We conjecture that, up to multiplication by an appropriate power of α, the right-hand side is always a polynomial with nonnegative coefficients in α. This actually strengthens a conjecture of Sahi, which asserts that $c^\lambda_{\mu\nu}$ is a quotient of two polynomials in α with nonnegative coefficients [**Sah11**].

CONJECTURE 4.1. For any partition λ,μ,ν with $|\lambda|\le|\mu|+|\nu|$, the quantity

$$g^\lambda_{\mu\nu} := H_\mu H_\nu H'_\lambda c^\lambda_{\mu\nu} \tag{4.2}$$

is a Laurent polynomial with nonnegative integer coefficients in α.

Here are a few examples of these coefficients

$$\begin{aligned}
g^{421}_{31,21} &= 8\alpha^5(9+97\alpha+294\alpha^2+321\alpha^3+131\alpha^4+12\alpha^5),\\
g^{541}_{41,41} &= 144\alpha^8(1+\alpha)^2(2+\alpha)(2+3\alpha)(1+4\alpha)(18+149\alpha+238\alpha^2+120\alpha^3),\\
g^{2111}_{2111,1111} &= \frac{144(\alpha+1)^2(\alpha+2)^2(\alpha+4)(2\alpha+3)^2}{\alpha}.
\end{aligned}$$

In the first example, $|\lambda|=|\mu|+|\nu|$, so this coefficient can also be expressed as $\langle J_{31}J_{21}, J_{421}\rangle$. This value was given as an example by Stanley (who credits Hanlon for the computation [**Sta89**, Section 8]). The other examples however are not expressible as such scalar products. The third line shows that $g^\lambda_{\mu\nu}$ is not always a polynomial in α. We attach[2] the values of all coefficients for $|\mu|, |\nu| \leq 6$ supporting the above conjecture.

5. A polynomiality result

PROPOSITION 5.1. *The expression* $\alpha^{|\mu|+|\nu|-|\lambda|-2} g^\lambda_{\mu\nu}$ *is a polynomial with rational coefficients in* α.

It is possible that multiplying by some smaller and/or more natural power of α suffices to get a polynomial, but we have no other results in this direction.

PROOF. Let us expand the Jack symmetric functions in the power-sum basis:

$$J_\lambda = \sum_\mu \theta^{(\alpha)}_\mu(\lambda)\, p_\mu. \tag{5.1}$$

We then introduce the following quantity

$$\vartheta^{(\alpha)}_\mu(\lambda) = \begin{cases} |\lambda|(|\lambda|-1)\cdots(|\lambda|-|\mu|+1)\dfrac{\theta^{(\alpha)}_{\mu 1^{|\lambda|-|\mu|}}(\lambda)}{\theta^{(\alpha)}_{1^{|\mu|}}(\lambda)} & \text{if } |\lambda| \geq |\mu| \\ 0 & \text{if } |\lambda| < |\mu|. \end{cases} \tag{5.2}$$

It was proved in [**Las08**] that there exists a linear isomorphism $\mathrm{sh}(\cdot)$ from usual symmetric functions to shifted symmetric ones so that

$$\mathrm{sh}(J_\lambda) = J^\#_\lambda, \quad \mathrm{sh}(p_\mu) = \alpha^{-(|\mu|-\ell(\mu))}\,\vartheta^{(\alpha)}_\mu.$$

(This morphism is denoted $f \mapsto f^\#$ in [**Las08**].) Applying this morphism to (5.1) gives

$$J^\#_\lambda = \sum_\mu \alpha^{-(|\mu|-\ell(\mu))}\,\theta^{(\alpha)}_\mu(\lambda)\,\vartheta^{(\alpha)}_\mu.$$

On the other hand, using self-duality of power-sums and Jack polynomials for the deformed Hall scalar product, (5.1) implies

$$p_\mu = \sum_\lambda \frac{\langle p_\mu, J_\lambda\rangle}{\langle J_\lambda, J_\lambda\rangle} J_\lambda = \sum_\lambda \frac{\theta^{(\alpha)}_\mu(\lambda)\langle p_\mu, p_\mu\rangle}{\langle J_\lambda, J_\lambda\rangle} J_\lambda = \alpha^{\ell(\mu)} z_\mu \left(\sum_\lambda \frac{\theta^{(\alpha)}_\mu(\lambda)}{\langle J_\lambda, J_\lambda\rangle} J_\lambda\right).$$

Applying sh, we get

$$\vartheta^{(\alpha)}_\mu = \alpha^{|\mu|} z_\mu \left(\sum_\lambda \frac{\theta^{(\alpha)}_\mu(\lambda)}{\langle J_\lambda, J_\lambda\rangle} J^\#_\lambda\right).$$

If we now let $d^\lambda_{\mu\nu}$ be defined via the relation $\vartheta^{(\alpha)}_\mu \vartheta^{(\alpha)}_\nu = \sum_\lambda d^\lambda_{\mu\nu} \vartheta^{(\alpha)}_\lambda$, we get

$$c^\lambda_{\mu\nu}\frac{H_\mu \cdot H_\nu}{H_\lambda} = \sum_{\mu',\nu',\lambda'} \alpha^{-(|\mu'|-\ell(\mu'))}\,\theta^{(\alpha)}_{\mu'}(\mu)\,\alpha^{-(|\nu'|-\ell(\nu'))}\,\theta^{(\alpha)}_{\nu'}(\nu)\cdot d^{\lambda'}_{\mu'\nu'}\alpha^{|\lambda'|} z_{\lambda'} \frac{\theta^{(\alpha)}_{\lambda'}(\lambda)}{\langle J_\lambda, J_\lambda\rangle},$$

where the sum runs over triplet of partitions (μ', ν', λ') of the same size as (μ, ν, λ). But from [**DF14**, Theorem 1.4], we get that coefficients $d^{\lambda'}_{\mu'\nu'}$ are polynomials in α

[2]To access the file with acrobat reader, click on view/navigation panels/attachments. Alternatively, you can download the source files from `https://arxiv.org/abs/1912.05203`.

with rational coefficients (beware that the normalization there is different). Furthermore, the quantities $\theta^{(\alpha)}_{\mu'}(\mu)$ are also polynomials in α by the Knop–Sahi theorem [**KS97**]. Together with the equality $\langle J_\lambda, J_\lambda\rangle = H_\lambda\, H'_\lambda$, we conclude that

$$\alpha^{|\mu|+|\nu|-|\lambda|-2}\, H_\mu\, H_\nu\, H'_\lambda c^\lambda_{\mu\nu} = \sum_{\mu',\nu',\lambda'} \alpha^{\ell(\mu')-1}\, \theta^{(\alpha)}_{\mu'}(\mu)\, \alpha^{\ell(\nu')-1}\, \theta^{(\alpha)}_{\nu'}(\nu)\cdot d^{\lambda'}_{\mu'\nu'} z_{\lambda'} \theta^{(\alpha)}_{\lambda'}(\lambda)$$

is a polynomial in α with rational coefficients. □

6. A recursion formula for $c^\lambda_{\mu\nu}$

In this section, we give initial conditions and a recursion formula, determining completely the coefficients $c^\lambda_{\mu\nu}$. We do not claim any novelty here. The recursion is already given in [**Sah11**], and the presentation here is an adaptation of the technique in [**MS99**], where the case $\alpha = 1$ was established.

First observe that, from the vanishing result by F. Knop and S. Sahi and a simple contradiction argument (see [**MS99**, top of page 4434] for the case $\alpha = 1$), it follows that $c^\lambda_{\mu\nu}$ is identically zero unless $\lambda \supseteq \mu$ and $\lambda \supseteq \nu$.

We continue with an easy proposition, that give simple formulas for the $c^\lambda_{\mu\lambda}$: this will be our initial conditions.

PROPOSITION 6.1. *We have that* $c^\lambda_{\mu\lambda} = P^\#_\mu(\lambda)$.

PROOF. This follows from evaluating the definition of $c^\lambda_{\mu\lambda}$ (4.1) in λ:

$$P^\#_\mu(\lambda)P^\#_\lambda(\lambda) = \sum_\rho c^\rho_{\mu\lambda} P^\#_\rho(\lambda).$$

Now using the vanishing condition in (3.2), we see that only the term $\rho = \lambda$ survives and from this we can easily deduce the statement. □

The next statement gives the recurrence relation. When T is a standard Young tableau, we set $\psi'_T(\alpha) := \psi_{T'}(1/\alpha)$, where T' is the transpose of T.

PROPOSITION 6.2 (The recursion formula). *Let* $\mu, \nu \subseteq \lambda$. *Then*

$$c^\lambda_{\mu\nu} = \frac{1}{|\lambda| - |\nu|}\left(\sum_{\nu\to\nu^+} \psi'_{\nu^+/\nu}(\alpha) c^\lambda_{\mu\nu^+} - \sum_{\lambda^-\to\lambda} \psi'_{\lambda/\lambda^-}(\alpha) c^{\lambda^-}_{\mu\nu}\right) \tag{6.1}$$

where the first sum is taken over all possible ways to add one box to the diagram ν, *and the second sum is over all ways to remove one box from* λ.

The remainder of the section is devoted to the proof of this proposition. We start with some lemmas.

LEMMA 6.3 (See [**OO97**]). *If* $\mu \subseteq \nu$ *then*

$$\frac{P^\#_\mu(\nu)}{P^\#_\nu(\nu)} = \frac{1}{(|\nu| - |\mu|)!} \sum_{T\in \mathrm{SYT}(\nu/\mu)} \psi'_T(\alpha),$$

where the sum is taken over all standard *Young tableaux of shape* ν/μ.

It will be convenient to define $H(\mu, \nu)$ as

$$H(\mu,\nu) = \frac{P^\#_\mu(\nu)}{P^\#_\nu(\nu)} = \frac{1}{(|\nu| - |\mu|)!} \sum_{T\in \mathrm{SYT}(\nu/\mu)} \psi'_T(\alpha)$$

and $H'(\mu,\nu) = (-1)^{|\nu|-|\mu|}H(\mu,\nu)$. It will be useful to keep in mind that $H(\mu,\mu) = H'(\mu,\mu) = 1$.

LEMMA 6.4. *We have the formula*

$$c^{\lambda}_{\mu\nu} = \sum_{\nu\subseteq\rho\subseteq\lambda} P^{\#}_{\mu}(\rho)H(\nu,\rho)H'(\rho,\lambda). \tag{6.2}$$

PROOF. We use induction over $|\lambda| - |\nu|$. The base case $\lambda = \nu$ is given by Proposition 6.1. Substituting $\mathbf{x} = \lambda$ in the definition of $c^{\lambda}_{\mu\nu}$ (4.1) and dividing both sides with $P^{\#}_{\lambda}(\lambda)$ together with Lemma 6.3, we obtain

$$P^{\#}_{\mu}(\lambda)\frac{P^{\#}_{\nu}(\lambda)}{P^{\#}_{\lambda}(\lambda)} = \sum_{\sigma\subseteq\lambda} c^{\sigma}_{\mu\nu}H(\sigma,\lambda). \tag{6.3}$$

Rewriting this now gives

$$c^{\lambda}_{\mu\nu} = P^{\#}_{\mu}(\lambda)H(\nu,\lambda) - \sum_{\sigma\subsetneq\lambda} c^{\sigma}_{\mu\nu}H(\sigma,\lambda). \tag{6.4}$$

The induction hypothesis now allows us to express (6.4) as

$$\begin{aligned} c^{\lambda}_{\mu\nu} &= P^{\#}_{\mu}(\lambda)H(\nu,\lambda) - \sum_{\sigma,\rho:\mu\subseteq\rho\subseteq\sigma\subsetneq\lambda} P^{\#}_{\mu}(\rho)H(\nu,\rho)H'(\rho,\sigma)H(\sigma,\lambda) \\ &= P^{\#}_{\mu}(\lambda)H(\nu,\lambda) - \sum_{\rho:\mu\subseteq\rho\subsetneq\lambda} P^{\#}_{\mu}(\rho)H(\nu,\rho)\left[\sum_{\sigma:\rho\subseteq\sigma\subsetneq\lambda} H'(\rho,\sigma)H(\sigma,\lambda)\right]. \end{aligned} \tag{6.5}$$

It now suffices to prove the following claim: For any $\rho \subseteq \lambda$,

$$\sum_{\sigma:\rho\subseteq\sigma\subseteq\lambda} H'(\rho,\sigma)H(\sigma,\lambda) = 0. \tag{6.6}$$

Indeed, if (6.6) holds, we can replace the sum in brackets in (6.5) by $-H'(\rho,\lambda)$ and we get (6.2) (the first term in (6.5) corresponds to the summand $\rho = \lambda$ in (6.2)).

Let us now prove (6.6). Using the definition of H, we have that the left-hand side of (6.6) is given by

$$\sum_{\sigma:\rho\subseteq\sigma\subseteq\lambda} \frac{(-1)^{|\sigma|-|\rho|}}{(|\sigma|-|\rho|)!(|\lambda|-|\sigma|)!}\left(\sum_{T_1\in\mathrm{SYT}(\sigma/\rho)} \psi'_{T_1}(\alpha)\right)\left(\sum_{T_2\in\mathrm{SYT}(\lambda/\sigma)} \psi'_{T_2}(\alpha)\right)$$

This triple sum over σ, T_1 and T_2 can be rewritten as follows. First, we sum over standard Young tableaux T of shape λ/ρ. This tableau T corresponds to a sequence $\rho = \sigma^0 \to \sigma^1 \to \cdots \to \sigma^l = \lambda$ which is the "concatenation" of T_1 and T_2. Then we should sum over σ that appear in this sequence, that is $\sigma = \sigma^k$ for some k between 0 and $|\lambda| - |\rho|$. Thus, we get:

$$\begin{aligned} \sum_{\rho\subseteq\sigma\subseteq\lambda} H'(\rho,\sigma)H(\sigma,\lambda) &= \sum_{T\in\mathrm{SYT}(\lambda/\rho)} \psi'_T(\alpha)\left[\sum_{\sigma=\sigma^k\in T} \frac{(-1)^{|\sigma|-|\rho|}}{(|\sigma|-|\rho|)!(|\lambda|-|\sigma|)!}\right] \\ &= \left(\sum_{T\in\mathrm{SYT}(\lambda/\rho)} \psi'_T(\alpha)\right)\left(\sum_{k=0}^{|\lambda|-|\rho|} \frac{(-1)^k}{k!(|\lambda|-|\rho|-k)!}a\right) \end{aligned}$$

But, from the binomial theorem, we know that

$$\sum_{k=0}^{|\lambda|-|\rho|} \frac{(-1)^k}{k!(|\lambda|-|\rho|-k)!} = 0.$$

Thus (6.6) is proved and this completes the proof of Lemma 6.4 by induction. □

We are now ready to prove Proposition 6.2.

PROOF OF PROPOSITION 6.2. We fix ρ with $\nu \subseteq \rho \subseteq \lambda$. First observe that, using the definition of H and H' we have

$$\sum_{\nu\to\nu^+} \psi'_{\nu^+/\nu} H(\nu^+,\rho) = (|\rho|-|\nu|)H(\nu,\rho) \text{ and}$$
$$\sum_{\lambda^-\to\lambda} \psi'_{\lambda/\lambda^-} H'(\rho,\lambda^-) = -(|\lambda|-|\rho|)H'(\rho,\lambda).$$

Multiplying the first by $H'(\rho,\lambda)$, the second by $H(\nu,\rho)$ and taking the difference gives

$$(6.7)\quad (|\lambda|-|\nu|)H(\nu,\rho)H'(\rho,\lambda) = \sum_{\nu\to\nu^+} \psi'_{\nu^+/\nu} H(\nu^+,\rho)H'(\rho,\lambda) - \sum_{\lambda^-\to\lambda} \psi'_{\lambda/\lambda^-} H(\nu,\rho)H'(\rho,\lambda^-).$$

Now, simply multiply both sides of the above equation (6.7) with $P^{\#}_{\mu}(\rho)$, and sum them for all ρ where $\mu \subseteq \rho \subseteq \lambda$. Using Lemma 6.4 then gives the recursion in Proposition 6.2. □

7. Some particular cases

A weaker form of the conjecture asserts that for any $\alpha > 0$ (and any triple of partitions λ,μ,ν), the coefficient $g^{\lambda}_{\mu,\nu}$, or equivalently $c^{\lambda}_{\mu,\nu}$, is nonnegative. This is obviously a strictly weaker statement since a polynomial may take nonnegative values on nonnegative real numbers but have some negative coefficients, e.g. $(\alpha-1)^2$. In this section, we prove that this weaker property holds whenever $|\lambda|-|\mu| \leq 1$.

PROPOSITION 7.1. *For fixed $\alpha \geq 0$, we have that $c^{\lambda}_{\mu\lambda} \geq 0$.*

PROOF. Using Proposition 6.1 and (3.1), it is enough to show that

$$(7.1)\qquad \sum_{T\in \mathrm{RSSYT}(\mu,n)} \psi_T \left(\prod_{s\in\mu} (\lambda_{T(s)} - a'(s) + l'(s)/\alpha) \right)$$

is nonnegative. Since ψ_T is nonnegative for all $\alpha > 0$, it is enough to show that for each fixed tableau T in the sum above, the product is either positive or 0.

Fix T and assume that for the box $s \in \mu$, $(\lambda_{T(s)} - a'(s) + l'(s)/\alpha) < 0$. If $s = (i,j)$, this means

$$\lambda_{T(i,j)} - (j-1) + (i-1)/\alpha < 0.$$

Since T is a reverse tableau, and λ is a partition, $\lambda_{T(i,j)} \geq \lambda_{T(1,j)}$. This implies that $\lambda_{T(1,j)} - (j-1) < 0$.

Now, we have that $\lambda_{T(1,l)} \geq \lambda_{T(1,l-1)}$ for any l such that $2 \leq l \leq \mu_1$. As a consequence, the sequence

$$\lambda_{T(1,j)} - j + 1,\ \lambda_{T(1,j-1)} - j + 2,\ \ldots,\ \lambda_{T(1,1)}.$$

can increase by at most one in each step. Since $\lambda_{T(1,1)} \geq 0$, we know that at least one of these numbers is 0.

Thus, every product in (7.1) which has a negative factor, also include a factor which is zero. Hence, $c^{\lambda}_{\mu\lambda} \geq 0$ for fixed $\alpha \geq 0$. □

PROPOSITION 7.2. *For fixed $\alpha \geq 0$ and $|\lambda| - |\nu| = 1$ we have that $c^{\lambda}_{\mu\nu} \geq 0$.*

PROOF. Using the recursion established above, we have that

$$c^{\lambda}_{\mu\nu} = \psi'_{\lambda/\nu} \left(c^{\lambda}_{\mu\lambda} - c^{\nu}_{\mu\nu} \right)$$

if $|\lambda| - |\nu| = 1$. Since $\psi'_{\lambda/\nu} \geq 0$ when $\alpha > 0$, we can use Proposition 6.1 and (3.1) and conclude that it suffices to show that the following is nonnegative, where the sum is over all reverse semi-standard Young tableaux of shape μ:

$$\sum_{T} \psi_T \left(\prod_{s\in\mu} (\lambda_{T(s)} - a'(s) + l'(s)/\alpha) - \prod_{s\in\mu} (\nu_{T(s)} - a'(s) + l'(s)/\alpha) \right). \tag{7.2}$$

The reasoning in Proposition 7.1, implies that if any of the factors in the second product is negative, then both products contains at least one factor which is zero. Thus, we can assume that all factors in both products are nonnegative.

Since λ and ν only differ in one coordinate, say i, we can factor the difference of the product, and use that $\lambda_i = \nu_i + 1$:

$$\left(\prod_{\substack{s\in\mu \\ T(s)\neq i}} \nu_{T(s)} - b(s) \right) \left(\prod_{\substack{s\in\mu \\ T(s)=i}} (\nu_i + 1 - b(s)) - \prod_{\substack{s\in\mu \\ T(s)=i}} (\nu_i - b(s)) \right)$$

where $b(s) = a'(s) + l'(s)/\alpha$. We assumed that all factors are nonnegative, so the first product is nonnegative. This fact also implies that the difference in the second parenthesis is nonnegative. □

REMARK 7.3. The above method does not generalize directly to $|\lambda| - |\nu| \geq 2$, since some tableaux give negative contributions in that case.

Let us come back to the original conjecture, stating that $g^{\lambda}_{\mu,\nu}$ is a Laurent polynomial in α with nonnegative coefficients. Even in the case $\nu = \lambda$, where we have $H_\mu H_\lambda H'_\lambda P^{\#}_\mu(\lambda)$, we are not aware of a formula for shifted Jack polynomials explaining the (Laurent) polynomiality of this quantity. In [**AF18**], we make the following stronger conjecture (supported by numerical data):

CONJECTURE 7.4. The expression $\alpha^{\ell(\mu)-1} H_\mu P^{\#}_\mu(x_1, \dots, x_n)$ has nonnegative coefficient in the basis

$$\left(\alpha^c (x_1 - x_2)_{b_1} \cdots (x_{n-1} - x_n)_{b_{n-1}} (x_n)_{b_n} \right)_{c,b_1,\dots,b_n \geq 0},$$

where $(x)_b$ is as usual the bth falling power of x, that is $x(x-1)\cdots(x-b+1)$.

This suggests that an analogue of Knop–Sahi combinatorial formula [**KS97**] could exist for shifted Jack polynomials.

References

[AF18] Per Alexandersson and Valentin Féray, *Shifted symmetric functions and multirectangular coordinates of Young diagrams*, J. Algebra **483** (2017), 262–305, DOI 10.1016/j.jalgebra.2017.03.036. MR3649821

[BZ88] A. D. Berenstein and A. V. Zelevinsky, *Tensor product multiplicities and convex polytopes in partition space*, J. Geom. Phys. **5** (1988), no. 3, 453–472, DOI 10.1016/0393-0440(88)90033-2. MR1048510

[DF14] Maciej Dołęga and Valentin Féray, *Gaussian fluctuations of Young diagrams and structure constants of Jack characters*, Duke Math. J. **165** (2016), no. 7, 1193–1282, DOI 10.1215/00127094-3449566. MR3498866

[KS96] Friedrich Knop and Siddhartha Sahi, *Difference equations and symmetric polynomials defined by their zeros*, Internat. Math. Res. Notices **10** (1996), 473–486, DOI 10.1155/S1073792896000311. MR1399412

[KS97] Friedrich Knop and Siddhartha Sahi, *A recursion and a combinatorial formula for Jack polynomials*, Invent. Math. **128** (1997), no. 1, 9–22, DOI 10.1007/s002220050134. MR1437493

[Las08] Michel Lassalle, *A positivity conjecture for Jack polynomials*, Math. Res. Lett. **15** (2008), no. 4, 661–681, DOI 10.4310/MRL.2008.v15.n4.a6. MR2424904

[Mac95] I. G. Macdonald, *Symmetric functions and Hall polynomials*, 2nd ed., Oxford Mathematical Monographs, The Clarendon Press, Oxford University Press, New York, 1995. With contributions by A. Zelevinsky; Oxford Science Publications. MR1354144

[Mol09] A. I. Molev, *Littlewood-Richardson polynomials*, J. Algebra **321** (2009), no. 11, 3450–3468, DOI 10.1016/j.jalgebra.2008.02.034. MR2510056

[MS99] Alexander I. Molev and Bruce E. Sagan, *A Littlewood-Richardson rule for factorial Schur functions*, Trans. Amer. Math. Soc. **351** (1999), no. 11, 4429–4443, DOI 10.1090/S0002-9947-99-02381-8. MR1621694

[OO97] A. Okounkov and G. Olshanski, *Shifted Jack polynomials, binomial formula, and applications*, Math. Res. Lett. **4** (1997), no. 1, 69–78, DOI 10.4310/MRL.1997.v4.n1.a7. MR1432811

[Ras04] Etienne Rassart, *Geometric approaches to computing Kostka numbers and Littlewood-Richardson coefficients*, ProQuest LLC, Ann Arbor, MI, 2004. Thesis (Ph.D.)–Massachusetts Institute of Technology. MR2717178

[Sah94] Siddhartha Sahi, *The spectrum of certain invariant differential operators associated to a Hermitian symmetric space*, Lie theory and geometry, Progr. Math., vol. 123, Birkhäuser Boston, Boston, MA, 1994, pp. 569–576, DOI 10.1007/978-1-4612-0261-5_21. MR1327549

[Sah11] Siddhartha Sahi, *Binomial coefficients and Littlewood-Richardson coefficients for Jack polynomials*, Int. Math. Res. Not. IMRN **7** (2011), 1597–1612, DOI 10.1093/imrn/rnq126. MR2806516

[Sta89] Richard P. Stanley, *Some combinatorial properties of Jack symmetric functions*, Adv. Math. **77** (1989), no. 1, 76–115, DOI 10.1016/0001-8708(89)90015-7. MR1014073

Department of Mathematics, Stockholm University, SE-106 91 Stockholm, Sweden
Email address: `per.w.alexandersson@gmail.com`

Université de Lorraine, CNRS, IECL, F-54000 Nancy, France
Email address: `valentin.feray@univ-lorraine.fr`

Proceedings of Symposia in Pure Mathematics
Volume **110**, 2024
https://doi.org/10.1090/pspum/110/02016

Complexity and asymptotics of structure constants

Greta Panova

"Thinking out of the black box"

ABSTRACT. Kostka, Littlewood-Richardson, Kronecker, and plethysm coefficients are fundamental quantities in algebraic combinatorics, yet many natural questions about them stay unanswered for more than 80 years. Kronecker and plethysm coefficients lack "nice formulas", a notion that can be formalized using computational complexity theory. Beyond formulas and combinatorial interpretations, we can attempt to understand their asymptotic behavior in various regimes, and inequalities they could satisfy. Understanding these quantities has also applications outside of combinatorics. On the one hand, the asymptotics of structure constants is closely related to understanding the [limit] behavior of vertex and tiling models in statistical mechanics. More recently, these structure constants have been involved in establishing computational complexity lower bounds and separation of complexity classes like VP vs VNP, the algebraic analogs of P vs NP in arithmetic complexity theory. Here we discuss the outstanding problems related to asymptotics, positivity, and complexity of structure constants focusing mostly on the Kronecker coefficients of the symmetric group and, less so, on the plethysm coefficients.

This expository paper is based on the talk presented at the Open Problems in Algebraic Combinatorics coneference in May 2022.

1. Introduction

Algebraic Combinatorics studies symmetries via their manifestations in discrete objects, connecting with areas like Representation Theory, Statistical Mechanics, Computational Complexity Theory, and Algebraic Geometry. Many of its problems arise from the need for a quantitative and explicit understanding of algebraic phenomena like group representations and decompositions into irreducible representations, dimension formulae for modules, intersection numbers in geometry, etc.

In its origin lies the representation theory of the symmetric group S_n and general linear group GL_N. Their irreducible modules are indexed by integer partitions, and their bases can be described via standard and semi-standard Young tableaux (SYTs and SSYTs). As a next step, it is natural to understand how tensor products of irreducible representations decompose into irreducible components. In the case of GL_N the multiplicities of isotypic components are the Littlewood-Richardson (LR)

2020 *Mathematics Subject Classification.* Primary 05E05, 05E10; Secondary 68Q15, 68Q25, 05A16, 60K35, 60C05.

The author has been partially supported by the NSF.

coefficients. For S_n the multiplicities are the Kronecker coefficients. The composition of irreducible GL modules decomposes into irreducibles with multiplicities given by the plethysm coefficients.

While no compact formula for the LR coefficients exists, they can be understood via their combinatorial interpretation, the LR rule, which gives them in the form of certain skew SSYTs. The LR rule was formulated in 1934, but it took 40 years to prove formally. Inspired by the LR coefficients, Murnaghan defined the Kronecker coefficients in 1938 and observed that computing them is a highly nontrivial task. In late 70s and 80s the research turned towards obtaining positive formulas. In 2000 Stanley stated as a major open problem in Algebraic Combinatorics the problem of finding a nonnegative combinatorial interpretation for the Kronecker coefficients, as well as the plethysm coefficients. However, the progress has been minimal.

Besides presenting fundamental natural questions needing answers, the Kronecker and plethysm coefficients also play an important role in algebraic complexity theory, specifically Geometric Complexity Theory (GCT). Algebraic complexity theory studies the complexity of computing formal polynomials using arithmetic circuits, and some of its main problems concern the separation of complexity classes $\mathsf{VP} \neq \mathsf{VNP}$, the algebraic analog of the $\mathsf{P} \neq \mathsf{NP}$ Millennium problem. GCT approaches these problems by studying the symmetries and geometric properties of the universal polynomials (usually determinant and permanent), passing onto the coordinate rings of GL orbit closures of these polynomials. Complexity lower bounds and separation of classes then reduces to understanding the GL-irreducible representations of these rings and comparing their multiplicities. Such multiplicities are closely related to LR, Kronecker, and plethysm coefficients, and separating complexity classes can be achieved by proving certain inequalities for such multiplicities.

Another application of Algebraic Combinatorics is in Statistical Mechanics, Asymptotic Representation Theory and Random Matrix Theory. Some of the early connections date to the remarkable story of the longest increasing subsequence of permutations, which has the same [Tracy-Widom] distribution as the maximal eigenvalue of a random Hermitian matrix and the height of an SYT from the Plancherel measure. Symmetric polynomials, and particularly Schur functions, appear extensively in the study of random lozenge tilings, exclusion processes and vertex models. Plane partitions themselves are lozenge tilings with certain boundaries. Conversely, probabilistic and asymptotic methods have been used to understand algebro-combinatorial quantities. In the lack of compact formulas and combinatorial interpretations, understanding the growth behavior of structure constants is a natural next step. Beyond the interaction with Statistical Mechanics, understanding their growth is closely related to their application in GCT mentioned above.

Paper structure. Here we will briefly define the relevant structure constants, give some background, and phrase the open problems of three separate aspects: positivity, complexity, asymptotics. For more background on Complexity Theory and its specific connections with Algebraic Combinatorics we refer to the companion paper [**Pan23**] and references therein.

This paper presents the professional view of the author and is centered around problems on which she has worked on. It has no aspirations to be a comprehensive survey on the topics.

2. Basic objects and notions

We will assume the reader is familiar with basic notions in Algebraic Combinatorics like integer partitions, Young tableaux and symmetric functions. For a concise recap of those see [**Pan23**]. For details on the combinatorial sides see [**Sta99**, **Mac95**] and for the representation theoretic aspects see [**Sag01**, **Ful97**].

2.1. Partitions and Young tableaux. We use standard notation from [**Mac95**] and [**Sta99**, §7] throughout the paper.

Let $\lambda = (\lambda_1, \lambda_2, \ldots, \lambda_\ell)$ be a *partition* of size $n := |\lambda| = \lambda_1 + \lambda_2 + \ldots + \lambda_\ell$, where $\lambda_1 \geq \lambda_2 \geq \ldots \geq \lambda_\ell \geq 1$. We write $\lambda \vdash n$ for this partition, and $\mathcal{P} = \{\lambda\}$ for the set of all partitions. The length of λ is denoted $\ell(\lambda) := \ell$. Denote by $p(n)$ the number of partitions $\lambda \vdash n$. Let $\lambda + \mu$ denote the partition $(\lambda_1 + \mu_1, \lambda_2 + \mu_2, \ldots)$

Special partitions include the *rectangular shape* $(a^b) = (a, \ldots, a)$, b times, the *hooks shape* $(k, 1^{n-k})$, the *two-row shape* $(n-k, k)$, and the *staircase shape* $\rho_\ell = (\ell, \ell - 1, \ldots, 1)$.

A *Young diagram* of *shape* λ is an arrangement of squares $(i, j) \subset \mathbb{N}^2$ with $1 \leq i \leq \ell(\lambda)$ and $1 \leq j \leq \lambda_i$. Let $\lambda \vdash n$. A *semistandard Young tableau* T of *shape* λ and *weight* α is an arrangement of α_k many integers k in squares of λ, which weakly increase along rows and strictly increase down columns, i.e. $T(i, j) \leq T(i, j+1)$ and $T(i, j) \leq T(i+1, j)$. For example, $\begin{array}{|c|c|c|c|c|} \hline 1&1&2&4&4 \\ \hline 2&2&3&5 \\ \cline{1-4} 4&5 \\ \cline{1-2} \end{array}$ is an SSYT of shape $\lambda = (5, 4, 2)$ and type $\alpha = (2, 3, 1, 3, 2)$. Denote by $\mathrm{SSYT}(\lambda, \alpha)$ the set of such tableaux, and $K(\lambda, \alpha) = \big|\mathrm{SSYT}(\lambda, \alpha)\big|$ the *Kostka number.* A *standard Young tableau* (SYT) of shape λ is an SSYT of type (1^n), and we have $f^\lambda := K_{\lambda, 1^n}$, which can be computed by the hook-length formula (HLF) of [**FRT54**].

2.2. Representations of S_n and GL_N. The irreducible representations of the *symmetric group* S_n are the *Specht modules* $\mathbb{S}_\lambda$ and are indexed by partitions $\lambda \vdash n$. A basis for $\mathbb{S}_\lambda$ can be indexed by the SYTs. In particular

$$\dim \mathbb{S}_\lambda = f^\lambda.$$

We have that $\mathbb{S}_{(n)}$ is the trivial representation assigning to every w the value 1 and $\mathbb{S}_{1^n}$ is the sign representation.

The *character* $\chi^\lambda(w)$ of $\mathbb{S}_\lambda$ can be computed via the *Murnaghan-Nakayama* rule. Let w have type α, i.e. it decomposes into cycles of lengths $\alpha_1, \alpha_2, \ldots, \alpha_k$. Then

$$\chi^\lambda(w) = \chi^\lambda(\alpha) = \sum_{T \in MN(\lambda; \alpha)} (-1)^{ht(T)},$$

where MN is the set of rim-hook tableaux of shape λ and type α, so that the entries are weakly increasing along rows and down columns, and all entries equal to i form a rim-hook shape (i.e. connected, no 2×2 boxes) of length α_i. The height of each rim-hook is one less than the number of rows it spans, and $ht(T)$ is the sum of all these heights. For examples

$$\begin{array}{|c|c|c|c|c|c|} \hline 1&1&2&3&3&3 \\ \hline 1&2&2&3&4 \\ \cline{1-5} 2&2&3&3&4 \\ \cline{1-5} \end{array}$$

is a Murnaghan-Nakayama tableau of shape $(6, 5, 5)$, type $(3, 5, 6, 2)$ and has height $ht(T) = 1 + 2 + 2 + 1 = 6$.

The irreducible polynomial representations of $GL_N(\mathbb{C})$ are the *Weyl modules* V_λ and are indexed by all partitions with $\ell(\lambda) \leq N$. Their characters are exactly the Schur functions $s_\lambda(x_1, \ldots, x_N)$, where $x_1, \ldots, x_N$ are the eigenvalues of $g \in GL_N(\mathbb{C})$.

2.3. Symmetric functions. Let $\Lambda[\boldsymbol{x}]$ be the ring of *symmetric functions* $f(x_1, x_2, \ldots)$, where the symmetry means that $f(\boldsymbol{x}) = f(\boldsymbol{x}_\sigma)$ for any permutation σ of the variables, and f is a formal power series. The ring Λ_n of homogeneous symmetric functions of degree n has several important bases: the *homogenous symmetric functions* h_λ, *elementary symmetric functions* e_λ, *monomial symmetric functions* m_λ , *power sum symmetric functions* p_λ and *Schur functions* s_λ. The Schur functions can be defined as the generating functions for SSYTs of shape λ

$$s_\lambda = \sum_{\mu \vdash n} K_{\lambda\mu} m_\mu,$$

but also using *Weyl's determinantal* formula

$$s_\lambda(x_1, \ldots, x_\ell) = \frac{\det[x_i^{\lambda_j + \ell - j}]_{i,j=1}^{\ell}}{\prod_{i<j}(x_i - x_j)}$$

and the *Jacobi-Trudi* identity

$$s_\lambda = \det[h_{\lambda_i - i + j}]_{i,j=1}^{\ell(\lambda)}.$$

The *Cauchy identity*

$$\sum_\lambda s_\lambda(\mathbf{x}) s_\lambda(\mathbf{y}) = \prod_{i,j} \frac{1}{1 - x_i y_j}$$

has a remarkable combinatorial proof given by the RSK correspondence between pairs of same shape SSYTs and matrices with nonnegative integer entries.

2.4. Computational Complexity. We refer to [**Aar16, Wig19, BCS97**] for details on Computational Complexity classes, and to [**Pak22+, Pan23**] and references therein for the connections with Algebraic Combinatorics.

A *decision problem* is a computational problem, for which the output is Yes or No. There are two major complexity classes P and NP, subject of the P vs NP Millennium problem. P is the class of decision problems, where given any input of size n (number of bits required to encode it), such that the answer can be obtained in *polynomial time* denoted by *poly*(n), i.e. there is a fixed k and an algorithm taking $O(n^k)$ many steps (elementary operations). NP is the class of decision problems, where if the answer is Yes, then it can be verified in polynomial time, i.e. there is a poly-time computable witness. Naturally, $\mathsf{P} \subset \mathsf{NP}$ and it is widely believed that $\mathsf{P} \neq \mathsf{NP}$. The classes FP and $\#\mathsf{P}$ are the counting analogues of P and NP. A *counting problem* is in $\#\mathsf{P}$ if it is the number of accepting paths of an NP Turing machine. In practice, these are counting problems where the answers are exponentially large sums of 0-1 valued functions M, each of which can be computed in $O(n^k)$ time for a fixed k:

$$\sum_{b \in \{0,1\}^m} M(b),$$

where $m = poly(n)$. We set the classes $\mathsf{GapP} = \{f - g | \, f, g \in \#\mathsf{P}\}$ and $\mathsf{GapP}_{\geq 0} = \{f - g | \, f, g \in \#\mathsf{P} \text{ and } f - g \geq 0\}$.

The algebraic complexity classes VP and VNP were introduced by Valiant [**V79a**], as the algebraic analogues of P and NP (we refer to [**Bür00a**] for formal definitions and properties). They concern the computation of polynomials $f \in \mathbb{F}[x_1, \dots, x_N]$ of degree $poly(N)$ using arithmetic circuits where the inputs are constants of arbitrary size from the field $\mathbb{F}$ and the formal variables $x_1, \dots, x_n$, the gates are $+, \times$, and the output is f. Then VP is the class of polynomials for which there is a fixed k and a circuit of size $O(n^k)$ computing the polynomial f. VNP is the class of polynomials f, such that there exist a k and $m = O(N^k)$ and a polynomial $g \in \mathsf{VP}$ in $N + m$ variables, such that

$$f(x_1, \dots, x_N) = \sum_{b \in \{0,1\}^m} g(x_1, \dots, x_N, b_1, \dots, b_m).$$

The main conjecture is that $\mathsf{VP} \neq \mathsf{VNP}$ and is closely related to $\mathsf{P} \neq \mathsf{NP}$, see e.g. [**Bür00b**]. As Valiant showed, for every polynomial $f \in \mathbb{C}[x_1, \dots, x_n]$ there exists a K and a $K \times K$ matrix $A := A_0 + \sum_{i=1}^{N} A_i x_i$ with $A_j \in \mathbb{C}^{K \times K}$ such that $\det A = f$. The smallest such K is the *determinantal complexity* of f denoted $\mathsf{dc}(f)$ and it is finite for every f. The $\mathsf{VP}_{\mathrm{ws}}$-universal polynomial is the determinant, in the sense that $f \in \mathsf{VP}_{\mathrm{ws}}$ iff $\mathsf{dc}(f) = poly(n)$, where $poly(n)$ denotes any fixed degree polynomial in n. Here $\mathsf{VP}_{\mathrm{ws}}$ is the class of polynomials with poly-sized Algebraic Branching Programs (a model between "formula" and "circuit"'). The class $\mathsf{VP}_{\mathrm{ws}}$ has been the main subject of study and is often interchangably used for VP. The classical universal VNP-complete polynomial is the permanent

$$\mathrm{per}_m [X_{ij}]_{i,j=1}^m = \sum_{\sigma \in S_m} \prod_{i=1}^{m} X_{i\sigma(i)}.$$

Thus, separating $\mathsf{VP}_{\mathrm{ws}}$ from VNP can be done by showing that $\mathsf{dc}(per_m) \gg poly(m)$.

The *Geometric Complexity Theory* (GCT) program of Mulmuley and Sohoni [**MS01, MS08**] compares the determinant and permanent by comparing their orbit closures under the action of GL_{n^2} on the variables $(X_{11}, \dots, X_{nn})$. If $\mathsf{dc}(\mathrm{per}_m) \leq n$, then $\overline{GL_{n^2}\mathrm{per}_m} \subset \overline{GL_{n^2}\det_n}$, and passing to the coordinate ring, this induces a GL_{n^2}-equivariant surjection

$$\mathbb{C}[\overline{GL_{n^2}\det_n}] \twoheadrightarrow \mathbb{C}[\overline{GL_{n^2}X_{11}^{n-m}\mathrm{per}_m}] \tag{2.1}$$

To show a lower bound for $\mathsf{dc}(\mathrm{per}_m)$, i.e. $\mathsf{dc}(\mathrm{per}_m) > n$, then we need to show that the above surjection does not happen. Decomposing these modules into irreducibles for any given degree d, we have

$$\mathbb{C}[\overline{\mathrm{GL}_{n^2}\det_n}]_d \simeq \bigoplus_{\lambda} V_\lambda^{\oplus \delta_{\lambda,d,n}}, \quad \text{and} \quad \mathbb{C}[\overline{\mathrm{GL}_{n^2}\mathrm{per}_m^n}]_d \simeq \bigoplus_{\lambda} V_\lambda^{\oplus \gamma_{\lambda,d,n,m}}. \tag{2.2}$$

Thus, to show that (2.1) is not possible, we can find an irreducible GL_{n^2} module V_λ, such that $\gamma_{\lambda,d,n,m} > \delta_{\lambda,d,n}$ for some d. Such λ's are called *multiplicity obstructions* and γ and δ are closely related to plethysm and Kronecker coefficients. When $\delta_{\lambda,d,n} = 0$ also, then λ is an *occurrence obstruction*. It was shown in [**BIP19**] that there are no occurrence obstructions for $n > m^{25}$, so we need to find multiplicity obstructions.

This is one of the main applications of Kronecker and plethysm coefficients and motivates our work. In particular, we have

$$\delta_{\lambda,d,n} \leq g(\lambda, n^d, n^d) \qquad \text{and} \qquad \gamma_{\lambda,d,n,m} \leq a_\lambda(d[n]).$$

See §3 for the definitions and [**Pan23**] for a brief overview of these connections.

3. Multiplicities and structure constants

Once the irreducible representations have been sufficiently understood, it is natural to consider what other representations can be formed by them and how such representations decompose into irreducibles.

In the case of $GL_N(\mathbb{C})$ these coefficients are the *Littlewood–Richardson coefficients* (LR) $c^{\lambda}_{\mu\nu}$ defined as the multiplicity of V_λ in $V_\mu \otimes V_\nu$, so

$$V_\mu \otimes V_\nu = \bigoplus_{\lambda} V_\lambda^{\oplus c^{\lambda}_{\mu\nu}}.$$

Via their characters, they can be equivalently defined as

$$s_\mu s_\nu = \sum_{\lambda} c^{\lambda}_{\mu\nu} s_\lambda, \qquad \text{and} \qquad s_{\lambda/\mu} = \sum_{\nu} c^{\lambda}_{\mu\nu} s_\nu.$$

While no nice product formula exists for their computation, they have a *combinatorial interpretation*, the so called *Littlewood-Richardson rule*, see [**Sta99**, App. 1].

THEOREM 3.1 (Littlewood-Richardson rule). *The Littlewood-Richardson coefficient $c^{\lambda}_{\mu\nu}$ is equal to the number of skew SSYT T of shape λ/μ and type ν, whose reading word is a ballot sequence.*

A ballot sequence means that in every prefix of the sequence the number of 1's$\geq$ the number of 2's $\geq$ number of 3's etc. For example, we have $c^{(6,4,3)}_{(3,1),(4,3,2)} = 2$ as there are two LR tableaux (SSYT of shape $(6,4,3)/(3,1)$ and type $(4,3,2)$ whose reading words are ballot sequences) $\begin{array}{cccccc} & & & 1 & 1 & 1 \\ & 1 & 2 & 2 & & \\ 2 & 3 & 3 & & & \end{array}$ with reading word 111221332 and $\begin{array}{cccccc} & & & 1 & 1 & 1 \\ & 2 & 2 & 2 & & \\ 1 & 3 & 3 & & & \end{array}$ with reading word 111222331.

The *Kronecker coefficients* $g(\lambda, \mu, \nu)$ of the symmetric group are the corresponding structure constants for the ring of S_n- irreducibles. Namely, S_n acts diagonally on the tensor product of two Specht modules and the corresponding module factors into irreducibles with multiplicities $g(\lambda, \mu, \nu)$

$$\mathbb{S}_\lambda \otimes \mathbb{S}_\mu = \bigoplus_{\nu} \mathbb{S}_\nu^{\oplus g(\lambda,\mu,\nu)}.$$

In terms of characters we can write them as

$$g(\lambda, \mu, \nu) = \langle \chi^\lambda \chi^\mu, \chi^\nu \rangle = \frac{1}{n!} \sum_{w \in S_n} \chi^\lambda(w) \chi^\mu(w) \chi^\nu(w). \tag{3.1}$$

The last formula shows that they are symmetric upon interchanging the underlying partitions $g(\lambda, \mu, \nu) = g(\mu, \nu, \lambda) = \cdots$ which motivates us to use such symmetric notation.

The *Kronecker product* $*$ on Λ is defined on the Schur basis by

$$s_\lambda * s_\mu = \sum_{\nu} g(\lambda, \mu, \nu) s_\nu,$$

and extended by linearity.

Inspired by the Littlewood-Richardson coefficients, Murnaghan defined them in 1938. In fact, he showed that, see [**Mur56**],

THEOREM 3.2 (Murnaghan). *For every λ, μ, ν, such that $|\lambda| = |\mu| + |\nu|$ we have that*

$$c^{\lambda}_{\mu\nu} = g((n - |\lambda|, \lambda), (n - |\mu|, \mu), (n - |\nu|, \nu)),$$

for sufficiently large n.

In particular, one can see that $n = 2|\lambda| + 1$ would work. Note that even when the sizes of the partitions do not add up, the RHS stabilizes as $n \to \infty$ to the *reduced Kronecker coefficients*, see 7.

Thanks to the Schur-Weyl duality, they can also be interpreted via Schur functions as

$$s_{\lambda}[\boldsymbol{xy}] = \sum_{\mu,\nu} g(\lambda, \mu, \nu) s_{\mu}(\boldsymbol{x}) s_{\nu}(\boldsymbol{y}),$$

where $\boldsymbol{xy} = (x_1y_1, x_1y_2, \ldots, x_2y_1, \ldots)$ is the vector of all pairwise products. In terms of GL representations they give us the dimension of the invariant space

$$g(\lambda, \mu, \nu) = \dim(V_{\mu} \otimes V_{\nu} \otimes V^*_{\lambda})^{GL_N \times GL_M},$$

where V_{μ} is considered as a GL_N-module, and V_{ν} as a GL_M-module and V^*_{λ} is a dual GL_{NM}-module. In other words, they are dimensions of highest weight vector spaces, and using this interpretation for them as dimensions of highest weight spaces one can show the following, see [**CHM07**]

THEOREM 3.3 (Semigroup property). *Let $(\alpha^1, \beta^1, \gamma^1)$ and $(\alpha^2, \beta^2, \gamma^2)$ be two partition triples, such that $|\alpha^1| = |\beta^1| = |\gamma^1|$ and $|\alpha^2| = |\beta^2| = |\gamma^2|$. Suppose that $g(\alpha^1, \beta^1, \gamma^1) > 0$ and $g(\alpha^2, \beta^2, \gamma^2) > 0$. Then*

$$g(\alpha^1 + \alpha^2, \beta^1 + \beta^2, \gamma^1 + \gamma^2) \geq \max\{g(\alpha^1, \beta^1, \gamma^1), g(\alpha^2, \beta^2, \gamma^2)\}.$$

From the original S_n characters definition we see that

$$g(\lambda, \mu, \nu) = g(\lambda', \mu', \nu)$$

since $\chi^{\lambda'}(w) = \chi^{\lambda}(w)\mathrm{sgn}(w)$. In particular, we have $g(\lambda, \mu, (n)) = \delta_{\lambda,\mu}$ for all λ and $g(\lambda, \mu, 1^n) = \delta_{\lambda,\mu'}$.

EXAMPLE 3.4. By the above observation we have that

$$h_k[\boldsymbol{xy}] = s_k[\boldsymbol{xy}] = \sum_{\lambda \vdash k} s_{\lambda}(\boldsymbol{x}) s_{\lambda}(\boldsymbol{y}).$$

Using the Jacobi-Trudi identity we can write

$$\begin{aligned} s_{2,1}[\boldsymbol{xy}] = h_2[\boldsymbol{xy}]h_1[\boldsymbol{xy}] - h_3[\boldsymbol{xy}] &= (s_2(\boldsymbol{x})s_2(\boldsymbol{y}) + s_{1,1}(\boldsymbol{x})s_{1,1}(\boldsymbol{y}))\, s_1(\boldsymbol{x})s_1(\boldsymbol{y}) \\ &\quad - s_3(\boldsymbol{x})s_3(\boldsymbol{y}) - s_{2,1}(\boldsymbol{x})s_{2,1}(\boldsymbol{y}) - s_{1,1,1}(\boldsymbol{x})s_{1,1,1}(\boldsymbol{y}) \\ = s_{2,1}(\boldsymbol{x})s_{2,1}(\boldsymbol{y}) + s_{2,1}(\boldsymbol{x})s_3(\boldsymbol{y}) + s_3(\boldsymbol{x})&s_{2,}(\boldsymbol{y}) + s_{1,1,1}(\boldsymbol{x})s_{2,1}(\boldsymbol{y}) + s_{2,1}(\boldsymbol{x})s_{1,1,1}(\boldsymbol{y}). \end{aligned}$$

So we see that $g((2,1),(2,1),(2,1)) = 1$.

The *plethysm coefficients* $a^{\lambda}_{\mu,\nu}$ are multiplicities of an irreducible GL representation in the composition of two GL representations. Namely, let $\rho^{\mu} : GL_N \to GL_M$ be one irreducible, and $\rho^{\nu} : GL_M \to GL_K$ be another. Then $\rho^{\nu} \circ \rho^{\mu} : GL_N \to GL_K$

is another representation of GL_N which has a character $s_\nu[s_\mu]$, which decomposes into irreducibles as

$$s_\nu[s_\mu] = \sum_\lambda a^\lambda_{\mu,\nu} s_\lambda.$$

Here the *plethystic* notation $f[g]$ means the evaluation of f over the monomials of g as variables: if $g = \boldsymbol{x}^{\alpha^1} + \boldsymbol{x}^{\alpha^2} + \cdots$, then $f[g] = f(\boldsymbol{x}^{\alpha^1}, \boldsymbol{x}^{\alpha^2}, \ldots)$.

EXAMPLE 3.5. We have that

$$\begin{aligned} s_{(2)}[s_{(1^2)}] = h_3[e_2] = h_2(x_1x_2, x_1x_3, \ldots) &= x_1^2x_2^2 + x_1^2x_2x_3 + 3x_1x_2x_3x_4 + \cdots \\ &= s_{2,2}(x_1, x_2, x_3, \ldots) + s_{1,1,1,1}(x_1, x_2, x_3, \ldots), \end{aligned}$$

so $a^{(2,2)}_{(2),(1,1)} = 1$ and $a^{(3,1)}_{(2),(1,1)} = 0$.

We will be particularly interested when $\mu = (d)$ or (1^d) which are the dth symmetric power Sym^d and the dth wedge power Λ^d, and $\nu = (n)$. We denote this plethysm coefficient by $a_\lambda(d[n]) := a^\lambda_{(d),(n)}$ and

$$h_d[h_n] = \sum_\lambda a_\lambda(d[n]) s_\lambda.$$

These coefficients are central to the GCT application and deserve special attention.

4. Combinatorial interpretations

"Combinatorial interpretation", albeit never formally defined in the literature, is assumed to mean a family of nice discrete objects whose cardinality gives the desired quantity. In practice, for quantities with combinatorial interpretation, computing them is in #P. On the other hand, being in #P does not necessarily give a "nice" combinatorial interpretation, but is certainly a starting point. All problems considered here are in GapP, which makes #P the natural class to consider in relation to "manifestly nonnegative" formulas. If such a problem is not in #P, then there could not be a reasonable nonnegative combinatorial interpretation within the given context. This paradigm is discussed at length in [**IP22, Pak22+, Pan23**].

Following the discovery of the Littlewood-Richardson rule in 1934, Murnaghan [**Mur38**] defined the Kronecker coefficients of S_n and computed various special cases, but quickly observed that this is no easy task. Interest in nonnegative combinatorial interpretation of these coefficients reemerged in the 80s, see [**Las79, GR85**], and was stated as Problem 10 in Stanley's list[1] of "Open Problems in Algebraic Combinatorics" [**Sta00**].

OPEN PROBLEM 4.1 (Stanley). Find a combinatorial interpretation of $g(\lambda, \mu, \nu)$, thereby combinatorially proving that they are nonnegative.

Over the years, there has been very little progress on the question. In 1989 Remmel determined $g(\lambda, \mu, \nu)$ when two of the partitions are hooks. In 1994 Remmel and Whitehead [**RW94**] determined $g(\lambda, \mu, \nu)$ when two of the partitions are two-rows, i.e. $\ell(\lambda), \ell(\mu) \leq 2$. This case was subsequently studied also in [**BMS15**]. In 2006 Ballantine and Orellana [**BO05**] determined a rule for $g(\lambda, \mu, \nu)$ when one partition is a two-row, e.g. $\mu = (n-k, k)$, and the first row of one of the others is large, namely $\lambda_1 \geq 2k-1$. The most general rule was determined by Blasiak [**B13**]

[1]See this for the original list and updates on the problems `https://mathoverflow.net/questions/349406/`

in 2012 when one partition is a hook, and this was later simplified by Blasiak and Liu [**BL18**], and further in [**Liu17**]. Other very special cases have been computed for multiplicity free Kronecker products by Bessenrodt-Bowman [**BB17**], Kroneckers corresponding to pyramids by Ikenmeyer-Mulmuley-Walter [**IMW17**], $g(m^k, m^k, (mk-n, n))$ as counting labeled trees by Pak-Panova see Remark 4.4, near rectangular partitions by Tewari in [**T15**], etc.

Similar questions pertain to the plethysm coefficients. The following problem is number 9 in Stanley's list [**Sta00**].

OPEN PROBLEM 4.2 (Stanley). Find a combinatorial interpretation for the plethysm coefficients $a^{\lambda}_{\mu,\nu}$.

A detailed survey on partial results and methods can be found in [**COSSZ22**].

Even the simple case for $a_\lambda(d[n]) = \langle s_\lambda, h_d[h_n]\rangle$ is not known in general unless $n = 2$. While, there is no direct connection between the Kronecker and plethysm coefficients, the case when $\ell(\lambda) = 2$ coincide, see [**PP14**], and actually give a combinatorial interpretation as explained below.

PROPOSITION 4.3. *We have that* $g(\lambda, n^d, n^d) = a_\lambda(d[n]) = p_k(n,d) - p_{k-1}(n,d)$ *for* $\lambda = (nd-k, k) \vdash nd$. *Here* $p_r(a,b) = \#\{\mu \vdash r : \mu_1 \le a, \ell(\mu) \le b\}$ *are the partitions of* r *which fit inside the* $a \times b$ *rectangle.*

REMARK 4.4. In particular, these are the coefficients in the q-binomials

$$\sum_r p_r(a,b) q^r = \binom{a+b}{a}_q := \prod_{i=1}^{a} \frac{(1-q^{i+b})}{1-q^i}.$$

The combinatorial proof by Kathy O'Hara of the unimodality of the $p_r(a,b)$ is what gives the cominatorial interpration for the particular Kronecker and plethysm coefficients as counting certain labeled trees, see [**Pan15**, p. 9].

To formalize these problems, we formulate them within the Computational Complexity framework. To set the stage with an example, we start with the Littlewood-Richardson coefficients.

ComputeLR:
`Input:` λ, μ, ν
`Output:` Value of $c^{\lambda}_{\mu\nu}$.

Here and below, the input can be encoded in *unary* or *binary*, and this could affect the complexity. By unary we mean that the input for each partition is written as $(\underbrace{11\ldots1}_{\lambda_1}, \underbrace{1\ldots1}_{\lambda_2}, \ldots)$, so the input size is $O(|\lambda|) = O(n)$. In binary, we encode each part λ_i in base 2, so the input size is $\log_2(\lambda_1) + \log_2(\lambda_2) + \cdots = O(\ell(\lambda)\log_2(\lambda_1))$.

By the Littlewood-Richardson rule we have that when the input is unary, ComputeLR$\in$ #P, as our witnesses are the skew λ/μ tableaux, which are at most $n^n = O(2^{n^2})$-many, and each can be checked to be an LR tableaux in time $O(n^2)$. The LR conditions can be translated into $O(\ell(\lambda)^2)$ many linear inequalities, which comprise the LR polytope. This description gives us that if the input is in binary we still have ComputeLR$\in$ #P.

Using a reduction to Knapsack for binary input, it is not difficult to see that

THEOREM 4.5 (Narayanan [**Nar06**]). *When the input* λ, μ, ν *is in binary, ComputeLR is* #P*-complete.*

When the input is in unary, we do not know whether the problem is still that hard.

CONJECTURE 4.6 (Pak-Panova 2020). *When the input is in unary we have that* $\mathsf{ComputeLR}$ *is* $\#\mathsf{P}$*-complete.*

None of the above has been achieved for the Kronecker and plethysm coefficients, however, due to the lack of any positive combinatorial formula. Mulmuley also conjectured that computing the Kronecker coefficients would be in $\#\mathsf{P}$, mimicking the Littlewood-Richardson coefficients.

ComputeKron:
`Input:` λ, μ, ν
`Output:` Value of $g(\lambda, \mu, \nu)$.

CONJECTURE 4.7 (Pak). COMPUTEKRON *is not in* $\#\mathsf{P}$ *under reasonable complexity theoretic assumptions like* PH *not collapsing.*

If the above is proven, that would make any solution to Open Problem 4.1 as unlikely as the polynomial hierarchy collapsing. Any reasonable combinatorial interpretation such as counting certain objects would show that the problem is in $\#\mathsf{P}$, since the objects would likely be verifiable in polynomial time. The problem stands for both unary and binary input.

Note that $\mathsf{ComputeKron} \in \mathsf{GapP}$ ([**BI08**]) as it is easy to write an alternating sum for its computation, for example using contingency arrays, see [**PP17b**].

The author's experience with Kronecker coefficients suggests that some particular families would be as hard as the general problem.

CONJECTURE 4.8 (Panova). $\mathsf{ComputeKron}$ *is in* $\#\mathsf{P}$ *when* $\ell(\lambda) = 2$ *if and only if* $\mathsf{ComputeKron}$ *is in* $\#\mathsf{P}$*. Likewise,* $\mathsf{ComputeKron}$ *is in* $\#\mathsf{P}$ *for* $\mu = \nu = (n^d)$ *and* $\lambda \vdash nd$ *as the input if and only if* $\mathsf{ComputeKron}$ *is in* $\#\mathsf{P}$*.*

Of course, the backward direction is clear, if the general problem is in $\#\mathsf{P}$, then also the special subproblems are in $\#\mathsf{P}$.

The plethysm coefficients can also be studied from this perspective.

ComputePleth:
`Input:` λ, μ, ν
`Output:` Value of $a^{\lambda}_{\mu,\nu}$.

OPEN PROBLEM 4.9. Determine whether $\mathsf{ComputePleth} \in \#\mathsf{P}$.

Using symmetric function identities, it is not hard to find an alternating formula for the plethysms and show that they are also in GapP, see [**FI20**]. They also show that deciding positivity is NP-hard and thus the computational problem is $\#\mathsf{P}$-hard. $\mathsf{ComputePleth}$ may not in be in $\#\mathsf{P}$ in the general case, but also when μ, ν are single row partitions. The coefficient $a_\lambda(d[n])$ in this case has special significance in GCT, see [**Pan23**].

Underlying all the representation theoretic multiplicities mentioned above are the characters of the symmetric group. For example, equation (3.1) expresses the Kronecker coefficients via characters, and the other structure constants can also be expressed in similar ways. This motivates us to understand them also computationally. As we shall see, they also provide a proof of concept for the idea that certain positive integral algebro-combinatorial quantities may indeed not have a combinatorial interpretation.

The characters satisfy some particularly nice identities coming from the orthogonality of the rows and columns of the character table in S_n. We have that

$$\sum_{\lambda \vdash n} \chi^\lambda(w)^2 = \prod_i i^{c_i} c_i!, \tag{4.1}$$

where c_i = number of cycles of length i in $w \in S_n$. When $w = \mathrm{id}$, we have that $\chi^\lambda(\mathrm{id}) = f^\lambda$, the number of SYTs and the identity is proven using the beautiful RSK bijection. The first step in this proof is to identify $(f^\lambda)^2$ as the number of pairs of SYTs of the same shape.

Could anything like that be done for equation (4.1)? The first step would be to understand what objects $\chi^\lambda(w)^2$ counts, does it have any positive combinatorial interpretation? We formulate it again using the CC paradigm as

ComputeCharSq:
`Input:` $\lambda, \alpha \vdash n$, unary.
`Output:` the integer $\chi^\lambda(\alpha)^2$.

THEOREM 4.10 ([**IPP22**]). *ComputeCharSq* $\notin$ #P *unless* $PH = \Sigma_2^{\mathsf{P}}$.

The last condition says "polynomial hierarchy collapses to the second level", which is almost as unlikely as P = NP. The widely believe complexity theoretic assumptions are that P $\neq$ NP and that PH does not collapse. This problem shows that some natural questions in Algebraic Combinatorics are indeed not in #P.

The proof of this Theorem follows from another complexity theoretic result we show, namely that deciding if $\chi^\lambda(\alpha) \neq 0$ is PH-hard as well, see end of §5.

5. Positivity problems

Motivated by other developments, further questions about the Kronecker coefficients have appeared. Following their work in [**HSTZ13**] on the square of the Steinberg character for finite groups of Lie type, the authors conjectured that a similar irreducible character should exist for the symmetric group and formulated the following.

CONJECTURE 5.1 (Tensor square conjecture). *For every $n \geq 9$ there exists a symmetric partition $\lambda \vdash n$, such that $\mathbb{S}_\lambda \otimes \mathbb{S}_\lambda$ contains every irreducible S_n module. In other words $g(\lambda, \lambda, \mu) > 0$ for all $\mu \vdash n$.*

Following this, Saxl conjectured a specific partition, namely the staircase, which would satisfy that.

CONJECTURE 5.2 (Saxl, see [**PPV16**]). *Let $\delta_k = (k, k-1, \dots, 1)$ be the staircase partition. Then $g(\delta_k, \delta_k, \mu) > 0$ for all k and $\mu \vdash \binom{k+1}{2}$.*

This conjecture has received a lot of attention partially due to its concreteness.

Positivity results were proved using a combination of three methods – the semigroup property constructing recursively positive triples from building blocks, explicit heighest weight constructions using the techniques in [**Ful00**], and an unusual comparison with characters, which was originally stated in by Bessenrodt and Behns [**BB04**] to show that $g(\lambda, \lambda, \lambda) > 0$ for $\lambda = \lambda'$, later generalized in [**PPV16**], and in its final form in [**PP17a**].

THEOREM 5.3 ([**PP17a**]). *Let $\lambda, \mu \vdash n$ and $\lambda = \lambda'$. Let $\hat\lambda = (2\lambda_1 - 1, 2\lambda_2 - 3, 3\lambda_3 - 5, \ldots)$ be the principal hooks of λ. Then*

$$g(\lambda, \lambda, \mu) \geq |\chi^\mu(\hat\lambda)|.$$

The technique of explicit highest weight vectors construction led to the following more general result.

THEOREM 5.4 ([**I15**]). *Let $\mu \vdash \binom{k+1}{2}$ be a partition comparable to ρ_k in the dominance order, i.e. $\mu \succ \rho_k$ or $\rho_k \succ \mu$. Then $g(\rho_k, \rho_k, \mu) > 0$.*

Using this result and generalizations, and clever combinations of the semigroup property 3.3, Luo and Sellke [**LS17**] proved that $\chi^{\rho_k} \otimes \chi^{\rho_k} \otimes \chi^{\rho_k} \otimes \chi^{\rho_k}$ contains every χ^μ, and that $\chi^{\rho_k} \otimes \chi^{\rho_k}$ contains "most" irreducible representations. In [**B18**], Bessenrodt proved that all double hooks, i.e. partitions with Durfee size 2, are contained in $\chi^{\rho_k} \otimes \chi^{\rho_k}$, and Li generalized to triple hooks [**L21**]. Harman and Ryba [**HR22**] showed that $\chi^{\rho_k} \otimes \chi^{\rho_k} \otimes \chi^{\rho_m}$ contains every irreducible representation.

The semigroup property is a powerful tool to create positive Kronecker triples by induction out of building blocks. However, creating a particular partition, like the staircase or a rectangle, out of building blocks is very restrictive and quickly encounters number theoretic issues. A rectangle can be "cut" only into other rectangles. This prompts us to enlarge the class of positive triples. In 2020, with Christine Bessenrodt we generalized Conjecture 5.1 as follows (and checked for partitions up to size 20).

CONJECTURE 5.5 (Bessenrodt-Panova 2020). *For every n there exists a $k(n)$, such that for every $\lambda \vdash n$ with $\lambda = \lambda'$ and $d(\lambda) > k_n$ which is not the square partition, we have $g(\lambda, \lambda, \mu) > 0$ for all $\mu \vdash n$.*

Here $d(\lambda) = \max\{i : \lambda_i \geq i\}$ is the Durfee square size of the partition. Partial progress on that conjecture will appear in the work of Chenchen Zhao.

Another positivity result, motivated by applications in GCT, was obtained in [**IP17**] as

LEMMA 5.6 ([**IP17**, Thm 1.10]). *Let $X := \{1, 1^2, 1^4, 1^6, 21, 31\}$, and let partition $\nu \notin X$. Denote $\ell := \max\{\ell(\nu) + 1, 9\}$, and suppose $r > 3\ell^{3/2}$, $s \geq 3\ell^2$, and $|\nu| \leq rs/6$. Let $\nu[N] := (N - |\nu|, \nu_1, \nu_2, \ldots)$. Then $g(s^r, s^r, \nu[rs]) > 0$.*

The tensor square conjectures raise the question on simply determining when $g(\lambda, \mu, \nu) > 0$. It is a consequence of representation theory that when $n > 2$ for every $\mu \vdash n$ there is a $\lambda \vdash n$, such that $g(\lambda, \lambda, \mu) > 0$, see [**Sta99**, Ex. 7.82] but even that has no combinatorial proof.

To understand why such problems are difficult, we resort to the computational complexity framework. For Kostka numbers, determining if $K_{\lambda,\mu} > 0$ is easy, it just requires us checking that $\lambda \succ \mu$ in the dominance order. For Littlewood-Richardson that's a lot less trivial despite the combinatorial rule.

LRPos:
`Input:` λ, μ, ν
`Output:` Is $c^\lambda_{\mu\nu} > 0$?

The proof of the Saturation Conjecture by Knutson and Tao showed that an LR coefficient is nonzero if and only if the corresponding hive polytope is nonempty. This polytope is a refinement of the Gelfand-Tsetlin polytope, and is defined by

$O(\ell(\lambda)^2)$ many inequalities. Showing that the polytope is nonempty is thus a linear programming problem, which can be solved in polynomial time, see [**MNS12**].

THEOREM 5.7. *We have that LRPos $\in$ P when the input is in binary (and unary).*

KronPos:
`Input:` λ, μ, ν
`Output:` Is $g(\lambda, \mu, \nu) > 0$?

In the early stages of GCT Mulmuley conjectured [**MS08**] that they would behave like the Littlewood-Richardson, so KronPos$\in$ P, which was recently disproved.

THEOREM 5.8 ([**IMW17**]). *When the input λ, μ, ν is in unary, KronPos is* NP*-hard.*

The proof uses the fact that in certain cases $g(\lambda, \mu, \nu)$ is equal to the number of pyramids with marginals λ, μ, ν and deciding if there is such a pyramid is NP-complete. However, the problem is not yet in NP: we do not have, for every λ, μ, ν, polynomial witnesses showing that $g(\lambda, \mu, \nu) > 0$ when this is the case. This is yet another open problem.

OPEN PROBLEM 5.9. Determine whether when the input λ, μ, ν is in unary, KronPos is in NP.

Needless to say, the problem would be [computationally] harder when the input is in binary.

Positivity of plethysm coefficients is even less understood. Plethysm coefficients $a_\lambda(d[n])$ are important in GCT and positivity of $a_\lambda(d[n])$ for partitions λ with long first rows was established in [**BIP19**]. In general, we do not expect the problem to be any easier.

PlethPos:
`Input:` λ, μ, ν
`Output:` Is $a^\lambda_{\mu,\nu} > 0$?

In [**FI20**], Fischer and Ikenmeyer showed that PlethPos is NP-hard. As there is no combinatorial interpretation and no nice positivity criteria in general we cannot say whether the problem is in NP.

OPEN PROBLEM 5.10. Determine whether PlethPos is in NP.

There is one major conjecture on plethysm coefficients.

CONJECTURE 5.11 (Foulkes). *Let $d > n$, then*

$$a_\lambda(d[n]) \geq a_\lambda(n[d])$$

for all $\lambda \vdash nd$.

This conjecture is related to the Alon-Tarsi conjecture, and has appeared in GCT-related research. In [**DIP19**] we proved it for some families of 3-row partitions.

As most of these quantities can be computed via symmetric group characters, it is natural to ask how easy it is to decide if a character is 0 or not.

CharVanish:
`Input:` $\lambda, \alpha \vdash n$, unary.
`Output:` Is $\chi^\lambda(\alpha) = 0$?

THEOREM 5.12 ([**IPP22**]). *We have that CharVanish is* $\mathsf{C_=P}$*-complete under many-to-one reductions.*

The class $\mathsf{C_=P}$ is defined formally as $[\mathsf{GapP} = 0]$, the class of deciding when two #P functions have equal values, and this is a class believed to be strictly larger than coNP and NP. Thus we cannot expect that deciding $\chi = 0$ would be in coNP and $\chi \neq 0$ in NP. This also leads to the proof of Proposition 5.3. See [**Pan23**] for a discussion on that topic and sketch of the proof.

6. Asymptotic algebraic combinatorics

Asymptotic Algebraic Combinatorics is a natural extension of Algebraic Combinatorics inspired by Probability and Asymptotic Representation Theory. Instead of searching for exact formulas, bijections, combinatorial interpretations, we are looking to understand quantities approximately. This is, indeed, what we could hope for with the given structure constants in the absence of nice formulas.

As mentioned in the Introduction, such problems have natural interplay with Statistical Mechanics and Integrable Probability. They would also be important in the search of multiplicity obstructions in GCT, see [**Pan23**].

6.1. Maximal multiplicities. The first step in such asymptotic analysis begins with estimating the range of possible values. In [**Sta99, Sta20**], Stanley observed that

$$\max_{\lambda \vdash n} \max_{\mu \vdash n} \max_{\nu \vdash n} g(\lambda, \mu, \nu) = \sqrt{n!}\, e^{-O(\sqrt{n})},$$

$$\max_{0 \le k \le n} \max_{\lambda \vdash n} \max_{\mu \vdash k} \max_{\nu \vdash n-k} c^{\lambda}_{\mu,\nu} = 2^{n/2 - O(\sqrt{n})}.$$

and asked

OPEN PROBLEM 6.1 (Stanley). Fix n. For which λ, μ, ν with $\lambda \vdash n$, is $c^{\lambda}_{\mu\nu}$ asymptotically maximal? For which λ, μ, ν is $g(\lambda, \mu, \nu)$ asymptotically maximal?

Using various symmetric function and representation theoretic identities involving sums of these coefficients, it was observed in [**PPY19a**] that such maximal partitions λ must also have asymptotically maximal dimensions f^{λ}. By the pioneering work of Vershik-Kerov [**VK**] and Loggan-Shepp [**LoS**], we know that such partitions approach certain explicit curve, which we refer to as the VKLS shape and VKLS partitions, and for all those we have $f^{\lambda} = \sqrt{n!} e^{-O(\sqrt{n})}$.

THEOREM 6.2 ([**PPY19a**]). *Let* $\{\lambda^{(n)} \vdash n\}$, $\{\mu^{(n)} \vdash n\}$, $\{\nu^{(n)} \vdash n\}$ *be three partition sequences, such that*

$$(*) \qquad g\big(\lambda^{(n)}, \mu^{(n)}, \nu^{(n)}\big) = \sqrt{n!}\, e^{-O(\sqrt{n})}.$$

Then $\lambda^{(n)}, \mu^{(n)}, \nu^{(n)}$ *have VKLS shape (i.e. are asymptotically maximal). Conversely, for every two VKLS sequences* $\{\lambda^{(n)} \vdash n\}$ *and* $\{\mu^{(n)} \vdash n\}$*, there exists a VKLS partition sequence* $\{\nu^{(n)} \vdash n\}$*, s.t. (*) holds.*

This statement shows "existence", the problem here is to exhibit a particular explicit family of partitions for which this asymptotics holds.

OPEN PROBLEM 6.3. Determine a family of partitions $\lambda^{(n)}, \mu^{(n)}, \nu^{(n)} \vdash n$, such that

$$g\big(\lambda^{(n)}, \mu^{(n)}, \nu^{(n)}\big) = \sqrt{n!}\, e^{-O(\sqrt{n})}$$

as $n \to \infty$.

We believe that for the staircase partitions $\rho_k = (k, k-1, \ldots, 1)$ the Kronecker coefficient $g(\rho_k, \rho_k, \rho_k)$ grows superexponentially, and this is one specific candidate which could achieve large, but not maximal, values. However, no exponential lower bounds are known in this case.

Similar analysis can be done for the Littlewood-Richardson coefficients. Again, the proofs can only derive "existence", but not exhibit any specific families.

THEOREM 6.4 ([**PPY19a**]). *Fix $0 < \theta < 1$ and let $k_n := \lfloor \theta n \rfloor$. Then:*

1. *for every VKLS partition sequence $\{\lambda^{(n)} \vdash n\}$, there exist VKLS partition sequences $\{\mu^{(n)} \vdash k_n\}$ and $\{\nu^{(n)} \vdash n - k_n\}$, s.t.*

$$(**) \qquad c^{\lambda^{(n)}}_{\mu^{(n)},\nu^{(n)}} = \binom{n}{k_n}^{1/2} e^{-O(\sqrt{n})},$$

2. *for all VKLS partition sequences $\{\mu^{(n)} \vdash k_n\}$ and $\{\nu^{(n)} \vdash n - k_n\}$, there exists a VKLS partition sequence $\{\lambda^{(n)} \vdash n\}$, s.t. $(**)$ holds,*

3. *for all VKLS partition sequences $\{\lambda^{(n)} \vdash n\}$ and $\{\mu^{(n)} \vdash k_n\}$, there exists a partition sequence $\{\nu^{(n)} \vdash n - k_n\}$, s.t.*

$$f^{\nu^{(n)}} = \sqrt{n!}\, e^{-O(n^{2/3}\log n)} \quad \text{and} \quad c^{\lambda^{(n)}}_{\mu^{(n)},\nu^{(n)}} = \binom{n}{k_n}^{1/2} e^{-O(n^{2/3}\log n)}.$$

We again ask when such maxima are achieved.

OPEN PROBLEM 6.5. Let $\theta \in (0, 1)$ be some fixed number and let $k_n := \lfloor \theta n \rfloor$. Find families of partitions $\{\lambda^{(n)} \vdash n\}$, $\{\mu^{(n)} \vdash k_n\}$ and $\{\nu^{(n)} \vdash n - k_n\}$, for which

$$c^{\lambda^{(n)}}_{\mu^{(n)},\nu^{(n)}} = \binom{n}{k_n}^{1/2} e^{-O(\sqrt{n})},$$

While asymptotics of Kronecker coefficients have barely been studied, Schur functions asymptotics in various regimes has been central to the field of Integrable Probability. It is closely related to the Harish-Chandra-Itzhykhson-Zuber integrals in random matrix theory. A step further takes us to Kostka and Littlewood-Richardson coefficients, which can also be estimated using the so-called elliptic integrals, see [**BGH**]. There, the problem of asymptotics is translated into an implicit variational analysis, and it still does not give explicit concrete answers.

There are currently no nontrivial asymptotic results of this kind for plethysm coefficients.

OPEN PROBLEM 6.6. Determine the maximal asymptotic value of the plethysm coefficients

$$\max_{\lambda \vdash n^2} a_\lambda(n[n]).$$

The only nontrivial bounds come from their relation to rectangular Kronecker coefficients, and in particular when λ is a two-row partition, see § 6.3. We expect that the maxima should be at least exponential. Another source of nontrivial lower bounds can be found in [**FI20**] where plethysms are compared to pyramids. Similar comparison was used to obtain a family of large Kronecker coefficients in [**PP20a, PP23**].

6.2. Bounded rows and diagonals. In the "maximal" regime, the partitions involved were close to the VKLS shape which is symmetric and has length $O(\sqrt{n})$ for partitions of n. The asymptotics changes significantly depending on the regimes of convergence for the partitions involved.

When the lengths of the partitions are bounded, formulas coming from symmetric functions and relating them to contingency tables become useful. We have the following upper bound.

THEOREM 6.7 ([**PP20a**]). *Let $\lambda, \mu, \nu \vdash n$ such that $\ell(\lambda) = \ell$, $\ell(\mu) = m$, and $\ell(\nu) = r$. Then:*

$$g(\lambda, \mu, \nu) \le \left(1 + \frac{\ell m r}{n}\right)^n \left(1 + \frac{n}{\ell m r}\right)^{\ell m r}.$$

In particular, when $\lambda = \mu = \nu = ((\ell^2)^\ell) \vdash \ell^3 = n$, we have $g(\lambda, \mu, \nu) \le 4^n$. Yet, a similar lower bound for that case is beyond reach.

Further analysis using Schur function techniques also gives bounds when the Durfee size (diagonal) of a partition is bounded, even if the length grows. We have the following, see [**PP23**].

THEOREM 6.8 ([**PP23**]). *Let $n, k \ge 1$, and let $\lambda, \mu, \nu \vdash n$, such that*

$$d(\lambda), d(\mu), d(\nu) \le k.$$

Then:

$$g(\lambda, \mu, \nu) \le \frac{1}{k^{8k^2} 2^{8k^3}}\, n^{4k^3 + 13k^2 + 31k}. \tag{6.1}$$

We also know that the actual maximal value is not much further from this bound. Let

$$A(n, k) := \max\left\{g(\lambda, \mu, \nu) \ : \ \lambda, \mu, \nu \vdash n \ \text{ and } \ \ell(\lambda), \ell(\mu), \ell(\nu) \le k\right\}.$$

THEOREM 6.9 ([**PP23**]). *For all $k \ge 1$, there is a constant $C_k > 0$, such that*

$$A(n, k) \ge C_k n^{k^3 - 3k^2 - 3k + 3} \qquad \text{for all} \quad n \ge 1. \tag{6.2}$$

However, we are nowhere close to finding asymptotics close to these maxima for concrete families of partitions.

OPEN PROBLEM 6.10. For every fixed k determine a family of partitions

$$\lambda^{(n)}, \mu^{(n)}, \nu^{(n)} \vdash n$$

with $\ell(\lambda^{(n)}), \ell(\mu^{(n)}), \ell(\nu^{(n)}) \le k$, such that

$$g(\lambda^{(n)}, \mu^{(n)}, \nu^{(n)}) = O(n^{(k-1)^3}).$$

Using the powerful technique of ACSV (Asymptotic Combinatorics in Several Variables), the cases of $k \le 4$ have been somewhat well understood and explicit asymptotics derived in [**MT**]. For arbitrary, but fixed, values of k though the ACSV is no longer applicable and we need other ways to compute this.

The same question pertains to the plethysm coefficients.

OPEN PROBLEM 6.11. Show that for every fixed k the maximal plethysm coefficient over partitions with length k grows polynomially in n. In particular, we have

$$\log \max_{\lambda \vdash n^2,\, \ell(\lambda) = k} a_\lambda(n[n]) = \Omega(\log(n))$$

It was establsihed in [**IP17**] that plethysms are bounded above by rectangular Kronecker coefficients,

$$a_\lambda(d[n]) \le g(\lambda, n^d, n^d)$$

for $\ell(\lambda) \le m^2$, $\lambda_1 \ge d(n-m)$ and d, n large enough so we are in the stable regime, i.e. the values of the plethysm and Kronecker are constant when only d, n and λ_1 grow. Thus, we would not expect that plethysm coefficients would have larger growth than Kroneckers.

6.3. Tight asymptotics. There are very few nontrivial cases where we have certain explicit formulas for Kronecker and plethysm coefficients and would be able to derive tight asymptotics. One such case is for rectangular Kronecker coefficients $g(\lambda, m^\ell, m^\ell)$ when λ has two rows. We have that, see [**PP17a**], for $\lambda = (m\ell - n, n)$,

$$g(m^\ell, m^\ell, \lambda) = p_n(\ell, m) - p_{n-1}(\ell, m) = a_\lambda(m[\ell]),$$

where $p_n(\ell, m) = \#\{\alpha \vdash n, \alpha_1 \le m, \ell(\alpha) \le \ell\}$ is the number of partitions of k whose Young diagram fits inside an $\ell \times m$ rectangle.

Using this partition counting interpretation, and deriving tight asymptotics for $p_n(m, \ell)$ using tilted geometric random variables, we obtain

THEOREM 6.12 ([**MPP19a**]). *Let* $A := \frac{\ell}{m}$ $B := \frac{n-1}{m^2}$. *Let* c, d *be solutions of a certian system of integral equations with parameters* A, B. *We have*

$$\begin{aligned} a_\lambda(m[\ell]) &= g(\lambda, m^\ell, m^\ell) \\ &= p_n(\ell, m) - p_{n-1}(\ell, m) \sim \frac{d}{m} p_{n-1}(\ell, m) \sim \frac{d\ e^{m[cA + 2dB - \log(1 - e^{-c-d})]}}{2\pi m^3 \sqrt{D}}. \end{aligned}$$

Note that some technical conditions for the theorem to hold also include that ℓ, m grow at the same rate, and also that $\frac{n}{m\ell}$ should be away from $\frac{1}{2}$. When $n \sim \frac{1}{2} m\ell$ we can no longer estimate the difference between p_n's so tightly and we need a different approach for the following

OPEN PROBLEM 6.13. Find tight asymptotics when $\lambda = (\lceil m^2/2 \rceil, \lfloor m^2/2 \rfloor)$ for $g(m^m, m^m, \lambda)$.

The best lower bound in this case comes from characters and is obtained in [**PP17a**].

Needless to say, any other explicit tight asymptotic results would be of great interest and, using semigroup property, could lead to improved lower bounds in other regimes.

7. Reduced Kronecker coefficients

Since the Kronecker coefficients quickly turned out to be difficult to understand, hopes turned towards the *stable Kronecker coefficients* defined as

$$\overline{g}(\alpha, \beta, \gamma) := \lim_{n \to \infty} g\big(\alpha[n], \beta[n], \gamma[n]\big), \qquad \alpha[n] := (n - |\alpha|, \alpha_1, \alpha_2, \ldots),\ n \ge |\alpha| + \alpha_1.$$

They generalize the Littlewood-Richardson coefficients, following Theorem 3.2 from [**Mur56**].

$$\overline{g}(\alpha, \beta, \gamma) = c^\alpha_{\beta\gamma} \quad \text{for} \quad |\alpha| = |\beta| + |\gamma|,$$

and were thereby called "extended Littlewood-Richardson coefficients" in [**Kir04**]. They have thus occupied an intermediate spot between the Littlewood-Richardson and ordinary Kronecker coefficients, and have been a subject of independent study and interest, see [**Mur38, Mur56, Bri93, Val99, Kir04, Kly04, BOR11, BDO15, CR15, Man15, SS16, IP17, PP20b, OZ21**]. Together with the ordinary Kronecker coefficients, the problem of finding a combinatorial interpretation has also been part of the agenda, see [**Sta00**].

The ordinary Kronecker coefficients can be expressed as a small alternating sum of reduced Kroneckers and reduced Kroneckers are certain sums of ordinary Kroneckers, see [**BOR11, BDO15**]. These relationships show that reduced Kroneckers are also #P-hard to compute, see [**PP20b**]. However, it could not be immediately deduced that their positivity is NP-hard, which prompted the following question.

OPEN PROBLEM 7.1. Given input λ, μ, ν (in unary), is deciding whether $\overline{g}(\lambda, \mu, \nu) > 0$ NP-hard? Determine whether computing $\overline{g}(\lambda, \mu, \nu)$ is #P-complete?

As we noted, the second part of that problem would follow if there is a combinatorial interpretation for them, or for the ordinary Kronecker coefficients.

Some special cases of combinatorial interpretations can be derived from the ordinary Kronecker coefficients. Separately, in [**CR15**] a combinatorial interpretation was given when μ, ν are rectangles and λ is one row, and other similar cases were derived in [**BO05**]. Methods to compute them, as well as for the ordinary Kronecker coefficients, have been developed in a series of papers, see [**BDO15, BOR11, OZ20, OZ21**].

Until recently, it was believed they are strictly better behaved than the ordinary Kroneckers, with the following conjecture appearing in [**Kir04, Kly04**]

CONJECTURE 7.2 (Kirillov, Klyachko). *The reduced Kronecker coefficients satisfy the saturation property:*

$$\overline{g}(N\alpha, N\beta, N\gamma) > 0 \quad \textit{for some } N \geq 1 \quad \Longrightarrow \quad \overline{g}(\alpha, \beta, \gamma) > 0.$$

Using the recent positivity results, in particular for rectangular Kroneckers from [**IP17**], this was also disproved.

THEOREM 7.3 ([**PP20b**]). *For all $k \geq 3$, the triple of partitions*

$$\left(1^{k^2-1}, 1^{k^2-1}, k^{k-1}\right)$$

is a counterexample to the Conjecture. For every partition γ s.t. $\gamma_2 \geq 3$, there are infinitely many pairs $(a, b) \in \mathbb{N}^2$ s.t. (a^b, a^b, γ) is a counterexample.

As shown in [**PP20b**], asymptotically, and complexity-wise they behave more similarly to the ordinary Kronecker than to the Littlewood-Richardson coefficients.

REMARK 7.4. In the course of writing this paper, the first part of Open Problem 7.1 was settled in [**IP23**], where it was shown that every Kronecker coefficient is equal to a [not much larger] reduced Kronecker coefficient:

$$g(\lambda, \mu, \nu) = \overline{g}(\nu_1^{\ell(\lambda)} + \lambda, \nu_1^{\ell(\mu)} + \mu, (\nu_1^{\ell(\lambda)+\ell(\mu)}, \nu)).$$

Thus the reduced Kronecker coefficients are just as hard as the Kroneckers, and the complexity results follow from§ 4. This also shows that the problems are essentially the same, and the reduced Kronecker coefficients are not easier in general.

8. Methods

While the concrete open problems inspire work in the area, the pertinent task is to create tools for understanding the Kronecker and plethysm coefficients. Of course, an ultimate tool could be the "combinatorial interpretation", but this itself has become the hardest problem to understand and its solution requires new tools. So far, as very little progress has been made in this direction, it is clear that we need something new.

Kronecker coefficients can be interpreted via S_n as multiplicities. Via Schur-Weyl duality, they are also dimensions of invariant spaces under GL action. The last interpretation is what gives the semigroup property, something that cannot be seen from the S_n perspective. On the other hand, the transpositional invariance $g(\lambda, \mu, \nu) = g(\lambda', \mu', \nu)$ is visible from their S_n definition. There are many more examples showing the variety of results one can get using different methods and the limitations of each one also. Perhaps the best tool would be able to unify all these views.

Symmetric functions have been one of the most useful tools in Algebraic Combinatorics, and is certainly the author's favorite. The Littlewood-Richardson coefficients are structure constants in the ring of symmetric functions and thus have been very successfully studied using Schur functions. But the Kronecker coefficients cannot be interpreted in that way. Motivated by that, a family of symmetric functions was developed in [**OZ20, OZ21**]. Symmetric functions have not been very useful in establishing positivity results, because the formulas involved are often alternating sums. However, they are useful for finding asymptotics and bounds as done in [**PP20a, PP23**] for example. Within integrable probability they have been a lot more successful, from Asymptotic Representation Theory, through lozenge tilings, to Random Matrices, see e.g. [**BG15, GP15, BGH**] and references therein.

Plethysm coefficients can also be studied using symmetric functions, yet again positivity properties cannot usually be derived this way. For example in [**DIP19**] symmetric functions were used to derive some explicit formulas for $a_\lambda(d[n])$ with λ a three row partition, and later this method generalized in [**FI20**] to exhibit an alternating formula and put them in GapP.

A technique to show positivity is the construction of explicit highest weight tableau evaluations, see [**Ful00**], which is easier for plethysms but also applicable for Kroneckers. For example, this technique was used in [**I15**] to prove Theorem 5.4, generalized in [**LS17**], and in [**BIP19**] to show positive plethysm coefficients.

Useful tools in Representation Theory are induction and restriction of modules, passing to the representation theory of other groups. For example, the character bound 5.3 was obtained by restricting to the alternating group $A_n \subset S_n$, see [**BB04, PPV16, PP17a**]. Similar methods were developed to extract positivity criteria working with representations over finite fields, see e.g. [**BBL**]. In another direction, diagram algebras can also be used to study Kronecker, see e.g. [**BDO15**], and plethysm coefficients, see e.g. [**COSSZ22, OSSZ**].

Another useful approach comes from discrete geometry. This is not at all surprising given that many objects in representation theory correspond to integer points in polytopes, e.g. SSYTs are integer points in Gelfand-Tsetlin polytopes, see e.g. [**DM06**] for some complexity connections also. Vallejo used such relations to develop methods of discrete tomography and find a lower bound for

Kronecker coefficients via pyramids, see [**Val97**, **Val99**, **IMW17**]. Kronecker coefficients are bounded above by 3d contingency tables with given 2d marginals, as seen in [**IMW17**, **PP20a**] to obtain upper bounds. Similar expressions tying them to Kostant partition functions were developed in [**MRS21**] and used in [**MT**] to find asymptotics.

Finally, Computational Complexity theory provides some tools besides the framework. The framework is to explain the hardness of the problems and what possible answers to expect. The tools come in the form of gadgets and basic hard/complete problems to reduce to like 3SAT, PerfectMatching, SetPartition etc. Embedding such a complete problem in our problem is an art, very similar to finding injections, see e.g. [**GJ79**] for some classical results. Other tools come from algebraic complexity and GCT, where the complexity of certain polynomials could imply inequalities between the relevant structure constants.

Despite the variety of questions, methods and applications, the Kronecker and plethysm coefficients are still a mysterious black box. It sometimes gives us presents. But we cannot see inside... just yet.

Acknowledgments

The author is grateful to her collaborators throughout the years and in particular Christian Ikenmeyer and Igor Pak for the extensive work on the subject of this paper. We would like to dedicate this to the memory of Christine Bessenrodt, whose deep and insightful work in the area has led a lot of the progress and with whom the author had the honor to collaborate briefly in 2020. We are grateful to the organizers of "Open Problems in Algebraic Combinatorics" for the inspiring conference and the opportunity to present these problems.

References

[Aar16] S. Aaronson, $\mathsf{P} \stackrel{?}{=} \mathsf{NP}$, in *Open problems in mathematics*, Springer, Cham, 2016, 1–122.

[BO05] Cristina M. Ballantine and Rosa C. Orellana, *A combinatorial interpretation for the coefficients in the Kronecker product $s_{(n-p,p)} * s_\lambda$*, Sém. Lothar. Combin. **54A** (2005/07), Art. B54Af, 29. MR2264933

[BGH] Serban Belinschi, Alice Guionnet, and Jiaoyang Huang, *Large deviation principles via spherical integrals*, Probab. Math. Phys. **3** (2022), no. 3, 543–625, DOI 10.2140/pmp.2022.3.543. MR4520314

[BB04] Christine Bessenrodt and Christiane Behns, *On the Durfee size of Kronecker products of characters of the symmetric group and its double covers*, J. Algebra **280** (2004), no. 1, 132–144, DOI 10.1016/j.jalgebra.2004.03.028. MR2081924

[BB17] Christine Bessenrodt and Christopher Bowman, *Multiplicity-free Kronecker products of characters of the symmetric groups*, Adv. Math. **322** (2017), 473–529, DOI 10.1016/j.aim.2017.10.009. MR3720803

[B18] Christine Bessenrodt, *Critical classes, Kronecker products of spin characters, and the Saxl conjecture*, Algebr. Comb. **1** (2018), no. 3, 353–369, DOI 10.5802/alco. MR3856528

[BBL] C. Bessenrodt, C. Bowman, and L. Sutton, *Kronecker positivity and 2-modular representation theory*, Trans. Amer. Math. Soc. Ser. B **8** (2021), 1024–1055, DOI 10.1090/btran/70. MR4350547

[BBR06] François Bergeron, Riccardo Biagioli, and Mercedes H. Rosas, *Inequalities between Littlewood-Richardson coefficients*, J. Combin. Theory Ser. A **113** (2006), no. 4, 567–590, DOI 10.1016/j.jcta.2005.05.002. MR2216453

[Bia98] Philippe Biane, *Representations of symmetric groups and free probability*, Adv. Math. **138** (1998), no. 1, 126–181, DOI 10.1006/aima.1998.1745. MR1644993

[B13] Jonah Blasiak, *Kronecker coefficients for one hook shape*, Sém. Lothar. Combin. **77** ([2016), Art. B77c, 40. MR3650220

[BL18] Jonah Blasiak and Ricky Ini Liu, *Kronecker coefficients and noncommutative super Schur functions*, J. Combin. Theory Ser. A **158** (2018), 315–361, DOI 10.1016/j.jcta.2018.02.007. MR3800131

[BMS15] Jonah Blasiak, Ketan D. Mulmuley, and Milind Sohoni, *Geometric complexity theory IV: nonstandard quantum group for the Kronecker problem*, Mem. Amer. Math. Soc. **235** (2015), no. 1109, x+160, DOI 10.1090/memo/1109. MR3338303

[BI18] M. Bläser and C. Ikenmeyer, *Introduction to geometric complexity theory*, Summer school lecture notes, 2018, 148 pp.; `https://tinyurl.com/nhe2wxvw`

[BDO15] C. Bowman, M. De Visscher, and R. Orellana, *The partition algebra and the Kronecker coefficients*, Trans. Amer. Math. Soc. **367** (2015), no. 5, 3647–3667, DOI 10.1090/S0002-9947-2014-06245-4. MR3314819

[BOR11] Emmanuel Briand, Rosa Orellana, and Mercedes Rosas, *The stability of the Kronecker product of Schur functions*, J. Algebra **331** (2011), 11–27, DOI 10.1016/j.jalgebra.2010.12.026. MR2774644

[Bri93] Michel Brion, *Stable properties of plethysm: on two conjectures of Foulkes*, Manuscripta Math. **80** (1993), no. 4, 347–371, DOI 10.1007/BF03026558. MR1243152

[BG15] Alexey Bufetov and Vadim Gorin, *Representations of classical Lie groups and quantized free convolution*, Geom. Funct. Anal. **25** (2015), no. 3, 763–814, DOI 10.1007/s00039-015-0323-x. MR3361772

[Bür00a] Peter Bürgisser, *Completeness and reduction in algebraic complexity theory*, Algorithms and Computation in Mathematics, vol. 7, Springer-Verlag, Berlin, 2000, DOI 10.1007/978-3-662-04179-6. MR1771845

[Bür00b] Peter Bürgisser, *Cook's versus Valiant's hypothesis*, Theoret. Comput. Sci. **235** (2000), no. 1, 71–88, DOI 10.1016/S0304-3975(99)00183-8. Selected papers in honor of Manuel Blum (Hong Kong, 1998). MR1765966

[BCS97] Peter Bürgisser, Michael Clausen, and M. Amin Shokrollahi, *Algebraic complexity theory*, Grundlehren der mathematischen Wissenschaften [Fundamental Principles of Mathematical Sciences], vol. 315, Springer-Verlag, Berlin, 1997. With the collaboration of Thomas Lickteig, DOI 10.1007/978-3-662-03338-8. MR1440179

[BI08] Peter Bürgisser and Christian Ikenmeyer, *The complexity of computing Kronecker coefficients* (English, with English and French summaries), 20th Annual International Conference on Formal Power Series and Algebraic Combinatorics (FPSAC 2008), Discrete Math. Theor. Comput. Sci. Proc., AJ, Assoc. Discrete Math. Theor. Comput. Sci., Nancy, 2008, pp. 357–368. MR2721467

[BIP19] Peter Bürgisser, Christian Ikenmeyer, and Greta Panova, *No occurrence obstructions in geometric complexity theory*, J. Amer. Math. Soc. **32** (2019), no. 1, 163–193, DOI 10.1090/jams/908. MR3868002

[BLMW11] Peter Bürgisser, J. M. Landsberg, Laurent Manivel, and Jerzy Weyman, *An overview of mathematical issues arising in the geometric complexity theory approach to* VP $\neq$ VNP, SIAM J. Comput. **40** (2011), no. 4, 1179–1209, DOI 10.1137/090765328. MR2861717

[CHM07] Matthias Christandl, Aram W. Harrow, and Graeme Mitchison, *Nonzero Kronecker coefficients and what they tell us about spectra*, Comm. Math. Phys. **270** (2007), no. 3, 575–585, DOI 10.1007/s00220-006-0157-3. MR2276458

[CDW12] Matthias Christandl, Brent Doran, and Michael Walter, *Computing multiplicities of Lie group representations*, 2012 IEEE 53rd Annual Symposium on Foundations of Computer Science—FOCS 2012, IEEE Computer Soc., Los Alamitos, CA, 2012, pp. 639–648. MR3186652

[COSSZ22] L. Colmenarejo, R. Orellana, F. Saliola, A. Schilling, M. Zabrocki, *The mystery of plethysm coefficients*, `arXiv:2208.07258`, 2022.

[CR15] Laura Colmenarejo and Mercedes Rosas, *Combinatorics on a family of reduced Kronecker coefficients* (English, with English and French summaries), C. R. Math. Acad. Sci. Paris **353** (2015), no. 10, 865–869, DOI 10.1016/j.crma.2015.07.012. MR3411213

[DM06] Jesús A. De Loera and Tyrrell B. McAllister, *On the computation of Clebsch-Gordan coefficients and the dilation effect*, Experiment. Math. **15** (2006), no. 1, 7–19. MR2229381

[D93] Yoav Dvir, *On the Kronecker product of S_n characters*, J. Algebra **154** (1993), no. 1, 125–140, DOI 10.1006/jabr.1993.1008. MR1201916

[DIP19] Julian Dörfler, Christian Ikenmeyer, and Greta Panova, *On geometric complexity theory: multiplicity obstructions are stronger than occurrence obstructions*, 46th International Colloquium on Automata, Languages, and Programming, LIPIcs. Leibniz Int. Proc. Inform., vol. 132, Schloss Dagstuhl. Leibniz-Zent. Inform., Wadern, 2019, pp. Art. No. 51, 14. MR3984868

[FI20] Nick Fischer and Christian Ikenmeyer, *The computational complexity of plethysm coefficients*, Comput. Complexity **29** (2020), no. 2, Paper No. 8, 43, DOI 10.1007/s00037-020-00198-4. MR4171310

[FRT54] J. S. Frame, G. de B. Robinson, and R. M. Thrall, *The hook graphs of the symmetric groups*, Canad. J. Math. **6** (1954), 316–324, DOI 10.4153/cjm-1954-030-1. MR62127

[Ful97] William Fulton, *Young tableaux*, London Mathematical Society Student Texts, vol. 35, Cambridge University Press, Cambridge, 1997. With applications to representation theory and geometry. MR1464693

[Ful00] William Fulton, *Eigenvalues, invariant factors, highest weights, and Schubert calculus*, Bull. Amer. Math. Soc. (N.S.) **37** (2000), no. 3, 209–249, DOI 10.1090/S0273-0979-00-00865-X. MR1754641

[GJ79] Michael R. Garey and David S. Johnson, *Computers and intractability: A guide to the theory of NP-completeness*, A Series of Books in the Mathematical Sciences, W. H. Freeman and Co., San Francisco, CA, 1979. MR519066

[GR85] A. M. Garsia and J. Remmel, *Shuffles of permutations and the Kronecker product*, Graphs Combin. **1** (1985), no. 3, 217–263, DOI 10.1007/BF02582950. MR951014

[GIP17] Fulvio Gesmundo, Christian Ikenmeyer, and Greta Panova, *Geometric complexity theory and matrix powering*, Differential Geom. Appl. **55** (2017), 106–127, DOI 10.1016/j.difgeo.2017.07.001. MR3724215

[GP15] Vadim Gorin and Greta Panova, *Asymptotics of symmetric polynomials with applications to statistical mechanics and representation theory*, Ann. Probab. **43** (2015), no. 6, 3052–3132, DOI 10.1214/14-AOP955. MR3433577

[HR22] Nate Harman and Christopher Ryba, *A tensor-cube version of the Saxl conjecture*, Algebr. Comb. **6** (2023), no. 2, 507–511. MR4591596

[HSTZ13] Gerhard Heide, Jan Saxl, Pham Huu Tiep, and Alexandre E. Zalesski, *Conjugacy action, induced representations and the Steinberg square for simple groups of Lie type*, Proc. Lond. Math. Soc. (3) **106** (2013), no. 4, 908–930, DOI 10.1112/plms/pds062. MR3056296

[I15] Christian Ikenmeyer, *The Saxl conjecture and the dominance order*, Discrete Math. **338** (2015), no. 11, 1970–1975, DOI 10.1016/j.disc.2015.04.027. MR3357782

[IMW17] Christian Ikenmeyer, Ketan D. Mulmuley, and Michael Walter, *On vanishing of Kronecker coefficients*, Comput. Complexity **26** (2017), no. 4, 949–992, DOI 10.1007/s00037-017-0158-y. MR3723353

[IP22] Christian Ikenmeyer and Igor Pak, *What is in #P and what is not?*, 2022 IEEE 63rd Annual Symposium on Foundations of Computer Science—FOCS 2022, IEEE Computer Soc., Los Alamitos, CA, 2022, pp. 860–871. MR4537263

[IPP22] Christian Ikenmeyer, Igor Pak, and Greta Panova, *Positivity of the symmetric group characters is as hard as the polynomial time hierarchy*, Proceedings of the 2023 Annual ACM-SIAM Symposium on Discrete Algorithms (SODA), SIAM, Philadelphia, PA, 2023, pp. 3573–3586, DOI 10.1137/1.9781611977554.ch136. MR4538090

[IP17] Christian Ikenmeyer and Greta Panova, *Rectangular Kronecker coefficients and plethysms in geometric complexity theory*, Adv. Math. **319** (2017), 40–66, DOI 10.1016/j.aim.2017.08.024. MR3695867

[IP23] C. Ikenemeyer and G. Panova, *All Kronecker coefficients are reduced Kronecker coefficients*, `arXiv:2305.03003`, 2023.

[Kir04] Anatol N. Kirillov, *An invitation to the generalized saturation conjecture*, Publ. Res. Inst. Math. Sci. **40** (2004), no. 4, 1147–1239. MR2105706

[Kly04] A. Klyachko, *Quantum marginal problem and representations of the symmetric group*, `arXiv:quant-ph/0409113`, 2004.

[KT99] Allen Knutson and Terence Tao, *The honeycomb model of* $\mathrm{GL}_n(\mathbf{C})$ *tensor products. I. Proof of the saturation conjecture*, J. Amer. Math. Soc. **12** (1999), no. 4, 1055–1090, DOI 10.1090/S0894-0347-99-00299-4. MR1671451

[Las79] A. Lascoux, *Produit de Kronecker des représentations du groupe symétrique* (French), Séminaire d'Algèbre Paul Dubreil et Marie-Paule Malliavin, 32ème année (Paris, 1979), Lecture Notes in Math., vol. 795, Springer, Berlin, 1980, pp. 319–329. MR582085

[L21] Xin Li, *Saxl conjecture for triple hooks*, Discrete Math. **344** (2021), no. 6, Paper No. 112340, 19, DOI 10.1016/j.disc.2021.112340. MR4222286

[Liu17] Ricky Ini Liu, *A simplified Kronecker rule for one hook shape*, Proc. Amer. Math. Soc. **145** (2017), no. 9, 3657–3664, DOI 10.1090/proc/13692. MR3665021

[LoS] B. F. Logan and L. A. Shepp, *A variational problem for random Young tableaux*, Advances in Math. **26** (1977), no. 2, 206–222, DOI 10.1016/0001-8708(77)90030-5. MR1417317

[LS17] Sammy Luo and Mark Sellke, *The Saxl conjecture for fourth powers via the semigroup property*, J. Algebraic Combin. **45** (2017), no. 1, 33–80, DOI 10.1007/s10801-016-0700-z. MR3591371

[Mac95] I. G. Macdonald, *Symmetric functions and Hall polynomials*, 2nd ed., Oxford Mathematical Monographs, The Clarendon Press, Oxford University Press, New York, 1995. With contributions by A. Zelevinsky; Oxford Science Publications. MR1354144

[Man15] Laurent Manivel, *On the asymptotics of Kronecker coefficients*, J. Algebraic Combin. **42** (2015), no. 4, 999–1025, DOI 10.1007/s10801-015-0614-1. MR3417256

[MPP19a] Stephen Melczer, Greta Panova, and Robin Pemantle, *Counting partitions inside a rectangle*, SIAM J. Discrete Math. **34** (2020), no. 4, 2388–2410, DOI 10.1137/20M1315828. MR4175829

[MRS21] Marni Mishna, Mercedes Rosas, and Sheila Sundaram, *Vector partition functions and Kronecker coefficients*, J. Phys. A **54** (2021), no. 20, Paper No. 205204, 29, DOI 10.1088/1751-8121/abf45b. MR4271276

[MT] M. Mishna, S. Trandafir, *Estimating and computing Kronecker Coefficients: a vector partition function approach*, `arXiv:2210.12128`, 2022.

[Mul11] K. D. Mulmuley, *Geometric complexity theory VI: The flip via positivity*, preprint available at `http://gct.cs.uchicago.edu/gct6.pdf`, 40 pp, 2011; cf. `arXiv:0704.0229`, 139 pp, 2007.

[Mul17] Ketan D. Mulmuley, *Geometric complexity theory V: Efficient algorithms for Noether normalization*, J. Amer. Math. Soc. **30** (2017), no. 1, 225–309, DOI 10.1090/jams/864. MR3556292

[MNS12] Ketan D. Mulmuley, Hariharan Narayanan, and Milind Sohoni, *Geometric complexity theory III: on deciding nonvanishing of a Littlewood-Richardson coefficient*, J. Algebraic Combin. **36** (2012), no. 1, 103–110, DOI 10.1007/s10801-011-0325-1. MR2927658

[MS01] Ketan D. Mulmuley and Milind Sohoni, *Geometric complexity theory. I. An approach to the P vs. NP and related problems*, SIAM J. Comput. **31** (2001), no. 2, 496–526, DOI 10.1137/S009753970038715X. MR1861288

[MS08] Ketan D. Mulmuley and Milind Sohoni, *Geometric complexity theory. II. Towards explicit obstructions for embeddings among class varieties*, SIAM J. Comput. **38** (2008), no. 3, 1175–1206, DOI 10.1137/080718115. MR2421083

[Mur38] F. D. Murnaghan, *The Analysis of the Kronecker Product of Irreducible Representations of the Symmetric Group*, Amer. J. Math. **60** (1938), no. 3, 761–784, DOI 10.2307/2371610. MR1507347

[Mur56] Francis D. Murnaghan, *On the Kronecker product of irreducible representations of the symmetric group*, Proc. Nat. Acad. Sci. U.S.A. **42** (1956), 95–98, DOI 10.1073/pnas.42.2.95. MR75213

[Nar06] Hariharan Narayanan, *On the complexity of computing Kostka numbers and Littlewood-Richardson coefficients*, J. Algebraic Combin. **24** (2006), no. 3, 347–354, DOI 10.1007/s10801-006-0008-5. MR2260022

[OSSZ] Rosa Orellana, Franco Saliola, Anne Schilling, and Mike Zabrocki, *Plethysm and the algebra of uniform block permutations*, Algebr. Comb. **5** (2022), no. 5, 1165–1203, DOI 10.5802/alco.243. MR4511162

[OZ20] Rosa Orellana and Mike Zabrocki, *A combinatorial model for the decomposition of multivariate polynomial rings as S_n-modules*, Electron. J. Combin. **27** (2020), no. 3, Paper No. 3.24, 18, DOI 10.37236/8935. MR4245137

[OZ21] Rosa Orellana and Mike Zabrocki, *Symmetric group characters as symmetric functions*, Adv. Math. **390** (2021), Paper No. 107943, 34, DOI 10.1016/j.aim.2021.107943. MR4295089

[Pak22+] I. Pak, *What is a combinatorial interpretation?*, in Open Problems in Algebraic Combinatorics, 189–258. Proc. Sympos. Pure Math., 110, American Mathematical Society, Providence, RI, 2024.

[PP13] Igor Pak and Greta Panova, *Strict unimodality of q-binomial coefficients* (English, with English and French summaries), C. R. Math. Acad. Sci. Paris **351** (2013), no. 11-12, 415–418, DOI 10.1016/j.crma.2013.06.008. MR3090120

[PP14] Igor Pak and Greta Panova, *Unimodality via Kronecker products*, J. Algebraic Combin. **40** (2014), no. 4, 1103–1120, DOI 10.1007/s10801-014-0520-y. MR3273403

[PP17a] Igor Pak and Greta Panova, *Bounds on certain classes of Kronecker and q-binomial coefficients*, J. Combin. Theory Ser. A **147** (2017), 1–17, DOI 10.1016/j.jcta.2016.10.004. MR3589885

[PP17b] Igor Pak and Greta Panova, *On the complexity of computing Kronecker coefficients*, Comput. Complexity **26** (2017), no. 1, 1–36, DOI 10.1007/s00037-015-0109-4. MR3620776

[PP20a] Igor Pak and Greta Panova, *Bounds on Kronecker coefficients via contingency tables*, Linear Algebra Appl. **602** (2020), 157–178, DOI 10.1016/j.laa.2020.05.005. MR4102189

[PP20b] Igor Pak and Greta Panova, *Breaking down the reduced Kronecker coefficients*, C. R. Math. Acad. Sci. Paris **358** (2020), no. 4, 463–468, DOI 10.5802/crmath.60. MR4134255

[PP23] Igor Pak and Greta Panova, *Durfee squares, symmetric partitions and bounds on Kronecker coefficients*, J. Algebra **629** (2023), 358–380, DOI 10.1016/j.jalgebra.2023.04.006. MR4585439

[PPV16] Igor Pak, Greta Panova, and Ernesto Vallejo, *Kronecker products, characters, partitions, and the tensor square conjectures*, Adv. Math. **288** (2016), 702–731, DOI 10.1016/j.aim.2015.11.002. MR3436397

[PPY19a] Igor Pak, Greta Panova, and Damir Yeliussizov, *On the largest Kronecker and Littlewood-Richardson coefficients*, J. Combin. Theory Ser. A **165** (2019), 44–77, DOI 10.1016/j.jcta.2019.01.008. MR3906648

[Pan15] G. Panova, *Kronecker coefficients: combinatorics, complexity and beyond*, AMS-EMS meeting, Porto, Portugal (2015), slides at `https://tinyurl.com/bdd4w3pj`

[Pan23] G. Panova, *Computational Complexity in Algebraic Combinatorics*, Current Developments in Mathematics, Harvard University, 2023.

[RW94] Jeffrey B. Remmel and Tamsen Whitehead, *On the Kronecker product of Schur functions of two row shapes*, Bull. Belg. Math. Soc. Simon Stevin **1** (1994), no. 5, 649–683. MR1315363

[Sag01] Bruce E. Sagan, *The symmetric group: Representations, combinatorial algorithms, and symmetric functions*, 2nd ed., Graduate Texts in Mathematics, vol. 203, Springer-Verlag, New York, 2001, DOI 10.1007/978-1-4757-6804-6. MR1824028

[SS16] Steven V. Sam and Andrew Snowden, *Proof of Stembridge's conjecture on stability of Kronecker coefficients*, J. Algebraic Combin. **43** (2016), no. 1, 1–10, DOI 10.1007/s10801-015-0622-1. MR3439297

[Sta99] Richard P. Stanley, *Enumerative combinatorics. Volume 1*, 2nd ed., Cambridge Studies in Advanced Mathematics, vol. 49, Cambridge University Press, Cambridge, 2012. MR2868112

[Sta00] Richard P. Stanley, *Positivity problems and conjectures in algebraic combinatorics*, Mathematics: frontiers and perspectives, Amer. Math. Soc., Providence, RI, 2000, pp. 295–319. MR1754784

[Sta20] R. P. Stanley, *Supplementary Excercies to* [**Sta99**], available at `http://www-math.mit.edu/~rstan/ec`.

[T15] Vasu V. Tewari, *Kronecker coefficients for some near-rectangular partitions*, J. Algebra **429** (2015), 287–317, DOI 10.1016/j.jalgebra.2015.01.018. MR3320625

[V79a] L. G. Valiant, *Completeness classes in algebra*, Conference Record of the Eleventh Annual ACM Symposium on Theory of Computing (Atlanta, Ga., 1979), ACM, New York, 1979, pp. 249–261. MR564634

[V79b] L. G. Valiant, *The complexity of computing the permanent*, Theoret. Comput. Sci. **8** (1979), no. 2, 189–201, DOI 10.1016/0304-3975(79)90044-6. MR526203

[Val97] Ernesto Vallejo, *Reductions of additive sets, sets of uniqueness and pyramids*, Discrete Math. **173** (1997), no. 1-3, 257–267, DOI 10.1016/S0012-365X(96)00134-3. MR1468852

[Val99] Ernesto Vallejo, *Stability of Kronecker products of irreducible characters of the symmetric group*, Electron. J. Combin. **6** (1999), Research Paper 39, 7, DOI 10.37236/1471. MR1725703

[VK] A. M. Vershik and S. V. Kerov, *Asymptotic behavior of the maximum and generic dimensions of irreducible representations of the symmetric group* (Russian), Funktsional. Anal. i Prilozhen. **19** (1985), no. 1, 25–36, 96. MR783703

[Wig19] Avi Wigderson, *Mathematics and computation: A theory revolutionizing technology and science*, Princeton University Press, Princeton, NJ, 2019. MR4205964

Department of Mathematics, University of Southern California, Los Angeles, California

URL: `https://sites.google.com/usc.edu/gpanova/`

Proceedings of Symposia in Pure Mathematics
Volume **110**, 2024
https://doi.org/10.1090/pspum/110/02021

Increasing subsequences and Kronecker coefficients

Jonathan Novak and Brendon Rhoades

A permutation π in the symmetric group $\mathfrak{S}_n$ is said to have an increasing subsequence of length k if there are numbers $1 \leq i_1 < \cdots < i_k \leq n$ such that

$$\pi(i_1) < \cdots < \pi(i_k).$$

Similarly, π has a decreasing subsequence of length l if we can find $1 \leq j_1 < \cdots < j_l \leq n$ with

$$\pi(j_1) > \cdots > \pi(j_l).$$

Increasing and decreasing subsequences in permutations have been the subject of sustained interest in combinatorics since 1935, when Erdős and Szekeres proved that, given any $k, l \in \mathbb{N}$, every permutation in $\mathfrak{S}_n$ contains either an increasing subsequence of length k or a decreasing subsequence of length l as soon as n exceeds $(k-1)(l-1)$. This is a permutation version of Ramsey's theorem in which the threshold between possible disorder and certain order is explicitly computable.

In view of the Erdős-Szekeres theorem, it is natural to wonder about the typical length of longest monotone subsequences in permutations. Since reversing a permutation interchanges its increasing and decreasing subsequences, we may focus on just one of the two orientations. Let LIS_n denote the length of the longest increasing subsequence in a uniformly random sample from $\mathfrak{S}_n$. What can be said about the distribution of LIS_n? This question has been answered in the $n \to \infty$ limit: we have both a Law of Large Numbers and a Central Limit Theorem for LIS_n. The LLN was was obtained by Vershik and Kerov in 1977, who showed that

$$\lim_{n\to\infty} \frac{\mathrm{LIS}_n}{\sqrt{n}} = 2,$$

the convergence being in probability. The corresponding CLT is a breakthrough 1999 result of Baik, Deift, and Johansson, who discovered that

$$\lim_{n\to\infty} \mathbb{P}\left(\frac{\mathrm{LIS}_n - 2\sqrt{n}}{n^{1/6}} \leq t\right) = F(t),$$

2020 *Mathematics Subject Classification.* Primary 05A05, 05E10.

Key words and phrases. increasing subsequences, log-concavity, coinvariant algebra.

The first author was partially supported by NSF Grant DMS-1812288 and a Lattimer Fellowship.

The second author was partially supported by NSF Grants DMS-1953781 and DMS-2246846.

where $F(t)$ is the Tracy-Widom distribution from random matrix theory. For an informative discussion of these results, together with pertinent references, we refer the reader to Stanley's 2006 ICM contribution [**7**].

The distribution of LIS_n at finite n remains poorly understood — even the question of unimodality has not been settled. Given that the distribution of most natural combinatorial statistics is unimodal, it is not unreasonable to conjecture that this holds true for LIS_n. The above limit theorems certainly suggest that this is the case, since they tell us that when n is large the distribution of LIS_n is concentrated in a small neighborhood of $2\sqrt{n}$. However, this does not rule out the possibility of irregular "noisy" behavior before the limit. In combinatorics and elsewhere, the source of unimodality is often a stronger property: log-concavity [**6**]. Conjecturally, this is the case for LIS_n.

CONJECTURE 1 (Chen [**2**]). *For any $n \geq 3$, the distribution of LIS_n is log-concave. That is, we have*

$$a_{n,k-1}a_{n,k+1} \leq a_{n,k}^2, \tag{1}$$

for all $2 \leq k \leq n-1$, where $a_{n,k}$ is the number of permutations in $\mathfrak{S}_n$ whose longest increasing subsequence has length k.

The inequality (1) is equivalent to the existence of an injection

$$\mathfrak{S}_{n,k-1} \times \mathfrak{S}_{n,k+1} \longrightarrow \mathfrak{S}_{n,k} \times \mathfrak{S}_{n,k}, \tag{2}$$

where $\mathfrak{S}_{n,k}$ is the set of permutations in $\mathfrak{S}_n$ whose longest increasing subsequence has length equal to k. It seems difficult to construct such an injection in any uniform way: tampering with a permutation alters its increasing subsequences in complicated ways which are difficult to track. Instead, one may invoke the Robinson-Schensted correspondence [**5**], which tells us that $a_{n,k}$ is equal to the number of pairs of standard Young tableaux on a common shape $\lambda \vdash n$ satisfying $\ell(\lambda) = k$. Thus (2) is equivalent to the existence of an injection

$$\mathfrak{P}_{n,k-1} \times \mathfrak{P}_{n,k+1} \longrightarrow \mathfrak{P}_{n,k} \times \mathfrak{P}_{n,k}, \tag{3}$$

where $\mathfrak{P}_{n,k}$ is the set of Robinson-Schensted pairs with n cells and k rows. Since operations on Young tableaux are easier to visualize than operations on permutations, such an injection may be easier to construct. Another possibility would be to use the hook-length formula $f^\lambda = n!/H_\lambda$ for the number f^λ of SYT of shape λ, which combined with (3) gives

$$\sum_{\substack{\lambda \vdash n \\ \ell(\lambda)=k-1}} \sum_{\substack{\mu \vdash n \\ \ell(\mu)=k+1}} \frac{1}{H_\lambda^2 H_\mu^2} \leq \sum_{\substack{\lambda \vdash n \\ \ell(\lambda)=k}} \sum_{\substack{\mu \vdash n \\ \ell(\mu)=k}} \frac{1}{H_\lambda^2 H_\mu^2}. \tag{4}$$

This inequality is equivalent to (1), but may be easier to work with. So far, neither of these strategies has been successfully implemented; see [**1**] for some partial results.

When faced with an intractable problem, there is nothing to lose and everything to gain by trying to solve an even harder problem. We submit that the "right" way to approach Conjecture 1 is to focus on the following stronger conjecture. Let $R(\mathfrak{S}_n)$ be the representation ring of $\mathfrak{S}_n$, i.e. the commutative ring generated by the isomorphism classes V^λ of irreducible complex representations of $\mathfrak{S}_n$ with operations being direct sum and tensor product of $\mathfrak{S}_n$-modules. Let us replace the sets $\mathfrak{S}_{n,k}$ appearing in (2) with the modules $V_{n,k} \in R(\mathfrak{S}_n)$ defined by

$$V_{n,k} = \bigoplus_{\substack{\lambda \vdash n \\ \ell(\lambda)=k}} f^\lambda V^\lambda.$$

Since $\dim V^\lambda = f^\lambda$, we have

$$\dim V_{n,k} = \sum_{\substack{\lambda \vdash n \\ \ell(\lambda)=k}} (f^\lambda)^2 = a_{n,k},$$

whence inequality (1) is equivalent to

$$\dim V_{n,k-1} \dim V_{n,k+1} \leq \dim V_{n,k} \dim V_{n,k}. \tag{5}$$

We propose the following strengthening of Conjecture 1.

CONJECTURE 2. *For any $n \geq 3$, there exists an $\mathfrak{S}_n$-equivariant injection*

$$V_{n,k-1} \otimes V_{n,k+1} \longrightarrow V_{n,k} \otimes V_{n,k}. \tag{6}$$

for all $2 \leq k \leq n-1$.

Conjecture 2 may equivalently be formulated as a numerical refinement of (4). To see this, we make use of the Kronecker coefficients, which linearize multiplication in $R(\mathfrak{S}_n)$:

$$V^\lambda \otimes V^\mu = \bigoplus_{\nu \vdash n} g^\nu_{\lambda\mu} V^\nu.$$

Decomposing the source and target in (6) into irreducibles, Conjecture 2 claims the existence of an injective $\mathfrak{S}_n$-module homomorphism

$$\bigoplus_{\nu \vdash n} \left(\sum_{\substack{\lambda \vdash n \\ \ell(\lambda)=k-1}} \sum_{\substack{\mu \vdash n \\ \ell(\mu)=k+1}} f^\lambda f^\mu g^\nu_{\lambda\mu} \right) V^\nu \longrightarrow \bigoplus_{\nu \vdash n} \left(\sum_{\substack{\lambda \vdash n \\ \ell(\lambda)=k}} \sum_{\substack{\mu \vdash n \\ \ell(\mu)=k}} f^\lambda f^\mu g^\nu_{\lambda\mu} \right) V^\nu.$$

Thus, by Schur's lemma and the hook-length formula, Conjecture 2 is equivalent to the following numerical inequality.

CONJECTURE 3. *For any $n \geq 3$ and $\nu \vdash n$, we have*

$$\sum_{\substack{\lambda \vdash n \\ \ell(\lambda)=k-1}} \sum_{\substack{\mu \vdash n \\ \ell(\mu)=k+1}} \frac{g^\nu_{\lambda\mu}}{H_\lambda H_\mu} \leq \sum_{\substack{\lambda \vdash n \\ \ell(\lambda)=k}} \sum_{\substack{\mu \vdash n \\ \ell(\mu)=k}} \frac{g^\nu_{\lambda\mu}}{H_\lambda H_\mu} \tag{7}$$

for all $2 \leq k \leq n-1$.

The inequality (4) is recovered from Conjecture 3 by summing (7) over all $\nu \vdash n$.

Yet another equivalent formulation of Conjecture 2 may be obtained by means of the Frobenius isomorphism

$$\text{Frob}\colon R(\mathfrak{S}_n) \longrightarrow \Lambda_n,$$

where Λ_n is the ring of homogeneous symmetric functions of degree n equipped with the Kronecker product. We recall that the Kronecker product in Λ_n is defined via bilinear extension of the rule

$$s_\lambda * s_\mu := \sum_{\nu \vdash n} g^\nu_{\lambda\mu} s_\nu, \tag{8}$$

where $\{s_\lambda\colon \lambda \vdash n\}$ is the Schur function basis. The Frobenius isomorphism is defined by $\mathrm{Frob}(V^\lambda) = s_\lambda$. In particular, the Frobenius image of the increasing subsequence module $V_{n,k} \in R(\mathfrak{S}_n)$ appearing in Conjecture 2 is the symmetric function $S_{n,k} \in \Lambda_n$ given by

$$S_{n,k} = \sum_{\substack{\lambda \vdash n \\ \ell(\lambda)=k}} f^\lambda s_\lambda.$$

Given symmetric functions $F, G \in \Lambda_n$, let us write $F \leq G$ if the difference $G - F$ is Schur positive, i.e. if the coefficients c_λ defined by

$$G - F = \sum_{\lambda \vdash n} c_\lambda s_\lambda$$

are nonnegative. Conjecture 2 may then be restated as follows.

CONJECTURE 4. *For any $n \geq 3$, we have*

$$S_{n,k-1} * S_{n,k+1} \leq S_{n,k} * S_{n,k}$$

for all $2 \leq k \leq n-1$.

Conjecture 4 is a useful equivalent formulation of Conjecture 2 in that Schur positivity is a well-developed topic in algebraic combinatorics. Conjecture 4 has been verified for $n \leq 15$ on a computer.

How could one go about proving Conjecture 2? Let us illustrate how a successful argument might look by outlining a representation-theoretic proof of a much simpler proposition: the log-concavity of binomial coefficients. From the hook-length formula, it is clear that the dimension of the irreducible representation of $\mathfrak{S}_n$ corresponding to the hook $\lambda = (n-k, 1^k)$ is given by $\binom{n-1}{k}$ for all $0 \leq k \leq n-1$:

$$\dim V^{(n-k,1^k)} = \binom{n-1}{k}. \tag{9}$$

The difference $\binom{n-1}{k}^2 - \binom{n-1}{k-1} \cdot \binom{n-1}{k+1}$ may therefore be expressed as a difference of dimensions of Kronecker products:

$$\binom{n-1}{k}^2 - \binom{n-1}{k-1} \cdot \binom{n-1}{k+1} = \dim(V^{(n-k,1^k)} \otimes V^{(k+1,1^{n-k-1})}) - \dim(V^{(n-k+1,1^{k-1})} \otimes V^{(k+2,1^{n-k-2})}). \tag{10}$$

The sequence of binomial coefficients with upper index $n-1$ will thus be certified log-concave if we can exhibit an $\mathfrak{S}_n$-module $V_{n,k}$ whose Frobenius image is

$$\mathrm{Frob}(V_{n,k}) = s_{(n-k,1^k)} * s_{(k+1,1^{n-k-1})} - s_{(n-k+1,1^{k-1})} * s_{(k+2,1^{n-k-2})}. \tag{11}$$

The required module was found by Kim and Rhoades [**4**]. Let $\theta_1, \ldots, \theta_n$, $\xi_1, \ldots, \xi_n$ be a list of $2n$ anticommuting variables and consider the exterior algebra

$$\wedge\{\Theta_n, \Xi_n\} := \wedge\{\theta_1, \ldots, \theta_n, \xi_1, \ldots, \xi_n\} \tag{12}$$

over $\mathbb{C}$ generated by these variables. This is a $\mathbb{C}$-vector space of dimension 2^{2n} which carries the bigrading

$$\wedge\{\Theta_n, \Xi_n\} = \bigoplus_{i,j=0}^{n} \wedge\{\Theta_n, \Xi_n\}_{i,j} \tag{13}$$

induced by considering the degree of the θ-variables and ξ-variables separately. In Physics, anticommuting variables are called "fermionic," and the relation $\theta_i^2 = 0$ corresponds to the Pauli Exclusion Principle: no two fermions may occupy the same state at the same time. Consider the diagonal action of $\mathfrak{S}_n$ on $\wedge\{\Theta_n, \Xi_n\}$, viz.

$$w \cdot \theta_i := \theta_{w(i)} \qquad w \cdot \xi_i := \xi_{w(i)} \qquad w \in \mathfrak{S}_n, \; 1 \leq i \leq n \tag{14}$$

and denote by $\langle \wedge\{\Theta_n, \Xi_n\}_+^{\mathfrak{S}_n} \rangle \subseteq \wedge\{\Theta_n, \Xi_n\}$ the two-sided ideal generated by $\mathfrak{S}_n$-invariants with vanishing constant term. The *fermionic diagonal coinvariant* ring is defined in [**4**] by

$$FDR_n := \wedge\{\Theta_n, \Xi_n\} / \langle \wedge\{\Theta_n, \Xi_n\}_+^{\mathfrak{S}_n} \rangle. \tag{15}$$

This is a doubly graded $\mathfrak{S}_n$-module and an anticommutative version of the *diagonal coinvariant ring* [**3**].

THEOREM 5. *(Kim-R. [**4**]) The (i,j)-graded piece $(FDR_n)_{i,j}$ is zero unless $i + j < n$. When $i + j < n$ we have*

$$\mathrm{Frob}((FDR_n)_{i,j}) = s_{(n-i,1^i)} * s_{(n-j,1^j)} - s_{(n-i+1,1^{i-1})} * s_{(n-j+1,1^{j-1})}, \tag{16}$$

where we interpret $s_{(n-i+1,1^{i-1})} * s_{(n-j+1,1^{j-1})} = 0$ if $i = 0$ or $j = 0$. In particular, for $0 \leq k \leq n-1$ if we set $V_{n,k} := (FDR_n)_{k,n-k-1}$ then $V_{n,k}$ has Frobenius image given by Equation (11).

Theorem 5 implies that the symmetric function in Equation (11) is Schur-positive, and taking vector space dimensions yields the log-concavity of the sequence $\binom{n-1}{0}, \binom{n-1}{1}, \ldots, \binom{n-1}{n-1}$.

References

[1] Miklós Bóna, Marie-Louise Lackner, and Bruce E. Sagan, *Longest increasing subsequences and log concavity*, Ann. Comb. **21** (2017), no. 4, 535–549, DOI 10.1007/s00026-017-0365-x. MR3721641

[2] W. Y. C. Chen, *Log-concavity and q-log-convexity conjectures on the longest increasing subsequences of permutations*, arXiv:0806.3392 (2008).

[3] Mark Haiman, *Vanishing theorems and character formulas for the Hilbert scheme of points in the plane*, Invent. Math. **149** (2002), no. 2, 371–407, DOI 10.1007/s002220200219. MR1918676

[4] Jongwon Kim and Brendon Rhoades, *Lefschetz theory for exterior algebras and fermionic diagonal coinvariants*, Int. Math. Res. Not. IMRN **4** (2022), 2906–2933, DOI 10.1093/imrn/rnaa203. MR4381935

[5] C. Schensted, *Longest increasing and decreasing subsequences*, Canadian J. Math. **13** (1961), 179–191, DOI 10.4153/CJM-1961-015-3. MR121305

[6] Richard P. Stanley, *Log-concave and unimodal sequences in algebra, combinatorics, and geometry*, Graph theory and its applications: East and West (Jinan, 1986), Ann. New York Acad. Sci., vol. 576, New York Acad. Sci., New York, 1989, pp. 500–535, DOI 10.1111/j.1749-6632.1989.tb16434.x. MR1110850

[7] Richard P. Stanley, *Increasing and decreasing subsequences and their variants*, International Congress of Mathematicians. Vol. I, Eur. Math. Soc., Zürich, 2007, pp. 545–579, DOI 10.4171/022-1/21. MR2334203

DEPARTMENT OF MATHEMATICS, UNIVERSITY OF CALIFORNIA, SAN DIEGO, LA JOLLA, CALIFORNIA, 92093-0112

Email address: jinovak@ucsd.edu

DEPARTMENT OF MATHEMATICS, UNIVERSITY OF CALIFORNIA, SAN DIEGO, LA JOLLA, CALIFORNIA, 92093-0112

Email address: bprhoades@ucsd.edu

Proceedings of Symposia in Pure Mathematics
Volume **110**, 2024
https://doi.org/10.1090/pspum/110/02022

Combinatorics and braid varieties

Nathan Williams

ABSTRACT. I summarize a framework for finding and proving interesting combinatorial formulas using braid varieties. The combinatorics comes from the Deodhar decomposition of these braid varieties, while the enumeration uses traces in Hecke algebras to compute the number of points in these varieties over finite fields. I give several examples of the success of this framework, and also give many open problems.

1. Introduction

1.1. Motivation. The initiated combinatorialist knows that nearly all [**42,47**] combinatorial objects are counted by the Catalan numbers

$$\mathsf{Cat}(n) := \frac{1}{n+1}\binom{2n}{n}.$$

It shouldn't have come as a surprise then that the top a-coefficient of the HOMFLY-PT polynomial of the $(n, n+1)$-torus knot is [**25, 30, 31**] (up to a power of q) the naive q-analogue of $\mathsf{Cat}(n)$:

$$\mathsf{Cat}(n;q) = \frac{1}{[n+1]}\begin{bmatrix}2n\\ n\end{bmatrix}_q.$$

More generally, the complete list of coefficients gives a q-analogue of the f-vector of the associahedron (the q-Kirkman numbers). But how and why do torus knots know about the Catalan numbers?

The satisfying answer is that Catalan objects index certain pieces of a space—called a *braid variety*—associated to torus knots. To sketch the idea, recall that any link can be represented as a braid closure (and that the HOMFLY-PT polynomial can be computed directly from this braid). The braid variety is a "thickening" of the braid into a collection of walks indexed by the braid in a building over the flag variety. The relation to the HOMFLY-PT polynomial of the original link is that its top a-coefficient is given by the number of points of this braid variety over a finite field with q elements.

But the braid variety also decomposes in a natural way (the *Deodhar decomposition*, relative to a base flag) into pieces that look like torus times affine. The pieces with no torus part (or smallest torus part)—that is, the only pieces that survive as $q \to 1$—often turn out to be indexed by well-known combinatorial objects. In

2020 *Mathematics Subject Classification.* Primary 05E15; Secondary 05E05, 05E10, 20C08.

the case of the $(n, n+1)$-torus knot, these pieces are indexed by a classical Catalan object: the noncrossing partitions.

1.2. Overview. In this note, I present a framework for finding and proving interesting combinatorial formulas using braid varieties, generalizing the story of torus knots and noncrossing partitions. This framework is not new [**14**]—I learned about it from Galashin and Lam's work [**19, 20**]—but its application to algebraic combinatorics is new, and has already proven successful in producing new results: recent joint work with Galashin, Lam, and Trinh resolved two decades-long open problems in Coxeter–Catalan combinatorics [**21**].

The framework is based on two objects giving the same q-polynomial:

- the number of points in a braid variety over a finite field $\mathbb{F}_q$ (Definition 2.1); and
- a trace of certain braids in a suitable Hecke algebra (Theorem 3.3).

The first item produces combinatorial objects via Deodhar decomposition and specializing q to 1 (Section 3.1); the second item allows the use of representation-theoretic techniques for proving enumerative formulas (Section 3.2). At different levels of generality, different techniques become available.

1.3. Summary. In Section 2, I recall the definition of braid varieties, which form the algebraic and geometric background of the framework. In Section 3, I show how to extract combinatorial objects and enumerative formulas from braid varieties.

For finite Weyl and Coxeter groups (Section 4), it is possible to compute everything in a case-by-case manner using an explicit decomposition of the Hecke algebra (Equation (4.1)), and there are many interesting combinatorial and representation-theoretic problems that should be easily resolved using these case-by-case methods. Special classes of elements in finite type (*periodic* elements) have favorable representation-theoretic properties that allow for uniform approaches using Lusztig's exotic Fourier transform, Springer theory, and graded modules of rational Cherednik algebras (Section 4.6). A special case has already led to the solution to two long-standing open problems in Coxeter–Catalan combinatorics in [**21**] (Theorem 4.2).

For affine Weyl groups (Section 5), the main tool available is Opdam's trace formula for translation elements [**41**] (Theorem 5.2). In this setting, the proposed framework recovers some Tesler matrix identities due to Haglund [**27**] (Theorem 5.8)—and we compile several other interesting conjectures of intermediate difficulty (Section 5). Some of these conjectures can be attacked with Opdam's trace formula, while others will require further tools.

For general Weyl groups of Kac–Moody groups (Section 6), we have only rather generic recursive and cluster-theoretic methods. These techniques are not as easy to apply as the specialized trace formulas above, but have the advantage that they provide a tool for studying braid varieties over $\mathbb{C}$, with the aim of computing mixed Hodge decompositions and providing a link to q, t-combinatorics (Section 7).

2. Braid varieties and the Deodhar decomposition

In this section, I briefly recall the construction and Deodhar decomposition of braid varieties.

2.1. Coxeter groups. Let (W, S) be a Coxeter system with simple reflections S and rank r. The *reflections* T of W are the conjugates of the simple reflections. For $w \in W$, the *length* (resp. *absolute length*) $\ell(w)$ (resp. $\ell_T(w)$) of w is the smallest integer $m \geq 0$ such that w can be expressed as a product of m simple reflections (resp. reflections). For $w \in W$ and $s \in S$, write $ws < w$ if $\ell(ws) < \ell(w)$ and $ws > w$ if $\ell(ws) > \ell(w)$. The *weak order* on W is the partial order formed by the transitive closure of these relations.

2.2. Braid varieties. Let G be a connected Kac-Moody group split over an algebraically closed field $\mathbb{F}$. Fix a pair of opposed $\mathbb{F}$-split Borel subgroups B_+, B_-. Let $T := B_+ \cap B_-$ be a split maximal torus of G, and write $W := N_G(T)/T$ for the *Weyl group* (a crystallographic Coxeter group). The coset space G/B_+, called the *flag variety* of G, parametrizes the set of all Borel subgroups of G and carries an action of G by conjugation–if $g \in G$ and $B \in \mathcal{B}$, then $g \cdot B := gBg^{-1}$.

Write $w \cdot B_+ := \dot{w} \cdot B_+$, where $\dot{w} \in G$ is any lift of $w \in W$ to $N_G(T)$. For any two Borels $B_1, B_2 \in \mathcal{B}$, there is a unique w such that $(B_1, B_2) = (g \cdot B_+, gw \cdot B_+)$ for some $g \in G$. In this case, write $B_1 \xrightarrow{w} B_2$ and say that (B_1, B_2) are in *relative position* w.

DEFINITION 2.1. Let $\mathsf{w} = (s_1, s_2, \ldots, s_m) \in S^m$ be a word in the simple reflections S of length m, and fix $u, v \in W$. Define the corresponding *braid variety* by

$$R_{u,\mathsf{w}}^{(v)}(\mathbb{F}) := \left\{ (vB_+ = B_0 \xrightarrow{s_1} B_1 \xrightarrow{s_2} \cdots \xrightarrow{s_m} B_m \xleftarrow{vuw_\circ} B_-) \mid B_i \in \mathcal{B} \text{ for all } i \right\}.$$

For $v = e$, write $R_{u,\mathsf{w}}(\mathbb{F}) := R_{u,\mathsf{w}}^{(v)}(\mathbb{F})$. When w is the reduced word of an element $w \in W$, we have the special cases that:

- $R_{e,\mathsf{w}}(\mathbb{F})$ is isomorphic to the Schubert cell $B_+\dot{w}B_+/B_+$, and
- $R_{u,\mathsf{w}}(\mathbb{F})$ recovers the open Richardson variety, defined as the intersection of a Schubert and opposite Schubert cells $(B_+\dot{w}B_+/B_+) \cap (B_-\dot{u}B_+/B_+)$.

It is natural to view the word w as an element of the braid group B_W, as there are isomorphisms between $R_{u,\mathsf{w}}^{(v)}(\mathbb{F})$ and $R_{u,\mathsf{w}'}^{(v)}(\mathbb{F})$ when w and w' are related by braid moves.

2.3. Distinguished subwords and the Deodhar decomposition. The variety $R_{u,\mathsf{w}}^{(v)}$ decomposes into manageable pieces using Deodhar's *distinguished subwords*, as I now explain; I closely follow the exposition from [**21**].

A *subword* of $\mathsf{w} = (s_1, s_2, \ldots, s_m) \in S^m$ is a sequence $\mathsf{u} = (u_1, u_2, \ldots, u_m)$ in which $u_i \in \{s_i, e\}$ for all i (the positions for which $u_i = e$ are called *skips*). For any such sequence, set $u_{(i)} = u_1u_2 \cdots u_i \in W$. If $u_{(m)} = u$, then we refer to u as a *u-subword* of w.

DEFINITION 2.2 ([**14, 37**]). For $u \in W$, a u-subword u of $\mathsf{w} \in S^m$ is *v-distinguished* if $vu_{(i)} \leq vu_{(i-1)}s_i$ for all i. Write $\mathcal{D}_{u,\mathsf{w}}^{(v)}$ for the set of v-distinguished u-subwords of w (and set $\mathcal{D}_{u,\mathsf{w}} = \mathcal{D}_{u,\mathsf{w}}^{(e)}$). Define

$$\mathrm{e}_{\mathsf{u}} := |\{i \in [m] \mid u_i = e\}|, \mathrm{d}_{\mathsf{u}} := |\{i \in [m] \mid vu_{(i)} < vu_{(i-1)}\}|, \text{ and } k_{u,w}^{(v)} := \min_{\mathbf{u} \in \mathcal{D}_{u,\mathsf{w}}^{(v)}} \mathrm{e}_{\mathsf{u}}.$$

Write $\mathcal{M}_{u,\mathsf{w}}^{(v)} := \{\mathbf{u} \in \mathcal{D}_{u,\mathsf{w}}^{(v)} \mid \mathrm{e}_{\mathsf{u}} = k_{u,w}^{(v)}\}$ for the set of *maximal v-distinguished u-subwords* (and set $\mathcal{M}_{u,\mathsf{w}} := \mathcal{M}_{u,\mathsf{w}}^{(e)}$).

I now explain some combinatorial facts from [**21**]. When $u = e$ is the identity, the minimal value $k^{(v)}_{u,w}$ is given by the absolute length $\ell_T(w)$ of w [**21**, Proposition 4.8]:

$$k^{(v)}_{e,w} = \ell_T(w). \tag{2.1}$$

For any $u \in W$, there is a characterization of distinguished subwords using *colored reflections*—that is, a pair $(t, l) \in T \times \mathbb{Z}_{\geq 0}$. Given a subword $\mathsf{u} = (u_1, u_2, \cdots, u_m)$ of a word $\mathsf{w} = (s_1, s_2, \ldots, s_m)$ and an index $j \in [m]$, define the colored reflection

$$t_j(\mathsf{u}) := (s_j^{u_{(j)}}, l_j), \quad \text{where} \quad l_j := \left|\{1 \leq i < j \mid s_i^{u_{(i)}} = s_j^{u_{(j)}} \text{ and } u_i \neq e\}\right|.$$

If $t = (s^u, l)$ is a colored reflection, write $\overline{t} = s^u \in T$ for the corresponding uncolored reflection.

DEFINITION 2.3 ([**21**, Definition 4.3]). If $\mathsf{w} = (s_1, s_2, \ldots, s_m)$ is a word of simple reflections and $\mathbf{u}$ is a subword of w, then set

$$\mathsf{inv}(\mathsf{u}) := (t_1(\mathbf{u}), t_2(\mathbf{u}), \ldots, t_m(\mathbf{u})).$$

Write $\mathsf{inv}_e(\mathsf{u})$ for the subsequence of $\mathsf{inv}(\mathsf{u})$ obtained by restricting to skips of u—that is, the indices j for which $u_j = e$. A subword is distinguished if and only if $\mathsf{inv}_e(\mathsf{u})$ only contains reflections with even color.

It turns out that $\mathsf{inv}_e(\mathsf{u})$ often encodes combinatorial objects. We finally arrive at the Deodhar decomposition of the braid variety.

THEOREM 2.4 ([**14, 37, 52**]). *Let W be a Weyl group. For a u-subword u of $\mathsf{w} = (s_1, s_2, \ldots, s_m)$ we let*

$$R^{(v)}_{\mathsf{u},\mathsf{w}}(\mathbb{F}) := \left\{(vB_+ = B_0 \xrightarrow{s_1} B_1 \xrightarrow{s_2} \cdots \xrightarrow{s_m} B_m \xleftarrow{vuw_\circ} B_-) \mid B_- \xrightarrow{vu_{(i)}w_\circ} B_i\right\}.$$

Then we have the Deodhar decomposition *of $R^{(v)}_{u,\mathsf{w}}(\mathbb{F})$:*

$$R^{(v)}_{u,\mathsf{w}}(\mathbb{F}) = \bigsqcup_{\mathsf{u} \in \mathcal{D}^{(v)}_{u,\mathsf{w}}} R^{(v)}_{\mathsf{u},\mathsf{w}}(\mathbb{F}) \quad \text{with} \quad R^{(v)}_{\mathsf{u},\mathsf{w}}(\mathbb{F}) \simeq (\mathbb{F}^*)^{\mathsf{e}_\mathsf{u}} \times \mathbb{F}^{\mathsf{d}_\mathsf{u}}. \tag{2.2}$$

EXAMPLE 2.5. Let $W = \mathfrak{S}_2 = \{e, s\}$. We find

$$\mathcal{D}_{e,(s,s,s)} = \{(e,e,e), (e,s,s), (s,s,e)\}$$

so that

$$\sum_{\mathsf{u} \in \mathcal{D}_{e,(s,s,s)}} (q-1)^{\mathsf{e}_\mathsf{u}} q^{\mathsf{d}_\mathsf{u}} = (q-1)^3 + 2(q-1)q = (q-1)(q^2+1).$$

3. Combinatorial objects and traces

The Deodhar decomposition of braid varieties over finite fields identifies maximal distinguished subwords as the combinatorial objects in the framework. To count these objects, we compute the number of points of a braid variety over a finite field as a trace in the corresponding Hecke algebra $\mathcal{H}_W$.

3.1. Combinatorial objects. Suppose that $\mathbb{F}_q$ is a finite field with q elements. Then by Theorem 2.4, we have

$$\left|R^{(v)}_{u,\mathsf{w}}(\mathbb{F}_q)\right| = \sum_{\mathsf{u}\in\mathcal{D}^{(v)}_{u,\mathsf{w}}} (q-1)^{\mathsf{e}_\mathsf{u}} q^{\mathsf{d}_\mathsf{u}}.$$

Writing $k = k^{(v)}_{u,w}$ (see Definition 2.2), we conclude by definition of $\mathcal{M}^{(v)}_{u,\mathsf{w}}$ that

$$\lim_{q\to 1} \frac{1}{(q-1)^k}\left|R^{(v)}_{u,\mathsf{w}}(\mathbb{F}_q)\right| = \left|\mathcal{M}^{(v)}_{u,\mathsf{w}}\right|. \tag{3.1}$$

If we can determine $k = k^{(v)}_{u,w}$ apriori (for example, by Equation (2.1) when $u = e$), then any technique for computing $\left|R^{(v)}_{u,\mathsf{w}}(\mathbb{F}_q)\right|$ allows us to deduce a formula for the combinatorial set $\mathcal{M}^{(v)}_{u,\mathsf{w}}$ of maximal distinguished subwords—in certain settings, we will even be able to identify $\mathcal{M}^{(v)}_{u,\mathsf{w}}$ with existing combinatorial objects.

PROBLEM 3.1. *Suppose* $\frac{1}{(q-1)^k}\left|R^{(v)}_{u,\mathsf{w}}(\mathbb{F}_q)\right|$ *has nonnegative coefficients.*

- *Relate the maximal distinguished subwords* $\mathcal{M}^{(v)}_{u,\mathsf{w}}$ *to existing combinatorial objects.*
- *Find a combinatorial statistic* stat *on* $\mathcal{M}^{(v)}_{u,\mathsf{w}}$ *so that*

$$\frac{1}{(q-1)^k}\left|R^{(v)}_{u,\mathsf{w}}(\mathbb{F}_q)\right| = \sum_{\mathbf{u}\in\mathcal{M}^{(v)}_{u,\mathsf{w}}} q^{\mathrm{stat}(\mathbf{u})}.$$

EXAMPLE 3.2. Continuing Example 2.5, we have $\mathcal{M}_{e,(s,s,s)} = \{(e,s,s),(s,s,e)\}$. Problem 3.1 asks for a way to assign a statistic so that

$$\frac{1}{(q-1)^k}\left|R^{(v)}_{u,\mathsf{w}}(\mathbb{F}_q)\right| = 1+q^2 = q^{\mathrm{stat}(e,s,s)} + q^{\mathrm{stat}(s,s,e)} = \sum_{\mathbf{u}\in\mathcal{M}_{u,\mathsf{w}}} q^{\mathrm{stat}(\mathbf{u})}.$$

3.2. Traces. Every braid w in simple reflections gives rise to a corresponding element T_w of the Hecke algebra $\mathcal{H}_W$, and it turns out that the point count $R^{(v)}_{u,\mathsf{w}}(\mathbb{F}_q)$ can be expressed in terms of T_w. Recall that the *Hecke algebra* of (W,S) is the $\mathbb{Z}[q^{\pm 1}]$-algebra $\mathcal{H}_W$ that is the quotient of the free algebra generated by symbols T_w for $w \in W$, modulo the relations

$$T_wT_s = \begin{cases} qT_{ws} + (q-1)T_w & \text{if } ws < w, \\ T_{ws} & \text{if } ws > w, \end{cases}$$

for all $w \in W$ and $s \in S$. For any word $\mathsf{w} = (s_1, s_2, \ldots, s_m)$, we set $T_\mathsf{w} := T_{s_1}T_{s_2}\cdots T_{s_m}$. Write $\tau : \mathcal{H}_W \to A$ for the *trace* defined linearly on the basis $\{T_w\}_{w\in W}$ by:

$$\tau(T_w^{-1}) := \begin{cases} 1 & w = e, \\ 0 & w \neq e \end{cases} \qquad \text{for } w \in W.$$

This trace gives us our main tool for computing the number of points in a braid variety.

THEOREM 3.3 ([**21**, Corollary 5.3]). *For any word* w *and* $u, v \in W$, *we have*

$$\left|R^{(v)}_{u,\mathsf{w}}(\mathbb{F}_q)\right| = q^{\ell(v)}\tau(T_{v^{-1}}^{-1} T_\mathsf{w} T_{vu}^{-1}).$$

4. Techniques, examples, and problems in finite reflection groups

In this section, I describe some techniques and problems in the setting of finite reflection groups.

4.1. Technique: traces and Schur elements. Let W be a finite Coxeter group and $\mathrm{Irr}(W)$ be the set of its irreducible characters. The Hecke algebra $\mathcal{H}_W$ decomposes similarly to the group algebra of W, with certain weights called *Schur elements* $\mathbf{s}_\tau(\chi_q)$:

$$\tau = \sum_{\chi \in \mathrm{Irr}(W)} \frac{1}{\mathbf{s}_\tau(\chi_q)} \chi_q. \tag{4.1}$$

This decomposition of $\mathcal{H}_W$, along with explicit identifications of the Schur elements, gives a general technique for computing $\left|R^{(v)}_{u,\mathsf{w}}(\mathbb{F}_q)\right|$ in a case-by-case manner for any finite Coxeter group as a sum over irreducible characters. Furthermore, when taking a union of braid varieties $R^{(v)}_{e,\mathsf{w}}$ over all elements $v \in W$, we have the following dramatic simplification of Theorem 3.3.

THEOREM 4.1 ([**21**, Section 6.6]). *For W finite,*

$$\sum_{v \in W} \left|R^{(v)}_{e,\mathsf{w}}(\mathbb{F}_q)\right| = \sum_{\chi \in \mathrm{Irr}(W)} \dim(\chi) \cdot \chi_q(T_\mathsf{w}).$$

There are many interesting examples, as we now discuss.

4.2. Example: Galashin and Lam's positroid Catalan combinatorics. In [**19, 20**], Galashin and Lam apply this framework in the context of type A rational Catalan combinatorics. I refer the reader to their excellent papers for a full description of their results and open problems.

4.3. Example: Rational noncrossing Coxeter–Catalan combinatorics. With Galashin, Lam, Trinh in [**21**], we solved two long-standing open problems in Coxeter–Catalan combinatorics using the above framework, building on results from [**19, 24, 50, 51**] and [**48**].

For any positive integer p coprime to h, set

$$\mathsf{Cat}_p(W;q) := \prod_{i=1}^{r} \frac{[p + (pe_i \bmod h)]}{[d_i]}, \tag{4.2}$$

where $d_1 \leq d_2 \leq \cdots \leq d_r$ are the *degrees* of W, $e_i = d_i - 1$ are the exponents, h is the Coxeter number, and where $0 \leq (pe_i \bmod h) < h$ is the integer in that range congruent to pe_i modulo h. (For well-generated finite complex reflection groups, $\mathsf{Cat}_p(W;q)$ is the graded character of the finite-dimensional irreducible representation $eL_{p/h}(\mathsf{triv})$ of the rational Cherednik algebra at the parameter p/h). Write $\mathrm{Cat}(W) := \mathrm{Cat}_{h+1}(W;1)$, which was previously known case-by-case to enumerate the noncrossing partitions.

The two open problems in Coxeter–Catalan combinatorics solved in [**21**] were:

- uniformly prove that noncrossing objects are enumerated by $\mathrm{Cat}(W)$.
- uniformly construct rational noncrossing objects enumerated by $\mathrm{Cat}_p(W;1)$.

The manuscript [**21**] also resolves the corresponding "parking" problems using a weighted sum of braid varieties over the Coxeter group.

THEOREM 4.2 ([**21**]). *For any (finite) Weyl group W of rank r and Coxeter number h, Coxeter word* c, *and integer p coprime to h,*

$$|R_{e,\mathsf{c}^p}(\mathbb{F}_q)| = (q-1)^r \mathsf{Cat}_p(W;q) \quad \textit{and} \quad \sum_{v\in W} q^{\ell(v)} \left| R^{(v)}_{e,\mathsf{c}^p}(\mathbb{F}_q) \right| = (q-1)^r [p]^r,$$

$$\mathcal{M}_{e,\mathsf{c}^p}(W)| = \mathsf{Cat}_p(W) \quad \textit{and} \quad \sum_{v\in W} \left| \mathcal{M}^{(v)}_{e,\mathsf{c}^p} \right| = p^r.$$

The maximal distinguished subwords are truly *noncrossing* objects—there is an easy uniform bijection between $\mathcal{M}_{e,\mathsf{c}^p}(W)$ and noncrossing partitions for $p = mh+1$ [**1, 48**], and between $\bigcup_{v\in W} \mathcal{M}^{(v)}_{e,\mathsf{c}^p}$ and the noncrossing parking functions for $p = h+1$ of Armstrong, Reiner, and Rhoades [**4**]. Both bijections use the colored inversions in $\mathsf{inv}_e(\mathsf{u})$ coming from the skips. In particular, using standard Coxeter–Catalan bijections [**48**], [**21**] gives the first uniform proof that the number of clusters in a finite-type cluster algebra is counted by $\mathsf{Cat}(W)$ [**18**, Theorem 1.9].

It would be nice to have combinatorial interpretations for these new, rational noncrossing objects. In the symmetric group, it is possible $\mathcal{M}_{e,\mathsf{c}^p}(W)$ agrees with existing models for rational noncrossing partitions [**5, 11**].

PROBLEM 4.3. *Develop graphical models for our rational noncrossing Catalan and parking objects (generalizing the usual depictions of noncrossing partitions in classical types).*

s_1	s_2	(14)	s_4	s_1	s_2	(25)	s_4	s_1	s_2	(34)	$(\ddot{45})$
s_1	s_2	(14)	s_4	(23)	s_2	s_3	(34)	s_1	(25)	s_3	s_4
s_1	s_2	(14)	(45)	s_1	s_2	(24)	s_4	s_1	s_2	(35)	s_4
s_1	(13)	s_3	s_4	s_1	s_2	(25)	s_4	(14)	s_2	s_3	(45)
s_1	(13)	s_3	s_4	s_1	(24)	s_3	s_4	$(\ddot{12})$	(25)	s_3	s_4
(12)	s_2	s_3	(25)	s_1	(14)	s_3	s_4	s_1	s_2	(35)	s_4
(12)	(23)	s_3	s_4	s_1	(14)	s_3	s_4	s_1	(25)	s_3	s_4

FIGURE 1. Figure taken from [**21**]. For $W = S_5$, Coxeter word $\mathsf{c} = (s_1, s_2, s_3, s_4)$, and $p = 3$, each row above depicts one of the seven maximal distinguished subwords counted by $\mathsf{Cat}_3(S_5)$. Skips have been replaced by the colored reflections in $\mathsf{inv}_e(\mathsf{u})$ (colors are denoted with dots).

Beyond maximal distinguished subwords, enumerative results are lacking (and the non-maximal distinguished subwords do not appear to be closely linked to the h- or f-vector of the associahedron).

PROBLEM 4.4. *Explore combinatorial and enumerative properties of non-maximal distinguished subwords.*

Preliminary calculations in type A suggest that these lower order terms have a deeper, perhaps not unexpected, connection to maps—just as there are the same number of maximal distinguished subwords and rooted bicolored unicellular maps of genus 0 on n edges, there appear to be the same number of distinguished subwords with $r+2$ skips as rooted bicolored unicellular maps of genus 1 on $n+2$ edges (given by $\frac{(2n+3)!}{6n!(n+1)!}$).

4.4. Problem: Rational nonnesting Catalan combinatorics. Coxeter–Catalan numbers naturally appear in a markedly different context: in the study of affine Weyl groups and affine Springer fibers. Specializing to crystallographic Coxeter groups, for p coprime to h, $\mathsf{Cat}_p(W)$ (uniformly) counts the number of coroot points inside a p-fold dilation of the fundamental alcove in the corresponding affine Weyl group [**29**, **49**]. For $p = h+1$, these coroot points are called *nonnesting partitions*, and are in bijection with order ideals in the root poset (or, equivalently, ad-nilpotent ideals in a Borel subalgebra of the corresponding complex simple Lie algebra). *Rational (nonnesting) parking functions* are the p^n alcoves inside this same dilation of the fundamental alcove.

Although nonnesting and noncrossing partitions have many similarities, finding a uniform bijection between the two sets has been an active and motivating area of research since the late 1990s [**7**, **44**]. The state of the art has now changed with our recent new definition of rational noncrossing objects—both noncrossing and nonnesting objects can be defined for Weyl groups for any p coprime to h.[1]

PROBLEM 4.5. *Let p be coprime to the Coxeter number h.*

- *Find a bijection between $\mathcal{M}_{e,c^p}$ and rational nonnesting partitions.*
- *Find a bijection between $\bigcup_{v\in W} \mathcal{M}^{(v)}_{e,c^p}$ and rational nonnesting parking functions.*

In [**53**] and for $p = h+1$, I conjectured exactly such a bijection between nonnesting and noncrossing objects for any Coxeter element and any finite Weyl group, suggesting that the root poset encodes a remarkable amount of information related to the corresponding Weyl group (compare with the duality between the heights of roots and the degrees). This conjectural bijection between noncrossing and nonnesting objects comes from mimicking walks on the W-associahedron—drawing inspiration from [**6**, **10**, **43**], this leads to remarkable conjectural (compatible) bijections from nonnesting partitions to clusters and noncrossing partitions (which have been exhaustively checked up to rank eight) [**53**, **54**].

PROBLEM 4.6. *Show the conjectured maps in* [**53**, **54**] *are bijections between nonnesting and noncrossing partitions. Use the new definition of rational noncrossing objects from* [**21**] *to extend them to the Fuss and rational levels of generality.*

A group of current and former LaCIM members and postdocs has recently made substantial progress on this problem, proving part of the conjectures by purely combinatorial means in type A, for all Coxeter elements [**15**]. The innovation is a realization of the Cambrian recurrence on nonnesting partitions.

4.5. Problem: Periodic elements. While Equation (4.1) enables brute-force computations of any traces for finite Coxeter groups, there are also more powerful specific tools available for sufficiently nice words w (the Coxeter words of Section 4.3 being a special case).

A braid $\mathsf{w} = \mathsf{s}_1 \cdots \mathsf{s}_k \in B^+(W)$ is called *periodic* if $\mathsf{w}^m = \mathsf{w}_\circ^{2p}$ for some p, m with $m \neq 0$; the *slope* of a periodic braid w is the ratio p/m. There a classification of periodic braids up to conjugacy using Springer theory as the dth roots of the full twist, where d is a regular number of W [**9**, **22**, **36**].

[1]Noncrossing objects are more general than nonnesting objects: noncrossing objects exist for any finite Coxeter group when p is coprime to h, and also exist for well-generated complex reflection groups for $p = mh+1$.

PROBLEM 4.7. *Find Catalan and parking-like formulas for braid varieties associated to powers of periodic elements* w^p*:*

$$|R_{e,\mathsf{w}^p}(\mathbb{F}_q)| \text{ and } \sum_{v\in W} q^{\ell(v)}\left|R^{(v)}_{e,\mathsf{w}^p}(\mathbb{F}_q)\right|.$$

If w *is periodic of order a regular number* d*, with reflection representation eigenvalues* $\{e^{2\pi e_i(w)/d}\}_{i=1}^n$*, then small computations suggest that*

$$|R_{e,\mathsf{w}^p}(\mathbb{F}_q)| = (q-1)^{\ell_T(w)}\prod_{i=1}^r \frac{[p+(pe_i(w) \bmod d)]}{[d_i]} C_w(q^p) \text{ and}$$

$$\sum_{v\in W} q^{\ell(v)}\left|R^{(v)}_{e,\mathsf{w}^p}(\mathbb{F}_q)\right| = (q-1)^{\ell_T(w)}[p]^n P_w(q^p),$$

for some polynomials $C_w(q)$ *and* $P_w(q)$ *with* $C_w(1) = P_w(1)$*.*

This should perhaps be compared with [**16**].

EXAMPLE 4.8. For type D_4 with regular number $d = 4$ and periodic braid $\mathsf{w} = \mathbf{s}_1\mathbf{s}_2\mathbf{s}_3\mathbf{s}_1\mathbf{s}_2\mathbf{s}_4$, we compute that

$$|R_{e,\mathsf{w}^p}(\mathbb{F}_q)| = (q-1)^4\frac{[p+1]^2[p+3]^2}{[2][4]^2[6]}(1+2q^{p+1}+2q^{p+3}+q^{2p+4}), \text{ and}$$

$$\sum_{v\in W} q^{\ell(v)}\left|R^{(v)}_{e,\mathsf{w}^p}(\mathbb{F}_q)\right| = (q-1)^4[p]^4(1+4q^p+q^{2p}).$$

Note that the order of $w = s_1s_2s_3s_1s_2s_4$ is 4, with reflection representation eigenvalues i^1 and i^3 (each with multiplicity 2).

4.6. Technique: Periodic elements. There is a concrete, systematic, and uniform approach to Problem 4.7, building on [**21**]. For all $\chi \in \mathrm{Irr}(W)$, the *fake* and *generic degrees* of χ are

$$\mathrm{Feg}_\chi(q) := \frac{(\chi, [\mathfrak{S}]_q)_W}{(1, [\mathfrak{S}]_q)_W} \text{ and } \mathrm{Deg}_\chi(q) := \frac{\mathbf{s}_\tau(1_q)}{\mathbf{s}_\tau(\chi_q)}.$$

It turns out that $\mathrm{Feg}_\chi(q) \in \mathbb{Z}[q]$ and $\mathrm{Deg}_\chi \in \mathbb{Q}_W[q]$; at $q = 1$, both polynomials specialize to the *degree* of χ. For W a finite Coxeter group and $\chi \in \mathrm{Irr}(W)$ an irreducible character, write $c(\chi) = \frac{1}{\dim(\chi)}\sum_{t\in T}\chi(t)$ for the *content* of χ [**51**]. (When $W = \mathfrak{S}_{n+1}$ is the symmetric group and the irreducible characters are indexed by integer partitions, this is the sum of the contents of all boxes in the partition.) If $\chi \in \mathrm{Irr}(W)$ and w is a periodic braid of slope $\nu \in \mathbb{Q}$, write $\sigma_\mathsf{w} = q^{-\ell(\mathsf{w})/2}T_\mathsf{w}$.

Then we have the following result on traces of periodic elements [**51**]: $\chi_q(\sigma_\mathsf{w}) = q^{\nu\, c(\chi)}\mathrm{Feg}_\chi(e^{2\pi i\nu})$, so that (up to a sign)

$$\begin{aligned}\tau(\sigma_\mathsf{w}) &= \frac{1}{\mathbf{s}_\tau(1_q)}\sum_{\chi\in\mathrm{Irr}(W)} q^{-\nu\, c(\chi)}\mathrm{Feg}_\chi(e^{2\pi i\nu})\mathrm{Deg}_\chi(q)\\ &= \frac{1}{\mathbf{s}_\tau(1_q)}\sum_{\chi\in\mathrm{Irr}(W)} q^{-\nu\, c(\chi)}\mathrm{Feg}_\chi(q)\mathrm{Deg}_\chi(e^{2\pi i\nu}),\end{aligned}$$

where the second, miraculous, equality follows from Lusztig's exotic Fourier transform. This last formula gives a dramatic simplification of the trace—even though periodic elements often have many vanishing characters, transferring the root of unity from fake degree to the generic degree typically gives a dramatic reduction in the number of terms vanishing in the sum.

It follows from the above (see also [**51**, Theorem 9.2.1] and Theorem 4.1) that

$$\sum_{v\in W} q^{\ell(v)}\left|R^{(v)}_{e,\mathbf{w}_\circ^2}(\mathbb{F})\right| = \sum_{\chi\in \mathrm{Irr}(W)} \dim(\chi)^2 q^{\mathrm{cont}(\chi)}.$$

More generally, writing $\mathrm{ord}(w)$ for the order of an element $w \in W$, the expression $\sum_{v\in W} q^{\ell(v)}\left|R^{(v)}_{e,\mathbf{w}^{\mathrm{ord}(w)}}(\mathbb{F})\right|$ appears to have positive coefficients.

4.7. Problem: Complex reflection groups. In this section, I propose an interesting extension of the results in [**21**] to *spetsial* complex reflection groups (that is, one of $G(d,1,r)$, $G(d,d,r)$, G_4, G_6, G_8, G_{14}, G_{23}, G_{24}, G_{25}, G_{26}, G_{27}, G_{28}, G_{29}, G_{30}, G_{32}, G_{33}, G_{34}, G_{35}, G_{36}, or G_{37} in the Shephard–Todd indexing of irreducible complex reflection groups [**45**]). Spetsial complex reflection groups still have a preferred set of "simple reflections", Coxeter elements, and a well-defined rational Catalan number [**23**]. Moreover, the notion of periodic elements naturally generalizes to well-generated complex reflection groups, and such elements have been classified. We may therefore consider Problem 4.7 in this context.

The first step in this direction has been recently carried out by Miller as part of his undergraduate honors thesis [**38**]. Subject to certain assumptions common in the field, he has performed a case-by-case computation of the trace $\tau(T^p_{\mathbf{c}})$ in the Hecke algebra of the spetsial complex reflection group [**12, 13**] (whose braid groups have Artin-like presentations, by Bessis); many of the favorable representation-theoretic properties of Coxeter elements that we used in Section 4.3 still carry over to the complex setting [**13**]. While we no longer have Lusztig's exotic Fourier transform uniformly nor a uniform definition of the Hecke algebra, Trinh has pointed out that there are partial results due to Lasy and Lacabanne in the infinite family [**33, 35**], and a reciprocity result due to Malle that has been used by Douvropoulos in a related setting [**17**].

THEOREM 4.9 (Miller [**38**]). *Let W be a spetsial complex reflection group and $\mathbf{c}$ a lift from W to B_W of a ζ_h-regular element such that $\mathbf{c}^h$ is the full twist. Then (up to a power of q) we have*

$$\tau(T^{-1}_{\mathbf{c}^p}) = (q-1)^r \prod_{i=1}^{r} \frac{[p+e_i(V^p)]}{[d_i]} = (q-1)^r \mathsf{Cat}_p(W;q),$$

where the $e_i(V^p)$ are the fake degrees of the p-th Galois twist of the reflection representation and the trace is taken in the Hecke algebra $\mathcal{H}_W$. If $\{T_w\}_{w\in W}$ is a basis of $\mathcal{H}_W$ adapted to the Wedderburn decomposition, then

$$\sum_{w\in W} \tau\left(T^\vee_w T_{\mathbf{c}^p} T_w\right) = (q-1)^r [p]^n.$$

EXAMPLE 4.10. The complex reflection group G_4 has rank $r = 2$, Coxeter number $h = 6$. Its reflection representation has fake degrees 3 and 5. We compute using GAP3 that

$$\tau(T_{\mathbf{c}^7}) = (q-1)^2(q^{12}+q^8+q^6+q^4+1) = (q-1)^2\frac{[7+3][7+5]}{[4][6]}.$$

It would also be interesting to extend this in the direction of Section 4.6 to the periodic elements of complex reflection groups. Although the Catalan piece of Theorem 4.9 does not hold for general complex reflection groups, Miller has recently observed that the parking part of Theorem 4.9 does seem to remain true.

The Deodhar decomposition allows us to build combinatorial models of braid varieties for general Coxeter groups. Besides the representation-theoretic computation above, there is the issue of finding the correct combinatorial definition of distinguished subwords for complex reflection groups—this is strongly dependent on the lift of the reflection group to the Hecke algebra, and so presents nontrivial technical challenges (beyond the case of the cyclic group).

PROBLEM 4.11. *Find a combinatorial description of the Deodhar decomposition for well-generated complex reflection groups.*

Because there is a reasonable lift of W to H_W in the case of cyclic groups, we can use the recurrence obtained by multiplying by T_s in the Hecke algebra to work out the combinatorics of distinguished subwords.

EXAMPLE 4.12. For general cyclic groups $G(m,1,1)$ of order m, we fix $\mathsf{w} = \mathbf{s}^k$, and a subword $\mathsf{u} = (u_1, \ldots, u_k)$ is a sequence of reflections with $u_i = s^j$. Writing $u_{(i)} = u_1 \cdots u_i$, u is *distinguished* when $u_{(k)} = e$ and

[1] if $u_{(i)} = e$, then u_{i+1} can be either
   (a) s^j for $0 \leq j < m-1$ (with weight $(1-q)$); or
   (b) s^{m-1} (with weight 1); and

[2] otherwise $u_{i+1} = s^{-1}$ (with weight q).

Some examples of distinguished subwords for the cyclic group $G(3,1,1)$ are illustrated in Figure 2.

5. Techniques, examples, and problems in affine Weyl groups

In this section, we consider affine Weyl groups $\widetilde{W}$.

5.1. Technique: Opdam's trace formula. Write Φ^+ for the positive roots of a simple Lie group, $Q = \bigoplus_{i=1}^r \mathbb{Z}\alpha_i$ for the *root lattice*, $Q^+ \subset Q$ for the positive span of the simple roots, and Λ for the weight lattice. Given $\lambda \in Q^+$, we express λ in the basis of fundamental weights as $\lambda = \sum_{i=1}^{n-1} a_i \lambda_i$ and define $\lambda_+ = \sum_{i:a_i>0} a_i\lambda_i$ and $\lambda_- = -\sum_{i:a_i<0} a_i\lambda_i$. For $x \in \Lambda$, write t_x for the translation in the extended affine Weyl group $\widehat{W}$.

DEFINITION 5.1. A *Kostant partition* $(a_\alpha)_{\alpha\in\Phi^+}$ for $\lambda \in Q^+$ is a sequence of nonnegative integers indexed by positive roots such that $\lambda = \sum_{\alpha\in\Phi^+} a_\alpha \alpha$. We denote the set of all Kostant partitions for λ by $K(\lambda)$.

Opdam proved the following trace formula, which—when combined with Theorem 3.3—is our main technique in this setting.

THEOREM 5.2 ([**41**, Cor. 1.18]). *Let* $[k]_q = \frac{(q-1)^2}{q}\frac{q^k - q^{-k}}{q-q^{-1}}$. *For* $\lambda \in Q^+$,

$$\tau(T_{t_{\lambda_-}} T_{t_{\lambda_+}}^{-1}) = q^{(\ell(t_{\lambda_-}) - \ell(t_{\lambda_+}))/2} \sum_{(a_\alpha)\in K(\lambda)} \prod_{\substack{\alpha\in\Phi^+ \\ a_\alpha>0}} [a_\alpha]_q.$$

s	s	s	s	wt
$\cdot$	$\cdot$	$\cdot$	$\cdot$	$(1-q)^4$
s	s^{-1}	$\cdot$	$\cdot$	$(1-q)^3q$
$\cdot$	s	s^{-1}	$\cdot$	$(1-q)^3q$
$\cdot$	$\cdot$	s	s^{-1}	$(1-q)^3q$
s	s^{-1}	s	s^{-1}	$(1-q)^2q^2$
s^2	s^{-1}	s^{-1}	$\cdot$	$(1-q)q^2$
$\cdot$	s^2	s^{-1}	s^{-1}	$(1-q)q^2$

s	s	s	s	s	wt
$\cdot$	$\cdot$	$\cdot$	$\cdot$	$\cdot$	$(1-q)^5$
s	s^{-1}	$\cdot$	$\cdot$	$\cdot$	$(1-q)^4q$
$\cdot$	s	s^{-1}	$\cdot$	$\cdot$	$(1-q)^4q$
$\cdot$	$\cdot$	s	s^{-1}	$\cdot$	$(1-q)^4q$
$\cdot$	$\cdot$	$\cdot$	s	s^{-1}	$(1-q)^4q$
s	s^{-1}	s	s^{-1}	$\cdot$	$(1-q)^3q^2$
s	s^{-1}	$\cdot$	s	s^{-1}	$(1-q)^3q^2$
$\cdot$	s	s^{-1}	s	s^{-1}	$(1-q)^3q^2$
s^2	s^{-1}	s^{-1}	$\cdot$	$\cdot$	$(1-q)^2q^2$
$\cdot$	s^2	s^{-1}	s^{-1}	$\cdot$	$(1-q)^2q^2$
$\cdot$	$\cdot$	s	s^{-1}	s^{-1}	$(1-q)^2q^2$
s^2	s^{-1}	s^{-1}	s	s^{-1}	$(1-q)q^3$
s	s^{-1}	s^2	s^{-1}	s^{-1}	$(1-q)q^4$

FIGURE 2. Fix $G = G(3,1,1)$, the cyclic group of order three. *Left:* the seven e-subwords of $\mathsf{w} = (s,s,s,s)$ for $p = 4$, with a total weight of $(1-q)(1+q^3) = (1-q)\mathrm{Cat}_4(G;q)$ (only the bottom two are maximal, and letters in a subword that are the identity have been replaced by a dot). *Right:* the 13 e-subwords of $\mathsf{w} = (s,s,s,s,s)$ for $p = 5$, with a total weight of $(1-q)(1+q^4) = \mathrm{Cat}_5(G;q)$ (only the bottom two are maximal).

5.2. Example: Haglund's formula and Tesler matrices. The results in this section were obtained in collaboration with Galashin and Lam. Let S_n act diagonally on the polynomial ring $\mathbb{C}[x_1,\ldots,x_n,y_1,\ldots,y_n]$, and write Q_n for its root lattice, Λ_n for its fundamental weights, and Φ_n^+ for its positive roots. The *quotient ring of diagonal coinvariants* DH_n is the quotient of this polynomial ring by the ideal generated by the invariants with no constant term; there is a more general S_n module DH_n^m depending on an integral parameter m. In [**27**], Haglund proved a remarkable formula for the bigraded (in x- and y-degree) Hilbert series of DH_n^m.

THEOREM 5.3 ([**27**, Corollary 1 and Theorem 3]). *Write* $[k]_{q,t} = (q-1)(1-t)\frac{q^k-t^k}{q-t}$ *and let* $\lambda = (m(n-1)+1)\lambda_{n-1} - (m-1)\lambda_1 \in Q_n^+$. *Then*

$$\mathrm{Hilb}(\mathrm{DH}_{n-1}^m;q,t) = \left(\frac{1}{(q-1)(t-1)}\right)^{n-1} \sum_{(a_\alpha)\in K(\lambda)} \prod_{\substack{\alpha\in\Phi_n^+\\ a_\alpha>0}} [a_\alpha]_{q,t}.$$

REMARK 5.4. Haglund stated Theorem 5.3 in terms of *Tesler matrices*, which are a simple combinatorial rephrasing of Kostant partitions. We choose to write the formula using Kostant partitions to mirror Theorem 5.2—note that since there are only $n-1$ simple roots in Φ_n^+, the formula for DH_{n-1}^m is written using Kostant partitions in Q_n^+.

REMARK 5.5. Galashin has pointed out that there is a similar formula due to Gorsky and Negut [**26**], which appears as [**28**, Theorem 32]—this formula sums over the same set of Kostant partitions and outputs the same q,t-polynomial, but extends to the rational case.

Let S_n be the symmetric group of order $n!$. The Weyl group of $\mathrm{GL}_n(\mathbb{F})$ is the group of *extended affine permutations* $\widehat{S}_n = \Lambda_n \ltimes S_n \simeq N_{\mathrm{GL}_n(\mathbb{F})}(\widehat{T})/\widehat{T}$. Given $\lambda \in Q_n^+$, we express λ in the basis Λ_n as $\lambda = \sum_{i=1}^{n-1} a_i\lambda_i$ and define $\lambda_+ = \sum_{i:a_i>0} a_i\lambda_i$ and $\lambda_- = -\sum_{i:a_i<0} a_i\lambda_i$. For $x \in \Lambda_n$, write t_x for the corresponding translation element of $\widehat{S}_n$.

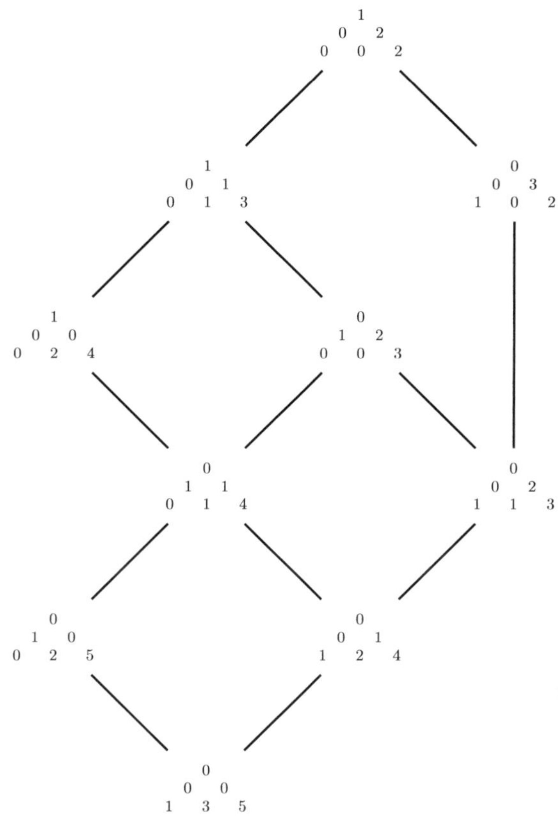

FIGURE 3. The set of Kostant partitions $K(\alpha_1 + 3\alpha_2 + 5\alpha_3) \in Q_4^+$, partially ordered using the canonical basis of the quantum group $U_q^+(\mathfrak{sl}_4)$. Coefficients are arranged by dominance ordering on positive roots.

EXAMPLE 5.6. The element $\lambda = \alpha_1 + 3\alpha_2 + 5\alpha_3 \in Q_4^+$ has ten possible Kostant partitions, partially ordered in Figure 3 according to the PBW basis terms appearing in the corresponding element of the canonical basis in the positive part of the quantum group $U_q^+(\mathfrak{sl}_4)$ (for the ordering $\alpha_1 < \alpha_{12} < \alpha_{123} < \alpha_2 < \alpha_{23} < \alpha_3$; it would be interesting to study this partial ordering in more detail—some recent progress appears to have been made by a group working with Dranowski—see also [**2**, **40**]). Since $\lambda = -\lambda_1 + 7\lambda_3$, we have $\lambda_+ = 7\lambda_3$ and $\lambda_- = \lambda_1$, so that $t_{\lambda_1} = \sigma \cdot s_3s_2s_1$ and $t_{7\lambda_3} = (\sigma^3 \cdot s_1s_2s_3)^7$.

EXAMPLE 5.7. As in Example 5.6 and Figure 3, $\lambda = -\lambda_1 + 7\lambda_3$ has ten Kostant partitions, while $\ell(t_{\lambda_1}) = 3$ and $\ell(t_{7\lambda_3}) = 21$. Taking the weighted sum over $K(\lambda)$, we have

$$\begin{aligned}\mathrm{tr}(T_{t_{\lambda_-}} T_{t_{\lambda_+}}^{-1}) = q^{(3-21)/2}\big(&[1][3][5] + [1][2][1][4] + [1][1][2][3] + [1][3][2] + [1][2][5]+ \\ &+ [1][1][1][4] + [1][2][3] + [2][1][4] + [1][1][1][3] + [1][2][2]\big) \\ = q^{-18}(q-1)^6[7]^2 &\end{aligned}$$

Since $[k]_q = [k]_{q,q^{-1}}$, we can specialize Theorem 5.3 using a result of Haiman to conclude the following.

THEOREM 5.8. *Fix the extended affine Weyl group* $\widehat{S}_n$, *and let* $u = t_{(m-1)\lambda_1}$ *and* $w = t_{(m(n-1)+1)\lambda_{n-1}}$. *Then*

$$|R_{u,\mathsf{w}}(\mathbb{F})| = (q-1)^{2(n-1)}\mathrm{Hilb}(\mathrm{DH}^m_{n-1}; q, q^{-1}).$$

PROOF. By [**32**, Lemmas A3 and A4] and [**21**, Corollary 5.3], the number of $\mathbb{F}_q$-points in the braid variety $R_{\widetilde{u},\widetilde{w}}(\mathbb{F})$ is given by the trace formula

$$q^{\ell(\widetilde{w})-\ell(\widetilde{u})}\mathrm{tr}(T_{\widetilde{u}}\, T_{\widetilde{w}}^{-1}).$$

Since $\widehat{u} = t_{(m-1)\lambda_1}$ and $\widehat{w} = t_{(m(n-1)+1)\lambda_n}$, we see that

$$\begin{aligned}\widehat{u} &= (\sigma s_{n-1}\cdots s_2 s_1)^{m-1} = \widetilde{u}\sigma^{m-1} \text{ and} \\ \widehat{w} &= (\sigma^{-1}s_1\cdots s_{n-2}s_{n-1})^{m(n-1)+1} = \widetilde{w}\sigma^{-m(n-1)-1} = \widetilde{w}\sigma^{-mn+(m-1)} = \widetilde{w}\sigma^{m-1},\end{aligned}$$

so that $T_{\widehat{u}} = T_{\widetilde{u}}T_{\sigma^{m-1}}$ and $T_{\widehat{w}}^{-1} = T_{\sigma^{m-1}}^{-1}T_{\widetilde{w}}^{-1}$. We conclude that $\mathrm{tr}(T_{\widehat{u}}\, T_{\widehat{w}^{-1}}) = \mathrm{tr}(T_{\widetilde{u}}\, T_{\widetilde{w}^{-1}})$, so that $\mathrm{tr}(T_{\widetilde{u}}\, T_{\widetilde{w}^{-1}})$ is given by Opdam's Theorem 5.2. We conclude the result by Haglund's Theorem 5.3. □

5.3. Example: MRWAC. When $m = 1$, we have the simplification that $u = e$—in this case, starting at the 2023 Minnesota Research Workshop in Algebraic Combinatorics, a group of researchers affiliated with the University of Minnesota and with the University of Texas at Dallas very recently worked out in [**8**] the combinatorics by considering the colored inversions in $\mathsf{inv}_e(\mathsf{u})$ coming from skips.

THEOREM 5.9 ([**8**]). *There is a bijection between maximal distinguished subwords of* $\mathsf{w} = (s_0, s_1, \ldots, s_{n-1})^{n-1}$ *in* $\widetilde{S}_n$ *and vertex-labeled trees with n-vertices.*

This combinatorial result gives a remarkably elaborate new proof of Cayley's formula for the number of such trees. Since the maximal distinguished subwords are naturally in bijection with trees, it makes sense to think of the braid variety $R_{e,t_{n\lambda_{n-1}}}(\mathbb{F})$ as a q-analogue of vertex-labeled trees.

EXAMPLE 5.10. Figure 4 gives an example of the correspondence proved in [**8**].

For $m = 1$, [**8**, Conjecture 8.6] gives a proposed decomposition of $R_{e,t_{n\lambda_{n-1}}}(\mathbb{F})$ into pieces indexed by permutations; this decomposition should match the decomposition of the noncrossing parking spaces from Section 4.3.

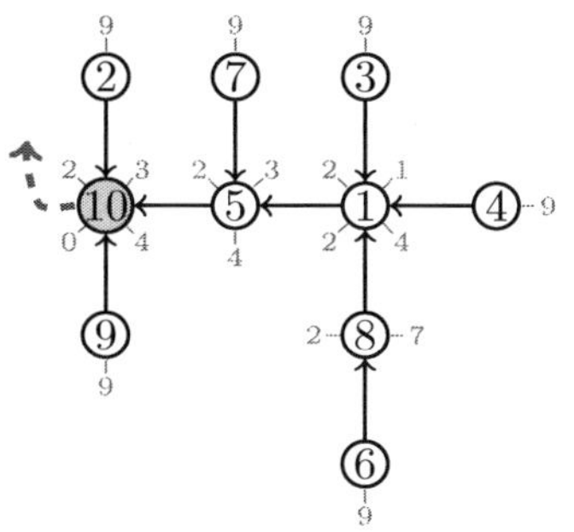

s_0	$((\overline{10}\,2))$	s_2	s_3	s_4	s_5	s_6	s_7	s_8
s_9	$((2\,10))$	s_1	s_2	$((\overline{10}\,5))$	s_4	$((57))$	s_6	s_7
s_8	s_9	s_0	s_1	s_2	s_3	$((\overline{7}5))$	s_5	s_6
$((\overline{5}1))$	s_8	$((13))$	s_0	s_1	s_2	s_3	s_4	s_5
s_6	s_7	$((\overline{3}1))$	$((14))$	s_0	s_1	s_2	s_3	s_4
s_5	s_6	s_7	$((\overline{4}1))$	s_9	s_0	s_1	$((18))$	s_3
s_4	s_5	s_6	s_7	s_8	$((\overline{8}6))$	s_0	s_1	s_2
s_3	s_4	s_5	s_6	s_7	$((68))$	s_9	$((\overline{8}1))$	s_1
$((15))$	s_3	s_4	s_5	$((5\,10))$	s_7	s_8	s_9	$((\overline{10}\,9))$
s_1	s_2	s_3	s_4	s_5	s_6	s_7	s_8	$((9\,10))$

FIGURE 4. Figure taken from [**8**]. *Left:* a cyclically-embedded vertex-labeled tree. *Right:* the corresponding maximal distinguished subword.

5.4. Problems: affine Weyl groups.

PROBLEM 5.11. *Extend the combinatorics of* [**8**] *to* $\mathcal{M}_{t_{(m-1)\lambda_1}, t_{(m(n-1)+1)\lambda_{n-1}}}$.

It appears that nice enumerations extend to other fundamental weights beyond λ_{n-1}. It should be possible evaluate the sum over the Kostant partitions using inductive methods similar to those Haglund used to establish Theorem 5.3. We propose a very conjectural extension of Theorem 4.1 to rational parking functions.

CONJECTURE 5.12. *Let n and m be relatively prime, and define the element*

$$c_{nm} = (s_{m-n+1}s_{m-n+2}\cdots s_{m-1}s_0 s_{m-n}s_{m-n-1}\cdots s_1)^n \in \widetilde{S}_m.$$

Then

$$|R_{e,c_{nm}}(\mathbb{F})| = (q-1)^{n+m-1}[m]^{n-1}.$$

This agrees with the computatations in Section 5.2 for $(n,m) = (n, n+1)$, since it is easily seen that $w_{n,n+1}$ produces a braid variety isomorphic to the one given by $t_{n\lambda_{n-1}}$. We have confirmed Conjecture 5.12 by computer for $(n,m) \in \{(3,5),(3,8),(4,7),(5,7)\}$; note that Opdam's formula does not apply because we don't get cancellation of the non-translation elements in this more general case.

The following conjecture is a restatement of a conjecture of Armstrong, Garsia, Haglund, Rhoades, and Sagan [**3**], originally stated in the language of Kostant partitions.

CONJECTURE 5.13 ([**3**, Conjecture 7.1]). *Fix $\widetilde{W} = \widetilde{S}_n$. Let $\lambda \in Q^+$ satisfy $\lambda = \sum_{i=1}^{n-1} a_i\alpha_i = t_{\lambda_+} - t_{\lambda_-}$ with $a_1 > a_2 > \cdots > a_{n-1} \geq a_n = 0$. Then*

$$\left|R_{t_{\lambda_-}, t_{\lambda_+}}(\mathbb{F})\right| = q^{(\ell(t_{\lambda_+})-\ell(t_{\lambda_-}))/2}(q-1)^{2n}\prod_{i=1}^{n-1}[(i+1)a_i - ia_{i+1}].$$

While we can express the left-hand side of the conjecture as a sum of Kostant partitions, the form of the right-hand side strongly suggests that there is another decomposition of the variety available. It would be interesting to extend the previous conjecture to other affine Weyl groups, and other affine Weyl group elements other than translations.

Write f_μ for the number of standard Young tableaux of shape μ.

CONJECTURE 5.14. *Fix $\widetilde{W} = \widetilde{S}_{2n}$ and take $2\lambda_n \in Q^+$. Then*

$$\left|R_{1,t_{2\lambda_n}}(\mathbb{F})\right| = \sum_{(a_\alpha)\in K(2\lambda_n)} \prod_{\substack{\alpha\in\Phi^+\\ a_\alpha>0}} [a_\alpha]_q = \sum_{\mu\vdash n} f_\mu^2 q^{2\left(c(\mu)+\binom{n}{2}\right)}.$$

We have confirmed Conjecture 5.14 up to $n = 5$. To use Opdam's formula, we need to compute the sum over $K(2\lambda_n)$—the number of Kostant partitions in $|K(2\lambda_n)|$ starts $1, 5, 86, 4274, 550919, \ldots$, and does not appear in the OEIS.

EXAMPLE 5.15. The six maximal Deodhar words for $\widetilde{S}_5$ and $\mathcal{M}_{e,2t_{\lambda_3}}$ are illustrated in Figure 5. A reduced word for $2t_{\lambda_3}$ is $[s_0, s_1, s_2, s_5, s_0, s_1, s_4, s_5, s_0, s_3, s_4, s_5, s_2, s_3, s_4, s_1, s_2, s_3]$, which we have split into two 3×3 squares—the left square containing the first nine letters and the right square containing the remaining nine letters. If one interprets the shaded letters as '0' and the unshaded letters as '1', then each pair of 3×3 squares appears to be the permutation matrix of a permutation, along with the permutation matrix of its inverse.

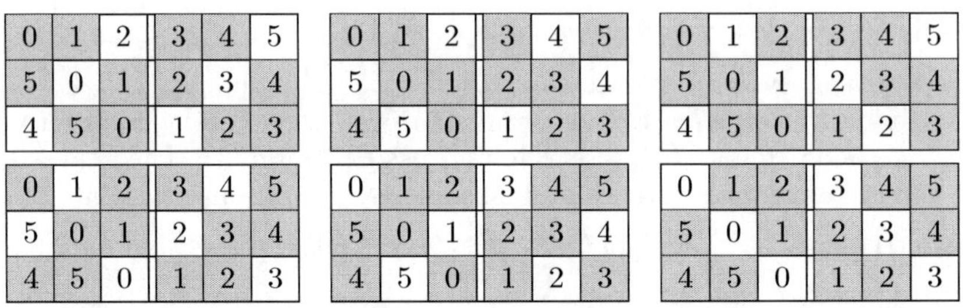

FIGURE 5. The six maximal Deodhar words for $\widetilde{S}_5$ and $\mathcal{M}_{e,2t_{\lambda_3}}$ appear to encode permutations (see also [**8**, Conjecture 8.6]).

6. Techniques and problems in Kac-Moody groups

In this section, we fix a general Kac–Moody Lie group G and describe general techniques that apply in this situation.

6.1. R-polynomials. Let W be the Weyl group of a Kac-Moody Lie group. There are two general techniques that persist at this level of generality. The first exploits a recursive structure—coming from the Hecke algebra—on the set of distinguished subwords to construct polynomials that count $\left|R^{(v)}_{u,\mathsf{w}}(\mathbb{F}_q)\right|$: setting $R^{(v)}_{u,\varnothing}(q) := R_{u,\varnothing}(q)$, for any word w and $s \in S$ we have

$$R^{(v)}_{u,\mathsf{ws}}(q) = \begin{cases} R^{(v)}_{us,\mathsf{w}}(q) & \text{if } vus < vu \\ qR^{(v)}_{us,\mathsf{w}}(q) + (q-1)R^{(v)}_{u,\mathsf{w}}(q) & \text{if } vus > vu. \end{cases} \tag{6.1}$$

THEOREM 6.1. *For arbitrary Weyl groups W, we have $R^{(v)}_{u,\mathsf{w}}(q) = \left|R^{(v)}_{u,\mathsf{w}}(\mathbb{F}_q)\right|$.*

Braid varieties arising from powers of Coxeter elements are likely varieties of interest for general W, and serve as analogues of rational noncrossing Catalan objects for general Coxeter groups.

6.2. Cluster varieties. There is a cluster-theoretic approach for computing mixed Hodge decompositions. For particular choices of u and w, it is possible to put a (locally acyclic) cluster structure on $R^{(v)}_{u,\mathsf{w}}(\mathbb{C})$. With this cluster structure, we may invoke technology of Lam and Speyer [**34, 39**] to recursively compute $|R^{(v)}_{u,\mathsf{w}}(\mathbb{F}_q)|$. This requires finding an artful sequence of mutations to isolate a separating edge, thereby enabling recursive arguments. This technique succeeds in small examples, but we require more systematic approaches using the specific properties of the quivers corresponding to the braid varieties of interest to handle large cases.

7. Cohomology and mixed Hodge structures

This section is speculative. Working over $\mathbb{C}$ now (rather than $\mathbb{F}$), we have the *Deligne splitting* of cohomology

$$H^k(R^{(v)}_{u,\mathsf{w}}(\mathbb{C})) = \bigoplus_{p=0}^{k}\bigoplus_{q=0}^{k} H^{k,(p,q)}(R^{(v)}_{u,\mathsf{w}}(\mathbb{C})).$$

Other than the long exact sequence for relative cohomology, we have only one tool. For particular choices of u and w—even, for example, in affine type—it is possible to put a (locally acyclic) cluster structure on $R^{(v)}_{u,\mathsf{w}}(\mathbb{C})$, from which we may conclude that $H^{k,(p,q)}(R^{(v)}_{u,\mathsf{w}}(\mathbb{C})) = 0$ for $p \neq q$. With this cluster structure, we may invoke technology of Lam and Speyer [**46**], building on work of Muller [**39**]. As in the previous section, the approach is to find an artful sequence of mutations to isolate a separating edge. Following [**19**], when $R^{(v)}_{u,\mathsf{w}}(\mathbb{C})$ has dimension d we define the *mixed Hodge polynomial*

$$\mathcal{P}(R^{(v)}_{u,\mathsf{w}}(\mathbb{C}); q, t) := \sum_{k,p\in\mathbb{Z}} q^{p-k/2} t^{(d-k)/2} \dim(H^{k,(p,p)}(R^{(v)}_{u,\mathsf{w}}(\mathbb{C}))).$$

PROBLEM 7.1. *Compute the mixed Hodge polynomial $\mathcal{P}(R^{(v)}_{u,\mathsf{w}}(\mathbb{C}); q, t)$ in all cases where the point count $|R^{(v)}_{u,\mathsf{w}}(\mathbb{F}_q)|$ has been established.*

This is already fascinating in the case of the Coxeter–Catalan varieties R_{e,c^p} and the parking varieties $\bigcup_{v\in W} R^{(v)}_{e,c^p}$—here, the mixed Hodge polynomials should compute the rational (W, q, t)-analogues of Catalan numbers and parking functions [**23**].

EXAMPLE 7.2. For $W = \mathfrak{S}_3$ and $\mathsf{w} = (s_1, s_2, s_1, s_2, s_1, s_2, s_1, s_2)$, we have six varieties in $\bigsqcup_{v\in W} R^{(v)}_{e,c^p}$ (one for each element v of W). Using the tables from [**34**], Lam has confirmed that, as cluster varieties, they are of type E_6, A_4 (twice), A_2 (twice), and A_0, giving the sum

$$(q^3 + q^2 t + qt + tq^2 + t^3) + 2(q^2 + qt + t^2) + 2(q + t) + 1,$$

which is the usual parking q, t-analogue of 4^2.

PROBLEM 7.3. *Find combinatorial statistics* stat_q *and* stat_t *on all distinguished subwords $\mathcal{D}^{(v)}_{u,\mathsf{w}}$ so that*

$$\mathcal{P}(R^{(v)}_{u,\mathsf{w}}(\mathbb{C}); q, t) = \sum_{\mathbf{u}\in\mathcal{D}^{(v)}_{u,\mathsf{w}}} [(q-1)(t-1)]^{\mathrm{e}_{\mathbf{u}}/2} q^{\mathrm{stat}_q(\mathbf{u})} t^{\mathrm{stat}_t(\mathbf{u})}.$$

When $\mathcal{P}(R_{u,\mathrm{w}}^{(v)}(\mathbb{C});q,t)$ *is a positive* q,t*-polynomial, find combinatorial statistics* stat'_q *and* stat'_t *on all maximal distinguished subwords* $\mathcal{M}_{u,\mathrm{w}}^{(v)}$ *so that*

$$\mathcal{P}(R_{u,\mathrm{w}}^{(v)}(\mathbb{C});q,t) = \sum_{\mathbf{u}\in\mathcal{M}_{u,\mathrm{w}}^{(v)}} q^{\mathrm{stat}'_q(\mathbf{u})}t^{\mathrm{stat}'_t(\mathbf{u})}.$$

Acknowledgments

It is an honor to be asked to contribute to this conference proceedings—especially as I am neither the most distinguished *Nathan*, nor the most distinguished *Williams* in algebraic combinatorics. Those in attendance for the final talk at Open Problems in Algebraic Combinatorics 2022 will observe that the content of this note does not exactly match the content of that talk (slides are available at [**55**], a recording is available for streaming [**56**]).

I thank my coauthors Thomas Lam, Pavel Galashin, and Minh-Tâm Trinh for their very patient explanations and generous insights, as well as the three Dennises at the University of Minnesota (that is, Reiner, Stanton, and White) for their mentorship. I thank Carlos Arreche, Colin Defant, and Weston Miller for their helpful comments on drafts of this note. I thank Ivan Marin for informative conversations on Hecke algebras of complex reflection groups.

References

[1] Drew Armstrong, *Generalized noncrossing partitions and combinatorics of Coxeter groups*, Mem. Amer. Math. Soc. **202** (2009), no. 949, x+159, DOI 10.1090/S0065-9266-09-00565-1. MR2561274

[2] Drew Armstrong, *Tesler matrices*, Talk slides at Bruce Saganfest, 2014.

[3] D. Armstrong, A. Garsia, J. Haglund, B. Rhoades, and B. Sagan, *Combinatorics of Tesler matrices in the theory of parking functions and diagonal harmonics*, J. Comb. **3** (2012), no. 3, 451–494, DOI 10.4310/JOC.2012.v3.n3.a7. MR3029443

[4] Drew Armstrong, Victor Reiner, and Brendon Rhoades, *Parking spaces*, Adv. Math. **269** (2015), 647–706, DOI 10.1016/j.aim.2014.10.012. MR3281144

[5] Drew Armstrong, Brendon Rhoades, and Nathan Williams, *Rational associahedra and noncrossing partitions*, Electron. J. Combin. **20** (2013), no. 3, Paper 54, 27, DOI 10.37236/3432. MR3118962

[6] Drew Armstrong, Christian Stump, and Hugh Thomas, *A uniform bijection between nonnesting and noncrossing partitions*, Trans. Amer. Math. Soc. **365** (2013), no. 8, 4121–4151, DOI 10.1090/S0002-9947-2013-05729-7. MR3055691

[7] Christos A. Athanasiadis, *On noncrossing and nonnesting partitions for classical reflection groups*, Electron. J. Combin. **5** (1998), Research Paper 42, 16, DOI 10.37236/1380. MR1644234

[8] Esther Banaian, Anh Trong Nam Hoang, Elizabeth Kelley, Weston Miller, Jason Stack, Carolyn Stephen, and Nathan Williams, *An elaborate new proof of Cayley's formula*, arXiv:2402.07798, 2024.

[9] David Bessis, *Finite complex reflection arrangements are* $K(\pi,1)$, Ann. of Math. (2) **181** (2015), no. 3, 809–904, DOI 10.4007/annals.2015.181.3.1. MR3296817

[10] David Bessis and Victor Reiner, *Cyclic sieving of noncrossing partitions for complex reflection groups*, Ann. Comb. **15** (2011), no. 2, 197–222, DOI 10.1007/s00026-011-0090-9. MR2813511

[11] Michelle Bodnar, *Rational noncrossing partitions for all coprime pairs*, J. Comb. **10** (2019), no. 2, 365–407, DOI 10.4310/JOC.2019.v10.n2.a10. MR3912220

[12] Michel Broué, *Introduction to complex reflection groups and their braid groups*, Lecture Notes in Mathematics, vol. 1988, Springer-Verlag, Berlin, 2010, DOI 10.1007/978-3-642-11175-4. MR2590895

[13] Michel Broué, Gunter Malle, and Raphaël Rouquier, *Complex reflection groups, braid groups, Hecke algebras*, J. Reine Angew. Math. **500** (1998), 127–190. MR1637497

[14] Vinay V. Deodhar, *On some geometric aspects of Bruhat orderings. I. A finer decomposition of Bruhat cells*, Invent. Math. **79** (1985), no. 3, 499–511, DOI 10.1007/BF01388520. MR782232

[15] Benjamin Dequêne, Gabriel Frieden, Alessandro Iraci, Florian Schreier-Aigner, Hugh Thomas, and Nathan Williams, *Charmed roots and the Kroweras complement*, `arXiv:2212.14831`, 2022.

[16] Theo Douvropoulos, Joel Brewster Lewis, and Alejandro H. Morales, *Hurwitz numbers for reflection groups II: Parabolic quasi-Coxeter elements*, J. Algebra **641** (2024), 648–715, DOI 10.1016/j.jalgebra.2023.11.015. MR4678298

[17] Theo Douvropoulos, *On enumerating factorizations in reflection groups*, Algebr. Comb. **6** (2023), no. 2, 359–385. MR4591591

[18] Sergey Fomin and Andrei Zelevinsky, *Y-systems and generalized associahedra*, Ann. of Math. (2) **158** (2003), no. 3, 977–1018, DOI 10.4007/annals.2003.158.977. MR2031858

[19] Pavel Galashin and Thomas Lam, *Positroids, knots, and q,t-Catalan numbers*, Sém. Lothar. Combin. **85B** (2021), Art. 54, 12. MR4311935

[20] Pavel Galashin, *Positroid Catalan numbers*, `arXiv:2104.05701`, 2021.

[21] Pavel Galashin, Thomas Lam, Minh-Tâm Quang Trinh, and Nathan Williams, *Rational noncrossing Coxeter-Catalan combinatorics*, `arXiv:2208.00121`, 2022.

[22] Owen Garnier, *Regular theory in complex braid groups*, J. Algebra **620** (2023), 534–557, DOI 10.1016/j.jalgebra.2022.12.031. MR4539101

[23] Iain G. Gordon and Stephen Griffeth, *Catalan numbers for complex reflection groups*, Amer. J. Math. **134** (2012), no. 6, 1491–1502, DOI 10.1353/ajm.2012.0047. MR2999286

[24] Iain G. Gordon and Stephen Griffeth, *Catalan numbers for complex reflection groups*, Amer. J. Math. **134** (2012), no. 6, 1491–1502, DOI 10.1353/ajm.2012.0047. MR2999286

[25] E. Gorsky, *q,t-Catalan numbers and knot homology*, Zeta functions in algebra and geometry, Contemp. Math., vol. 566, Amer. Math. Soc., Providence, RI, 2012, pp. 213–232, DOI 10.1090/conm/566/11222. MR2858925

[26] Eugene Gorsky and Andrei Neguţ, *Refined knot invariants and Hilbert schemes* (English, with English and French summaries), J. Math. Pures Appl. (9) **104** (2015), no. 3, 403–435, DOI 10.1016/j.matpur.2015.03.003. MR3383172

[27] J. Haglund, *A polynomial expression for the Hilbert series of the quotient ring of diagonal coinvariants*, Adv. Math. **227** (2011), no. 5, 2092–2106, DOI 10.1016/j.aim.2011.04.013. MR2803796

[28] James Haglund, *Catalan paths and q,t-enumeration*, Handbook of Enumerative Combinatorics (2015), 679–751.

[29] Mark D. Haiman, *Conjectures on the quotient ring by diagonal invariants*, J. Algebraic Combin. **3** (1994), no. 1, 17–76, DOI 10.1023/A:1022450120589. MR1256101

[30] Vaughan Jones, *The Jones polynomial for dummies*, lecture notes, 2014.

[31] V. F. R. Jones, *Hecke algebra representations of braid groups and link polynomials*, Ann. of Math. (2) **126** (1987), no. 2, 335–388, DOI 10.2307/1971403. MR908150

[32] David Kazhdan and George Lusztig, *Representations of Coxeter groups and Hecke algebras*, Invent. Math. **53** (1979), no. 2, 165–184, DOI 10.1007/BF01390031. MR560412

[33] Abel Lacabanne, *Fourier matrices for $G(d,1,n)$ from quantum general linear groups*, J. Algebra **586** (2021), 433–466, DOI 10.1016/j.jalgebra.2021.06.034. MR4291489

[34] Thomas Lam and David E. Speyer, *Cohomology of cluster varieties I: Locally acyclic case*, Algebra Number Theory **16** (2022), no. 1, 179–230, DOI 10.2140/ant.2022.16.179. MR4384567

[35] Trafim Lasy, *Special Markov traces and Gomi's formula for reflection groups*, PhD thesis, University Denis Diderot, 2012.

[36] Eon-Kyung Lee and Sang-Jin Lee, *Periodic elements in Garside groups*, J. Pure Appl. Algebra **215** (2011), no. 10, 2295–2314, DOI 10.1016/j.jpaa.2010.12.011. MR2793937

[37] R. J. Marsh and K. Rietsch, *Parametrizations of flag varieties*, Represent. Theory **8** (2004), 212–242, DOI 10.1090/S1088-4165-04-00230-4. MR2058727

[38] Weston Miller, *Rational Catalan numbers for complex reflection groups*, `arXiv:2310.12354`, 2023.

[39] Greg Muller, *$\mathcal{A} = \mathcal{U}$ for locally acyclic cluster algebras*, SIGMA Symmetry Integrability Geom. Methods Appl. **10** (2014), Paper 094, 8, DOI 10.3842/SIGMA.2014.094. MR3261850

[40] Jason O'Neill, *On the poset and asymptotics of Tesler matrices*, Electron. J. Combin. **25** (2018), no. 2, Paper No. 2.4, 27, DOI 10.37236/6877. MR3799422

[41] Eric M. Opdam, *A generating function for the trace of the Iwahori-Hecke algebra*, Studies in memory of Issai Schur (Chevaleret/Rehovot, 2000), Progr. Math., vol. 210, Birkhäuser Boston, Boston, MA, 2003, pp. 301–323. MR1985730

[42] Igor Pak, *History of Catalan numbers*, arXiv:1408.5711, 2014.

[43] Dmitri I. Panyushev, *On orbits of antichains of positive roots*, European J. Combin. **30** (2009), no. 2, 586–594, DOI 10.1016/j.ejc.2008.03.009. MR2489252

[44] Victor Reiner, *Non-crossing partitions for classical reflection groups*, Discrete Math. **177** (1997), no. 1-3, 195–222, DOI 10.1016/S0012-365X(96)00365-2. MR1483446

[45] G. C. Shephard and J. A. Todd, *Finite unitary reflection groups*, Canad. J. Math. **6** (1954), 274–304, DOI 10.4153/cjm-1954-028-3. MR59914

[46] David Speyer and Hugh Thomas, *Acyclic cluster algebras revisited*, Algebras, quivers and representations, Abel Symp., vol. 8, Springer, Heidelberg, 2013, pp. 275–298, DOI 10.1007/978-3-642-39485-0_12. MR3183889

[47] Richard P. Stanley, *Catalan numbers*, Cambridge University Press, New York, 2015, DOI 10.1017/CBO9781139871495. MR3467982

[48] Christian Stump, Hugh Thomas, and Nathan Williams, *Cataland: Why the fuss?* (English, with English and French summaries), 28th International Conference on Formal Power Series and Algebraic Combinatorics (FPSAC 2016), Discrete Math. Theor. Comput. Sci. Proc., BC, Assoc. Discrete Math. Theor. Comput. Sci., Nancy, [2016] ©2016, pp. 1123–1134. MR4111284

[49] Ruedi Suter, *The number of lattice points in alcoves and the exponents of the finite Weyl groups*, Math. Comp. **67** (1998), no. 222, 751–758, DOI 10.1090/S0025-5718-98-00919-3. MR1443124

[50] Minh-Tam Trinh, *Algebraic Braids and Geometric Representation Theory*, ProQuest LLC, Ann Arbor, MI, 2020. Thesis (Ph.D.)–The University of Chicago. MR4132362

[51] Minh-Tâm Quang Trinh, *From the Hecke category to the unipotent locus*, arXiv:2106.07444v2, 2021.

[52] Ben Webster and Milen Yakimov, *A Deodhar-type stratification on the double flag variety*, Transform. Groups **12** (2007), no. 4, 769–785, DOI 10.1007/s00031-007-0061-8. MR2365444

[53] Nathan Williams, *Cataland*, ProQuest LLC, Ann Arbor, MI, 2013. Thesis (Ph.D.)–University of Minnesota. MR3193161

[54] Nathan Williams, *Bijactions in Cataland* (English, with English and French summaries), 26th International Conference on Formal Power Series and Algebraic Combinatorics (FPSAC 2014), Discrete Math. Theor. Comput. Sci. Proc., AT, Assoc. Discrete Math. Theor. Comput. Sci., Nancy, 2014, pp. 597–608. MR3466406

[55] Nathan Williams, *Rational noncrossing Coxeter-Catalan combinatorics*, http://www.samuelfhopkins.com/OPAC/files/slides/n_williams.pdf, 2022, Slides; accessed: November 29, 2022.

[56] Nathan Williams, *Rational noncrossing Coxeter-Catalan combinatorics*, https://youtu.be/17BVk4WiX9U, 2022, Recorded talk; accessed: November 29, 2022.

University of Texas at Dallas, Dallas, Texas
Email address: nathan.williams1@utdallas.edu

Proceedings of Symposia in Pure Mathematics
Volume **110**, 2024
https://doi.org/10.1090/pspum/110/02015

Neighborly spheres and transversal numbers

Isabella Novik and Hailun Zheng

1. Introduction

The goal of this paper is to advertise several old and new problems related to the number of simplicial spheres (Section 3), the number of neighborly simplicial spheres (Section 4), and the number of centrally symmetric simplicial spheres that are cs-neighborly (Section 5). Neighborly spheres provide maximizers for the upper bound type problems. They are also natural candidates for having large transversal numbers. Thus, in the second part of the paper (Section 6), we discuss problems as well as some results aimed at understanding transversal numbers of simplicial spheres. In particular, we estimate transversal ratios of some of the neighborly spheres presented in Sections 4 and 5 and conclude that for $k \geq 2$, there exist families of simplicial $2k$-spheres such that their transversal ratios are at least $2/5 - o(1)$ as the number of vertices tends to infinity. (The previously known bound was $1/(k+1)$.)

The tools developed and the results proved in the last decade suggest that significant progress on some of these problems might be just around the corner. In short, we believe that these problems deserve to be looked at right now!

2. Basic definitions

We assume that the reader is familiar with basics of convex polytopes and simplicial complexes. However, for completeness sake, below we review some of the definitions, especially those related to simplicial complexes. For all undefined terminology related to polytopes, the reader is referred to Ziegler's book [**Zie95**].

In this paper we only consider finite simplicial complexes. A *simplicial complex* Δ on a finite vertex set V is a collection of subsets of V that is closed under inclusion, i.e., if $F \in \Delta$ and $G \subset F$, then $G \in \Delta$. We also usually assume that every singleton is an element of Δ, that is, $\{v\} \in \Delta$ for all $v \in V$. The elements of

---

2020 *Mathematics Subject Classification.* Primary 05E45, 52B05, 52B70, 57Q15; Secondary 05C65.

Key words and phrases. simplicial spheres, neighborly spheres, centrally symmetric spheres, transversal number.

The research of the first author was partially supported by NSF grant DMS-1953815 and by Robert R. & Elaine F. Phelps Professorship in Mathematics.

The research of the second author was partially supported by a postdoctoral fellowship from ERC grant 716424 - CASe.

V are called *vertices* while the elements of Δ are called *faces*. For a vertex v, we abuse notation and write $v \in \Delta$ instead of $\{v\} \in \Delta$.

We say that $F \in \Delta$ is an *i-face* or an *i-dimensional face* if $|F| = i+1$. The *dimension* of Δ is defined as $\dim \Delta := \max\{\dim F : F \in \Delta\}$. The number of i-faces is denoted by $f_i(\Delta)$. We refer to 0-faces as *vertices*, 1-faces as *edges*, and the maximal under inclusion faces as *facets*. We say that Δ is *pure* if all facets of Δ have the same dimension. For instance, the complex whose faces are the empty set, the vertices $v_1, \dots, v_n$, and the edges $\{v_i, v_{i+1}\}$ for all $1 \leq i \leq n-1$, is a pure 1-dimensional complex known as a *path*. The notation we adopt for this complex is $(v_1, v_2, \dots, v_n)$.

Two examples of simplicial complexes worth mentioning are the *simplex on* V, denoted $\overline{V}$, and the *boundary complex* of the simplex $\overline{V}$, denoted $\partial\overline{V}$. The former complex consists of *all* subsets of V while the latter complex consists of all subsets of V but V itself.

An important operation on simplicial complexes is that of joins: If Δ and Γ are simplicial complexes on disjoint vertex sets V and V', then the *join* of Δ and Γ is the simplicial complex $\Delta * \Gamma = \{\sigma \cup \tau \ : \ \sigma \in \Delta \text{ and } \tau \in \Gamma\}$ with vertex set $V \cup V'$. A special case is the *cone* over Δ with apex v, $\Delta * v$, defined as the join of Δ and the simplex $\overline{\{v\}}$.

A *polytope* is the convex hull of finitely many points in a Euclidean space. One example is given by a (geometric) simplex: the convex hull of affinely independent points. A *face* of a polytope P is the intersection of P with any supporting hyperplane. A polytope P is *simplicial* if all facets of P are simplices.

There is a natural way to associate with every simplicial complex Δ its *geometric realization*, $\|\Delta\|$: it is built out of geometric simplices in a way that every two simplices intersect along a common (possibly empty) face and the collection of vertex sets of faces of $\|\Delta\|$ is Δ. We say that Δ is a *simplicial* $(d-1)$*-sphere* if $\|\Delta\|$ is homeomorphic to a $(d-1)$-sphere; we also say that Δ is a simplicial d-ball if $\|\Delta\|$ is homeomorphic to a d-ball. For instance, $\overline{V}$ is a simplicial ball, while $\partial\overline{V}$ as well as the boundary complex of any simplicial polytope is a simplicial sphere.

Many balls and spheres we discuss in this paper are *piecewise-linear*; we call them *PL balls* and *PL spheres*, for short. A PL d-ball is a simplicial complex PL homeomorphic to a (geometric) d-simplex. Similarly, a PL $(d-1)$-sphere is a simplicial complex PL homeomorphic to the boundary complex of a d-simplex. We refer the reader to [**Hud69**] and [**Lic99**] for some basics on PL topology.

A simplicial $(d-1)$-sphere Δ is called k-neighborly if every k vertices of Δ form the vertex set of a face. The boundary complex of a d-simplex is d-neighborly. On the other hand, it is not hard to see (for instance, from Dehn-Sommerville relations [**Kle64, Som27**]) that a simplicial $(d-1)$-sphere with at least $d+2$ vertices cannot be more than $\lfloor d/2 \rfloor$-neighborly. Furthermore, $\lfloor d/2 \rfloor$-neighborly simplicial $(d-1)$-spheres with arbitrarily many vertices do exist. One famous example is given by the boundary complex of the cyclic polytope whose construction is outlined below.

The *moment curve* $M = M_d : \mathbb{R} \to \mathbb{R}^d$ is defined by $M(t) := (t, t^2, \dots, t^d)$. Let $t_1 < t_2 < \cdots < t_n$ be any $n \geq d+1$ distinct real numbers. The *cyclic polytope*, $C(d,n)$ is the convex hull of the points $M(t_1), \dots, M(t_n)$. The cyclic polytope has several amazing properties [**Gal63, Zie95**]: it is a simplicial d-polytope with n vertices; its combinatorial type is independent of the choice of $t_1, \dots, t_n$; and it is $\lfloor d/2 \rfloor$-neighborly. That the combinatorial type of $C(d,n)$ is independent of the choice of $t_1, \ldots, t_n$ is a consequence of a beautiful result by Gale [**Gal63**] that provides a complete characterization of the facets of $C(d,n)$; it is known as the *Gale evenness condition.*

The interest in $\lfloor d/2 \rfloor$-neighborly simplicial $(d-1)$-spheres comes in part from the celebrated Upper Bound Theorem due to Stanley [**Sta75**]. It asserts that in the class of all simplicial $(d-1)$-spheres with n vertices, any $\lfloor d/2 \rfloor$-neighborly sphere (e.g., the boundary complex of the cyclic polytope) simultaneously maximizes all the face numbers.

Another polytope we will encounter in this paper is the *cross-polytope* $\mathcal{C}_d^*$ defined as the convex hull of $\{\pm e_1, \pm e_2, \ldots, \pm e_d\}$, where $e_1, \ldots, e_d$ are the endpoints of the standard basis in $\mathbb{R}^d$. As an abstract simplicial complex, the boundary complex $\partial \mathcal{C}_d^*$ of $\mathcal{C}_d^*$ is the collection of all subsets of $\{\pm 1, \ldots, \pm d\}$ that contain at most one vertex from each pair $\{\pm i\}$.

We close this section with a quick review of the asymptotic notations. Let $f(n)$ and $g(n)$ be two functions from the set of non-negative integers to itself. We say that $g(n) = O(f(n))$ if there exists a positive constant C such that $g(n) \leq Cf(n)$ for all sufficiently large n. Similarly, $g(n) = \Omega(f(n)$, if there exists a positive constant C' such that $g(n) \geq C'f(n)$ for all sufficiently large n. Finally, we say that $g(n) = \Theta(f(n))$ if both $g(n) = O(f(n))$ and $g(n) = \Omega(f(n))$ hold.

3. How many simplicial spheres are there?

Denote by $s(d,n)$ the number of simplicial $(d-1)$-spheres with n *labeled* vertices. How large is $s(d,n)$?

We first discuss the case of $d = 3$. It is well known that $s(3,n) = 2^{\Theta(n \log n)}$. In fact, it is even known that the number of *combinatorial types* of simplicial 2-spheres with n vertices is asymptotically exponential in n. This follows from works of Tutte [**Tut62, Tut80**], Brown [**Bro63**], and Richmond and Wormald [**RW82**] along with Steinitz' theorem [**Zie95**, Chapter 4] which implies that all simplicial (and even polyhedral) 2-spheres are boundary complexes of 3-polytopes.

Assume now that $d \geq 4$. The currently best known bounds are:

THEOREM 3.1. *For $d \geq 4$,*

$$2^{\Omega(n^{\lfloor d/2 \rfloor})} \leq s(d,n) \leq 2^{O(n^{\lfloor d/2 \rfloor} \log n)}.$$

For comparison, we should mention that there are much fewer polytopes. By works of Goodman and Pollack [**GP86**] and Alon [**Alo86**], for $d \geq 4$, the number of d-polytopes (simplicial and non-simplicial) with n labeled (or unlabeled) vertices is $2^{\Theta(n \log n)}$.

Let's take a closer look at Theorem 3.1.[1] The upper bound on $s(d,n)$ essentially follows from Stanley's Upper Bound Theorem, see [**Kal88**, Section 4.2]. The lower bound for odd values of d is due to Kalai [**Kal88**]. In fact, Kalai proved that for *all* $d \geq 5$ (even and odd), $s(d,n) \geq 2^{\Omega(n^{\lfloor \frac{d-1}{2} \rfloor})}$. For $d = 4$, Pfeifle and Ziegler [**PZ04**] showed that $s(4,n) \geq 2^{\Omega(n^{5/4})}$. The lower bound of Theorem 3.1 for even values of d is due to Nevo, Santos, and Wilson [**NSW16**]. As the reader will notice, the

[1] Note that although we count the number of spheres with *labeled* vertices (i.e., we count complexes up to equality), since $n! = 2^{O(n \log n)}$, the bounds of Theorem 3.1 also apply to the number of combinatorial types of simplicial $(d-1)$-spheres on n vertices, i.e., to the number of spheres counted up to isomorphism.

exponents in the lower and upper bounds differ by the factor of $\log n$. This brings us to

PROBLEM 3.2. Find tighter bounds on $\log s(n,d)$. Is $\log s(d,n) = \Theta(n^{\lfloor d/2 \rfloor})$ or is it closer to $\Theta(n^{\lfloor d/2 \rfloor} \log n))$?

The lower bounds in Theorem 3.1 are obtained by ingenious constructions. Kalai's construction of many $(2k-2)$-spheres [**Kal88**] (for $k \geq 3$) is based on squeezed balls — certain full-dimensional subcomplexes of the boundary complex of the cyclic polytope $C(2k,n)$. The boundary complexes of squeezed balls, called *squeezed spheres*, provide us with $2^{\Omega(n^{k-1})}$ simplicial $(2k-2)$-spheres with n vertices. One of constructions by Nevo, Santos, and Wilson [**NSW16**] (at least in the 3-dimensional case) is also based on the cyclic polytope. Furthermore, all the constructions in [**Kal88, NSW16, PZ04**] are PL spheres, and, as was proved by Lee [**Lee00**], all of Kalai's squeezed spheres are even shellable. Since there are many non-shellable (and even non-constructible) simplicial spheres [**HZ00**], there seems to be large unexplored terrain for potential additional constructions. One may even speculate that the value of $s(d,n)$ is closer to the upper bound in Problem 3.2. To this end, it is worth noting that there are constructions of simplicial $(d-1)$-spheres, $d \geq 6$, that are *not PL* and have only $d+12$ vertices [**BL00**].

A related problem, where much less is known, is

PROBLEM 3.3. For $d \geq 4$, how many combinatorial types of simplicial $(d-1)$-spheres with N facets are there? Is the number of such spheres at most exponential in N, that is, is it bounded from above by $(C_d)^N$, where C_d is some constant that depends only on d?

This problem, for $d = 4$, was raised by Ambjørn, Durhuus, and Jonsson [**ADJ91**] and advertised by Gromov [**Gro00**, pp. 156–157]. For 2-spheres, the answer to the second question of Problem 3.3 is indeed affirmative, see [**Tut62, Tut80**]. For $d \geq 4$, Benedetti and Ziegler [**BZ11**] proved that the number of shellable (and, in fact, even the number of locally constructible) $(d-1)$-spheres with N facets is at most exponential in N. Some additional results on exponential growth were recently established by Adiprasito and Benedetti [**AB20**]. However, the general case of Problem 3.3 remains wide open. It is also worth mentioning that if Δ is a simplicial $(d-1)$-sphere with n vertices, then $f_{d-1}(\Delta) \leq O(n^{\lfloor d/2 \rfloor})$. (This is a simple consequence of the Upper Bound Theorem [**Sta75**].) Thus, an affirmative answer to Problem 3.3 would imply that

$$s(d,n) \leq n! \cdot \sum_{N=1}^{O(n^{\lfloor d/2 \rfloor})} (C_d)^N = 2^{O(n^{\lfloor d/2 \rfloor})}, \quad \text{and hence, that } s(d,n) = 2^{\Theta(n^{\lfloor d/2 \rfloor})}.$$

4. How many neighborly simplicial spheres are there?

Since $\lfloor d/2 \rfloor$-neighborly simplicial d-polytopes and $(d-1)$-spheres serve as maximizers in the Upper Bound Theorem [**McM70, Sta75**], one might be tempted to think that the property of being $\lfloor d/2 \rfloor$-neighborly is quite rare. This, however, is not the case: Shemer [**She82**] introduced a sewing construction and used it to produce many neighborly polytopes. The combinatorial part of the sewing construction can be roughly described as follows: one starts with a neighborly PL sphere Δ with n vertices, finds in Δ a full-dimensional PL ball B that satisfies certain nice properties, and then produces a new neighborly sphere Δ' with $n+1$ vertices by replacing the ball B with another ball — the cone over the boundary of B, where the cone vertex is a new vertex.[2]

Padrol [**Pad13**] extended and generalized Shemer's technique to produce even more neighborly polytopes: he constructed on the order of $n^{dn/2}$ polytopes of dimension d with n (labeled) vertices that are $\lfloor d/2 \rfloor$-neighborly. Until very recently, this lower bound was the best lower bound not only on the number of neighborly polytopes, but also on the total number of d-polytopes with n (labeled) vertices. The current best lower bound on the total number of polytopes is on the order of n^{dn}, see [**PPS24**].

Let $\mathrm{sn}(d,n)$ be the number of $\lfloor d/2 \rfloor$-neighborly simplicial $(d-1)$-spheres with n *labeled* vertices. Motivated by Shemer's results, Kalai [**Kal88**] proposed the following bold conjecture:

CONJECTURE 4.1. *For all* $d \geq 4$, $\lim_{n\to\infty}(\log \mathrm{sn}(d,n)/\log s(d,n)) = 1$.

The current best lower bound on $\mathrm{sn}(d,n)$ is due to Novik and Zheng [**NZ24**] who proved that for $d \geq 5$,

$$\mathrm{sn}(d,n) \geq 2^{\Omega(n^{\lfloor (d-1)/2 \rfloor})}.$$

The construction of [**NZ24**] relies on Kalai's squeezed balls [**Kal88**]. More precisely, it relies on *differences* of appropriately chosen squeezed balls, called *relative squeezed balls*, see [**NZ24**, Theorem 3.1(1)] and Section 6.1 below for precise conditions. Such balls are $(2k-1)$-dimensional subcomplexes of $\partial C(2k,n)$ — the boundary complex of $C(2k,n)$. The desired $(k-1)$-neighborly $(2k-2)$-spheres are obtained as the boundaries of relative squeezed balls while the desired k-neighborly $(2k-1)$-spheres are obtained by the sewing construction where a relative squeezed ball B is deleted from $\partial C(2k,n)$ and replaced by the cone over the boundary of B.

While Conjecture 4.1 is likely out of reach at the moment, in view of the result of Nevo, Santos, and Wilson [**NSW16**], the following problem might be more accessible:

PROBLEM 4.2. Let $d \geq 4$. Are there $2^{\Omega(n^{\lfloor d/2 \rfloor})}$ simplicial $(d-1)$-spheres on n vertices that are $\lfloor d/2 \rfloor$-neighborly?

5. How many centrally symmetric spheres are there?

A polytope $P \subset \mathbb{R}^d$ is *centrally symmetric* (or *cs* for short) if $P = -P$. For instance, the cross-polytope is cs. Similarly, a simplicial sphere Δ is *centrally symmetric* (or *cs*) if the vertex set of Δ is endowed with a *free involution* α that induces a free involution on the set of all nonempty faces of Δ. In more detail, for every nonempty face $F \in \Delta$, the following holds:

$$\alpha(F) \in \Delta, \quad \alpha(F) \neq F, \quad \text{and } \alpha(\alpha(F)) = F.$$

We call vertices v and $\alpha(v)$ *antipodal*; for brevity we write $-v$ instead of $\alpha(v)$.

Note that if Δ is cs, then v and $-v$ never form an edge (as otherwise such an edge would be fixed by the involution). This leads to an adjusted notion of

[2]Since Shemer's goal was to produce neighborly *polytopes* rather than spheres, he also needed to ensure that this ball B can be chosen in a way that allows to place the new vertex beyond all facets in B and beneath the rest of facets.

neighborliness for cs simplicial spheres: a cs simplicial sphere Δ is called *cs-k-neighborly* if every subset of k vertices of Δ that does not contain a pair of antipodal vertices is the vertex set of a face.

While cs-neighborliness of cs polytopes is very restricted (for example, a cs d-polytope with 2^d or more vertices cannot be even cs-2-neighborly [**LN06**, **Nov18**]), cs $(d-1)$-spheres that are cs-$\lfloor d/2 \rfloor$-neighborly and have an arbitrary large number of vertices do exist. The following result was proved by Jockusch [**Joc95**] for $d = 4$ and by Novik and Zheng [**NZ20**] for all d:

THEOREM 5.1. *For all values of $d \geq 4$ and $n \geq d$, there exists a cs PL $(d-1)$-sphere with $2n$ vertices, Δ_n^{d-1}, that is cs-$\lfloor d/2 \rfloor$-neighborly.*

The construction uses the following cs analog of combinatorial sewing. It starts with the boundary complex of the cross-polytope: $\Delta_d^{d-1} := \partial \mathcal{C}_d^*$. At each step of the construction, it is shown that there exists a pair of antipodal $(d-1)$-balls $\pm B$ in Δ_{n-1}^{d-1} that share no common facets and have other nice properties. The complex Δ_n^{d-1} is then obtained from Δ_{n-1}^{d-1} by introducing two new vertices, n and $-n$, and replacing B and $-B$ with the cones over the boundary complexes of B and $-B$: $\partial B * n$ and $\partial(-B) * (-n)$, respectively; see Definition 6.7 below for more details.

The significance of Theorem 5.1 is that exactly as in the case of spheres without symmetry assumption, in the class of cs simplicial $(d-1)$-spheres with $2n$ vertices, a cs-$\lfloor d/2 \rfloor$-neighborly sphere simultaneously maximizes all the face numbers. This follows from work of Adin [**Adi91**] and Stanley (unpublished). Also, as in the case without symmetry assumption, it is very natural to consider the following problems:

PROBLEM 5.2. Let $d \geq 4$. Denote by $\mathrm{cs}(d, n)$ the number of cs simplicial $(d-1)$-spheres with $2n$ labeled vertices. How large is $\mathrm{cs}(d, n)$?

PROBLEM 5.3. Let $d \geq 4$. How many cs simplicial $(d-1)$-spheres with $2n$ vertices are cs-$\lfloor d/2 \rfloor$-neighborly? Can one construct $2^{\Omega(n^{\lfloor (d-1)/2 \rfloor})}$ such spheres, or, more optimistically, $2^{\Omega(n^{\lfloor d/2 \rfloor})}$ such spheres?

An additional impetus for Problem 5.3 comes from a result of McMullen–Shephard [**MS68**] asserting that for $d \geq 4$, a cs $(d-1)$ sphere that is cs-$\lfloor d/2 \rfloor$-neighborly and has more than $2d + 2$ vertices cannot be realized as the boundary of a cs d-polytope. In contrast with the situation for spheres without symmetry assumption, not much is known about this problem: for $d = 4$, there are at least $\Omega(2^n)$ pairwise *non-isomorphic* cs 3-spheres with $2n$ vertices that are cs-2-neighborly, see [**NZ22**, Section 7]. For each $d = 2k > 4$ and $n \gg 0$, only two non-isomorphic constructions are available at present (see [**NZ22**, Thms. 5.5 and 5.7]). One of these two spheres is Δ_n^{2k-1}. The other one, which we denote by Λ_n^{2k-1}, is the link of one of the edges in Δ_{n+2}^{2k+1}.

As for Problem 5.2, we claim that $\mathrm{cs}(d, n)$ satisfies the inequalities of Theorem 3.1, that is,

$$2^{\Omega(n^{\lfloor d/2 \rfloor})} \leq \mathrm{cs}(d, n) \leq 2^{O(n^{\lfloor d/2 \rfloor} \log n)}.$$

We provide a sketch of the proof. The upper bound follows from the fact that $\mathrm{cs}(d, n) \leq s(d, n)$. It remains to prove the lower bound for $d = 2k$. (For $d = 2k + 1$, simply count suspensions of cs $(2k-1)$-spheres.) Assume that $n = mk$. By [**NZ22**, Proposition 4.3], a cs simplicial $(2k-1)$-sphere Λ_{2n}^{2k-1} from the family of cs-k-neighborly spheres mentioned above satisfies the following property: it contains an isomorphic image of the ball that is the join of k paths

$$(1, 2, \ldots, m),\ (m+1, m+2, \ldots, 2m), \ldots, \quad \text{and } ((k-1)m+1, \ldots, km).$$

Denote this isomorphic image by I. By symmetry, Λ_{2n}^{2k-1} also contains $-I$. We now apply Construction 3 from [**NSW16**] to I, that is, we retriangulate I in such a way that the resulting ball has $O(n)$ labeled vertices and its boundary is the same as ∂I. Let $\mathcal{T}$ be the set of new balls. As shown in [**NSW16**], the size of $\mathcal{T}$ is $2^{\Omega(n^{\lfloor d/2 \rfloor})}$.

Consider the prism $\partial I \times [0,1]$. Its facets are prisms over simplices. Triangulate this polyhedral complex without introducing new vertices. For each $B \in \mathcal{T}$, let $\Gamma(B)$ be this triangulation of $\partial I \times [0,1]$ with a copy of B glued to the top, i.e., to $\partial I \times \{1\} = \partial B \times \{1\}$, along the boundary of B. Then $\Gamma(B)$ is a PL ball whose boundary is ∂I (the bottom of the prism). The sphere obtained from Λ_{2n}^{2k-1} by replacing I with $\Gamma(B)$ and $-I$ with $-\Gamma(B)$ is then a cs simplicial $(2k-1)$-sphere with $O(n)$ vertices. In this way, we obtain $2^{\Omega(n^{\lfloor d/2 \rfloor})}$ cs $(2k-1)$-spheres with $O(n)$ vertices.

6. The transversal numbers of simplicial spheres

Let $H = (V, E)$ be a hypergraph. A *transversal* of H is a subset T of V that intersects all edges of H. The *transversal number* $\tau(H)$ of H is the minimum cardinality of transversals of H. Transversal numbers of interesting families of graphs and hypergraphs have been studied extensively, see, for example, [**Alo90, AKMM02**].

In this section, we restrict our discussion to transversal numbers of hypergraphs that arise from simplicial spheres. Given a simplicial sphere Δ, we define a hypergraph $H(\Delta) = (V, E)$, where V is the vertex set of Δ, and E is the set of facets of Δ. Then $\tau(H(\Delta))$ is the minimum number of vertices needed so that the union of their (closed) vertex stars is the entire complex Δ, i.e., this union does not miss any facet of Δ. In addition to the transversal number $\tau(H(\Delta))$, we are also interested in the *transversal ratio* $\mu(\Delta) := \tau(H(\Delta))/f_0(\Delta)$.

PROBLEM 6.1. Let $d \geq 4$. How does the function

$$\mu_d(n) := \max\{\mu(\Delta) : \Delta \text{ is a simplicial } (d-1)\text{-sphere with } n \text{ vertices}\}$$

behave when n is large? Find the value of $\mu_d := \limsup_{n\to\infty} \mu_d(n)$.

In [**BDL24**], it is shown that $\mu_3 = 1/2$: $\mu_3 \leq 1/2$ follows from the four color theorem and $\mu_3 \geq 1/2$ follows from the existence of an infinite family of 2-spheres with transversal ratio $1/2$. The problem remains wide open for $d \geq 4$. Several lower bounds on μ_d proved in [**BDL24**] can be summarized as follows: $\mu_4 \geq 11/21$, while for $k \geq 3$, $\mu_{2k} \geq 1/2$ and $\mu_{2k-1} \geq 1/k$. The proof of the bound $\mu_4 \geq 11/21$ begins with a construction of a 2-neighborly PL 3-sphere with 21 vertices whose transversal number is 11; it is found with computer help. The bound $\mu_{2k} \geq 1/2$ is obtained by considering the family of the boundary complexes of the cyclic polytopes $\{\partial C(2k, n) : n \geq 2k+1\}$. This approach does not work for even-dimensional spheres as the transversal number of $\partial C(2k-1, n)$ is equal to two independently of n and k. The bound $\mu_{2k-1} \geq 1/k$ is verified by looking at the family of the boundary complexes of stacked $(2k-1)$-polytopes. This discussion leads to the following weaker version of Problem 6.1:

PROBLEM 6.2. For a fixed $d \geq 4$, is μ_d bounded away from 1? Is $\limsup_{d\to\infty} \mu_d$ strictly smaller than 1 or is it equal to 1?

A simplicial sphere is *flag* if it is the clique complex of its graph. Under the metric that assigns the same length $\pi/2$ to all edges of the sphere, the flag spheres have particularly nice geometric properties; for example, the vertex stars are geodesically convex, see [**Gro87**]. This motivates our final problem.

PROBLEM 6.3. Let $d \geq 3$. Find the value of $\lambda_d := \limsup_{n\to\infty} \lambda_d(n)$, where

$$\lambda_d(n) := \max\{\mu(\Delta) : \Delta \text{ is a flag } (d-1)\text{-sphere with } n \text{ vertices}\}.$$

Is $\lambda_d < \mu_d$?

In the rest of the paper, we obtain some estimates on transversal ratios. Our main result is that for all odd $d \geq 5$, $\mu_d \geq 2/5$. It is curious to note that most of the large transversal ratios mentioned above are attained by neighborly spheres. This suggests we consider various families of neighborly spheres to find better lower bounds on μ_d. To start, we notice that $\mu(\Delta) \geq 1/4$ for *any* 2-neighborly 3-sphere Δ with n vertices: by the 2-neighborliness of Δ and the Dehn–Sommerville relations, $f_3(\Delta) = f_1(\Delta) - f_0(\Delta) = n(n-3)/2$. Now, the link of each vertex is a 2-sphere with $n-1$ vertices. Thus, each vertex belongs to $2(n-1)-4 = 2(n-3)$ facets. It follows that $\mu(\Delta) \geq f_3(\Delta)/2(n-3)n = 1/4$. (In fact, by using the inclusion-exclusion principle, one can even improve this lower bound to $\frac{3}{2} - \sqrt{\frac{3}{2}}$; we invite the reader to check this fact.)

As promised, we now turn to proving that $\mu_d \geq 2/5$ for all odd $d \geq 5$. We achieve this by estimating the transversal ratios of two classes of highly neighborly spheres that we encountered in Sections 4 and 5, respectively: the family of relative squeezed spheres and the family of cs neighborly spheres $\{\Delta_n^{d-1} : n \geq d\}$.

6.1. The traversal numbers of relative squeezed spheres. We start by reviewing definitions of squeezed balls and relative squeezed balls; the reader is referred to [**Kal88, NZ24**] for more details.

Let $k \geq 1$, and for $1 \leq m \leq n$, let $[m,n]$ denote the set $\{m, m+1, \dots, n\}$. The poset $\mathcal{F}_{2k}^{[m,n]}$ is defined as follows: as a set, it consists of the following $2k$-sets (which are in fact facets of $\partial C(2k,n)$):

$$\big\{\{i_1, i_1+1, i_2, i_2+1, \dots, i_k, i_k+1\} : m \leq i_1,\ i_k+1 \leq n,\ i_j \leq i_{j+1}-2\ \forall 1 \leq j \leq k-1\big\},$$

ordered by the standard partial order $\leq_p$ (the product order on $\mathbb{N}^{2k}$). Given an antichain S in $\mathcal{F}_{2k}^{[1,n]}$, one can define $B(S)$ to be the pure simplicial complex whose facets are the elements of the order ideal of $\mathcal{F}_{2k}^{[1,n]}$ generated by S. It is a theorem of Kalai [**Kal88**] that for any antichain S, $B(S)$ is a PL $(2k-1)$-ball, called a *squeezed ball*; its boundary, $\partial B(S)$, is then a PL $(2k-2)$-sphere, called a *squeezed sphere.*

Given an antichain S in $\mathcal{F}_{2k}^{[1,n]}$, one can consider another (potentially empty) antichain

$$S - \mathbf{1}_{2k} := \big\{\{x_1-1, x_1, x_2-1, x_2, \dots, x_k-1, x_k\} :$$
$$\{x_1, x_1+1, x_2, x_2+1, \dots, x_k, x_k+1\} \in S,\ x_1 > 1\big\}.$$

Let B_S denote the pure simplicial complex whose facets are all facets of $B(S)$ that are not facets of $B(S-\mathbf{1}_{2k})$. It is proved in [**NZ24**] that for any antichain S, B_S is a PL $(2k-1)$-ball, and hence, ∂B_S is a PL $(2k-2)$-sphere. We call B_S a *relative squeezed ball* and ∂B_S a *relative squeezed sphere*. An important property of these complexes is that if S contains the element $[1,2]\cup[n-2k+3,n]$, then ∂B_S is a $(k-1)$-neighborly sphere with vertex set $[1,n]$, see [**NZ24**, Theorem 3.1].

For the rest of this subsection we fix $k\geq 3$ and we let A_n denote the following antichain in $\mathcal{F}_{2k}^{[1,n]}$; to simplify notation we write A instead of A_n when n is fixed or understood from context:

$$A = A_n := \{F_1, F_2, \ldots, F_{\lfloor \frac{n}{2}\rfloor - k+1}\}, \qquad \text{where } F_i = [i,i+1]\cup[n-2k+4-i, n-i+1].$$

In particular, $F_1 = [1,2]\cup[n-2k+3,n]$, and so ∂B_{A_n} is $(k-1)$-neighborly. Our goal is to estimate the transversal ratios of ∂B_{A_n} as $n\to\infty$, see Proposition 6.6. To do so, it is useful to describe some of the facets of $\partial B_A = \partial B_{A_n}$ (for a fixed n). This is done in the following lemma.

LEMMA 6.4. *Let $k\geq 3$ and let $n\geq 2k+1$. The complex ∂B_A contains the following facets:*

(1) $[i,i+1]\cup H\cup\{n-i+1\}$, *where* $1\leq i\leq\lfloor\frac{n}{2}\rfloor-k+1$ *and* $H\in\mathcal{F}_{2k-4}^{[i+2,n-i]}$.
(2) $[i,i+1]\cup H\cup\{n-i-1\}$, *where* $1\leq i\leq\lfloor\frac{n}{2}\rfloor-k+1$ *and* $H\in\mathcal{F}_{2k-4}^{[i+2,n-i-2]}$.
(3) $\{i+1\}\cup H\cup[n-i,n-i+1]$, *where* $1\leq i\leq\lfloor\frac{n}{2}\rfloor-k+1$ *and* $H\in\mathcal{F}_{2k-4}^{[i+2,n-i-1]}$.
(4) $\{i\}\cup H\cup[n-i-1,n-i]$, *where* $1\leq i\leq\lfloor\frac{n}{2}\rfloor-k+1$ *and* $H\in\mathcal{F}_{2k-4}^{[i+2,n-i-2]}$.
(5) $\{1\}\cup H\cup[n-1,n]$, *where* $H\in\mathcal{F}_{2k-4}^{[2,n-2]}$.
(6) $\{\lfloor\frac{n}{2}\rfloor-k+2\}\cup H$, *where* $H\in\mathcal{F}_{2k-2}^{[\lfloor\frac{n}{2}\rfloor-k+3,\lceil\frac{n}{2}\rceil+k]}$.

PROOF. It suffices to show that every item in the list is contained in a unique facet of B_A. For $1\leq i\leq\lfloor\frac{n}{2}\rfloor-k+1$, the set $[i,i+1]\cup G$ is a facet of B_A if and only if

$$G\leq_p[n-2k+4-i,n-i+1],\quad G\not\leq_p[n-2k+2-i,n-i-1];$$

that is, it is a facet of B_A if and only if $G=H\cup[n-i,n-i+1]$, where $H\in\mathcal{F}_{2k-4}^{[i+2,n-i-1]}$, or $G=H\cup[n-i-1,n-i]$, where $H\in\mathcal{F}_{2k-4}^{[i+2,n-i-2]}$. This shows that the only facet of B_A that contains the set $F=[i,i+1]\cup H\cup\{n-i+1\}$, where $H\in\mathcal{F}_{2k-4}^{[i+2,n-i]}$, is the facet $F\cup\{m\}$, where m is the maximum of $[i+2,n-i]\backslash H$. Similarly, the only facet of B_A that contains the second item in the list (for a given H) is $[i,i+1]\cup H\cup[n-i-1,n-i]$.

By shifting the indices, a facet of B_A with minimal elements $\{i+1,i+2\}$ must have maximal elements $\{n-i-2,n-i-1\}$ or $\{n-i-1,n-i\}$. Similarly, a facet of B_A with minimal elements $\{i-1,i\}$ must have maximal elements $\{n-i,n-i+1\}$ or $\{n-i+1,n-i+2\}$. This shows that the third and fourth items (for a given H) are also contained in unique facets of B_A, namely, in the facets $[i,i+1]\cup H\cup[n-i,n-i+1]$ and $[i,i+1]\cup H\cup[n-i-1,n-i]$, respectively. The fifth and sixth cases are similar: the only facet of B_A that contains $F=\{1\}\cup H\cup[n-1,n]$, where $H\in\mathcal{F}_{2k-4}^{[2,n-2]}$, is $F\cup\{m\}$ where m is the minimum of $[2,n-2]\backslash H$, while the only facet that contains $F'=\{\lfloor\frac{n}{2}\rfloor-k+2\}\cup H$, where $H\in\mathcal{F}_{2k-2}^{[\lfloor\frac{n}{2}\rfloor-k+3,\lceil\frac{n}{2}\rceil+k]}$, is $F'\cup\{\lfloor n/2\rfloor-k+1\}$. □

To establish Proposition 6.6, we need one additional lemma; its proof follows from the proof of [**BDL24**, Proposition 3.5].

LEMMA 6.5. *Any transversal of* $\mathcal{F}_{2k}^{[1,n]}$ *has size at least* $\lceil \frac{n}{2} \rceil - k$.

PROPOSITION 6.6. *Let* $k \geq 3$. *Then* $\liminf_{n\to\infty} \mu(\partial B_{A_n}) \geq 2/5$. *In particular,* $\mu_{2k-1} \geq 2/5$.

PROOF. Fix $n \gg 0$. Let T be a transversal of ∂B_A and let T^c be the complement of T in $[1, n]$.

Let i be the smallest number in $[1, \lceil \frac{n}{2} \rceil - k + 1]$ such that T^c contains at least one of the following sets: $\{i, i+1, n-i+1\}$, $\{i, i+1, n-i-1\}$, $\{i+1, n-i, n-i+1\}$ and $\{i, n-i-1, n-i\}$. If no such i exists, let $i = \lceil \frac{n}{2} \rceil$. By Lemma 6.4 (see the first four items), $T' := T \cap [i+2, n-i-2]$ must be a transversal of $\mathcal{F}_{2k-4}^{[i+2,n-i-2]}$. Let $\tilde{T} := T \cap ([1, i] \cup [n-i, n])$. By our choice of i, $\tilde{T}$ must be a transversal of the following collection $\mathcal{A}_i$ of 3-sets

$$\mathcal{A}_i := \cup_{j=1}^{i-1} \{\{j, j+1, n-j+1\}, \{j, j+1, n-j-1\}, \\ \{j+1, n-j, n-j+1\}, \{j, n-j-1, n-j\}\}.$$

Since $T \supset \tilde{T} \cup T'$ and since by Lemma 6.5, $|T'| \geq \frac{1}{2}(n-2i) - O(k)$, to complete the proof, it suffices to show that $|\tilde{T}| \geq \frac{2}{5} \cdot 2i - O(1)$. To do so, write $\tilde{T}^c \cap [1, i]$ as the union of maximal disjoint intervals $[j_1, j_1'] \cup [j_2, j_2'] \cup \cdots \cup [j_m, j_m']$, where $j_\ell' + 1 < j_{\ell+1}$ for all ℓ, and let $L_\ell = [j_\ell, j_\ell']$. Using that $\tilde{T}$ is a transversal of $\mathcal{A}_i$, we now find a lower bound on $|\tilde{T}|$ in terms of $\left|\tilde{T}^c \cap [1, i]\right|$, the sizes of these intervals, and i. There are the following possible cases for an interval $L_\ell = [j_\ell, j_\ell']$:

- $j_\ell' - j_\ell \geq 2$. Then $R_\ell := [n-j_\ell', n-j_\ell+1]$ is a subset of $\tilde{T}$ and $|R_\ell| = |L_\ell|+1$.
- $j_\ell' - j_\ell = 1$. Then $R_\ell := \{n-j_\ell', n-j_\ell+1\}$ is a subset of $\tilde{T}$ and $|R_\ell| = 2 = |L_\ell|$.
- $j_\ell' = j_\ell$. Then (assuming $1 < j_\ell < i$), $\tilde{T}$ contains at least one element in $[n-j_\ell-1, n-j_\ell]$ and one element in $[n-j_\ell+1, n-j_\ell+2]$, and so $R_\ell := \tilde{T} \cap [n-j_\ell-1, n-j_\ell+2]$ has size at least $2 = |L_\ell| + 1$.

Since for every $\ell > 1$, $j_\ell \geq j_{\ell-1}' + 2$, it follows that the intervals $[n-j_\ell', n-j_\ell+1]$ and $[n-j_{\ell-1}', n-j_{\ell-1}+1]$ are disjoint. Consequently, if both $L_{\ell-1}$ and L_ℓ are *not singletons* (i.e., $j_\ell' > j_\ell$ and $j_{\ell-1}' > j_{\ell-1}$), then $R_{\ell-1}$ and R_ℓ are disjoint.

What happens if L_ℓ (for $\ell > 1$) is a singleton? There are two possible cases. If $L_{\ell-1}$ is also a singleton, then by the third bullet point, $R_{\ell-1} \cup R_\ell$ contains at least one element in each of the following pairwise disjoint intervals: $[n-j_\ell-1, n-j_\ell]$, $[n-j_\ell+1, n-j_{\ell-1}]$, and $[n-j_{\ell-1}+1, n-j_{\ell-1}+2]$. Hence, $|R_{\ell-1} \cup R_\ell| \geq 3 = (|L_{\ell-1}|+1) + (|L_\ell|+1) - 1$. Similarly, if $L_{\ell-1}$ is not a singleton, then $R_{\ell-1}$ and R_ℓ have at most one element in common (namely, $n - j_\ell + 2$; this can happen only if $j_\ell - j_{\ell-1}' = 2$), and so in this case, $|R_{\ell-1} \cup R_\ell| \geq |R_{\ell-1}| + |R_\ell| - 1$. A similar discussion applies to R_ℓ and $R_{\ell+1}$.

Let m be the total number of intervals. Let t be the number of intervals with $j' - j \geq 2$ (i.e., intervals of size ≥ 3), x the number of singleton intervals, y the number of pairs $(L_{\ell-1}, L_\ell)$ where exactly one of $L_{\ell-1}, L_\ell$ is a singleton, and z the number of pairs $(L_{\ell-1}, L_\ell)$ where both $L_{\ell-1}$ and L_ℓ are singletons. The three bullet points and the above discussion imply that

$$|\tilde{T} \cap [n-i, n]| \geq \left|\bigcup_{\ell=1}^{m} R_\ell\right| \geq \left(\sum_{\ell=1}^{m} |L_\ell|\right) + t + x - y - z = \left|\tilde{T}^c \cap [1, i]\right| + t + x - y - z.$$

Adding $|\tilde{T} \cap [1,i]|$ to the outer parts of this inequality, we conclude that $|\tilde{T}| \geq i + (t + x - y - z)$.

It follows from our definitions that $y \leq m - 1$ and $x - z \geq y/2$, and so to minimize $t + x - y - z$, we need to maximize y and minimize t. This implies that the lower bound on $|\tilde{T}|$ is the smallest when the sequence of intervals L_ℓ in $\tilde{T}^c \cap [1,i]$ has the following properties: there are no intervals of size ≥ 3 (i.e., $t = 0$), singletons and 2-element intervals alternate (i.e., $z = 0$ and $y = m-1$), and the gaps between the intervals are as small as possible, that is, every two adjacent intervals are separated by only one element, $j_1 \leq 2$, and $j'_m \geq i - 2$ (i.e., $x = i/5 - O(1)$ and $y = 2i/5 - O(1)$). Hence $|\tilde{T}| \geq i + (i/5 - 2i/5) - O(1) = \frac{2}{5} \cdot 2i - O(1)$, as desired. (For example, $\tilde{T}^c \cap [1,i]$ could be $[1,2] \cup \{4\} \cup [6,7] \cup \{9\} \cup \ldots$ in which case $\tilde{T} \cap [1,i] = \{3, 5, 8, 10, \ldots\}$.) $\square$

6.2. The transversal number of Δ_n^d. Our goal in this section is to estimate the transversal ratio of the family of cs PL d-spheres Δ_n^d. We begin by reviewing some relevant definitions, see [**NZ20, NZ22**] for more details. If Γ and Δ are pure simplicial complexes of the same dimension and $\Gamma \subset \Delta$, then $\Delta \backslash \Gamma$ denotes the subcomplex of Δ generated by facets of Δ that are not facets of Γ.

The vertex set of Δ_n^d is $V_n := [-1, -n] \cup [1, n]$. The definition of Δ_n^d is by interlaced recursion that introduces not only Δ_n^d, but also certain PL d-balls $B_n^{d,i}$ that for $i \leq \lceil d/2 \rceil - 1$ are contained in Δ_n^d.

DEFINITION 6.7. Let $d \geq 1$, $i \leq \lceil d/2 \rceil$, and $n \geq d + 1$ be integers. Define Δ_n^d and $B_n^{d,i}$ inductively as follows:

- For the initial cases, let Δ_n^1 be the cycle $(1, 2, \ldots, n, -1, -2, \ldots, -n, 1)$ and let Δ_{d+1}^d be the boundary complex of the $(d+1)$-dimensional cross-polytope C_{d+1}^*; define $B_n^{d,j} := \emptyset$ if $j < 0$, and $B_n^{1,0} := \overline{\{-1, n\}}$. (In particular, $B_n^{1,j} \subseteq \Delta_n^1$ for all $j \leq 0$.)
- If Δ_m^{d-1} and $B_m^{d-1,i} \subseteq \Delta_m^{d-1}$ are already defined for all $i \leq \lfloor (d-1)/2 \rfloor$ and $m \geq d$, then for $d = 2k$ and $n \geq 2k$, define $B_n^{2k-1,k} := \Delta_n^{2k-1} \backslash B_n^{2k-1,k-1}$; in addition, for all $n \geq d + 1$ and $i \leq \lfloor d/2 \rfloor$, define

$$B_n^{d,i} := \left(B_{n-1}^{d-1,i} * n \right) \cup \left((-B_{n-1}^{d-1,i-1}) * (-n) \right).$$

- If Δ_n^d is already defined, then define Δ_{n+1}^d as the complex obtained from Δ_n^d by replacing the subcomplex $B_n^{d,\lceil d/2 \rceil - 1}$ with $\partial B_n^{d,\lceil d/2 \rceil - 1} * (n+1)$ and $-B_n^{d,\lceil d/2 \rceil - 1}$ with $\partial(-B_n^{d,\lceil d/2 \rceil - 1}) * (-n-1)$.

A big portion of [**NZ20**] is devoted to proving that Δ_n^d and $B_n^{d,i}$ are well-defined, that Δ_n^d is a cs PL sphere while $B_n^{d,i}$ are PL balls, and that Δ_n^d is cs-$\lceil d/2 \rceil$-neighborly.

To estimate the transversal number of Δ_n^d, we need an explicit description of some of the facets of Δ_n^d. For $d = 3$ and $d = 4$, the complete lists of facets are recorded in [**NZ22**, Corollary 3.2] and [**Pfe24**, Proof of Theorem 5.13], respectively. For any $d = 2k - 1 \geq 5$, a large collection of facets of Δ_n^{2k-1} was identified in [**NZ22**, Lemma 3.5]; the following result states a special case of this lemma.

We say that a $2k$-set $G = \{p_1, \dots, p_{2k}\} \subset V_n$ *satisfies property* P_n if

(1) $1 \le |p_1| < |p_2| < \cdots < |p_{2k}| \le n$,
(2) the numbers p_{2i-1} and p_{2i} have the same sign for $1 \le i \le k$, and
(3) $|p_2| - |p_1| = 1$ while $|p_{2i}| - |p_{2i-1}| = 2$ for $2 \le i \le k$.

LEMMA 6.8. *Any $2k$-set $G \subset V_n$ that satisfies property P_n is a facet of*

$$\Delta_n^{2k-1} \backslash \pm B_n^{2k-1,k-1}.$$

The next lemma provides an analogous result for Δ_n^{2k}.

LEMMA 6.9. *Let $k \ge 2$ and $n < m$. If F is of the form $\{p_1, \dots, p_{2k}, p_{2k+1}\}$, where*

(1) $\{p_{2k-1}, p_{2k}, p_{2k+1}\} = \{n-2, n-1, n+1\}$ *or* $\{n-2, n, n+1\}$, *and*
(2) $G = \{p_1, \dots, p_{2k-2}\}$ *has property* P_{n-3},

then F and $-F$ are facets of $\Delta_m^{2k} \backslash \pm B_m^{2k,k-1}$.

PROOF. According to Lemma 6.8, the statement would follow if we show that for every facet G of $\Delta_{n-3}^{2k-3} \backslash \pm B_{n-3}^{2k-3,k-2}$, both $F_1 := G \cup \{n-2, n-1, n+1\}$ and $F_2 := G \cup \{n-2, n, n+1\}$ are facets of $\Delta_m^{2k} \backslash \pm B_m^{2k,k-1}$. So consider such G and let $i \in \{1, 2\}$.

The maximum of F_i is $n+1$. Hence F_i is a facet of Δ_m^{2k} if

$$F_i \in \partial B_n^{2k,k-1} * (n+1) \backslash \pm B_{n+1}^{2k,k-1}.$$

That is, F_i is a facet of Δ_m^{2k} if

$$F_i \backslash (n+1) \in \Gamma_n := \partial B_n^{2k,k-1} \backslash (B_n^{2k-1,k-1} \cup B_n^{2k-1,k-2}).$$

By [**NZ20**, Section 3, page 7],

$$\begin{aligned} B_n^{2k,k-1} &= \left(B_{n-2}^{2k-2,k-1} * (n-1, n)\right) \cup \left(-B_{n-2}^{2k-2,k-2} * (n, -n+1, -n)\right) \\ &\quad \cup \left(B_{n-2}^{2k-2,k-3} * (-n, n-1)\right); \end{aligned}$$

on the other hand, the facets of $B_n^{2k-1,k-1} \cup B_n^{2k-1,k-2}$ always contain one of $\pm(n-1)$ and one of $\pm n$. Thus, computing the boundary of $B_n^{2k,k-1}$ and keeping in mind that (see [**NZ20**, Lemma 3.3]) $B_{n-2}^{2k-2,k-3} \subset -B_{n-2}^{2k-2,k-2} \subset B_{n-2}^{2k-2,k-1}$, we conclude that

$$\begin{aligned} \Gamma_n \supseteq &\left((B_{n-2}^{2k-2,k-1} \backslash B_{n-2}^{2k-2,k-3}) * (n-1)\right) \cup \left((B_{n-2}^{2k-2,k-1} \backslash (-B_{n-2}^{2k-2,k-2})) * (n)\right) \\ &\cup \left((-B_{n-2}^{2k-2,k-2} \backslash B_{n-2}^{2k-2,k-3}) * (-n)\right) =: S_1 \cup S_2 \cup S_3. \end{aligned}$$

The result follows since if G is a facet of

$$\Delta_{n-3}^{2k-3} \backslash \pm B_{n-3}^{2k-3,k-2} = B_{n-3}^{2k-3,k-1} \backslash (-B_{n-3}^{2k-3,k-2}) \subset B_{n-3}^{2k-3,k-1} \backslash B_{n-3}^{2k-3,k-3},$$

then

$$F_1 \backslash (n+1) = G \cup \{n-2, n-1\} \in S_1 \subset \Gamma_n \text{ and } F_2 \backslash (n+1) = G \cup \{n-2, n\} \in S_2 \subset \Gamma_n.$$

□

While the following lemma will not be used in this paper, we state it for completeness.

LEMMA 6.10. *Let $k \ge 2$ and and let G be a $(2k-2)$-set that has property P_{n-3}. Then $F = G \cup \{n-2, n-1, n\}$ is a facet of $B_n^{2k,k-1}$.*

PROOF. This follows immediately from the fact that
$B_n^{2k,k-1} \supseteq B_{n-3}^{2k-3,k-1} * \overline{\{n-2,n-1,n\}}$ and $G \in \Delta_{n-3}^{2k-3} \backslash \pm B_{n-3}^{2k-3,k-2} \subset B_{n-3}^{2k-3,k-1}$. □

We are now in a position to establish the main result of this subsection.

PROPOSITION 6.11. *Let $k \geq 2$. Then*

$$\liminf_{n\to\infty} \mu(\Delta_n^{2k}) \geq \frac{2}{5} \quad \textit{while} \quad \liminf_{n\to\infty} \mu(\Delta_n^{2k-1}) \geq \frac{1}{2}.$$

In particular, $\mu_{2k+1} \geq 2/5$ and $\mu_{2k} \geq 1/2$.

PROOF. We first discuss the case of $d = 2k$. Let $T = T_+ \cup T_-$ be a transversal of Δ_n^d, where T_+ and T_- consist of positive and negative vertices, respectively. It suffices to show that $|T^+| \geq \frac{2}{5}n + O(k)$. (By symmetry, this will also imply that $|T^-| \geq \frac{2}{5}n + O(k)$.) To do so, write $T_+ = \cup_{p=1}^m I_p$ as the union of maximal disjoint intervals. Similarly, write the complement of T_+ in $[1,n]$ as the union of maximal disjoint intervals $T_+^c = \cup_{p=1}^{m'} J_p$. Let $[a_1,b_1]$ be the first non-singleton interval in T_+^c. Then $|T^+ \cap [1,a_1-1]| \geq \frac{1}{2}(a_1-1)$. Hence to complete the proof, it suffices to show that $|T_+ \cap [a_1,n]| \geq \frac{2}{5}(n-a_1) - O(k)$. This will follow from the next two claims.

CLAIM 1. The intervals in T_+^c of size ≥ 4 together occupy at most $O(k)$ elements.

PROOF. Let $[a_\ell, b_\ell]$ be the last interval of size ≥ 4 in T_+^c, and let $[a_2,b_2] < \cdots < [a_{\ell-1}, b_{\ell-1}]$ be the intervals of size ≥ 3 in T_+^c that lie between $[a_1,b_1]$ and $[a_\ell,b_\ell]$. Define the following sets:

$$\begin{aligned}
A_1 &= \{a_1, a_1+1\} \cup \bigcup_{i=1}^{\lfloor \frac{b_1-a_1-1}{3} \rfloor} \{a_1+3i-1, a_1+3i+1\} \\
&= \{a_1, a_1+1, a_1+2, a_1+4, \dots\} \subset [a_1,b_1], \\
A_j &= \bigcup_{i=0}^{\lfloor \frac{b_j-a_j-2}{3} \rfloor} \{a_j+3i, a_j+3i+2\} \\
&= \{a_j, a_j+2, a_j+3, a_j+5, \dots\} \subset [a_j,b_j], \;\; 2 \leq j \leq \ell-1, \\
A_\ell &= \Big(\bigcup_{i=2}^{\lfloor \frac{b_\ell-a_\ell}{3} \rfloor} \{b_\ell-3i, b_\ell-3i+2\} \Big) \cup \{b_\ell-3, b_\ell-2, b_\ell\} \\
&= \{\dots, b_\ell-3, b_\ell-2, b_\ell\} \subset [a_\ell, b_\ell].
\end{aligned}$$

Note that $\sum_{j=1}^{\ell} |A_j| < 2k$ as otherwise, by Lemma 6.9, the set $\cup_{j=1}^{\ell} A_j \subset T_+^c$ would contain a facet of Δ_n^{2k}. But such a facet would be disjoint from T contradicting our assumption that T is a transversal of Δ_n^{2k}. Thus $\ell \leq k-1$ and

$$\left\lfloor \frac{b_1-a_1+2}{3} \right\rfloor + \left\lfloor \frac{b_\ell - a_\ell}{3} \right\rfloor + \sum_{j=2}^{\ell-1} \left\lfloor \frac{b_j-a_j+1}{3} \right\rfloor \leq k-1.$$

We conclude that $\sum_{j=1}^{\ell} \big|[a_j,b_j]\big| = \ell + \sum_{j=1}^{\ell}(b_j-a_j) = O(k)$, as claimed. □

CLAIM 2. The number of singleton intervals in $T^+ \cap [b_1+1, n]$ that are adjacent on the right to an interval in T_+^c of size ≥ 2 is at most $k-1$.

PROOF. If there were at least k such intervals, say, $\{y_1\}, \dots, \{y_k\}$, then, by Lemma 6.9,

$$F := \{a_1, a_1+1\} \cup \big(\bigcup_{i=2}^{k-1} \{y_i - 1, y_i + 1\} \big) \cup \{y_k - 1, y_k + 1, y_k + 2\}$$

would be a facet of Δ_n^{2k} that is disjoint from T. This is impossible since T is a transversal of Δ_n^{2k}. □

To finish the proof of the proposition, observe that the above two claims imply that there is a set C of size $O(k)$ such that $[a_1, n] \backslash C$ is the disjoint union of pairs of adjacent intervals (I, J) where $I \subset T_+$, $J \subset T_+^c$ and such that $|I| \geq 2, |J| \leq 3$ or $|I| = |J| = 1$. In particular, $T_+ \cap [a_1, n] \geq \frac{2}{5}(n - a_1) - O(k)$.

The case of $d = 2k+1$ is very similar: the main difference is that in this case, using Lemma 6.8 and taking $[a_\ell, b_\ell]$ to be the last interval of size ≥ 3, allows to strengthen Claim 1 to the statement that *the intervals in T_+^c of size ≥ 3 together occupy at most $O(k)$ elements.* □

We would like to point out that $\lim_{n\to\infty} \mu(\Delta_n^3) = 1/2$ while $\lim_{n\to\infty} \mu(\Delta_n^4) = 2/5$. Indeed, by looking at the complete list of facets of Δ_n^3 and Δ_n^4 (see [**NZ22**, Corollary 3.2] and [**Pfe24**, Proof of Theorem 5.13], respectively), one can easily check that $T = (\pm\{1, 3, 5, 7, \dots\} \cap V_n) \cup \{\pm n\}$ is a transversal of $H(\Delta_n^3)$; similarly, $T = (\pm\{1, 2, 6, 7, 11, 12, \dots\} \cap V_n) \cup \{\pm n\}$ is a transversal of $H(\Delta_n^4)$.

To summarize the discussion of this section, the following lower bounds on μ_d are known at present: $\mu_3 = 1/2$, $\mu_4 \geq 11/21$, and for $k \geq 3$, $\mu_{2k} \geq 1/2$ while $\mu_{2k-1} \geq 2/5$. For $d > 3$, no upper bounds on μ_d, except for the trivial one $\mu_d \leq 1$, are known. Shedding any new light on possible values of μ_d would be of great interest; so will be shedding any light on possible values of λ_d from Problem 6.3. At the moment nothing at all is known about these numbers.

Acknowledgments

We are grateful to Rowan Rowlands and Seunghun Lee for their comments on the preliminary version of this paper.

References

[Adi91] R. M. Adin, *Combinatorial structure of simplicial complexes with symmetry*, Ph.D. thesis, Hebrew University, Jerusalem, 1991.

[AB20] Karim Adiprasito and Bruno Benedetti, *A Cheeger-type exponential bound for the number of triangulated manifolds*, Ann. Inst. Henri Poincaré D **7** (2020), no. 2, 233–247, DOI 10.4171/AIHPD/85. MR4109834

[Alo86] Noga Alon, *The number of polytopes, configurations and real matroids*, Mathematika **33** (1986), no. 1, 62–71, DOI 10.1112/S0025579300013875. MR859498

[Alo90] Noga Alon, *Transversal numbers of uniform hypergraphs*, Graphs Combin. **6** (1990), no. 1, 1–4, DOI 10.1007/BF01787474. MR1058542

[AKMM02] Noga Alon, Gil Kalai, Jiří Matoušek, and Roy Meshulam, *Transversal numbers for hypergraphs arising in geometry*, Adv. in Appl. Math. **29** (2002), no. 1, 79–101, DOI 10.1016/S0196-8858(02)00003-9. MR1921545

[ADJ91] Jan Ambjørn, Bergfinnur Durhuus, and Thórdur Jónsson, *Three-dimensional simplicial quantum gravity and generalized matrix models*, Modern Phys. Lett. A **6** (1991), no. 12, 1133–1146, DOI 10.1142/S0217732391001184. MR1115607

[BZ11] Bruno Benedetti and Günter M. Ziegler, *On locally constructible spheres and balls*, Acta Math. **206** (2011), no. 2, 205–243, DOI 10.1007/s11511-011-0062-2. MR2810852

[BL00] Anders Björner and Frank H. Lutz, *Simplicial manifolds, bistellar flips and a 16-vertex triangulation of the Poincaré homology 3-sphere*, Experiment. Math. **9** (2000), no. 2, 275–289. MR1780212

[BDL24] Joseph Briggs, Michael Gene Dobbins, and Seunghun Lee, *Transversals and Colorings of Simplicial Spheres*, Discrete Comput. Geom. **71** (2024), no. 2, 738–763, DOI 10.1007/s00454-023-00520-1. MR4700324

[Bro63] William G. Brown, *Enumeration of non-separable planar maps*, Canadian J. Math. **15** (1963), 526–545, DOI 10.4153/CJM-1963-056-7. MR150759

[Gal63] David Gale, *Neighborly and cyclic polytopes*, Proc. Sympos. Pure Math., Vol. VII, Amer. Math. Soc., Providence, RI, 1963, pp. 225–232. MR152944

[GP86] Jacob E. Goodman and Richard Pollack, *Upper bounds for configurations and polytopes in* $\mathbf{R}^d$, Discrete Comput. Geom. **1** (1986), no. 3, 219–227, DOI 10.1007/BF02187696. MR861891

[Gro87] M. Gromov, *Hyperbolic groups*, Essays in group theory, Math. Sci. Res. Inst. Publ., vol. 8, Springer, New York, 1987, pp. 75–263, DOI 10.1007/978-1-4613-9586-7_3. MR919829

[Gro00] Misha Gromov, *Spaces and questions*, Geom. Funct. Anal. **Special Volume** (2000), 118–161. GAFA 2000 (Tel Aviv, 1999). MR1826251

[HZ00] Masahiro Hachimori and Günter M. Ziegler, *Decompositons of simplicial balls and spheres with knots consisting of few edges*, Math. Z. **235** (2000), no. 1, 159–171, DOI 10.1007/s002090000129. MR1785077

[Hud69] J. F. P. Hudson, *Piecewise linear topology*, W. A. Benjamin, Inc., New York-Amsterdam, 1969. University of Chicago Lecture Notes prepared with the assistance of J. L. Shaneson and J. Lees. MR248844

[Joc95] William Jockusch, *An infinite family of nearly neighborly centrally symmetric* 3-*spheres*, J. Combin. Theory Ser. A **72** (1995), no. 2, 318–321, DOI 10.1016/0097-3165(95)90070-5. MR1357779

[Kal88] Gil Kalai, *Many triangulated spheres*, Discrete Comput. Geom. **3** (1988), no. 1, 1–14, DOI 10.1007/BF02187893. MR918176

[Kle64] Victor Klee, *A combinatorial analogue of Poincaré's duality theorem*, Canadian J. Math. **16** (1964), 517–531, DOI 10.4153/CJM-1964-053-0. MR189039

[Lee00] C. W. Lee, *Kalai's squeezed spheres are shellable*, Discrete Comput. Geom. **24** (2000), no. 2-3, 391–396, DOI 10.1007/s004540010043. The Branko Grünbaum birthday issue. MR1758058

[Lic99] W. B. R. Lickorish, *Simplicial moves on complexes and manifolds*, Proceedings of the Kirbyfest (Berkeley, CA, 1998), Geom. Topol. Monogr., vol. 2, Geom. Topol. Publ., Coventry, 1999, pp. 299–320, DOI 10.2140/gtm.1999.2.299. MR1734414

[LN06] Nathan Linial and Isabella Novik, *How neighborly can a centrally symmetric polytope be?*, Discrete Comput. Geom. **36** (2006), no. 2, 273–281, DOI 10.1007/s00454-006-1235-1. MR2252105

[McM70] P. McMullen, *The maximum numbers of faces of a convex polytope*, Mathematika **17** (1970), 179–184, DOI 10.1112/S0025579300002850. MR283691

[MS68] P. McMullen and G. C. Shephard, *Diagrams for centrally symmetric polytopes*, Mathematika **15** (1968), 123–138, DOI 10.1112/S0025579300002473. MR238180

[NSW16] Eran Nevo, Francisco Santos, and Stedman Wilson, *Many triangulated odd-dimensional spheres*, Math. Ann. **364** (2016), no. 3-4, 737–762, DOI 10.1007/s00208-015-1232-x. MR3466849

[Nov18] Isabella Novik, *From acute sets to centrally symmetric 2-neighborly polytopes*, SIAM J. Discrete Math. **32** (2018), no. 3, 1572–1576, DOI 10.1137/18M1164202. MR3824165

[NZ20] Isabella Novik and Hailun Zheng, *Highly neighborly centrally symmetric spheres*, Adv. Math. **370** (2020), 107238, 16, DOI 10.1016/j.aim.2020.107238. MR4106648

[NZ22] Isabella Novik and Hailun Zheng, *New families of highly neighborly centrally symmetric spheres*, Trans. Amer. Math. Soc. **375** (2022), no. 6, 4445–4475, DOI 10.1090/tran/8631. MR4419065

[NZ24] Isabella Novik and Hailun Zheng, *Many neighborly spheres*, Math. Ann. **388** (2024), no. 1, 969–984, DOI 10.1007/s00208-022-02538-x. MR4693951

[Pad13] Arnau Padrol, *Many neighborly polytopes and oriented matroids*, Discrete Comput. Geom. **50** (2013), no. 4, 865–902, DOI 10.1007/s00454-013-9544-7. MR3138140

[PPS24] Arnau Padrol, Eva Philippe, and Francisco Santos, *Many regular triangulations and many polytopes*, Math. Ann. **389** (2024), no. 1, 745–763, DOI 10.1007/s00208-023-02652-4. MR4735961

[Pfe24] Julian Pfeifle, *Positive Plücker tree certificates for non-realizability*, Exp. Math. **33** (2024), no. 1, 69–85, DOI 10.1080/10586458.2021.1994487. MR4713863

[PZ04] Julian Pfeifle and Günter M. Ziegler, *Many triangulated 3-spheres*, Math. Ann. **330** (2004), no. 4, 829–837, DOI 10.1007/s00208-004-0594-2. MR2102314

[RW82] L. B. Richmond and N. C. Wormald, *The asymptotic number of convex polyhedra*, Trans. Amer. Math. Soc. **273** (1982), no. 2, 721–735, DOI 10.2307/1999938. MR667170

[She82] Ido Shemer, *Neighborly polytopes*, Israel J. Math. **43** (1982), no. 4, 291–314, DOI 10.1007/BF02761235. MR693351

[Som27] D. M. Y. Sommerville, *The relations connecting the angle-sums and volume of a polytope in space of n dimensions*, Proceedings of the Royal Society of London. Series A, Containing papers of a mathematical and physical character **115** (1927), no. 770, 103–119.

[Sta75] Richard P. Stanley, *The upper bound conjecture and Cohen-Macaulay rings*, Studies in Appl. Math. **54** (1975), no. 2, 135–142, DOI 10.1002/sapm1975542135. MR458437

[Tut62] W. T. Tutte, *A census of planar triangulations*, Canadian J. Math. **14** (1962), 21–38, DOI 10.4153/CJM-1962-002-9. MR130841

[Tut80] W. T. Tutte, *On the enumeration of convex polyhedra*, J. Combin. Theory Ser. B **28** (1980), no. 2, 105–126, DOI 10.1016/0095-8956(80)90059-3. MR572468

[Zie95] Günter M. Ziegler, *Lectures on polytopes*, Graduate Texts in Mathematics, vol. 152, Springer-Verlag, New York, 1995, DOI 10.1007/978-1-4613-8431-1. MR1311028

Department of Mathematics, University of Washington, Seattle, Washington 98195-4350

Email address: `novik@uw.edu`

Department of Mathematical Sciences, University of Copenhagen, Universitesparken 5, 2100 Copenhagen, Denmark

Email address: `zhenghailun1990@gmail.com`

Proceedings of Symposia in Pure Mathematics
Volume **110**, 2024
https://doi.org/10.1090/pspum/110/02023

Realizing simplicial complexes as the boundary of the totally non-negative part of a variety

Hugh Thomas

ABSTRACT. We raise the question of when a simplicial complex can be recovered as the boundary of the totally non-negative part of a certain variety defined based on either the simplicial complex itself or on a "compatibility degree" function on pairs of vertices. We present some non-trivial examples with origins in combinatorics and representation theory, and briefly discuss the motivation from the study of scattering amplitudes.

1. The simplest form of the combinatorial problem

Let Δ be a simplicial complex on a ground set V. That is to say, we suppose that Δ is a collection of subsets of V, such that for any $F \in \Delta$, and $G \subseteq F$, we have $G \in \Delta$. We do not assume that all the elements of V appear in Δ.

Suppose further that Δ is *flag*, that is to say, the minimal non-faces of Δ are of size at most two. In other words, $F \in \Delta$ if and only if all the subsets of F of size at most two are in Δ.

Introduce a variable u_x for each $x \in V$, and consider the following system of equations, one for each $x \in V$:

$$u_x + \prod_{\{x,y\} \notin \Delta} u_y = 1 \tag{1.1}$$

Define $X(\Delta)$ to be the solutions to this system of equations over $\mathbb{C}$. Define $X_{\geq 0}(\Delta)$ to be the solutions with all u_x real and non-negative. We call this the totally non-negative part of $X(\Delta)$.

$\Sigma(\Delta)$ is a collection of subsets of V, describing the combinatorics of the boundary $X_{\geq 0}(\Delta)$, defined as follows. We say that $F \in \Sigma(\Delta)$ if there is some point $p \in X_{\geq 0}(\Delta)$ such that $u_x(p) = 0$ if and only if $x \in F$.

The question is: when does the combinatorics of the boundary of $X_{\geq 0}(\Delta)$ agree with Δ?

PROBLEM 1.1. *For what V and Δ is it the case that $\Sigma(\Delta) = \Delta$?*

Let us look at some examples where this holds.

2020 *Mathematics Subject Classification.* Primary 05E45, 16G10.

I acknowledge with gratitude financial support from NSERC Discovery Grant RGPIN-2022-03960 and the Canada Research Chairs program, grant number CRC-2021-00120.

EXAMPLE 1.2. Let $V = \{1, 2\}$, and let $\Delta = \{\emptyset, \{1\}, \{2\}\}$. The two equations from (1.1) are both $x_1 + x_2 = 1$. The totally positive part of the solutions is a line segment, and the boundary recovers Δ.

EXAMPLE 1.3. Let $V = \{1, 2, 3, 4, 5\}$. Let Δ be a five-cycle. It is less straightforward to show that $\Sigma(\Delta) = \Delta$.

EXAMPLE 1.4. Fix a positive integer $N \geq 4$. Let V be the set of diagonals of an N-gon, and let Δ be the simplicial complex consisting of non-crossing collections of diagonals. The maximal faces of Δ then correspond to triangulations of the N-gon, and Δ is the (simplicial) associahedron. Then it is shown in [**3**] that $\Sigma(\Delta) = \Delta$. When $N = 4, 5$, this recovers the previous two examples.

We should remark that in general, we do not know that $X(\Delta)$ is reduced or irreducible, so the term "variety" in the title of this paper is intended in a loose sense.

2. A more complicated version of the combinatorial problem

As before, let V be a finite vertex set. Define a *compatibility degree* $c : V \times V \to \mathbb{Z}_{\geq 0}$ such that $c(x, y) = 0$ if and only if $c(y, x) = 0$. If $c(x, y) = 0 = c(y, x)$, we say that x and y are compatible.

Define $\Delta(c)$ to be the simplicial complex consisting of subsets of V which are pairwise compatible, and such that every element is compatible with itself.

Now consider the following system of equations, one for each $x \in V$:

$$u_x + \prod_{y \in V} u_y^{c(y,x)} = 1. \tag{2.1}$$

Write $X(c)$ for the solutions to (2.1), and similarly $X_{\geq 0}(c)$ and $\Sigma(c)$.

We recover the system (1.1) if we define the compatibility degree of x and y to be zero if $\{x, y\}$ is a face of Δ and 1 otherwise. The following is an example with a different choice of c.

EXAMPLE 2.1. Let N be even, and consider the N-gon with the action of 180-degree rotation. Let $\widetilde{V}$ be the set of equivalence classes of diagonals under rotation. For $\widetilde{x}, \widetilde{y} \in \widetilde{V}$, define $\widetilde{c}(\widetilde{y}, \widetilde{x})$ to be the total number of intersections of all diagonals from $\widetilde{y}$ with one diagonal from $\widetilde{x}$. The faces of $\Delta(\widetilde{c})$ are in natural bijection with the partial triangulations of the N-gon which are symmetric under 180-degree rotation. (Note that this example of $\widetilde{c}$ does not satisfy $\widetilde{c}(\widetilde{x}, \widetilde{y}) = \widetilde{c}(\widetilde{y}, \widetilde{x})$ for all $\widetilde{x}, \widetilde{y}$.)

As in Example 1.4, we can define V to be the set of all diagonals of the N-gon, and Δ to be the collections of non-crossing diagonals. Then (1.1) defines $X(\Delta)$. We observe that

$$X(\widetilde{c}) \simeq X(\Delta) \cap \{p \mid u_x(p) = u_{\bar{x}}(p) \mid \forall x \in \Delta\},$$

where we write $\bar{x}$ for the 180-degree rotation of x.

It now follows $\Sigma(\widetilde{c}) = \Delta(\widetilde{c})$.

PROBLEM 2.2. *For what compatibility degree functions c is it the case that $\Sigma(c) = \Delta(c)$?*

3. Examples from representation theory

Let A be a finite-dimensional algebra over a field k, with n isomorphism classes of simple modules, $S_1, \dots, S_n$. Let P_i be the projective cover of S_i, and I_i its injective hull. (A good introduction to the representation theory of finite-dimensional algebras is [**4**].)

Let V_A be the set of indecomposable A-modules, together with n additional objects, which we denote $P_i[1]$.

(There is a slightly more sophisticated framework in which we could work, which is a certain extension-closed subcategory of the homotopy category $K^b(\operatorname{proj} A)$. The category $K^b(\operatorname{proj} A)$ is triangulated, and in that context, $P_i[1]$ is the shift of P_i, but for simplicity we will use the simpler framework in which $P_i[1]$ is just a formal object, much as one adds n negative roots to the set of positive roots in order to account for the initial variables in a cluster algebra.)

To define the compatibility degree, we need another preliminary definition. Given an indecomposable A-module M, Let τM be the Auslander–Reiten translation of M. We also define $\tau(P_i[1]) = I_i$, and $\operatorname{Hom}(P_i[1], M) = 0$ for any module M.

Now define the compatibility degree of two objects of V_A by saying

$$c_A(X, Y) = \dim \operatorname{Hom}(X, \tau Y) + \dim \operatorname{Hom}(Y, \tau X).$$

$\Delta(c_A)$ is the abstract simplicial complex corresponding to the g-fan, as studied in [**1, 5, 12**].

We have the following theorem, which will be established in [**2**].

THEOREM 3.1. *Let A be a finite-dimensional algebra of finite representation type. Then $\Sigma(c_A) = \Delta(c_A)$.*

Note that in this case, if a module M has $\operatorname{Hom}(M, \tau M) \neq 0$, then the corresponding vertex does not appear in any face of $\Delta(c_A)$, but the corresponding variable still plays a role.

EXAMPLE 3.2. Fix a positive integer n. Let Q be the quiver with arrows α_i from i to $i+1$ and β_i from $i+1$ to i, for $1 \leq i \leq n-1$. Let $\mathbb{C}Q$ be the path algebra of this quiver, and let I be the ideal generated by $\alpha_i\beta_i$ and $\beta_i\alpha_i$ for $1 \leq i \leq n-1$. Let $A = \mathbb{C}Q/I$. It follows from combining the main result of [**11**] in the type A_n case with [**6**, Theorem 1.15] that $\Delta(c_A)$ is the type A_n Coxeter complex, which is dual to the permutohedron.

4. The geometrical problem

Beyond the combinatorial problem considered so far, we can also consider the following geometrical upgrade of the problem. Suppose that Δ can be realized as the boundary of a polytope, and let $P^\vee(\Delta)$ be the dual polytope.

PROBLEM 4.1.

(1) *For what Δ, which can be realized as a simplicial polytope, is it the case that $X_{\geq 0}(\Delta)$ has a face-preserving diffeomorphism to $P^\vee(\Delta)$?*
(2) *For what compatibility degree c, such that $\Delta(c)$ can be realized as a simplicial polytope, is it the case that $X_{\geq 0}(c)$ has a face-preserving diffeomorphism to $P^\vee(\Delta(c))$?*

A positive answer is shown in [**3**] for the c's that give rise to *cluster configuration spaces.* This includes Example 1.4 and Example 2.1. More precisely, the results from [**3**] cover exactly the examples in Section 3 which come from Q an orientation of a simply-laced Dynkin diagram, together with those that come from the natural extension of Section 3 to non-simply-laced Dynkin diagrams, via the theory of species. (For the theory of species, see, for example, [**7**].)

Currently no results are known beyond these cases.

5. Motivation

A system of equations equivalent to those coming from Example 1.4 were studied in 1969 by physicists Koba and Nielsen [**8–10**]. As in that example, let V be the set of diagonals of an N-gon, and Δ the collections of non-crossing diagonals.

Koba and Nielsen defined the "tree string amplitude" as an integral over $X_{\geq 0}(\Delta)$. The structure of the boundary of $X_{\geq 0}(\Delta)$ is important because of its implication for factorization properties of the integral.

This integral should be viewed as being associated to a disk; in perturbative quantum field theory one then expect higher corrections to the tree string amplitude associated to surfaces of higher genus.

As in Section 3, one can associate algebras to these surfaces, via the correspondence between triangulated surfaces and gentle algebras [**13**]. However, unlike in Section 3, these algebras are not of finite representation type, i.e., they have an infinite number of indecomposable modules.

PROBLEM 5.1. *Formulate an analogue of* (2.1) *which extends the set-up of Section* 3 *to algebras having infinitely many indecomposable modules.*

This problem is intentionally somewhat vague. One possible approach would be to consider the variables u_x taking values in a ring of power series, in which the products (now infinite) in (2.1) have a chance to converge.

The goal is to define integrals providing higher loop corrections to the Koba–Nielsen tree string amplitude.

Acknowledgments

I would like to acknowledge with thanks my collaborators Nima Arkani-Hamed, Hadleigh Frost, Song He, Thomas Lam, Pierre-Guy Plamondon, and Giulio Salvatori, from whom I have learned a great deal. I am very grateful to Open Problems in Algebraic Combinatorics for the opportunity to have presented a version of this paper at the conference.

References

[1] T. Aoki, A. Higashitani, O. Iyama, R. Kase, Y. Mizuno, *Fans and polytopes in tilting theory*, `arXiv:2203.15213`, 2022.

[2] N. Arkani-Hamed, H. Frost, P.-G. Plamondon, G. Salvatori and H. Thomas, *A representation-theoretic approach to the u-equations* (in preparation).

[3] Nima Arkani-Hamed, Song He, and Thomas Lam, *Cluster configuration spaces of finite type*, SIGMA Symmetry Integrability Geom. Methods Appl. **17** (2021), Paper No. 092, 41, DOI 10.3842/SIGMA.2021.092. MR4324836

[4] Ibrahim Assem, Daniel Simson, and Andrzej Skowroński, *Elements of the representation theory of associative algebras. Vol. 1*, London Mathematical Society Student Texts, vol. 65, Cambridge University Press, Cambridge, 2006. Techniques of representation theory, DOI 10.1017/CBO9780511614309. MR2197389

[5] Laurent Demonet, Osamu Iyama, and Gustavo Jasso, *τ-tilting finite algebras, bricks, and g-vectors*, Int. Math. Res. Not. IMRN **3** (2019), 852–892, DOI 10.1093/imrn/rnx135. MR3910476

[6] Laurent Demonet, Osamu Iyama, Nathan Reading, Idun Reiten, and Hugh Thomas, *Lattice theory of torsion classes: beyond τ-tilting theory*, Trans. Amer. Math. Soc. Ser. B **10** (2023), 542–612, DOI 10.1090/btran/100. MR4579952

[7] Bangming Deng, Jie Du, Brian Parshall, and Jianpan Wang, *Finite dimensional algebras and quantum groups*, Mathematical Surveys and Monographs, vol. 150, American Mathematical Society, Providence, RI, 2008, DOI 10.1090/surv/150. MR2457938

[8] Z. Koba and H. Nielsen, *Reaction amplitude for n-mesons a generalization of the Veneziano–Bardakçi–Ruegg–Virasoro model*, Nuclear Physics B 10 (1969) 633–655.

[9] Z. Koba and H. Nielsen, *Manifestly crossing-invariant parametrization of n-meson amplitude*, Nuclear Physics B 12 (1969), 517–536.

[10] Zirô Koba and Holger Bech Nielsen, *Generalized Veneziano model from the point of view of manifestly crossing-invariant parametrization*, Z. Physik **229** (1969), 243–263. MR339708

[11] Yuya Mizuno, *Classifying τ-tilting modules over preprojective algebras of Dynkin type*, Math. Z. **277** (2014), no. 3-4, 665–690, DOI 10.1007/s00209-013-1271-5. MR3229959

[12] Y. Mizuno, *Shard theory for g-fans*, arXiv:2212.10745, 2022.

[13] Yann Palu, Vincent Pilaud, and Pierre-Guy Plamondon, *Non-kissing and non-crossing complexes for locally gentle algebras*, J. Comb. Algebra **3** (2019), no. 4, 401–438, DOI 10.4171/jca/35. MR4046035

LACIM, Université du Québec à Montréal, Canada
Email address: thomas.hugh_r@uqam.ca

Proceedings of Symposia in Pure Mathematics
Volume **110**, 2024
https://doi.org/10.1090/pspum/110/02006

Order polynomial product formulas and poset dynamics

Sam Hopkins

ABSTRACT. We survey all known examples of finite posets whose order polynomials have product formulas, and we propose the heuristic that these are the same posets with good dynamical behavior. Here the dynamics in question are the actions of promotion on the linear extensions of the poset and rowmotion on the P-partitions of the poset.

1. Introduction

A *linear extension* of a poset is a linear (i.e., total) ordering of its elements extending the partial ordering. Perhaps the single most important numerical invariant associated to a finite[1] poset P is its number $e(P)$ of linear extensions. Computing the number of linear extensions of a poset is a #P-complete problem [**BW91**], which means that we cannot hope for a "good" formula for this number in general. Nevertheless, especially within the context of algebraic combinatorics, there is great interest in obtaining good formulas – in the best cases, product formulas – for the number of linear extensions of special families of posets. The most famous such product formula is the celebrated hook-length formula [**FRT54**] for the number of linear extensions of a poset of Young diagram shape. There are also hook-length formulas for shifted shapes, and for rooted forests; and more generally, the d-complete posets of Proctor [**Pro14, KY19, NO19**], which include Young diagram shapes, shifted shapes, and rooted forests, have hook-length formulas counting their linear extensions.

Here we will concentrate on a finer combinatorial invariant of a poset P than its number of linear extensions: namely, its order polynomial. A *P-partition of height m* is a weakly order-preserving map $P \to \{0, 1, \ldots, m\}$; the *order polynomial* $\Omega_P(m)$ of P counts the number of P-partitions of height m.[2] The order polynomial is a polynomial in m of degree $\#P$, with leading coefficient equal to $e(P)/\#P!$. Posets with product formulas for their order polynomials are much rarer than posets with product formulas enumerating their linear extensions.

2020 *Mathematics Subject Classification.* Primary 06A07; Secondary 05E10, 05A15, 05E18, 05A19.

The author was supported in part by NSF grant #1802920.

[1] All posets we consider will be finite and we will drop this adjective from now on.

[2] Traditionally, as in [**Sta12**, Chapter 3], P-partitions are order-reversing and the order polynomial is what is $\Omega_P(m-1)$ in our notation. Our changes are superficial but lead to a cleaner presentation.

We will review all known examples of posets with product formulas for their order polynomials below. However, our goal is not just to survey families of posets with order polynomial product formulas, but also to advertise an apparently powerful heuristic which says that:

the posets with order polynomial product formulas are the same as the posets with good dynamical behavior.

Let us explain what we mean by poset dynamics. There are several constructions from algebraic combinatorics of interesting invertible operators acting on objects associated to a poset (e.g., linear extensions or P-partitions). And while these operators are defined for any poset, they tend to have good behavior (e.g., a small, predictable order and regular orbit structure) only for a select few families of posets. Here we will focus on two such operators: *promotion* and *rowmotion*.

Promotion is an invertible operator acting on the linear extensions of any poset. It was first defined and studied by Schützenberger [**Sch72**], in conjunction with a related involutive operator called evacuation. Schützenberger's initial motivation for studying these operators was the relation between evacuation and the RSK algorithm [**Sch63**] (see [**Sta99**, Chapter 7, §A1.2] for a modern account of this relation).

Rowmotion is an invertible operator acting on the order ideals of any poset, which has been studied by a number of authors over several decades [**BS74, DF90, FDF93, CFDF95**], with a renewed interest especially in the last 10 or so years because of a surprising connection to the combinatorics of root systems [**Pan09, SW12, Wil13**]. Einstein and Propp [**EP21**] gave a piecewise-linear extension of rowmotion to the entire order polytope $\mathcal{O}(P)$ of a poset P. By identifying the points in $\frac{1}{m}\mathbb{Z}^P \cap \mathcal{O}(P)$ with P-partitions of height m, we thus obtain an action of rowmotion on these P-partitions. It is this (piecewise-linear) action of rowmotion on P-partitions which we will mostly be concerned with.

Then, the heuristic we are proposing is more precisely that the following three properties of a poset P are related and tend to occur simultaneously:

(1) the order polynomial $\Omega_P(m)$ has a product formula;
(2) promotion acting on the linear extensions of P has good behavior;
(3) rowmotion acting on the P-partitions of height m has good behavior, for all m.

These three properties are not perfectly correlated, in that we have examples of posets which satisfy some but not all of them (although we know of no counterexamples to (3) $\Rightarrow$ (2) $\Rightarrow$ (1)). Nevertheless, we do think this heuristic is powerful, and we remark that it is powerful "in both directions": that is, both for finding posets with good dynamical behavior (see, e.g., [**HR22**]), and for finding posets with order polynomial product formulas (see, e.g., [**HL21**]).

1.1. A more detailed account of the heuristic. Let us give a more detailed account of what (1), (2), and (3) mean, and how they are related.

For $\Omega_P(m)$ to have a product formula, ideally all of its roots should be integers. There are also some interesting examples where the roots are *half*-integers, so we will consider this acceptable as well. Furthermore, P should come in a family, with certain numerical parameters attached to it, and we should be able to write $\Omega_P(m)$ in a simple way as a product of rational expressions involving these parameters.

Since $e(P) = \#P! \cdot \lim_{m\to\infty} \frac{\Omega_P(m)}{m^{\#P}}$, whenever we have a product formula for $\Omega_P(m)$ we also have one for $e(P)$.

As it turns out, we also always seem to get very nice q-analogs when P has an order polynomial product formula. Define

$$\Omega_P(m;q) := \prod_{\alpha} \frac{(1-q^{\kappa(m-\alpha)})}{(1-q^{-\kappa\alpha})},$$

where the product is over all roots α of $\Omega_P(m)$, with multiplicity, and κ is 1 if these roots are all integers and 2 if they are half-integers. Then, miraculously, in the examples we observe that $\Omega_P(m;q)$ is (for nonnegative integers m) a polynomial in q with nonnegative integer coefficients, which at $q=1$ is equal to $\Omega_P(m)$. Similarly, define

$$e(P;q) := \prod_{j=1}^{\#P}(1-q^{j\kappa}) \cdot \lim_{m\to\infty} \Omega_P(m;q) = \prod_{j=1}^{\#P}(1-q^{j\kappa}) \prod_{\alpha} \frac{1}{(1-q^{-\kappa\alpha})}.$$

Again, $e(P;q)$ is miraculously a q-analog of $e(P)$ in the sense that it is a polynomial in q with nonnegative integer coefficients which at $q=1$ is equal to $e(P)$. (See [**Sta90**] for discussion of when q-expressions of these forms are polynomials with nonnegative integer coefficients.)

Let $\mathcal{L}(P)$ denote the linear extensions of P and $\mathrm{Pro}\colon \mathcal{L}(P) \to \mathcal{L}(P)$ denote promotion. When we formally define promotion and go over the basics below, we will see why $\mathrm{Pro}^{\#P}$ is the "right power" of promotion to look at to find good behavior. For promotion to have good behavior, ideally $\mathrm{Pro}^{\#P}$ is the identity. There are also some interesting examples where $\mathrm{Pro}^{\#P}$ is a non-identity involutive poset automorphism: these are exactly the examples where the roots of $\Omega_P(m)$ are half-integers.

Moreover, whenever promotion of $\mathcal{L}(P)$ has good behavior, $e(P;q)$ is apparently a cyclic sieving polynomial for the action of promotion. We will review the cyclic sieving phenomenon later, but for now the important remark is that if a polynomial with an expression as a product of ratios of q-numbers is a cyclic sieving polynomial for some cyclic action, then this action has a very regular orbit structure: for instance, this means there is a product formula enumerating every symmetry class.

Let $\mathcal{PP}^m(P)$ denote the height m P-partitions and $\mathrm{Row}\colon \mathcal{PP}^m(P) \to \mathcal{PP}^m(P)$ denote rowmotion. Rowmotion only ever has good behavior when P is graded (i.e., all maximal chains of P have the same length). For a graded poset P we use $r(P)$ to denote the rank of P (i.e., the length of a maximal chain); we will see below why $\mathrm{Row}^{r(P)+2}$ is the "right power" of rowmotion to look at. For rowmotion to have good behavior, ideally $\mathrm{Row}^{r(P)+2}$ is the identity; but there are also interesting examples where it is an involutive automorphism, and again this happens when the roots of $\Omega_P(m)$ are half-integers. Moreover, when rowmotion of $\mathcal{PP}^m(P)$ has good behavior, $\Omega_P(m;q)$ is apparently a cyclic sieving polynomial for this action.

1.2. Whence all this? Where does this heuristic comes from, and why might these properties of a poset be related? The short answer is that these properties are indicative of some connection of the poset to *algebra*, especially, the representation theory of Lie algebras, Lie groups, Weyl groups, etc.

For instance, it often happens that $\mathcal{PP}^m(P)$ indexes a basis of an irreducible representation of a Lie algebra, in which case we can compute $\Omega_P(m)$ using the

Weyl dimension formula. Similarly, the q-analog $\Omega_P(m;q)$ can be obtained via a q-Weyl dimension formula for the principal specialization of the corresponding character.

Furthermore, the actions of promotion and rowmotion routinely have nice algebraic models as well. In the simplest cases, there are in fact diagrammatic models where the action is realized as rotation; but there are also examples where sophisticated tools from algebra like crystals and canonical bases are required. Indeed, these models are part of what make promotion and rowmotion so fascinating. We will review all known models of promotion and rowmotion below.

For the families of posets which have a direct connection to algebra, it is desirable to prove results uniformly, that is, without relying on classification theorems. We should also note that for some posets (such as the "chain of V's"), the relevant algebra has apparently yet to be uncovered.

1.3. The open problems. As we will see in the subsequent sections, simply by carrying out the program in Section 1.1 for the posets known to have an order polynomial product formula, we obtain many intriguing dynamical conjectures. But our heuristic also presents a few "meta-problems":

PROBLEM 1.1. Find formal relations between the properties (1), (2), and (3).

PROBLEM 1.2. Find more examples of posets satisfying (1), (2), and (3).

PROBLEM 1.3. Find a unified algebraic explanation for the good (enumerative and dynamical) behavior of the families of posets considered here.

Regarding Problem 1.1, we should remark that while there are several papers which study connections between promotion and rowmotion (see, e.g., [**SW12, DPS17, DSV19, BSV21, BSV24**]), we know of none which discusses a direct connection between promotion of $\mathcal{L}(P)$ and rowmotion of $\mathcal{PP}^m(P)$.

1.4. Acknowledgments. I thank Ira Gessel, Soichi Okada, Rebecca Patrias, Robert Proctor, Victor Reiner, Martin Rubey, Jessica Striker, Bruce Westbury, and Nathan Williams for useful discussion.

2. The posets

In this section we introduce the posets which have order polynomial product formulas. We assume the reader is familiar with poset basics as laid out for instance in [**Sta12**, Chapter 3]. All the properties of posets we are interested in decompose in a natural way over disjoint unions, so we will only consider connected posets. These properties also evidently translate directly from a poset P to its dual P^*. In fact, these properties (at least conjecturally) translate from a poset to any other poset with an isomorphic comparability graph (see [**Hop22**]). Therefore, we will not separately list posets with isomorphic comparability graphs to the ones below.

2.1. Shapes and shifted shapes. Several of the families of posets which have order polynomial product formulas will be Young diagram shapes or shifted shapes.

We assume the reader is familiar with the basics concerning partitions, shapes, and so on. We view a shape as a poset on its boxes with the partial order where $u \leq v$ means box u is weakly northwest of box v. The poset objects associated

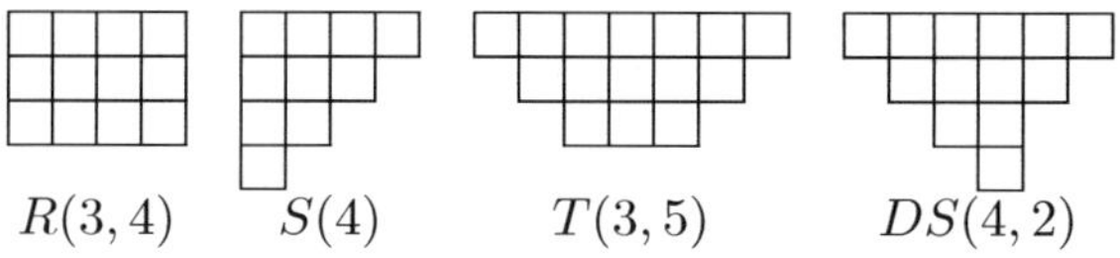

FIGURE 1. Examples of the families of shapes.

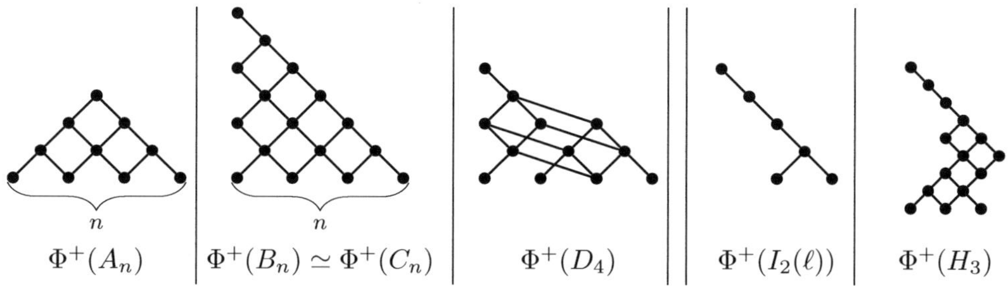

FIGURE 2. Left: some crystallographic root posets. Right: the non-crystallographic root posets of coincidental type.

to shapes have different traditional names: e.g., a linear extension is a standard Young tableau, a P-partition is a plane partition, etc.

We now define the relevant families of shapes. The *rectangle* $R(a, b)$ is the shape for the partition (b^a) using multiplicity notation. The *staircase* $S(n)$ is the shape for the partition $(n, n-1, \ldots, 1)$. The *shifted trapezoid* $T(a, b)$ is the shifted shape for the strict partition $(a+b-1, a+b-3, a+b-5, \ldots, b-a+1)$. The *shifted double staircase* $DS(n, k)$ is the shifted shape for the strict partition $(n, n-1, \ldots, 1) + (k, k-1, \ldots, 1)$. Observe that $T(n, n) = DS(n, n-1)$ and $T(n, n+1) = DS(n, n)$. Figure 1 depicts examples of these shapes.

We also define the arithmetic progression $AP(M, d, \ell)$ to be the shape for the partition $(M-d, M-2d, \ldots, M-\ell d)$. Observe that $R(a, b) = AP(b, 0, a)$ and $S(n) = AP(n+1, 1, n)$.

2.2. Root posets. The root posets are a very interesting family of posets coming from Lie theory, and some of them (specifically, the root posets of coincidental type) have order polynomial product formulas. We give only a very cursory account of root posets here; for a detailed account see [**Wil13**] or [**HPPW20**, §8].

Let Φ be an irreducible crystallographic root system of rank n, and W its Weyl group. We use Φ^+ to denote the positive roots of Φ. We view Φ^+ as a poset, called the *root poset*, where the partial order is $\alpha \leq \beta$ if and only if $\beta - \alpha$ is a nonnegative sum of simple roots. The poset Φ^+ has n minimal elements (the simple roots) and a unique maximal root (the highest root). It is a graded poset (with the rank function given by height). And it contains important numerical information about W. For instance, if $\Phi^+_0, \Phi^+_1, \ldots, \Phi^+_{r(\Phi^+)}$ are the ranks of this poset, then $(\#\Phi^+_0, \#\Phi^+_1, \ldots, \Phi^+_{r(\Phi^+)})$ is a partition and its conjugate partition is $(d_n - 1, d_{n-1} - 1, \ldots, d_1 - 1)$, where $d_1 \leq d_2 \leq \cdots \leq d_n$ are the degrees of W. In particular $r(P) + 2 = d_n = h$ is the Coxeter number of W. Some of these crystallographic root posets are depicted on the left in Figure 2.

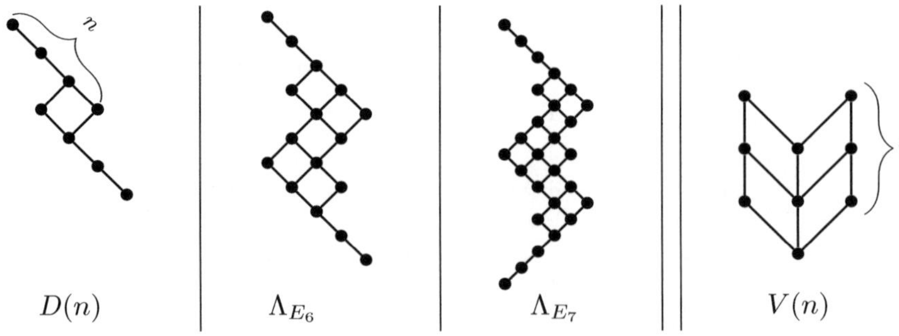

FIGURE 3. Left: the other minuscule posets besides the rectangle and shifted staircase. Right: the "chain of V's."

Now let Φ be an irreducible non-crystallographic root system. One could naively apply the same definition of partial order to Φ^+, but it would fail to have the desirable features discussed in the last paragraph. Armstrong [**Arm09**] gave an *ad hoc* construction of root posets $\Phi^+(I_2(\ell))$ and $\Phi^+(H_3)$ which do have these desirable features; these are depicted on the right in Figure 2. For H_4, the other non-crystallographic root system, there are either many or no analogous root posets, depending on exactly which properties one chooses [**CS15**].

Among all the complex reflection groups, a special sub-class are the so-called "coincidental types," which are those whose degrees form an arithmetic progression. For the finite Coxeter groups, these are the types A_n, $B_n \simeq C_n$, $I_2(\ell)$, and H_3. Their corresponding root posets are the *root posets of coincidental type*. These are all depicted in Figure 2. Observe that $\Phi^+(A_n) \simeq S(n)^*$ and $\Phi^+(B_n) \simeq T(n,n)^*$.

For Φ^+ a crystallographic root poset, there is a canonical involutive poset automorphism $\delta\colon \Phi^+ \to \Phi^+$ defined by $\delta(\alpha) := -w_0(\alpha)$ where $w_0 \in W$ is the longest element of the Weyl group. For the cases that concern us: δ is the reflection across the vertical axis of symmetry for $\Phi^+(A_n)$; and δ is the identity for $\Phi^+(B_n)$. By convention, we define a poset automorphism $\delta\colon \Phi^+ \to \Phi^+$ for Φ^+ a non-crystallographic root poset of coincidental type by: δ swaps the minimal elements of $\Phi^+(I_2(\ell))$ if ℓ is odd, and is the identity if ℓ is even; and δ is the identity for $\Phi^+(H_3)$.

2.3. Minuscule posets. The minuscule posets are another family of posets coming from Lie theory, with many remarkable properties. Again, we give only a cursory account; see [**HPPW20**, §4] for a detailed treatment.

Let $\mathfrak{g}$ be a simple Lie algebra over $\mathbb{C}$, with Φ its root system and W its Weyl group. A non-zero (integral, dominant) weight λ of $\mathfrak{g}$ is said to be minuscule if the Weyl group acts transitively on the weights of the corresponding highest weight irreducible representation V^λ. In this case, the Weyl orbit $W\lambda$ is a distributive lattice, where the partial order is again given by root order: i.e., $\nu \leq \mu$ means $\mu - \nu$ is a nonnegative sum of simple roots. The *minuscule poset* corresponding to the minuscule weight λ is the poset of join irreducible elements of the distributive lattice $W\lambda$.

If λ is minuscule, then it must be equal to some fundamental weight ω_i, and we can also describe the corresponding minuscule poset P as the order filter in Φ^+ generated by the corresponding simple root α_i: i.e., $P = \{\alpha \in \Phi^+ \colon \alpha \geq \alpha_i\}$. The minuscule poset P is always graded, and has a unique minimal and a unique maximal element. Furthermore, it has a canonical involutive poset anti-automorphism $\iota\colon P \to P^*$ which is induced from the action of multiplication by w_0 on $W\lambda$.

The minuscule posets, up to isomorphism, have been classified: they are the *rectangle* $R(a,b)$, the *shifted staircase* $DS(n,0)$, the *"propeller poset"* $D(n)$, and two *exceptional posets* $\Lambda_{E_6}, \Lambda_{E_7}$ coming from the types E_6 and E_7. The last three of these are depicted in Figure 3.

2.4. The "chain of V's". The final family of posets with an order polynomial product formula is the *"chain of V's"*: $V(n) :=$  $\times [n]$, the Cartesian product of the 3-element "V"-shaped poset  and the n-element chain $[n]$ (see the right of Figure 3). It was first studied by Kreweras and Niederhausen [**KN81**]. It has a rather different structure than the other examples: for instance, in the other examples each element covers and is covered by at most two elements, but this is not true for $V(n)$. Let us use $\delta\colon V(n) \to V(n)$ to denote the involutive poset automorphism which is reflection across the vertical axis of symmetry of the "V."

3. Order polynomial product formulas

In this section we review the order polynomial product formulas for the posets introduced in Section 2, and briefly explain where these formulas come from.

3.1. Symmetry classes of plane partitions. The origin of all these kind of product formulas is MacMahon's investigation of plane partitions, and the subsequent investigation of plane partitions symmetry classes. See [**Kra16**] for a complete history.

An $a \times b$ plane partition of height m is an $a \times b$ array $\pi = (\pi_{i,j})_{\substack{1\leq i\leq a,\\ 1\leq j\leq b}}$ of nonnegative integers which is weakly decreasing in rows and columns and for which the largest entry is at most m. We denote the set of such plane partitions by $\mathcal{PP}^m(a\times b)$. Observe that $\mathcal{PP}^m(a\times b)$ is exactly the same as $\mathcal{PP}^m(R(a,b))$.

We define the size of a plane partition $\pi \in \mathcal{PP}^m(a\times b)$ by $|\pi| := \sum_{\substack{1\leq i\leq a,\\ 1\leq j\leq b}} \pi_{i,j}$. MacMahon [**Mac04**] obtained the following celebrated product formula for the size generating function of plane partitions:

$$\sum_{\pi\in\mathcal{PP}^m(a\times b)} q^{|\pi|} = \prod_{i=1}^{a}\prod_{j=1}^{b} \frac{(1-q^{m+i+j-1})}{(1-q^{i+j-1})}$$

This is the q-analog $\Omega_P(m;q)$ from Section 1.1 for $P = R(a,b)$.

The order polynomials for other posets beyond $R(a,b)$ arise when considering symmetries of plane partitions. The symmetries relevant to us are transposition $\mathrm{Tr}\colon \mathcal{PP}^m(n\times n) \to \mathcal{PP}^m(n\times n)$ given by $\mathrm{Tr}(\pi)_{i,j} := \pi_{j,i}$, and complementation $\mathrm{Co}\colon \mathcal{PP}^m(a\times b) \to \mathcal{PP}^m(a\times b)$ given by $\mathrm{Co}(\pi)_{i,j} := m - \pi_{a+1-i,b+1-j}$.

Plane partitions $\pi \in \mathcal{PP}^m(n\times n)$ with $\mathrm{Tr}(\pi) = \pi$ are called symmetric; they are evidently in bijection with $\mathcal{PP}^m(DS(n,0))$. MacMahon [**Mac99**] conjectured, and Andrews [**And78**] and Macdonald [**Mac79**] proved, the following product formula for the size generating function of symmetric plane partitions:

$$\sum_{\pi\in\mathcal{PP}^m(n\times n),\mathrm{Tr}(\pi)=\pi} q^{|\pi|} = \prod_{1\leq i<j\leq n} \frac{(1-q^{2(i+j+m-1)})}{(1-q^{2(i+j-1)})} \cdot \prod_{i=1}^{n} \frac{(1-q^{2i+m-1})}{(1-q^{2i-1})}.$$

There is a second q-analog for symmetric plane partitions as well. Namely, for a plane partition $\pi \in \mathcal{PP}^m(n \times n)$ define $|\pi|' := \sum_{1 \leq i \leq j \leq n} \pi_{i,j}$. Then, Bender and Knuth [**BK72**] conjectured, and Gordon [**Gor83**], Andrews [**And77**], and Macdonald [**Mac79**] proved:

$$\sum_{\pi \in \mathcal{PP}^m(n\times n), \mathrm{Tr}(\pi)=\pi} q^{|\pi|'} = \prod_{1 \leq i \leq j \leq n} \frac{(1-q^{i+j+m-1})}{(1-q^{i+j-1})}.$$

It is this second q-analog which is the $\Omega_P(m;q)$ from Section 1.1 for $P = DS(n,0)$.

Plane partitions $\pi \in \mathcal{PP}^{2m}(n \times n)$ with $\mathrm{Tr}(\pi) = \mathrm{Co}(\pi)$ are called transpose-complementary; they are in bijection with $\mathcal{PP}^m(S(n))$. These were first enumerated by Proctor (see Theorem 3.5).

Plane partitions $\pi \in \mathcal{PP}^{2m}(n \times n)$ with $\mathrm{Tr}(\pi) = \pi$ and $\mathrm{Co}(\pi) = \pi$ are called symmetric self-complementay; they are in bijection with $\mathcal{PP}^m(T(\lfloor n/2 \rfloor, \lceil n/2 \rceil))$. These were again first enumerated by Proctor (see Theorem 3.4).

3.2. Minuscule posets. Let P be a poset. For a P-partition $\pi \in \mathcal{PP}^m(P)$, define its size to be $|\pi| := \sum_{p \in P} \pi(p)$. Let $F_P(m;q) := \sum_{\pi \in \mathcal{PP}^m(P)} q^{|\pi|}$ denote the size generating function of these P-partitions. The basic theory of P-partitions (see, e.g., [**Sta12**, §3.15.2]) says that

$$(1-q)(1-q^2)\cdots(1-q^{\#P}) \cdot \lim_{m\to\infty} F_{P^*}(m;q) = \sum_{L \in \mathcal{L}(P)} q^{\mathrm{maj}(L)},$$

the major index generating function of linear extensions of P (with respect to any fixed natural labeling). We remark that every d-complete poset P has a product formula for $\lim_{m\to\infty} F_{P^*}(m;q)$ [**Pro14, KY19, NO19**].

THEOREM 3.1 (Proctor [**Pro84**]). *Let P be a minuscule poset. Then*

$$F_P(m;q) = F_{P^*}(m;q) = \prod_{p \in P} \frac{(1-q^{m+r(p)+1})}{(1-q^{r(p)+1})},$$

where $r\colon P \to \mathbb{N}$ is the rank function of P.

Theorem 3.1 gives the q-analog $\Omega_P(m;q)$ from Section 1.1 for P a minuscule poset. Theorem 3.1 is due to Proctor [**Pro84**] (although observe that for $R(a,b)$ and $DS(n,0)$ it is equivalent to results just mentioned in Section 3.1). To prove this theorem he used Standard Monomial Theory, which explains that $\mathcal{PP}^m(P)$ indexes a basis of the representation $V^{m\lambda}$ when P is the minuscule poset corresponding to the minuscule weight λ. Proctor moreover conjectured that the minuscule posets are the *only* posets which have a product formula for $F_P(m;q)$ of this form.

3.3. Root posets. Let W be a finite Coxeter group of rank n. Define the q-W-Catalan number by

$$\mathrm{Cat}(W;q) := \prod_{i=1}^{n} \frac{(1-q^{h+d_i})}{(1-q^{d_i})}$$

where $d_1, \ldots, d_n$ are the degrees of W and h its Coxeter number. It is uniformly known that $\mathrm{Cat}(W;q)$ is a polynomial with nonnegative integer coefficients (see [**BR11**]).

An *order ideal* of a poset P is a downwards-closed subset $I \subseteq P$ (i.e., a subset for which $y \in I$ and $x \leq y \in P$ implies $x \in I$). We use $\mathcal{J}(P)$ to denote the order ideals of P. We have a natural identification $\mathcal{J}(P) \simeq \mathcal{PP}^1(P)$ where an order ideal corresponds to the indicator function of its complement.

THEOREM 3.2 (Cellini–Papi [**CP02**], Haiman [**Hai94**]). *Let Φ be a crystallographic root system and W its Weyl group. Then $\#\mathcal{J}(\Phi^+) = \mathrm{Cat}(W; q := 1)$.*

The proof of Theorem 3.2 is uniform; however, there is no known statistic for order ideals of which $\mathrm{Cat}(W;q)$ is the generating function.

Now let W be a finite Coxeter group of coincidental type. Define the *q-W-multi-Catalan number* by

$$\mathrm{Cat}(W, m; q) := \prod_{j=0}^{m-1} \prod_{i=1}^{n} \frac{(1 - q^{h+d_i+2j})}{(1 - q^{d_i+2j})}.$$

It is known, in a case-by-case fashion, that $\mathrm{Cat}(W, m; q)$ is a polynomial in q with nonnegative integer coefficients (for instance, for Type A this follows from [**Pro90**, Theorem 1, Case 'CGI']; for other types it can be deduced from consideration of the poset's minuscule doppelgänger in the sense of Section 3.4).

The multi-Catalan numbers are **not** the same as the more well-known Fuss–Catalan numbers. They first appeared, with this name, in the paper of Ceballos–Labbé–Stump [**CLS14**] which studied multi-triangulations and the multi-cluster complex. Our interest in these numbers is, however, the following:

THEOREM 3.3 ([**Pro83, Pro90, Wil13**]). *Let Φ be a coincidental type root system and W its corresponding Coxeter group. Then $\Omega_{\Phi^+}(m) = \mathrm{Cat}(W, m; q := 1)$.*

Theorem 3.3 gives the q-analog $\Omega_P(m;q)$ from Section 1.1 for P a root poset of coincidental type. The proof of Theorem 3.3 is case-by-case, with the difficult cases of $\Phi^+(A_n)$ and $\Phi^+(B_n)$ proved by Proctor [**Pro90, Pro83**] using representations of the symplectic group. The cases $\Phi^+(I_2(\ell))$ and $\Phi^+(H_3)$ were checked by Williams [**Wil13**].

3.4. Doppelgängers.

Following [**HPPW20**], we call a pair of posets

$$(P, Q) \in \{(R(a,b), T(a,b)^*), (DS(5,0), \Phi^+(H_3)), (D(\ell), \Phi^+(I_2(2\ell)))\}$$

a *minuscule doppelgänger pair*.

THEOREM 3.4 ([**Pro83**]). *Let (P, Q) be a minuscule doppelgänger pair. Then we have $\Omega_P(m) = \Omega_Q(m)$ for all $m \geq 1$.*

Since minuscule posets have product formulas for their order polynomials (Theorem 3.1), Theorem 3.4 says their doppelgängers do too. The difficult case of Theorem 3.4, the rectangle/trapezoid pair, was yet again established by Proctor [**Pro83**] using representations of the symplectic group. The other cases are an easy check.

In [**HPPW20**], the authors gave a uniform bijection between $\mathcal{PP}^m(P)$ and $\mathcal{PP}^m(Q)$ (and also a uniform bijection between $\mathcal{L}(P)$ and $\mathcal{L}(Q)$) for any minuscule doppelgänger pair (P, Q). In [**Hop22**] it was suggested that minuscule doppelgänger pairs are "very similar." We will see more similarities below.

3.5. Other examples.

THEOREM 3.5 (Proctor). *For $P = AP(M, d, \ell)$ an arithmetic progression,*

$$\Omega_P(m) = \prod_{\substack{(i,j)\in P,\\ \ell+c(i,j)\leq M-id}} \frac{m+\ell+c(i,j)}{\ell+c(i,j)} \cdot \prod_{\substack{(i,j)\in P,\\ \ell+c(i,j)>M-id}} \frac{(d+1)m+\ell+c(i,j)}{\ell+c(i,j)},$$

where $c(i,j) := j - i$ is the content of the box $(i,j) \in P$.

Theorem 3.5 is due to Proctor. The case $d = 1$ has an interpretation in terms of the representation theory of the symplectic group (see [**Pro88**]), and the minuscule case $d = 0$ has an interpretation in terms of the representation theory of the general linear group. However, for general d, Proctor's proof instead manipulates a determinantal formula due to MacMahon (see [**Sta99**, Exercise 7.101]).

THEOREM 3.6 (Kreweras–Niederhausen [**KN81**]). *For the "chain of V's,"*

$$\Omega_{V(n)}(m) = \frac{\prod_{i=1}^{n}(m+1+i)\prod_{i=1}^{2n}(2m+i+1)}{(n+1)!(2n+1)!}.$$

Theorem 3.6 is due to Kreweras and Niederhausen [**KN81**]. Their proof uses some basic P-partition theory together with a lot of clever algebraic manipulation and recurrences. Earlier, Kreweras [**Kre65**] obtained the product formula for $e(V(n))$, which has an interpretation in terms of a 3 candidate ballot problem.

THEOREM 3.7 (Hopkins–Lai [**HL21**]; Okada [**Oka21b**]). *For the shifted double staircase,*

$$\Omega_{DS(n,k)}(m) = \prod_{1\leq i\leq j\leq n} \frac{m+i+j-1}{i+j-1} \cdot \prod_{1\leq i\leq j\leq k} \frac{m+i+j}{i+j}.$$

The proof of Theorem 3.7 in [**HL21**] is based on the theory of lozenge tilings of the triangular lattice and the Kuo condensation recurrence technique. In [**Oka21b**], Okada proves Theorem 3.7 using "intermediate symplectic group" [**Pro88**] characters; in fact, he obtains a q-analog $\Omega_{DS(n,k)}(m;q)$.

4. Promotion of linear extensions

In this section we survey the posets which have good behavior of promotion.

4.1. Definitions and basics. Let P be a poset on n elements. For our purposes it is best to represent a linear extension of P as a list $(p_1, p_2, \ldots, p_n)$ of the elements of P, each appearing once, for which $p_i \leq p_j$ implies that $i \leq j$. We then define for each $i = 1, \ldots, n-1$ the *Bender–Knuth involution* $\tau_i \colon \mathcal{L}(P) \to \mathcal{L}(P)$ by

$$\tau_i(p_1, \ldots, p_n) := \begin{cases} (p_1, \ldots, p_{i-1}, p_{i+1}, p_i, p_{i+2}, \ldots, p_n) & \text{if } p_i,\ p_{i+1} \text{ incomparable;} \\ (p_1, \ldots, p_n) & \text{otherwise.} \end{cases}$$

Promotion $\mathrm{Pro} \colon \mathcal{L}(P) \to \mathcal{L}(P)$ is the following composition of these τ_i:

$$\mathrm{Pro} := \tau_{n-1} \circ \tau_{n-2} \circ \cdots \circ \tau_1.$$

Evacuation $\mathrm{Evac} \colon \mathcal{L}(P) \to \mathcal{L}(P)$ is the following composition of the τ_i:

$$\mathrm{Evac} := (\tau_1) \circ (\tau_2 \circ \tau_1) \circ \cdots \circ (\tau_{n-2} \circ \cdots \circ \tau_2 \circ \tau_1) \circ (\tau_{n-1} \circ \cdots \circ \tau_2 \circ \tau_1).$$

There is a duality $\mathcal{L}(P) \to \mathcal{L}(P^*)$ which sends $L = (p_1, \ldots, p_n)$ to $L^* = (p_n, \ldots, p_1)$. Dual evacuation $\mathrm{Evac}^* \colon \mathcal{L}(P) \to \mathcal{L}(P)$ is defined by $\mathrm{Evac}^*(L) := \mathrm{Evac}(L^*)^*$.

The following are the basic results concerning promotion and evacuation established by Schützenberger [**Sch72**]; see also the presentation of Stanley [**Sta09**]:

PROPOSITION 4.1 (Schützenberger [**Sch72**]). *For any poset P,*

- Evac *and* Evac* *are both involutions;*
- Evac $\circ$ Pro $=$ Pro^{-1} $\circ$ Evac*;*
- Pro$^{\#P}$ $=$ Evac* $\circ$ Evac.

Proposition 4.1 explains why Pro$^{\#P}$ is the "right" power of promotion to look at. As mentioned in Section 1.1, Pro$^{\#P}$ is ideally the identity, but we will also see interesting examples where it is a non-identity involutive poset automorphism. Let us remark that if Pro$^{\#P}$ is a poset automorphism, then it must be an involution. Indeed, suppose Pro$^{\#P}$ $= \delta$ is an automorphism; then by conjugating $\delta =$ Evac$^* \circ$ Evac by Evac we get $\delta =$ Evac $\circ$ Evac*, since evacuation commutes with any automorphism; in other words, we have $\delta = \delta^{-1}$, as claimed.

4.2. Models. We now review models (diagrammatic, algebraic, etc.) for promotion and evacuation for certain families of posets. These models lead to a precise understanding of the order and orbit structure of these operators.

4.2.1. *Rotation of noncrossing matchings and webs.* A noncrossing matching of $[2n]$ is a partition of $[2n]$ into blocks of size 2 for which there is no pair of crossing blocks. D. White observed that promotion of standard Young tableaux of $2 \times n$ rectangular shape corresponds to rotation of noncrossing matchings of $[2n]$ (see for instance [**Rho10**, §8]).

Webs are a class of planar graphs Kuperberg introduced to study the invariant theory of Lie algebras. Khovanov and Kuperberg [**KK99**] (see also Tymoczko [**Tym12**]) defined a bijection between standard Young tableaux of $3 \times n$ rectangular shape and a subset of $\mathfrak{sl}_3$-webs. Petersen, Pylyavskyy, and Rhoades [**PPR09**] showed that under this bijection, promotion of tableaux corresponds to rotation of webs. Russell [**Rus13**] (see also Patrias [**Pat19**]) showed that rotation for a broader class of $\mathfrak{sl}_3$-webs corresponds to promotion of certain *semi*standard tableaux of $3 \times n$ rectangular shape (see Section 5.2.3). Finally, Hopkins and Rubey [**HR22**] showed that linear extensions of $V(n)$ can be encoded as certain 3-edge-colored $\mathfrak{sl}_3$-webs for which promotion again corresponds to rotation.

In these correspondences between linear extensions and diagrams it is also possible to show that evacuation corresponds to reflection across a diameter (see, e.g., [**PP23**]), so that the full dihedral action is apparent.

4.2.2. *Rotation of reduced words and Edelman–Greene-style bijections.* Let W be a finite Coxeter group with Φ its root system. For a reduced word $s_{\alpha_{i_1}} s_{\alpha_{i_2}} \cdots s_{\alpha_{i_p}}$ of the longest word $w_0 \in W$, we define its (twisted) rotation to be $s_{\alpha_{i_2}} \cdots s_{\alpha_{i_p}} s_{\delta(\alpha_{i_1})}$, which is again a reduced word of w_0.

For W of coincidental type, there is an equivariant bijection between linear extensions of Φ^+ under promotion and reduced words of w_0 under rotation. For Type A, this is due to Edelman and Greene [**EG87**]. For Type B, it was established by Haiman [**Hai89**] and Kraśkiewicz [**Kra89**]. For $I_2(\ell)$ and H_3 it was shown by Williams [**Wil13**].

We believe that evacuation should correspond to (twisted) reflection of the reduced word under these Edelman–Greene-style bijections, but we know of nowhere in the literature where this is explicitly stated.

4.2.3. *Kazhdan–Lusztig cell representations.* The theory of Kazhdan–Lusztig cells gives a canonical basis for any irreducible representation of the symmetric group S_n. Rhoades [**Rho10**] showed that for an irreducible symmetric group representation of rectangular shape, the action of the long cycle $c = (1, 2, \ldots, n)$ corresponds to promotion of standard tableaux in the Kazhdan–Lusztig basis. Rhoades's result built on an earlier results of Berenstein–Zelevinsky [**BZ96**] and Stembridge [**Ste96**], who showed that for *any* irreducible symmetric group representation, the action of the longest word $w_0 = n(n-1)\cdots 1$ corresponds (up to sign) to evacuation of standard tableaux in the Kazhdan–Lusztig basis.

4.2.4. *Crystals and cactus group actions.* The cactus group action on the tensor product of crystals for simple Lie algebra representations, as defined by Henriques and Kamnitzer [**HK06**], gives rise to notions of evacuation and promotion acting on the corresponding highest weight words of weight zero: see [**FK14, Wes16, PRW20**]. In the case of V being the vector representation of $\mathfrak{sl}_k$, this cactus group promotion for weight zero highest weight words of $V^{\otimes kn}$ corresponds to promotion of standard Young tableaux of $k \times n$ rectangular shape, and similarly for evacuation.

4.2.5. *The Wronski map and monodromy.* For a list $f_1(z), f_2(z), \ldots, f_d(z)$ of d linearly independent polynomials in $\mathbb{C}[z]$ of degree at most $n-1$, their Wronskian is the determinant of the $d \times d$ matrix whose rows are the derivatives $f_j^{(i)}(z)$ of these polynomials for $i = 0, \ldots, d-1$. Up to scale, the Wronskian depends only the linear span of the $f_j(z)$. We thus obtain the Wronski map from the Grassmannian $\mathrm{Gr}(d, n)$ to projective space $\mathbb{P}^{n-1}$.

Standard Young tableaux of $d \times (n-d)$ rectangular shape index the fibers of the Wronski map. In [**Pur13**], Purbhoo showed that a certain monodromy action for the Wronski map corresponds to promotion for these tableaux. Moreover, in [**Pur18**], Purbhoo showed that by restricting to those preimages which lie in either the orthogonal or Lagrangian Grassmannian (under a certain embedding of these Type B/C Grassmannians into the usual Type A Grassmannian), one can similarly obtain the action of promotion on standard tableaux of shifted staircase or staircase shape.

4.2.6. *Evacuation of minuscule posets.* The extension of Schützenberger's theory of jeu de taquin [**Sch77**] to cominuscule Schubert calculus due to Thomas and Yong [**TY16**, Lemma 5.2] implies that evacuation of a minuscule poset has a simple description in terms of its canonical anti-automorphism:

THEOREM 4.2 ([**Sch77, TY16**]). *For a minuscule poset P and $L \in \mathcal{L}(P)$, we have* $\mathrm{Evac}(L) = \iota(L)^*$.

4.2.7. *Doppelgängers bijections.* The minuscule doppelgänger pairs have the same orbit structure of promotion:

THEOREM 4.3 (Haiman [**Hai89, Hai92**]). *For (P, Q) a minuscule doppelgänger pair, there is a bijection between $\mathcal{L}(P)$ and $\mathcal{L}(Q)$ which commutes with promotion.*

The bijection for the difficult case of Theorem 4.3, the rectangle/trapezoid pair, is due to Haiman [**Hai89, Hai92**]. The other cases are an easy check.

4.3. Order. We now review the posets P for which $\mathrm{Pro}^{\#P}$ can be described.

THEOREM 4.4 ([**Sch77, EG87, Hai92**]). *For $P = R(a,b), T(a,b), DS(n,k)$, $\mathrm{Pro}^{\#P}$ is the identity. For $P = S(n)$, $\mathrm{Pro}^{\#P}$ is transposition.*

The case of Theorem 4.4 for the rectangle $R(a,b)$ follows from Schützenberger's theory of jeu de taquin [**Sch77**]. The case of the staircase $S(n)$ is due to Edelman–Greene [**EG87**]. The cases of $T(a,b)$ and $DS(n,k)$ are due to Haiman [**Hai92**], who developed a method which recaptures the rectangle and staircase cases as well. In fact, Haiman and Kim [**HK92**] showed that the *only* shapes and shifted shapes for which $\mathrm{Pro}^{\#P}$ is the identity or transposition are those appearing in Theorem 4.4.

THEOREM 4.5. *For P a minuscule poset,* $\mathrm{Pro}^{\#P}$ *is the identity.*

Theorem 4.5 follows, for instance, from Theorem 4.2.

THEOREM 4.6. *For $P = \Phi^+$ a root poset of coincidental type,* $\mathrm{Pro}^{\#P} = \delta$.

Theorem 4.6 follows, for instance, from the reduced word bijections (see the discussion in Section 4.2.2).

THEOREM 4.7 (Hopkins–Rubey [**HR22**]). *For $P = V(n)$,* $\mathrm{Pro}^{\#P} = \delta$.

4.4. Orbit structure. We now discuss the orbit structure of promotion for the posets with good promotion behavior.

The cyclic sieving phenomenon of Reiner–Stanton–White [**RSW04**] provides a very compact way to record the orbit structure of a cyclic group action. Recall that if X is a combinatorial set, $C = \langle c\rangle$ is a cyclic group of order n acting on X, and $f(q) \in \mathbb{N}[q]$ is a polynomial with nonnegative integer coefficients, then we say that the triple $(X, C, f(q))$ exhibits *cyclic sieving* if for all integers k,

$$\#\{x \in X : c^k(x) = x\} = f(q := \zeta^k),$$

where $\zeta := e^{2\pi i/n}$ is a primitive nth root of unity. As mentioned in Section 1.1, cyclic sieving phenomena (CSPs) where the polynomial has a simple product formula are especially valuable, because they imply a product formula for every symmetry class.

For the minuscule posets there is a beautiful such CSP:

THEOREM 4.8 ([**Rho10, Pur18, SSD14**]). *Let P be a minuscule poset. Then* $(\mathcal{L}(P), \langle \mathrm{Pro}\rangle, f(q))$ *exhibits cyclic sieving, where*

$$f(q) := \sum_{L \in \mathcal{L}(P)} q^{\mathrm{maj}(L)} = (1-q)(1-q^2)\cdots(1-q^{\#P}) \cdot \prod_{p\in P} \frac{1}{(1-q^{r(p)+1})}.$$

The case of Theorem 4.8 for the rectangle $R(a,b)$ is due to Rhoades [**Rho10**]. The exceptional cases $\Lambda_{E_6}, \Lambda_{E_7}$ are a finite check which is easily carried out by computer. The case $D(n)$ is trivial. For the shifted staircase $DS(n,0)$, Purbhoo [**Pur18**, Theorem 5.1(i)] gave an interpretation of this fixed point count in terms of ribbon tableaux; and in [**SSD14**] it was verified that this matches the CSP evaluation. It would be preferable to have a uniform proof of Theorem 4.8.

For the other posets with good promotion behavior, we have conjectural CSPs:

CONJECTURE 4.9. *Let Φ be a rank n root system of coincidental type, W its Coxeter group, and h its Coxeter number. Let $C = \langle c\rangle \simeq \mathbb{Z}/nh\mathbb{Z}$ act on $\mathcal{L}(\Phi^+)$ via $c(L) := \mathrm{Pro}(L)$. Then $(\mathcal{L}(\Phi^+), C, f(q))$ exhibits cyclic sieving, where*

$$f(q) := (1-q^2)(1-q^4)\cdots(1-q^{nh}) \cdot \lim_{m\to\infty} \mathrm{Cat}(W,m;q) \in \mathbb{N}[q].$$

The cases $\Phi = B_n, H_3, I_2(2\ell)$ of Conjecture 4.9 follow from Theorems 4.3 and 4.8. The case $\Phi = I_2(2\ell+1)$ is trivial. Hence, the only open case is $\Phi = A_n$, for which N. Williams conjectured this CSP in a different but equivalent form c. 2010. Purbhoo [**Pur18**, Theorem 5.1(ii)] again gave an interpretation of this fixed point count in terms of ribbon tableaux, so possibly this conjecture could be resolved as in [**SSD14**].

CONJECTURE 4.10 (Hopkins–Rubey [**HR22**]). *For all $n \geq 1$, the rational expression*

$$f(q) := \frac{\prod_{i=1}^{3n}(1-q^{2i})}{\prod_{i=2}^{2n+1}(1-q^{i})\prod_{i=2}^{n+1}(1-q^{2i})}$$

is in $\mathbb{N}[q]$, and $(\mathcal{L}(V(n)), \langle \mathrm{Pro} \rangle, f(q))$ exhibits cyclic sieving.

CONJECTURE 4.11. *For all $1 \leq k \leq n$, the rational expression*

$$f(q) := \frac{\prod_{i=1}^{n(n+1)/2+k(k+1)/2}(1-q^{i})}{\prod_{1\leq i\leq j\leq n}(1-q^{i+j-1})\prod_{1\leq i\leq j\leq k}(1-q^{i+j})}$$

is in $\mathbb{N}[q]$, and $(\mathcal{L}(DS(n,k)), \langle \mathrm{Pro} \rangle, f(q))$ exhibits cyclic sieving.

Finally, we note that for evacuation acting on the linear extensions $\mathcal{L}(P)$ of *any* poset P, there is a CSP using the (co)major index generating function [**Sta09**, §3].

5. Rowmotion of order ideals and P-partitions

In this section we survey the posets which have good behavior of rowmotion.

5.1. Definitions and basics. Let P be a poset. As discussed in Section 1, rowmotion was originally defined as an action on $\mathcal{J}(P)$. But, following [**EP21**], we will right away define it as a piecewise-linear action on $\mathcal{PP}^m(P)$ for any m. For $p \in P$ we define the *(piecewise-linear) toggle* $\tau_p \colon \mathcal{PP}^m(P) \to \mathcal{PP}^m(P)$ by

$$\tau_p(\pi)(q) := \begin{cases} \pi(q) & \text{if } p \neq q; \\ \min(\{\pi(r) \colon p \lessdot r\}) + \max(\{\pi(r) \colon r \lessdot p\}) - \pi(p) & \text{if } p = q, \end{cases}$$

where $\min(\varnothing) := m$ and $\max(\varnothing) := 0$. *Rowmotion* $\mathrm{Row} \colon \mathcal{PP}^m(P) \to \mathcal{PP}^m(P)$ is then the following composition of these τ_i:

$$\mathrm{Row} := \tau_{p_1} \circ \tau_{p_2} \circ \cdots \circ \tau_{p_n},$$

where $(p_1, \ldots, p_n)$ is any linear extension of P. The traditional case of order ideal rowmotion is recovered via the identification $\mathcal{J}(P) \simeq \mathcal{PP}^1(P)$.

Rowmotion only ever has good behavior when P is graded (indeed, the name "rowmotion" indicates toggling "row-by-row," i.e., "rank-by-rank"). So from now on assume P is graded and $P_0, P_1, \ldots, P_{r(P)}$ are its ranks. Define $\tau_i := \prod_{p \in P_i} \tau_p$ for $i = 0, \ldots, r(P)$ (these toggles all commute, so this product makes sense). Observe that $\mathrm{Row} = \tau_0 \circ \cdots \circ \tau_{r(P)-1} \circ \tau_{r(P)}$. In analogy with promotion/evacuation, let us then define *rowvacuation* $\mathrm{Rvac} \colon \mathcal{PP}^m(P) \to \mathcal{PP}^m(P)$ by

$$\mathrm{Rvac} := (\tau_{r(P)}) \circ (\tau_{r(P)-1} \circ \tau_{r(P)}) \circ \cdots \circ (\tau_1 \circ \cdots \circ \tau_{r(P)-1} \circ \tau_{r(P)}) \circ (\tau_0 \circ \cdots \circ \tau_{r(P)-1} \circ \tau_{r(P)})$$

There is a duality $\mathcal{PP}^m(P) \to \mathcal{PP}^m(P^*)$ given by $\pi^*(p) := m - \pi(p)$. Dual rowvacuation $\mathrm{Rvac}^* \colon \mathcal{PP}^m \to \mathcal{PP}^m(P)$ is defined by $\mathrm{Rvac}^*(\pi) := \mathrm{Rvac}(\pi^*)^*$.

The arguments from [**Sta09**] which establish Proposition 4.1 are very formal (they only use that the τ_i are involutions and that τ_i and τ_j commute if $|i-j| \geq 2$) and apply here as well, so that we have:

PROPOSITION 5.1. *For any graded poset P,*

- Rvac *and* Rvac^* *are both involutions;*
- $\mathrm{Rvac} \circ \mathrm{Row} = \mathrm{Row}^{-1} \circ \mathrm{Rvac}$*;*
- $\mathrm{Row}^{r(P)+2} = \mathrm{Rvac}^* \circ \mathrm{Rvac}$.

Proposition 5.1 explains why $\mathrm{Row}^{r(P)+2}$ is the "right" power of rowmotion to look at. Again, ideally $\mathrm{Row}^{r(P)+2}$ is the identity, but in some interesting cases it is an involutive poset automorphism. The same argument as for promotion shows that if $\mathrm{Row}^{r(P)+2}$ is a poset automorphism, it must be an involution.

N.B.: often "left-to-right" toggling of order ideals or P-partitions is studied, as opposed to the "top-to-bottom" toggling of rowmotion. This left-to-right toggling is routinely called "promotion," but to avoid confusion with promotion of linear extensions, we will not use that term. In practice, the techniques of [**SW12**] can always be used to show that left-to-right toggling is conjugate to top-to-bottom toggling.

5.2. Models. We now review models for rowmotion and rowvacuation for certain families of posets. As with the models for promotion and evacuation, these models give us a precise understanding of the order and orbit structure.

5.2.1. *Rotation of binary words, and parabolic cosets of Weyl groups.* Under the "Stanley–Thomas word" bijection (see [**Sta09, PR15**]), rowmotion of $\mathcal{J}(R(a,b))$ corresponds to rotation of binary words with a 1's and b 0's. Extending this description, Rush and Shi [**RS13**] showed that if P is the minuscule poset corresponding to the minuscule weight λ, then under the natural isomorphism $\mathcal{J}(P) \simeq W/W_J$, where W_J is the parabolic subgroup of the Weyl group W stabilizing λ, the action of rowmotion is conjugate to the action of a Coxeter element $c \in W$.

5.2.2. *Kreweras complementation for noncrossing partitions, in all types.* A noncrossing partition of $[n]$ is a set partition of $[n]$ for which there is no pair of crossing blocks. The noncrossing partitions form a lattice, and the Kreweras complementation is an operator acting on this lattice which has order $2n$ (its square is rotation).

Now let W be a finite Coxeter group and Φ its root system. By analogy, the noncrossing partitions for W are the elements less than some fixed Coxeter element $c \in W$ in absolute order, and Kreweras complementation is the operator $w \mapsto cw^{-1}$ on these noncrossing partitions. Proving conjectures of Panyushev [**Pan09**] and Bessis–Reiner [**BR11**], Amstrong, Stump, and Thomas [**AST13**] showed that when W is a Weyl group, rowmotion of $\mathcal{J}(\Phi^+)$ is in equivariant bijection with Kreweras complementation of the noncrossing partitions of W (and for non-crystallographic types, see [**CS15**]).

In [**DH21**], rowvacuation of $\mathcal{J}(\Phi^+)$ is studied. In particular, it is shown there that the bijection of Amstrong–Stump–Thomas between $\mathcal{J}(\Phi^+)$ and the noncrossing partitions of W transports the action of $\mathrm{Row}^{-1} \circ \mathrm{Rvac}$ to the involution $w \mapsto gw^{-1}g^{-1}$ (for the appropriate involution $g \in W$ depending on c). For the diagrammatic models of noncrossing partitions, this involution is a reflection across a diameter. In [**HJ22**] it is shown that, in Type A, rowvacuation of $\mathcal{J}(\Phi^+)$ is the same as the "Lalanne–Kreweras involution" on Dyck paths.

5.2.3. *Promotion and evacuation of semistandard Young tableaux.* A semistandard Young tableau (SSYT) of shape λ is a filling of the boxes of λ with positive integers that is weakly increasing in rows and strictly increasing in columns. Let $\mathrm{SSYT}(\lambda, k)$ denote the set of semistandard tableaux of shape λ with entries at most k. For $i = 1, \ldots, k-1$, the ith Bender–Knuth involution is an operator on $\mathrm{SSYT}(\lambda, k)$ which exchanges the number of i's and $(i+1)$'s in a tableau. Promotion and evacuation can then be defined as operators on $\mathrm{SSYT}(\lambda, k)$ which are the appropriate compositions of these Bender–Knuth involutions.

Promotion and evacuation of semistandard Young tableaux are well-studied actions, with several algebraic guises (e.g., see Section 5.2.4 below). In particular, promotion of SSYTs of rectangular shape is used in the combinatorial definition of affine Type A crystals, or more precisely, the so-called "Kirillov–Reshetikhin crystals;" see [**Shi02, BST10**]. In some sense promotion corresponds to the cyclic symmetry of the affine Dynkin diagram.

It is known that rowmotion acting on $\mathcal{PP}^m(R(a,b))$ is in equivariant bijection with promotion acting on $\mathrm{SSYT}(m^a, a+b)$ (see, e.g., [**KB95**] and [**Hop20**, Appendix A]). And likewise for rowvacuation and evacuation.

5.2.4. *Canonical bases, from quantum groups and cluster algebras.* The theory of quantum groups gives canonical bases for Lie group representations. Rhoades [**Rho10**] (see also [**Rus21**]) showed that for an irreducible representation of the general linear group of rectangular shape, the action of the long cycle on the dual canonical basis corresponds to the action of promotion on rectangular SSYTs. Again, Rhoades's result built on an earlier results of Berenstein–Zelevinsky [**BZ96**] and Stembridge [**Ste96**], who showed that for *any* irreducible representation of the general linear group, the action of the longest word corresponds (up to sign) to evacuation of tableaux in the dual canonical basis. Thanks to the discussion in Section 5.2.3 above, this means that these actions are conjugate to romotion and rowvacuation of $\mathcal{PP}^m(R(a,b))$.

The theory of cluster algebras also gives canonical bases. Shen and Weng [**SW20**] showed that the action of the cyclic shift on the theta basis of the coordinate ring of the Grassmannian – which is a canonical basis coming from its structure as a cluster algebra – is also conjugate to the action of rowmotion on $\mathcal{PP}^m(R(a,b))$.

Recently, Gao, Lam, and Xu [**GLX22**] introduced the grove algebra, a version of the coordinate ring of the Lagrangian Grassmannian. They conjectured the existence of an electrical canonical basis for this grove algebra, which comes with a cyclic action. We believe that this cyclic action on electrical canonical basis elements should be in equivariant bijection with rowmotion of $\mathcal{PP}^m(\Phi^+(A_n))$.

5.2.5. *Crystals and cactus group actions.* Recall the cactus group promotion of highest weight words of weight zero for tensor products of crystals discussed in Section 4.2.4. As explained in [**PRW20**, Example 2.4], for tensor products of the spin representation of the spin group, these weight zero highest weight words correspond to fans of Dyck paths, which are in bijection with $\mathcal{PP}^m(\Phi^+(A_n))$. Moreover, using the techniques of [**PRW20**] it can be shown that cactus group promotion of these words is in equivariant bijection with rowmotion of $\mathcal{PP}^m(\Phi^+(A_n))$, and similarly for evacuation and rowvacuation. Promotion of fans of Dyck paths is studied in detail in the recent paper [**PPSS22**].

5.2.6. *Quiver representations and reflection functors.* In [**GPT23**], Garver, Patrias, and Thomas studied minuscule posets and their P-partitions from the perspective of quiver representations. Fixing a quiver Q whose underlying graph is a Dynkin diagram, and a node i of this Dynkin diagram corresponding to a minuscule weight ω_i, they showed that the Jordan form of a generic nilpotent endomorphism gives a bijection from representations X of Q with support at i to P-partitions for the corresponding minuscule poset P. Moreover, they described the piecewise-linear toggles in terms of reflection functors. In this way, they were able to analyze rowmotion for P-partitions of minuscule posets using quiver representations.

5.2.7. *Rowvacuation of minuscule posets.* Rowvacuation of a minuscule poset has a simple description in terms of its canonical anti-automorphism:

THEOREM 5.2 ([**GR15**, **Oka21a**]). *For a minuscule poset P and $\pi \in \mathcal{PP}^m(P)$, we have* $\mathrm{Rvac}(\pi) = \iota(\pi)^*$.

The case $P = R(a,b)$ of Theorem 5.2 was proved by Grinberg–Roby [**GR15**]; the rest of the theorem was proved, in a case-by-case manner, by Okada [**Oka21a**]. Those authors described their results in terms of "reciprocity" of rowmotion, but it is easy to translate their results to this statement about rowvacuation.

5.2.8. *Doppelgängers bijections.* We conjectured that minuscule doppelgänger pairs have the same orbit structure of rowmotion:

CONJECTURE 5.3 ([**Hop22**]). *For (P, Q) a minuscule doppelgänger pair, there is a bijection between $\mathcal{PP}^m(P)$ and $\mathcal{PP}^m(Q)$ which commutes with rowmotion.*

The case $m = 1$ of Conjecture 5.3 was proved in [**DWYWZ22**], using the bijection of [**HPPW20**]. In the very recent preprint [**JL23**], Johnson and Liu proved the main case of Conjecture 5.3 involving the rectangle-trapezoid doppelgänger pair.

5.2.9. *Symmetry classes of plane partitions.* Grinberg and Roby [**GR15**] explained how rowmotion for the three "triangular" posets $DS(n,0)$, $\Phi^+(A_n)$, and $\Phi^+(B_n)$ can be understood by imposing symmetries on the rectangle:

LEMMA 5.4 (Grinberg–Roby [**GR15**]; see also [**Hop20**, §5]).

- *There is a* Row*-equivariant bijection between $\mathcal{PP}^m(DS(n,0))$ and the subset $\pi \in \mathcal{PP}^m(n \times n)$ with* $\mathrm{Tr}(\pi) = \pi$.
- *There is a* Row*-equivariant bijection between $\mathcal{PP}^m(\Phi^+(A_n))$ and the subset of $\pi \in \mathcal{PP}^{2m}((n+1)\times(n+1))$ with* $\mathrm{Row}^{n+1}(\pi) = \mathrm{Tr}(\pi)$.
- *There is a* Row*-equivariant bijection between $\mathcal{PP}^m(\Phi^+(B_n))$ and the subset of $\pi \in \mathcal{PP}^{2m}(2n \times 2n)$ with* $\mathrm{Tr}(\pi) = \pi$ *and* $\mathrm{Row}^{2n}(\pi) = \pi$.

We note that a similar "triangle-into-rectangle" embedding, but for linear extensions rather than P-partitions, was studied by Pon and Wang [**PW11**].

5.3. Order. We now review the posets P for which $\mathrm{Row}^{r(P)+2}$ acting on $\mathcal{PP}^m(P)$ can be described.

THEOREM 5.5 ([**GR15**, **GPT23**, **Oka21a**]). *For P a minuscule poset, we have that* $\mathrm{Row}^{r(P)+2}$ *is the identity.*

THEOREM 5.6 ([**GR15**]). *For $P = \Phi^+$ a root poset of coincidental type, we have* $\mathrm{Row}^{r(P)+2} = \delta$.

Theorems 5.5 and 5.6 were essentially proved, in a case-by-case fashion, by Grinberg–Roby [**GR15**]; the only case they could not address was Λ_{E_7}, which was resolved in [**GPT23**] and [**Oka21a**].

Conjecture 5.3 and Theorem 5.5 together imply that the trapezoid should have $\mathrm{Row}^{r(P)+2}$ equal to the identity (this was also conjecture by N. Williams [**GR15**, Conjecture 75]).

The only other poset which apparently has good (piecewise-linear) rowmotion behavior is $V(n)$:

CONJECTURE 5.7. *For* $P = V(n)$, *we have* $\mathrm{Row}^{r(P)+2} = \delta$.

In the very recent preprint [**Ade23**], Adenbaum has proved Conjecture 5.7.

5.4. Orbit structure. There are very nice conjectural CSPs for all the posets with good P-partition rowmotion behavior:

CONJECTURE 5.8 ([**Hop22**]). *Let* P *be a minuscule poset. Then the triple* $(\mathcal{PP}^m(P), \langle \mathrm{Row} \rangle, F_P(m;q))$ *exhibits cyclic sieving.*

CONJECTURE 5.9 ([**Hop22**]). *Let* Φ *be a root system of coincidental type,* W *its Coxeter group, and* h *its Coxeter number. Let* $C = \langle c \rangle \simeq \mathbb{Z}/2h\mathbb{Z}$ *act on* $\mathcal{PP}^m(\Phi^+)$ *via* $c(\pi) := \mathrm{Row}(\pi)$. *Then* $(\mathcal{PP}^m(\Phi^+), C, \mathrm{Cat}(W, m; q))$ *exhibits cyclic sieving.*

The case $m = 1$ of Conjecture 5.8 was proved by Rush–Shi [**RS13**]. The case $m = 1$ of Conjecture 5.9 was proved by Armstrong–Stump–Thomas [**AST13**] (and in fact they showed this for *any* root system; see also [**CS15**]). The case of Conjecture 5.8 for the rectangle $R(a, b)$ was proved by Rhoades [**Rho10**] (and later, Shen–Weng [**SW20**]). All other cases of these conjectures are open. We note the Type A case of Conjecture 5.9 was conjectured also by J. Propp c. 2016.

Via Lemma 5.4, the main remaining cases of Conjectures 5.8 and 5.9 can be translated into statements about the numbers of fixed points of various subgroups of $\langle \mathrm{Row}, \mathrm{Tr} \rangle$ acting on $\mathcal{PP}^m(n \times n)$ being counted by CSP-type evaluations. In [**Hop20**], it was shown, building off the work of Rhoades [**Rho10**], that the number of fixed points of any *element* of $\langle \mathrm{Row}, \mathrm{Tr} \rangle$ acting on $\mathcal{PP}^m(n \times n)$ is given by a CSP-type evaluation of polynomial with a nice product formula.

Conjectures 5.3 and 5.8 together describe the orbit structure of rowmotion for minuscule doppelgängers (and their claims are consistent with Conjecture 5.9).

Finally, for $V(n)$ we conjecture:

CONJECTURE 5.10. *For all* $n, m \geq 1$, *the rational expression*

$$f(q) := \prod_{i=2}^{2n+1} \frac{(1 - q^{2m+i})}{(1 - q^i)} \cdot \prod_{i=2}^{n+1} \frac{(1 - q^{2m+2i})}{(1 - q^{2i})}$$

is in $\mathbb{N}[q]$, *and* $(\mathcal{PP}^m(V(n)), \langle \mathrm{Pro} \rangle, f(q))$ *exhibits cyclic sieving.*

References

[Ade23] Ben Adenbaum, *On the order of P-strict promotion on* $V \times [\ell]$, arXiv:2308.08757, 2023.

[And77] George E. Andrews, *Plane partitions. II. The equivalence of the Bender-Knuth and MacMahon conjectures*, Pacific J. Math. **72** (1977), no. 2, 283–291. MR472108

[And78] George E. Andrews, *Plane partitions. I. The MacMahon conjecture*, Studies in foundations and combinatorics, Adv. Math. Suppl. Stud., vol. 1, Academic Press, New York-London, 1978, pp. 131–150. MR520557

[Arm09] Drew Armstrong, *Generalized noncrossing partitions and combinatorics of Coxeter groups*, Mem. Amer. Math. Soc. **202** (2009), no. 949, x+159, DOI 10.1090/S0065-9266-09-00565-1. MR2561274

[AST13] Drew Armstrong, Christian Stump, and Hugh Thomas, *A uniform bijection between nonnesting and noncrossing partitions*, Trans. Amer. Math. Soc. **365** (2013), no. 8, 4121–4151, DOI 10.1090/S0002-9947-2013-05729-7. MR3055691

[BK72] Edward A. Bender and Donald E. Knuth, *Enumeration of plane partitions*, J. Combinatorial Theory Ser. A **13** (1972), 40–54, DOI 10.1016/0097-3165(72)90007-6. MR299574

[BR11] David Bessis and Victor Reiner, *Cyclic sieving of noncrossing partitions for complex reflection groups*, Ann. Comb. **15** (2011), no. 2, 197–222, DOI 10.1007/s00026-011-0090-9. MR2813511

[BS74] A. E. Brouwer and A. Schrijver, *On the period of an operator, defined on antichains*, Mathematisch Centrum, Afdeling Zuivere Wiskunde, ZW 24/74, Mathematisch Centrum, Amsterdam, 1974. MR349497

[BST10] Jason Bandlow, Anne Schilling, and Nicolas M. Thiéry, *On the uniqueness of promotion operators on tensor products of type A crystals*, J. Algebraic Combin. **31** (2010), no. 2, 217–251, DOI 10.1007/s10801-009-0182-3. MR2592077

[BSV21] Joseph Bernstein, Jessica Striker, and Corey Vorland, *P-strict promotion and B-bounded rowmotion, with applications to tableaux of many flavors*, Comb. Theory **1** (2021), Paper No. 8, 46, DOI 10.5070/C61055369. MR4396213

[BSV24] Joseph Bernstein, Jessica Striker, and Corey Vorland, *P-strict promotion and Q-partition rowmotion: the graded case*, European J. Combin. **115** (2024), Paper No. 103776, 24.

[BW91] Graham Brightwell and Peter Winkler, *Counting linear extensions*, Order **8** (1991), no. 3, 225–242, DOI 10.1007/BF00383444. MR1154926

[BZ96] Arkady Berenstein and Andrei Zelevinsky, *Canonical bases for the quantum group of type A_r and piecewise-linear combinatorics*, Duke Math. J. **82** (1996), no. 3, 473–502, DOI 10.1215/S0012-7094-96-08221-6. MR1387682

[CFDF95] P. J. Cameron and D. G. Fon-Der-Flaass, *Orbits of antichains revisited*, European J. Combin. **16** (1995), no. 6, 545–554, DOI 10.1016/0195-6698(95)90036-5. MR1356845

[CLS14] Cesar Ceballos, Jean-Philippe Labbé, and Christian Stump, *Subword complexes, cluster complexes, and generalized multi-associahedra*, J. Algebraic Combin. **39** (2014), no. 1, 17–51, DOI 10.1007/s10801-013-0437-x. MR3144391

[CP02] Paola Cellini and Paolo Papi, *Ad-nilpotent ideals of a Borel subalgebra. II*, J. Algebra **258** (2002), no. 1, 112–121, DOI 10.1016/S0021-8693(02)00532-X. Special issue in celebration of Claudio Procesi's 60th birthday. MR1958899

[CS15] Michael Cuntz and Christian Stump, *On root posets for noncrystallographic root systems*, Math. Comp. **84** (2015), no. 291, 485–503, DOI 10.1090/S0025-5718-2014-02841-X. MR3266972

[DF90] Michel-Marie Deza and Komei Fukuda, *Loops of clutters*, Coding theory and design theory, Part I, IMA Vol. Math. Appl., vol. 20, Springer, New York, 1990, pp. 72–92, DOI 10.1007/978-1-4613-8994-1_7. MR1047873

[DH21] Colin Defant and Sam Hopkins, *Symmetry of Narayana numbers and rowvacuation of root posets*, Forum Math. Sigma **9** (2021), Paper No. e53, 24, DOI 10.1017/fms.2021.47. MR4291955

[DPS17] Kevin Dilks, Oliver Pechenik, and Jessica Striker, *Resonance in orbits of plane partitions and increasing tableaux*, J. Combin. Theory Ser. A **148** (2017), 244–274, DOI 10.1016/j.jcta.2016.12.007. MR3603321

[DSV19] Kevin Dilks, Jessica Striker, and Corey Vorland, *Rowmotion and increasing labeling promotion*, J. Combin. Theory Ser. A **164** (2019), 72–108, DOI 10.1016/j.jcta.2018.12.007. MR3894190

[DWYWZ22] Quang Vu Dao, Julian Wellman, Calvin Yost-Wolff, and Sylvester W. Zhang, *Rowmotion orbits of trapezoid posets*, Electron. J. Combin. **29** (2022), no. 2, Paper No. 2.29, 19, DOI 10.37236/9769. MR4423786

[EG87] Paul Edelman and Curtis Greene, *Balanced tableaux*, Adv. in Math. **63** (1987), no. 1, 42–99, DOI 10.1016/0001-8708(87)90063-6. MR871081

[EP21] David Einstein and James Propp, *Combinatorial, piecewise-linear, and birational homomesy for products of two chains*, Algebr. Comb. **4** (2021), no. 2, 201–224, DOI 10.5802/alco.139. MR4244370

[FDF93] D. G. Fon-Der-Flaass, *Orbits of antichains in ranked posets*, European J. Combin. **14** (1993), no. 1, 17–22, DOI 10.1006/eujc.1993.1003. MR1197471

[FK14] Bruce Fontaine and Joel Kamnitzer, *Cyclic sieving, rotation, and geometric representation theory*, Selecta Math. (N.S.) **20** (2014), no. 2, 609–625, DOI 10.1007/s00029-013-0144-4. MR3177928

[FRT54] J. S. Frame, G. de B. Robinson, and R. M. Thrall, *The hook graphs of the symmetric groups*, Canad. J. Math. **6** (1954), 316–324, DOI 10.4153/cjm-1954-030-1. MR62127

[GLX22] Yibo Gao, Thomas Lam, and Zixuan Xu, *Electrical networks and the grove algebra*, Canad. J. Math. (in press). DOI 10.4153/S0009414X24000105.

[Gor83] Basil Gordon, *A proof of the Bender-Knuth conjecture*, Pacific J. Math. **108** (1983), no. 1, 99–113. MR709701

[GPT23] Alexander Garver, Rebecca Patrias, and Hugh Thomas, *Minuscule reverse plane partitions via quiver representations*, Selecta Math. (N.S.) **29** (2023), no. 3, Paper No. 37, 48, DOI 10.1007/s00029-023-00831-4. MR4581740

[GR15] Darij Grinberg and Tom Roby, *Iterative properties of birational rowmotion II: rectangles and triangles*, Electron. J. Combin. **22** (2015), no. 3, Paper 3.40, 49, DOI 10.37236/4335. MR3414186

[Hai89] Mark D. Haiman, *On mixed insertion, symmetry, and shifted Young tableaux*, J. Combin. Theory Ser. A **50** (1989), no. 2, 196–225, DOI 10.1016/0097-3165(89)90015-0. MR989194

[Hai92] Mark D. Haiman, *Dual equivalence with applications, including a conjecture of Proctor*, Discrete Math. **99** (1992), no. 1-3, 79–113, DOI 10.1016/0012-365X(92)90368-P. MR1158783

[Hai94] Mark D. Haiman, *Conjectures on the quotient ring by diagonal invariants*, J. Algebraic Combin. **3** (1994), no. 1, 17–76, DOI 10.1023/A:1022450120589. MR1256101

[HJ22] Sam Hopkins and Michael Joseph, *The birational Lalanne-Kreweras involution*, Algebr. Comb. **5** (2022), no. 2, 227–265, DOI 10.5802/alco.201. MR4426438

[HK92] Mark D. Haiman and Dongsu Kim, *A characterization of generalized staircases*, Discrete Math. **99** (1992), no. 1-3, 115–122, DOI 10.1016/0012-365X(92)90369-Q. MR1158784

[HK06] André Henriques and Joel Kamnitzer, *Crystals and coboundary categories*, Duke Math. J. **132** (2006), no. 2, 191–216, DOI 10.1215/S0012-7094-06-13221-0. MR2219257

[HL21] Sam Hopkins and Tri Lai, *Plane partitions of shifted double staircase shape*, J. Combin. Theory Ser. A **183** (2021), Paper No. 105486, 29, DOI 10.1016/j.jcta.2021.105486. MR4268722

[Hop20] Sam Hopkins, *Cyclic sieving for plane partitions and symmetry*, SIGMA Symmetry Integrability Geom. Methods Appl. **16** (2020), Paper No. 130, 40, DOI 10.3842/SIGMA.2020.130. MR4184618

[Hop22] Sam Hopkins, *Minuscule doppelgängers, the coincidental down-degree expectations property, and rowmotion*, Exp. Math. **31** (2022), no. 3, 946–974, DOI 10.1080/10586458.2020.1731881. MR4477416

[HPPW20] Zachary Hamaker, Rebecca Patrias, Oliver Pechenik, and Nathan Williams, *Doppelgängers: bijections of plane partitions*, Int. Math. Res. Not. IMRN **2** (2020), 487–540, DOI 10.1093/imrn/rny018. MR4073193

[HR22] Sam Hopkins and Martin Rubey, *Promotion of Kreweras words*, Selecta Math. (N.S.) **28** (2022), no. 1, Paper No. 10, 38, DOI 10.1007/s00029-021-00714-6. MR4346507

[JL23] Joseph Johnson and Ricky Ini Liu, *Plane partitions and rowmotion on rectangular and trapezoidal posets*, `arXiv:2311.07133`, 2023.

[KB95] A. N. Kirillov and A. D. Berenstein, *Groups generated by involutions, Gel′fand-Tsetlin patterns, and combinatorics of Young tableaux*, Algebra i Analiz **7** (1995), no. 1, 92–152; English transl., St. Petersburg Math. J. **7** (1996), no. 1, 77–127. MR1334154

[KK99] Mikhail Khovanov and Greg Kuperberg, *Web bases for* sl(3) *are not dual canonical*, Pacific J. Math. **188** (1999), no. 1, 129–153, DOI 10.2140/pjm.1999.188.129. MR1680395

[KN81] G. Kreweras and H. Niederhausen, *Solution of an enumerative problem connected with lattice paths*, European J. Combin. **2** (1981), no. 1, 55–60, DOI 10.1016/S0195-6698(81)80020-0. MR611931

[Kra89] Witold Kraśkiewicz, *Reduced decompositions in hyperoctahedral groups* (English, with French summary), C. R. Acad. Sci. Paris Sér. I Math. **309** (1989), no. 16, 903–907. MR1055219

[Kra16] C. Krattenthaler, *Plane partitions in the work of Richard Stanley and his school*, The mathematical legacy of Richard P. Stanley, Amer. Math. Soc., Providence, RI, 2016, pp. 231–261, DOI 10.1090//mbk/100/14. MR3618037

[Kre65] Germain Kreweras, *Sur une classe de problemes de dénombrement liés au treillis des partitions des entiers*, Cahiers du B.U.R.O. **6** (1965), 5–105.

[KY19] Jang Soo Kim and Meesue Yoo, *Hook length property of d-complete posets via q-integrals*, J. Combin. Theory Ser. A **162** (2019), 167–221, DOI 10.1016/j.jcta.2018.11.003. MR3873874

[Mac99] Percy Alexander MacMahon, *Partitions of numbers whose graphs possess symmetry*, Trans. Cambridge Philos. Soc. **17** (1899), 149–170.

[Mac79] I. G. Macdonald, *Symmetric functions and Hall polynomials*, Oxford Mathematical Monographs, The Clarendon Press, Oxford University Press, New York, 1979. MR553598

[Mac04] Percy A. MacMahon, *Combinatory analysis. Vol. I, II (bound in one volume)*, Dover Phoenix Editions, Dover Publications, Inc., Mineola, NY, 2004. Reprint of *An introduction to combinatory analysis* (1920) and *Combinatory analysis. Vol. I, II* (1915, 1916). MR2417935

[NO19] Hiroshi Naruse and Soichi Okada, *Skew hook formula for d-complete posets via equivariant K-theory*, Algebr. Comb. **2** (2019), no. 4, 541–571, DOI 10.5802/alco.54. MR3997510

[Oka21a] Soichi Okada, *Birational rowmotion and Coxeter-motion on minuscule posets*, Electron. J. Combin. **28** (2021), no. 1, Paper No. 1.17, 30, DOI 10.37236/9557. MR4245250

[Oka21b] Soichi Okada, *Intermediate symplectic characters and shifted plane partitions of shifted double staircase shape*, Comb. Theory **1** (2021), Paper No. 10, 42.

[Pan09] Dmitri I. Panyushev, *On orbits of antichains of positive roots*, European J. Combin. **30** (2009), no. 2, 586–594, DOI 10.1016/j.ejc.2008.03.009. MR2489252

[Pat19] Rebecca Patrias, *Promotion on generalized oscillating tableaux and web rotation*, J. Combin. Theory Ser. A **161** (2019), 1–28, DOI 10.1016/j.jcta.2018.07.005. MR3861768

[PP23] Rebecca Patrias and Oliver Pechenik, *Tableau evacuation and webs*, Proc. Amer. Math. Soc. Ser. B **10** (2023), 341–352, DOI 10.1090/bproc/191. MR4648350

[PPR09] T. Kyle Petersen, Pavlo Pylyavskyy, and Brendon Rhoades, *Promotion and cyclic sieving via webs*, J. Algebraic Combin. **30** (2009), no. 1, 19–41, DOI 10.1007/s10801-008-0150-3. MR2519848

[PPSS22] Joseph Pappe, Stephan Pfannerer, Anne Schilling, and Mary Claire Simone, *Promotion and growth diagrams for fans of Dyck paths and vacillating tableaux*, J. Algebra **655** (2024), 794–842. https://doi.org/10.1016/j.jalgebra.2023.07.038.

[PR15] James Propp and Tom Roby, *Homomesy in products of two chains*, Electron. J. Combin. **22** (2015), no. 3, Paper 3.4, 29, DOI 10.37236/3579. MR3367853

[Pro83] Robert A. Proctor, *Shifted plane partitions of trapezoidal shape*, Proc. Amer. Math. Soc. **89** (1983), no. 3, 553–559, DOI 10.2307/2045516. MR715886

[Pro84] Robert A. Proctor, *Bruhat lattices, plane partition generating functions, and minuscule representations*, European J. Combin. **5** (1984), no. 4, 331–350, DOI 10.1016/S0195-6698(84)80037-2. MR782055

[Pro88] Robert A. Proctor, *Odd symplectic groups*, Invent. Math. **92** (1988), no. 2, 307–332, DOI 10.1007/BF01404455. MR936084

[Pro90] Robert A. Proctor, *New symmetric plane partition identities from invariant theory work of De Concini and Procesi*, European J. Combin. **11** (1990), no. 3, 289–300, DOI 10.1016/S0195-6698(13)80128-X. MR1059559

[Pro14] Robert A. Proctor, *d-Complete posets generalize Young diagrams for the hook product formula: Partial presentation of proof*, RIMS Kokyuroku **1913** (2014), 120–140.

[PRW20] Stephan Pfannerer, Martin Rubey, and Bruce Westbury, *Promotion on oscillating and alternating tableaux and rotation of matchings and permutations*, Algebr. Comb. **3** (2020), no. 1, 107–141, DOI 10.5802/alco.87. MR4068745

[Pur13] Kevin Purbhoo, *Wronskians, cyclic group actions, and ribbon tableaux*, Trans. Amer. Math. Soc. **365** (2013), no. 4, 1977–2030, DOI 10.1090/S0002-9947-2012-05676-5. MR3009651

[Pur18] Kevin Purbhoo, *A marvellous embedding of the Lagrangian Grassmannian*, J. Combin. Theory Ser. A **155** (2018), 1–26, DOI 10.1016/j.jcta.2017.08.012. MR3741418

[PW11] Steven Pon and Qiang Wang, *Promotion and evacuation on standard Young tableaux of rectangle and staircase shape*, Electron. J. Combin. **18** (2011), no. 1, Paper 18, 18, DOI 10.37236/505. MR2770123

[Rho10] Brendon Rhoades, *Cyclic sieving, promotion, and representation theory*, J. Combin. Theory Ser. A **117** (2010), no. 1, 38–76, DOI 10.1016/j.jcta.2009.03.017. MR2557880

[RS13] David B. Rush and XiaoLin Shi, *On orbits of order ideals of minuscule posets*, J. Algebraic Combin. **37** (2013), no. 3, 545–569, DOI 10.1007/s10801-012-0380-2. MR3035516

[RSW04] V. Reiner, D. Stanton, and D. White, *The cyclic sieving phenomenon*, J. Combin. Theory Ser. A **108** (2004), no. 1, 17–50, DOI 10.1016/j.jcta.2004.04.009. MR2087303

[Rus13] Heather M. Russell, *An explicit bijection between semistandard tableaux and non-elliptic sl_3 webs*, J. Algebraic Combin. **38** (2013), no. 4, 851–862, DOI 10.1007/s10801-013-0428-y. MR3119361

[Rus21] David B. Rush, *Restriction of global bases and Rhoades's theorem*, Adv. Math. **384** (2021), Paper No. 107725, 19, DOI 10.1016/j.aim.2021.107725. MR4242901

[Sch63] M. P. Schützenberger, *Quelques remarques sur une construction de Schensted* (French), Math. Scand. **12** (1963), 117–128, DOI 10.7146/math.scand.a-10676. MR190017

[Sch72] M. P. Schützenberger, *Promotion des morphismes d'ensembles ordonnés* (French), Discrete Math. **2** (1972), 73–94, DOI 10.1016/0012-365X(72)90062-3. MR299539

[Sch77] M.-P. Schützenberger, *La correspondance de Robinson* (French), Combinatoire et représentation du groupe symétrique (Actes Table Ronde CNRS, Univ. Louis-Pasteur Strasbourg, Strasbourg, 1976), Lecture Notes in Math., Vol. 579, Springer, Berlin-New York, 1977, pp. 59–113. MR498826

[Shi02] Mark Shimozono, *Affine type A crystal structure on tensor products of rectangles, Demazure characters, and nilpotent varieties*, J. Algebraic Combin. **15** (2002), no. 2, 151–187, DOI 10.1023/A:1013894920862. MR1887233

[SSD14] Sarah Sekheri, Stéphanie Schank, and Fatiha Djermane, *Un CSP sur des tableaux décalés en forme d'escalier*, Unpublished report, available online at `https://lacim.uqam.ca/wp-content/uploads/Rapport/SekheriSchankDjermane2014.pdf`, 2014.

[Sta90] Dennis Stanton, *Fake Gaussian sequences*, Unpublished note, available online at `http://www-users.math.umn.edu/~stant001/PAPERS/macmahon.pdf`, 1997.

[Sta99] Richard P. Stanley, *Enumerative combinatorics. Vol. 2*, Cambridge Studies in Advanced Mathematics, vol. 62, Cambridge University Press, Cambridge, 1999. With a foreword by Gian-Carlo Rota and appendix 1 by Sergey Fomin, DOI 10.1017/CBO9780511609589. MR1676282

[Sta09] Richard P. Stanley, *Promotion and evacuation*, Electron. J. Combin. **16** (2009), no. 2, Special volume in honor of Anders Björner, Research Paper 9, 24.

[Sta12] Richard P. Stanley, *Enumerative combinatorics. Volume 1*, 2nd ed., Cambridge Studies in Advanced Mathematics, vol. 49, Cambridge University Press, Cambridge, 2012. MR2868112

[Ste96] John R. Stembridge, *Canonical bases and self-evacuating tableaux*, Duke Math. J. **82** (1996), no. 3, 585–606, DOI 10.1215/S0012-7094-96-08224-1. MR1387685

[SW12] Jessica Striker and Nathan Williams, *Promotion and rowmotion*, European J. Combin. **33** (2012), no. 8, 1919–1942, DOI 10.1016/j.ejc.2012.05.003. MR2950491

[SW20] Linhui Shen and Daping Weng, *Cyclic sieving and cluster duality of Grassmannian*, SIGMA Symmetry Integrability Geom. Methods Appl. **16** (2020), Paper No. 067, 41, DOI 10.3842/SIGMA.2020.067. MR4126699

[TY16] Hugh Thomas and Alexander Yong, *Cominuscule tableau combinatorics*, Schubert calculus—Osaka 2012, Adv. Stud. Pure Math., vol. 71, Math. Soc. Japan, [Tokyo], 2016, pp. 475–497, DOI 10.2969/aspm/07110475. MR3644832

[Tym12] Julianna Tymoczko, *A simple bijection between standard* $3 \times n$ *tableaux and irreducible webs for* $\mathfrak{sl}_3$, J. Algebraic Combin. **35** (2012), no. 4, 611–632, DOI 10.1007/s10801-011-0317-1. MR2902703

[Wes16] Bruce W. Westbury, *Invariant tensors and the cyclic sieving phenomenon*, Electron. J. Combin. **23** (2016), no. 4, Paper 4.25, 40, DOI 10.37236/4569. MR3577672

[Wil13] Nathan Williams, *Cataland*, ProQuest LLC, Ann Arbor, MI, 2013. Thesis (Ph.D.)–University of Minnesota. MR3193161

DEPARTMENT OF MATHEMATICS, HOWARD UNIVERSITY, WASHINGTON, DISTRICT OF COLUMBIA
Email address: samuelfhopkins@gmail.com

Proceedings of Symposia in Pure Mathematics
Volume **110**, 2024
https://doi.org/10.1090/pspum/110/02013

An invitation to positive geometries

Thomas Lam

ABSTRACT. This short introduction to positive geometries, targeted at a mathematical audience, is based on my talk at OPAC 2022.

Positive geometries are certain semialgebraic spaces, equipped with a distinguished meromorphic form called the canonical form [**ABL**]. Examples of positive geometries include polytopes, positive parts of toric varieties, totally nonnegative spaces, and Grassmann polytopes and amplituhedra.

Positive geometries were first introduced in theoretical physics. Canonical forms of positive geometries are used to write formulae for scattering amplitudes, analytic functions that are used to compute probabilities in particle scattering experiments. Roughly speaking, different positive geometries correspond to different quantum field theories. There is a flourishing and well-developed industry expanding the zoo of positive geometries and related physical processes. We refer the reader to the recent surveys [**FL**, **HT**] for more on the physics of positive geometries, and to [**EH**, **HP**] for textbook introductions to scattering amplitudes.

The mathematical study of positive geometries is still in its infancy. In this short note, we give a brief introduction to positive geometries, with a mathematical audience, especially an algebraic or geometric combinatorialist, in mind. The course webpage [**CGA**] contains many additional references for further exploration.

1. The polytope canonical form

We begin by illustrating the main ideas with the example of convex projective polytopes. A *projective polytope* is a subspace $P \subset \mathbb{P}^d(\mathbb{R})$ of projective space such that there exists a hyperplane $H \subset \mathbb{P}^d$ and a linear identification $\mathbb{P}^d \setminus H \cong \mathbb{R}^d$ such that $P \subset \mathbb{P}^d \setminus H \cong \mathbb{R}^d$ is a Euclidean convex polytope. An orientation of a projective polytope is an orientation of its interior $\mathrm{Int}(P)$, which is always a manifold. The following theorem is a reformulation of the statement that polytopes are positive geometries [**ABL**, Section 6].

THEOREM 1 (Residue definition). *For each full-dimensional oriented projective polytope $P \subset \mathbb{P}^d$ there exists a rational d-form $\Omega(P)$ on $\mathbb{P}^d$, with poles only along facet hyperplanes, and these poles are simple, and such that $\Omega(P)$ is uniquely*

2020 *Mathematics Subject Classification.* Primary 14P10, 05E14.

The author was supported by the National Science Foundation under grant DMS-1953852.

determined by the recursive properties:

(1) *the canonical form of a point* pt $= \mathbb{P}^0$ *is* $\Omega(\mathrm{pt}) = \pm 1$ *depending on the orientation,*

(2) *for any facet* $F \subset P$, *we have* $\mathrm{Res}_H\, \Omega(P) = \Omega(F)$, *where* H *is the hyperplane spanned by the facet* F.

The differential form $\Omega(P)$ is called the *canonical form* of P. The notation $\mathrm{Res}_H\, \Omega(P)$ denotes a residue, defined in (8); we illustrate the computation in some examples below.

In the following examples, if $P \subset \mathbb{R}^d$ is a Euclidean polytope, then we consider it a projective polytope via the natural inclusion $\mathbb{R}^d = \{(1 : \mathbf{x}) \mid \mathbf{x} \in \mathbb{R}^d\} \subset \mathbb{P}^d$. Let $P = [a, b] \subset \mathbb{R} \subset \mathbb{P}^1(\mathbb{R})$ be a closed interval with one of the two orientations. Then the canonical form is (up to a sign) given by

$$\Omega([a,b]) = \frac{dx}{x-a} - \frac{dx}{x-b}, \tag{1}$$

which has residue $+1$ at $x = a$ and -1 at $x = b$.

EXAMPLE 1. Let P be the quadrilateral with vertices $(0,0), (2,0), (0,1), (1,2)$ in $\mathbb{R}^2$ (Fig. 1(a)). Then

$$\Omega(P) = \frac{4 + 4x - y}{xy(1+x-y)(4-2x-y)} dxdy. \tag{2}$$

As required, $\Omega(P)$ has simple poles along each of the four facet hyperplanes

$$x = 0, \qquad y = 0, \qquad 1 + x - y = 0, \qquad 4 - 2x - y = 0 \tag{3}$$

of P. The numerator will be explained in Theorem 5 below. Let us first take the residue along $x = 0$. Since

$$\Omega(P) = \frac{dx}{x} \wedge \frac{4 + 4x - y}{y(1+x-y)(4-2x-y)} dy,$$

we have, using (1),

$$\mathrm{Res}_{x=0}(\Omega(P)) = \frac{4-y}{y(1-y)(4-y)} dy = \frac{1}{y(1-y)} dy = \frac{1}{y} dy - \frac{1}{y-1} dy = \Omega([0,1]).$$

Now, let us take the residue along the hyperplane $H = \{f = 1 + x - y = 0\}$. Since $df/f = (dx - dy)/(1 + x - y)$, we can write

$$\Omega(P) = \frac{df}{f} \wedge \frac{4 + 4x - y}{xy(4-2x-y)} dx,$$

Making the substitution $y = x + 1$, we get

$$\mathrm{Res}_H(\Omega(P)) = \left(\frac{4 + 4x - y}{xy(4-2x-y)} dx\right)_{y \mapsto 1+x} = \frac{3 + 3x}{x(x+1)(3-3x)} dx = \frac{1}{x(1-x)} dx,$$

which again by (1) is the canonical form of the edge connecting the vertices $(0,1)$ and $(1,2)$. We invite the reader to verify that the residues along the two other facets are correct.

Note that changing the orientation of P negates the canonical form $\Omega(P)$. Henceforth, we omit the adjective "oriented" from our theorem statements, implicitly assuming that our polytopes have been equipped with (compatible) orientations.

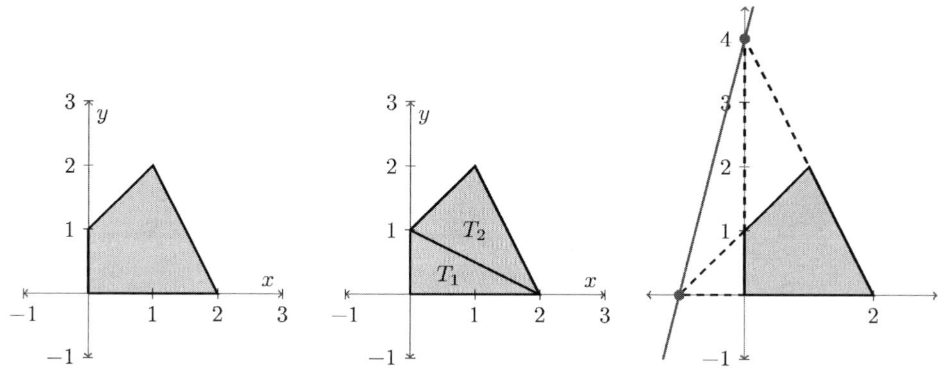

(a) A quadrilateral P. (b) Triangulation into T_1, T_2. (c) The adjoint line of P.

FIGURE 1. Our running example.

The uniqueness of $\Omega(P)$ follows from the fact that $\mathbb{P}^d$ has no nonzero holomorphic d-forms. The quickest proof of the existence of $\Omega(P)$ is via triangulations ([**ABL**, Section 3]).

THEOREM 2 (Additivity for subdivisions). *Suppose that a projective polytope P is subdivided into polytopes $T_1, T_2, \dots, T_r$. Then*

$$\Omega(P) = \sum_{i=1}^{r} \Omega(T_r).$$

Since every polytope has a triangulation, to construct canonical forms, it suffices to construct the canonical form of a simplex. Let

$$\Delta^d := \{(X_0 : X_1 : \cdots : X_d) \in \mathbb{P}^d \mid X_i \geq 0\} \subset \mathbb{P}^d$$

denote the standard projective d-simplex. Every face of Δ^d is again a standard simplex, so it is easy to verify that

$$\Omega(\Delta^d) = \frac{dx_1}{x_1} \wedge \cdots \wedge \frac{dx_d}{x_d}, \qquad x_i := \frac{X_i}{X_0}$$

satisfies the recursion in Theorem 1.

EXAMPLE 2. Let P be as in Example 1. Then P can be triangulated as in Fig. 1(b). The two canonical forms are

(4)
$$\Omega(T_1) = \frac{2}{xy(2-x-2y)} dxdy, \quad \Omega(T_2) = \frac{9}{(1+x-y)(4-2x-y)(2-x-2y)} dxdy$$

which can be calculated by finding a projective transformation $g \in \mathrm{PGL}(3) \cong \mathrm{Aut}(\mathbb{P}^2)$ sending T_1 (or T_2) to the standard simplex Δ^2 and computing the pullback $g^*(\Omega(\Delta^2))$. More explicitly, suppose the triangle T has facets $a_i + b_i x + c_i y = 0$ for $i = 1, 2, 3$. Then letting g be the 3×3 matrix with columns (a_i, b_i, c_i), we have

$$\Omega(T) = \pm \frac{\det(g)}{\prod_i (a_i + b_i x + c_i y)}.$$

It is easily verified that indeed $\Omega(T_1) + \Omega(T_2)$ recovers $\Omega(P)$ in (2), in agreement with Theorem 2.

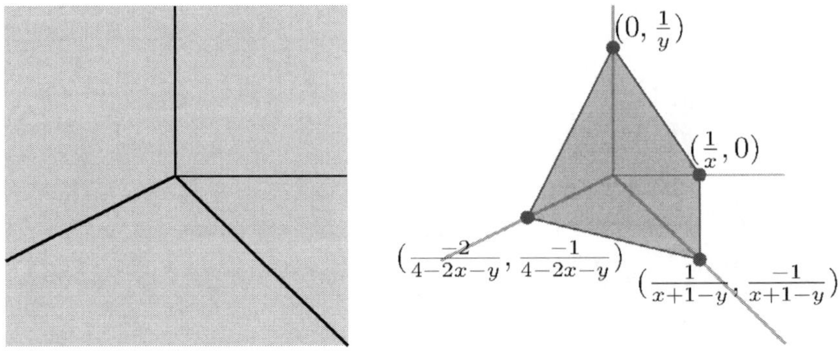

FIGURE 2. Left: the normal fan of P. Right: the polar polytope $(P-(x,y))^\vee$.

We give an explicit formula for $\Omega(P)$. For a subset $S \subset \mathbb{R}^d$, the polar set $S^\vee$ is defined by

$$S^\vee := \{\mathbf{x} \in \mathbb{R}^d \mid \mathbf{x}.\mathbf{y} \geq -1 \text{ for all } \mathbf{y} \in S\}.$$

THEOREM 3 (Dual volume [**ABL**, Section 7.4]). *Suppose $P \subset \mathbb{R}^d \subset \mathbb{P}^d$ is full-dimensional. Then*

$$\Omega(P)(\mathbf{x}) = \mathrm{Vol}((P-\mathbf{x})^\vee) d^d\mathbf{x}. \tag{5}$$

where $\mathbf{x} \in \mathrm{Int}(P)$. Here, Vol *denotes the Euclidean volume, normalized so that the unit simplex has volume* 1.

The function $\mathrm{Vol}((P-\mathbf{x})^\vee)$ is defined when $\mathbf{x} \in \mathrm{Int}(P)$. It analytically continues to a rational function on $\mathbb{R}^d$. While (5) depends on (compatible) choices of inner product and measure, we emphasize that the definition (Theorem 1) does not depend on any metric notions.

EXAMPLE 3. The polar polytope $(P-x)^\vee$ is illustrated in Fig. 2. The vertices of $(P-x)^\vee$ lie on the rays of the (inner) normal fan of P. Intersecting $(P-x)^\vee$ with each maximal cone of the normal fan we obtain a triangle. Summing the volumes of these triangles gives

$$\begin{aligned} &\frac{1}{xy} + \frac{1}{1+x-y} + \frac{3}{(4-2x-y)(x+1-y)} + \frac{2}{y(4-2x-y)} \\ &\quad = \frac{4+4x-y}{xy(1+x-y)(4-2x-y)}, \end{aligned}$$

agreeing with Theorem 3.

REMARK 1. The rational function $\mathrm{Vol}((P-\mathbf{x})^\vee)$ is a special case of the *dual mixed volume* function. Let $P_1, P_2, \ldots, P_r \subset \mathbb{R}^d$ be Euclidean polytopes. The *mixed volume* is the polynomial

$$V_{\mathbf{P}}(\mathbf{x}) := \mathrm{Vol}(x_1P_1 + \cdots + x_rP_r)$$

where $P+Q$ denotes the Minkowski sum of polytopes P and Q. We define the *dual mixed volume* function by

$$V_{\mathbf{P}}^\vee(\mathbf{x}) := \mathrm{Vol}((x_1P_1 + \cdots + x_rP_r)^\vee).$$

As before, this is defined when $0 \in \operatorname{Int}(x_1P_1+\cdots+x_rP_r)$ and extended by analytic continuation. The function $V^\vee(\mathbf{x})$ is a rational function. Choosing $P_i = -\mathbf{e}_i$ and $P_{d+1} = P$, and setting $x_{d+1} = 1$, one sees that $\operatorname{Vol}((P-\mathbf{x})^\vee)$ is a special case of $V^\vee(\mathbf{x})$. See also [**AHL21a**] for an appearance of the dual mixed volume function.

Define the d-form $\langle \mathbf{X} d^d \mathbf{X}\rangle$ on $\mathbb{R}^{d+1}$,

$$\langle \mathbf{X} d^d \mathbf{X}\rangle := d! \sum_{i=0}^{d} (-1)^i X_i dX_0 \wedge \cdots \wedge \widehat{dX_i} \wedge \cdots \wedge dX_d.$$

Under the tautological rational map $\tau : \mathbb{R}^{d+1} \dashrightarrow \mathbb{P}^d$, the pullback $\tau^*(\Omega)$ of a rational d-form Ω on $\mathbb{P}^d$ can be written as $\tau^*(\Omega) = \frac{P(\mathbf{X})}{Q(\mathbf{X})}\langle \mathbf{X} d^d \mathbf{X}\rangle$, where $\frac{P(\mathbf{X})}{Q(\mathbf{X})}$ is a rational function, homogeneous of degree $-d-1$.

For a cone $C \subset \mathbb{R}^{d+1}$, define the dual cone $C^\vee$ by

$$C^\vee = \{\mathbf{x} \in \mathbb{R}^{n+1} \mid \mathbf{x}.\mathbf{y} \geq 0 \text{ for all } \mathbf{y} \in C\}.$$

THEOREM 4 (Laplace transform [**ABL**, Section 7.4]). *Let $P \subset \mathbb{P}^d$ be a projective polytope, and let $C(P) \subset \mathbb{R}^{d+1}$ be the (pointed, convex, polyhedral) cone over P. Then*

$$\tau^*\Omega(P) = \frac{1}{d!}\left(\int_{C^\vee} e^{-\mathbf{X}.\mathbf{Y}} d^{d+1}\mathbf{Y}\right)\langle \mathbf{X} d^d \mathbf{X}\rangle$$

where $C^\vee \subset \mathbb{R}^{d+1}$ is the dual cone, and $\mathbf{X}, \mathbf{Y}$ are vectors in $\mathbb{R}^{d+1}$. Here, the integral converges when $\mathbf{X} \in C(P)$, and is extended by analytic continuation to $\mathbb{R}^{d+1}$.

EXAMPLE 4. Continuing Example 1, the cone $C(P)$ over P is given by

$$C(P) = \operatorname{Cone}\left(\begin{bmatrix}1\\0\\0\end{bmatrix}, \begin{bmatrix}1\\2\\0\end{bmatrix}, \begin{bmatrix}1\\1\\2\end{bmatrix}, \begin{bmatrix}1\\0\\1\end{bmatrix}\right) \tag{6}$$

with dual cone

$$C^\vee = \operatorname{Cone}\left(\mathbf{a}_1 = \begin{bmatrix}0\\1\\0\end{bmatrix}, \mathbf{a}_2 = \begin{bmatrix}1\\1\\-1\end{bmatrix}, \mathbf{a}_3 = \begin{bmatrix}4\\-2\\-1\end{bmatrix}, \mathbf{a}_4 = \begin{bmatrix}0\\0\\1\end{bmatrix}\right)$$

whose generators correspond to the four facets (3). The cone $C^\vee$ can be triangulated $C^\vee = C_1 \cup C_2$ into two simplicial cones $C_1 = \operatorname{Cone}(\mathbf{a}_1, \mathbf{a}_2, \mathbf{a}_4)$ and $C_2 = \operatorname{Cone}(\mathbf{a}_2, \mathbf{a}_3, \mathbf{a}_4)$, and we compute that

$$\begin{aligned}
\int_{C_1} e^{-\mathbf{X}.\mathbf{Y}} d^{d+1}\mathbf{Y} &= \frac{|\det(\mathbf{a}_1, \mathbf{a}_2, \mathbf{a}_4)|}{X_1(X_0+X_1-X_2)X_2} = \frac{1}{X_1(X_0+X_1-X_2)X_2} \\
\int_{C_2} e^{-\mathbf{X}.\mathbf{Y}} d^{d+1}\mathbf{Y} &= \frac{|\det(\mathbf{a}_2, \mathbf{a}_3, \mathbf{a}_4)|}{(X_0+X_1-X_2)(4X_0-2X_1-X_2)X_2} \\
&= \frac{6}{(X_0+X_1-X_2)(4X_0-2X_1-X_2)X_2}.
\end{aligned} \tag{7}$$

Summing then specializing to $(X_0, X_1, X_2) = (1, x, y)$, we obtain (2), agreeing with Theorem 4.

We remark that the calculation of Example 4 comes from a triangulation of the dual cone (or dual polytope) and is qualitatively quite different from the one in Theorem 3 (e.g., unlike (4), both rational functions in (7) have poles belonging to the facets of P). Indeed, the volume of a polytope P can be obtained by triangulating P or by triangulating the dual of P, a phenomenon sometimes known as Filliman duality [**Fil**].

Let P be a full-dimensional polytope in $\mathbb{P}^d$ with f facets. We define the *adjoint hypersurface* $A_P \subset \mathbb{P}^d$ as follows. Define $\mathcal{H}_P$ to be the projective hyperplane arrangement consisting of all facet hyperplanes of P. We say that $\mathcal{H}_P$ is *simple* if through any point in $\mathbb{P}^d$ pass at most d hyperplanes of $\mathcal{H}_P$. The *residual arrangement* $\mathcal{R}_P$ consists of all linear subspaces that are intersections of hyperplanes in $\mathcal{H}_P$ that do not contain faces of P.

When $\mathcal{H}_P$ is simple, Kohn and Ranestad [**KR**] proved that there is a unique hypersurface A_P in $\mathbb{P}^d$ of degree $f-d-1$ which vanishes along the residual arrangement $\mathcal{R}_P$. In general (for P not necessarily simple), we write $P = \lim_{t\to 0} P_t$ for polytopes P_t such that $\mathcal{H}_{P_t}$ is simple for $t > 0$, and define the adjoint hypersurface $A_P := \lim_{t\to 0} A_{P_t}$. In the following, we abuse notation by also using A_P to denote a polynomial whose vanishing set is the adjoint hypersurface. See [**War**] for a different definition of adjoint.

The following result connects the canonical form with adjoint hypersurfaces; see the learning seminar notes of Gaetz [**Gae**].

THEOREM 5 (Zeros equal adjoint hypersurface). *Suppose $P \subset \mathbb{R}^d \subset \mathbb{P}^d$. Then*

$$\Omega(P) = \alpha \cdot \frac{A_P}{\prod_{facets\ H} H} d^d\mathbf{x}$$

for some nonzero constant α.

Theorem 5 can be proved by triangulating P and using Theorem 3.

We remark that any rational form $\mathbb{P}^d$ has degree $-d-1$ since $\omega_{\mathbb{P}^d} \simeq \mathcal{O}_{\mathbb{P}^d}(-d-1)$. Since all the poles of $\Omega(P)$ are linear, this agrees with the degree of A_P being equal to $f-d-1$.

If P is a simplex, then the adjoint hypersurface is empty, A_P is a constant, and $\Omega(P)$ has no zeroes.

EXAMPLE 5. The rational 2-form (2) has four poles, which are the four facets of P. The numerator vanishes along the line connecting $(-1,0)$ to $(0,4)$, the adjoint hypersurface of P. See Fig. 1(c).

Let $\pi : X \dashrightarrow Y$ be a rational map between complex algebraic varieties of the same dimension. In the following, we shall use the notion of the *pushforward* $\pi_*\omega$ of a rational differential form ω on X; see [**ABL**, (4.1)].

THEOREM 6 (Pushforward [**ABL**, Section 7.3]). *Suppose P has vertices $\mathbf{W}_1, \mathbf{W}_2, \dots, \mathbf{W}_m \in \mathbb{P}^d$ viewed as vectors in $\mathbb{R}^{d+1}$ in the following. Let*

$$\mathbf{V}_1 = (1, \mathbf{v}_1), \mathbf{V}_2 = (1, \mathbf{v}_2), \dots, \mathbf{V}_m = (1, \mathbf{v}_m) \in \mathbb{Z}^{d+1}$$

be integer vectors with first coordinate equal to 1. Assume that $\mathbf{W}_i$ and $\mathbf{V}_i$ have the same oriented matroid, that is,

$$\operatorname{sign}\det(\mathbf{W}_{i_1}, \dots, \mathbf{W}_{i_{d+1}}) = \operatorname{sign}\det(\mathbf{V}_{i_1}, \dots, \mathbf{V}_{d+1}) \in \{-1, 0, 1\}$$

for all $\{i_1, \dots, i_{d+1}\}$. Define the rational map $\Phi : \mathbb{P}^d \to \mathbb{P}^d$ by

$$\Phi : (1 : \mathbf{z}) = (1 : z_1 : \cdots : z_d) \longmapsto \sum_{i=1}^m \mathbf{z}^{\mathbf{v}_i}\mathbf{W}_i \in \mathbb{P}^d.$$

Then

$$\Omega(P) = \Phi_*\Omega(\Delta^d).$$

In Theorem 6, the rational map Φ has degree equal to the normalized volume of the polytope with vertices $\mathbf{V}$, and maps $\operatorname{Int}(\Delta^d)$ homeomorphically to $\operatorname{Int}(P)$. Theorem 6 also has an interpretation in terms of toric varieties; see Section 3.3.

EXAMPLE 6. Choose $\mathbf{W}_1, \mathbf{W}_2, \mathbf{W}_3, \mathbf{W}_4 \in \mathbb{R}^3$ to be the generators of $C(P)$ in (6). We will choose $\mathbf{V}$ to be a unit square, with $\mathbf{V}_1 = (1,0,0), \mathbf{V}_2 = (1,1,0), \mathbf{V}_3 = (1,1,1), \mathbf{V}_4 = (1,0,1)$, which clearly has the same oriented matroid as $\mathbf{W}$. The map Φ, restricted to $\mathbb{R}^2 = \{(z_1, z_2)\}$ is given by

$$\Phi(z_1, z_2) = \begin{bmatrix}1\\0\\0\end{bmatrix} + z_1\begin{bmatrix}1\\2\\0\end{bmatrix} + z_1z_2\begin{bmatrix}1\\1\\2\end{bmatrix} + z_2\begin{bmatrix}1\\0\\1\end{bmatrix}$$

$$= \begin{bmatrix}1 + z_1 + z_1z_2 + z_2\\ 2z_1 + z_1z_2\\ 2z_1z_2 + z_2\end{bmatrix} \approx \begin{bmatrix}1\\ \frac{2z_1+z_1z_2}{1+z_1+z_1z_2+z_2}\\ \frac{2z_1z_2+z_2}{1+z_1+z_1z_2+z_2}\end{bmatrix}.$$

Using a symbolic computation package, one computes that there are two solutions to

$$(x, y) = (\frac{2z_1 + z_1z_2}{1 + z_1 + z_1z_2 + z_2}, \frac{2z_1z_2 + z_2}{1 + z_1 + z_1z_2 + z_2}),$$

given by

$$(z_1^{(1)}, z_2^{(1)}) = \mathbf{z}^{(1)} = \left(\frac{\sqrt{2(x-2)y + (x+2)^2 + y^2} - 3x - y + 2}{2(2x + y - 4)},\right.$$
$$\left. -\frac{\sqrt{2(x-2)y + (x+2)^2 + y^2} + x - 3y + 2}{2(x - y + 1)}\right)$$

$$(z_1^{(2)}, z_2^{(2)}) = \mathbf{z}^{(2)} = \left(-\frac{\sqrt{2(x-2)y + (x+2)^2 + y^2} + 3x + y - 2}{2(2x + y - 4)},\right.$$
$$\left. \frac{\sqrt{2(x-2)y + (x+2)^2 + y^2} - x + 3y - 2}{2(x - y + 1)}\right)$$

The differential form $\Phi_*(\frac{dz_1dz_2}{z_1z_2})$ is given by summing over the solutions $\mathbf{z}^{(1)}$ and $\mathbf{z}^{(2)}$. Explicitly, for $i = 1, 2$, define the Jacobian

$$J^{(i)}(x, y) = \det\left(\begin{bmatrix}\frac{\partial z_1^{(i)}}{\partial x} & \frac{\partial z_1^{(i)}}{\partial y}\\ \frac{\partial z_2^{(i)}}{\partial x} & \frac{\partial z_2^{(i)}}{\partial y}\end{bmatrix}\right).$$

Then

$$\Phi_*\left(\frac{dz_1dz_2}{z_1z_2}\right) = \left(\frac{J^{(1)}(x,y)}{z_1^{(1)}z_2^{(1)}} + \frac{J^{(2)}(x,y)}{z_1^{(2)}z_2^{(2)}}\right) dxdy$$

which recovers (2). Note that neither term in the summation is a rational form!

2. Definition of positive geometry

Let X be complex d-dimensional irreducible algebraic variety, ω a meromorphic d-form on X, and $H \subset X$ an (irreducible) hypersurface on X. Assume that ω has at most simple poles on H. Then the residue $\operatorname{Res}_H \omega$ is the $(d-1)$-form on H

defined as follows. Let f be a local coordinate such that f vanishes to order one on H. Write

$$\omega = \frac{df}{f} \wedge \eta + \eta'$$

for a $(d-1)$-form η and a d-form η', both without poles along H. Then the restriction

$$\operatorname{Res}_H \omega := \eta|_H \tag{8}$$

is a well-defined $(d-1)$-form on H, not depending on the choices of f, η, η'.

Henceforth, we assume that X is a complex d-dimensional irreducible algebraic variety defined over $\mathbb{R}$. We equip the real points $X(\mathbb{R})$ with the analytic topology. Let $X_{\geq 0} \subset X(\mathbb{R})$ be a closed semialgebraic subset such that the interior $X_{>0} = \operatorname{Int}(X_{\geq 0})$ is an oriented d-manifold, and the closure of $X_{>0}$ recovers $X_{\geq 0}$. Let $\partial X_{\geq 0}$ denote the boundary $X_{\geq 0} \setminus X_{\geq 0}$ and let ∂X denote the Zariski closure of $\partial X_{\geq 0}$. Let $C_1, C_2, \ldots, C_r$ be the irreducible components of ∂X. (We assume that X is generically smooth along each C_i, but see Remark 3.) Let $C_{i,\geq 0}$ denote the closures of the interior of $C_i \cap X_{\geq 0}$ in $C_i(\mathbb{R})$. The spaces $C_{1,\geq 0}, C_{2,\geq 0}, \ldots, C_{r,\geq 0}$ are called the boundary components, or facets of $X_{\geq 0}$.

DEFINITION 1. We call $(X, X_{\geq 0})$ a *positive geometry* if there exists a unique nonzero rational d-form $\Omega(X, X_{\geq 0})$, called the *canonical form*, satisfying the recursive axioms:

(1) If $d = 0$, then $X = X_{\geq 0} = \text{pt}$ is a point and we define $\Omega(X, X_{\geq 0}) = \pm 1$ depending on the orientation.
(2) If $d > 0$, then we require that $\Omega(X, X_{\geq 0})$ has poles only along the boundary components C_i, these poles are simple, and for each $i = 1, 2, \ldots, r$, we have

$$\operatorname{Res}_{C_i} \Omega(X, X_{\geq 0}) = \Omega(C_i, C_{i,\geq 0}). \tag{9}$$

REMARK 2. While orientations are suppressed in our notation and statements, we always insist that the orientation on a boundary component $C_{i,>0}$ is induced by that of $X_{>0}$.

We call $(X, X_{\geq 0})$ *normal* if X is a normal variety and each boundary component $(C_i, C_{i,\geq 0})$ is a normal positive geometry.

REMARK 3. If X is not normal, one or more irreducible components $C \subset \partial X$ may belong to the singular locus of X. In this case, the definition of boundary component and residue should be modified as follows. Let $\pi : \tilde{X} \to X$ be the blowup of X along the codimension one subvariety C. Define $\tilde{X}_{\geq 0} = \overline{\pi^{-1}(X_{>0} \setminus C)}$ to be the closure of the preimage under π of the part of $X_{>0}$ that does not belong to C. Note that π is an isomorphism away from C, so $\pi^{-1}(X_{>0} \setminus C)$ and $X_{>0} \setminus C$ are diffeomorphic manifolds. We then define the boundary components of $(X, X_{\geq 0})$ over C as before: let the irreducible components of the Zariski-closure of $(\tilde{X}_{\geq 0} \setminus \operatorname{Int}(\tilde{X}_{\geq 0})) \cap \pi^{-1}(C)$ be $\tilde{C}_1, \ldots, \tilde{C}_r$ and let $\tilde{C}_{i,\geq 0}$ denote the closures of the interior of $\tilde{C}_i \cap \tilde{X}_{\geq 0}$ in $\tilde{C}_i(\mathbb{R})$. In Definition 1, we use $(\tilde{C}_i, \tilde{C}_{i,\geq 0})$ for $i = 1, 2, \ldots, r$ in place of $(C, C_{\geq 0})$ and in (9) we take the residue of $\pi^* \Omega(X, X_{\geq 0})$ along each $\tilde{C}_i$.

For example, let X be a rational curve with a node $p \in X(\mathbb{R})$, and let $\pi : \mathbb{P}^1 \to X$ be the normalization where $a, b \in \mathbb{P}^1(\mathbb{R})$ both map to p. Let $X_{\geq 0} := \pi([a, b])$ be the image of the closed interval from a to b. Then $\partial X_{\geq 0} = \{p\}$. Then $(X, X_{\geq 0})$ is a positive geometry with canonical form the 1-form on X that pullsback to (1) on $\mathbb{P}^1$. This example appears as a boundary of the positive geometry in Fig. 4(a).

For brevity, we may refer to $X_{\geq 0}$ as a positive geometry, and $\Omega(X_{\geq 0})$ its canonical form. For $(X, X_{\geq 0})$ to be a positive geometry, X cannot have nonzero holomorphic d-forms. Furthermore, if X does not have nonzero holomorphic d-forms, then the uniqueness of the canonical form is immediate: the difference of any two such forms would be holomorphic.

Any positive geometry $(C, C_{\geq 0})$ encountered in the recursion Definition 1 is called a *face* of the positive geometry $X_{\geq 0}$.

For $d = 1$, we must have that X is a rational curve. If $(X, X_{\geq 0})$ is normal, then we have $X \cong \mathbb{P}^1$. Any finite union of closed intervals in $\mathbb{P}^1(\mathbb{R})$ is a positive geometry. For an interval $[a, b]$ (in some affine chart), the canonical form is given by (1). The canonical form of a disjoint union $\bigcup_i [a_i, b_i]$ of closed intervals is the sum $\sum_i \Omega([a_i, b_i])$.

Note that $\mathbb{P}^1(\mathbb{R})$ is not a positive geometry. We have $\partial\mathbb{P}^1(\mathbb{R}) = \emptyset$, but there is no 1-form on $\mathbb{P}^1$ with no poles.

Disjoint unions of positive geometries in the same ambient projective algebraic variety are again positive geometries.

3. Examples of positive geometries

It follows from Theorem 1 that polytopes are positive geometries.

3.1. Simplex-like positive geometries. With Theorem 5 in mind, we call a positive geometry $X_{\geq 0}$ *simplex-like* if $\Omega(X_{\geq 0})$ has no zeroes. Simplex-like positive geometries are particularly simple since their canonical forms are almost defined without using the recursion of Definition 1. Namely, let Ω and Ω' be two rational forms, with simple poles along ∂X, and no other poles or zeroes. Then the ratio Ω/Ω' is a regular function on the projective variety X, and therefore a constant. Thus the canonical form is defined up to a scalar without any requirement on the residues.

If $(X, X_{\geq 0})$ is a normal simplex-like positive geometry, then ∂X is an anti-canonical divisor in X. See [**KLS14**, Section 5] for related discussion.

It is also convenient to introduce a slight weakening of the simplex-like condition. We say that two positive geometries $(X, X_{\geq 0})$ and $(Y, Y_{\geq 0})$ are *birationally isomorphic* if there is a birational isomorphism $f : X \dashrightarrow Y$, inducing a diffeomorphism $X_{\geq 0} \cong Y_{\geq 0}$ which respects the face stratification. For example, X could be a blow up of Y along a subvariety that does not intersect $X_{\geq 0}$.

We say that $(X, X_{\geq 0})$ is *birationally simplex-like* if it is birationally isomorphic to a simplex-like positive geometry.

3.2. Dimension 2. Let us first consider normal positive geometries $(X, X_{\geq 0})$ with $X = \mathbb{P}^2$. The normality condition implies that each boundary component is a smooth rational curve, and thus either a line or a conic. Indeed, let $R \subset \mathbb{P}^2(\mathbb{R})$ be a region that that is bounded by a simple closed curve γ, which is a union of finitely many pieces each of which is either a line segment or a part of a conic. Most such regions R are positive geometries. For example, R could be a convex polygon, or a non-convex polygon, or half of a disk; see Fig. 3.

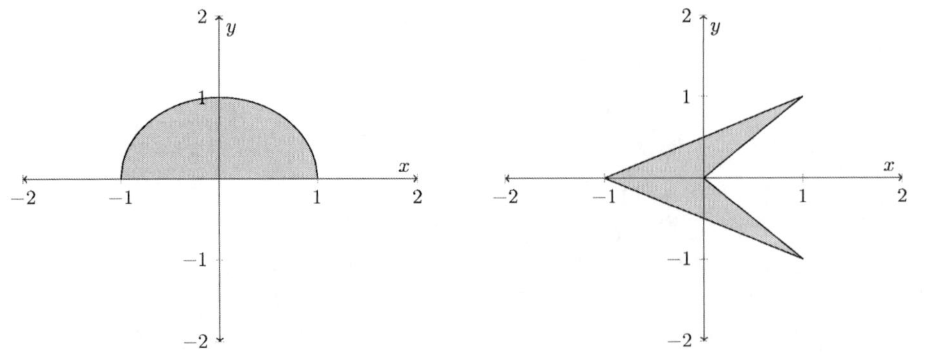

(a) $\Omega = \dfrac{2}{y(x^2+y^2-1)}dxdy.$ (b) $\Omega = \dfrac{6x+2}{(x-2y+1)(x-y)(x+y)(x+2y+1)}dxdy.$

FIGURE 3. Some positive geometries in the plane that are not convex polygons. (a): upper half unit disk. (b): nonconvex 4-gon.

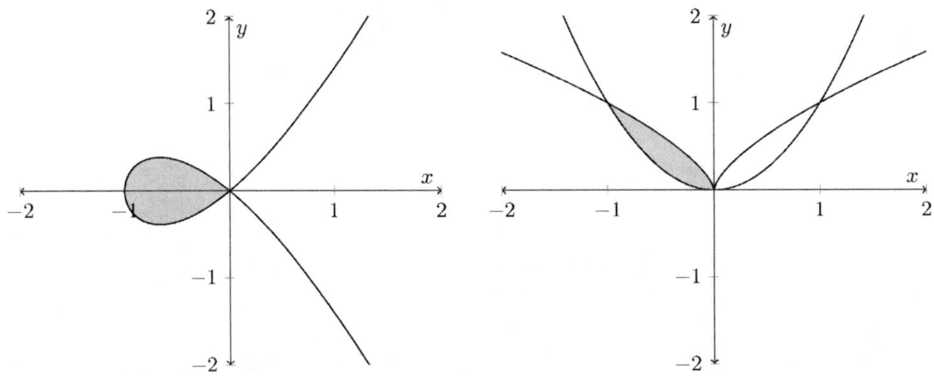

(a) $\Omega = \dfrac{1}{x^3+x^2-y^2}dxdy.$ (b) $\Omega = \dfrac{x^2+xy+y^2+x}{(x^2-y)(x^2-y^3)}dxdy.$

FIGURE 4. Some rational polypols. (a): boundary nodal cubic with equation $y^2 = x^2(x+1)$. (b): boundary curves $y = x^2$ and $y^3 = x^2$.

On the other hand, let R be the closed unit disk. Then R is *not* a positive geometry, since the boundary is a circle, the whole of $\mathbb{P}^1(\mathbb{R})$, which is not a one-dimensional positive geometry.

The list grows if we allow $(X, X_{\geq 0})$ to be non-normal. A large zoo of examples arise from the theory of *rational polypols* in the work of Kohn, Piene, Ranestad, Rydell, Shapiro, Sinn, Sorea, and Telen [**KPRRSSST**]. See Fig. 4 for some examples.

Unfortunately, a classification of positive geometries in arbitrary dimensions seems out of reach.

3.3. Toric varieties. Let X_P be the normal, projective toric variety associated to a d-dimensional lattice polytope P. Then it has a natural positive part

$X_{P,>0}$, defined to be the positive part $\mathbb{R}^d_{>0}$ of the torus $T(\mathbb{R}) \subset T$ that sits inside X_P. We define the nonnegative part $X_{P,\geq 0} := \overline{X_{P,>0}}$.

THEOREM 7 ([**ABL**]). *$(X_P, X_{P,\geq 0})$ is a normal, simplex-like positive geometry.*

The canonical form is the natural form $\Omega = \mathrm{dlog}\, x_1 \wedge \cdots \wedge \mathrm{dlog}\, x_n$ on T, extended to a rational top-form on X_P.

While the combinatorics of faces for the two positive geometries, the polytope P, and the positive toric variety $X_{P,\geq 0}$ are identical, the algebraic geometry is quite different. In particular, the canonical form $\Omega(X_{P,\geq 0})$ has no zeroes while $\Omega(P)$ typically has a very interesting zero set (Theorem 5). From this perspective, toric varieties are simpler positive geometries than polytopes.

A *morphism* $f : (X, X_{\geq 0}) \to (Y, Y_{\geq 0})$ of postive geometries is a rational map $f : X \dashrightarrow Y$ which restricts to a diffeomorphism $f|_{X_{>0}} : X_{>0} \to Y_{>0}$. In [**ABL**], we formulated the heuristic that canonical forms should pushforward under morphisms of positive geometries. Theorem 6 is an example of this for a morphism from a toric variety to a polytope. More precisely, the rational map $\Phi : \mathbb{P}^d \dashrightarrow \mathbb{P}^d$ of in Theorem 6 can be factored into the rational maps

$$\alpha : \mathbb{P}^d \dashrightarrow \mathbb{P}^{m-1} \qquad (1 : z_1 : \cdots : z_d) \longmapsto (\mathbf{z}^{\mathbf{v}_1} : \cdots : \mathbf{z}^{\mathbf{v}_m})$$

$$\beta : \mathbb{P}^{m-1} \dashrightarrow \mathbb{P}^d \qquad (u_1 : u_2 : \cdots : u_m) \longmapsto \sum_{i=1}^m u_i \mathbf{W}_i.$$

The (closure of the) image of α is the toric variety $X_{\mathbf{V}}$ associated to $\mathbf{V}$. The linear map β is a morphism of positive geometries from $X_{\mathbf{V},\geq 0}$ to the polytope P. In the case that $\mathbf{V} = \mathbf{W}$, the map β is called the *algebraic moment map.*

3.4. Totally nonnegative Grassmannian and flag varieties. Let G be a split real reductive group and $P \subset G$ a parabolic subgroup. Lusztig [**Lus98**] has defined the totally nonnegative part $(G/P)_{\geq 0}$ of the partial flag variety G/P. The following result was stated as an expectation in [**ABL**].

THEOREM 8. *$(G/P, (G/P)_{\geq 0})$ is a normal, simplex-like positive geometry.*

For completeness, we give a proof of Theorem 8 in Appendix A.

The face stratification of $(G/P)_{\geq 0}$ is well studied; see [**Rie99, Lus98**]. We have a decomposition $G/P = \bigsqcup_{[v,w]_P} \mathring{\Pi}_v^w$ where the *open projected Richardson varieties* $\mathring{\Pi}_v^w$ [**KLS14**] are indexed by equivalence classes of P-Bruhat intervals. Denote by $\Pi_v^w = \overline{\mathring{\Pi}_v^w}$ the (closed) projected Richardson variety. The faces of $(G/P)_{\geq 0}$ are the positive geometries $(\Pi_v^w, (\Pi_v^w)_{\geq 0})$, where $(\Pi_v^w)_{\geq 0} := \Pi_v^w \cap (G/P)_{\geq 0}$.

The case of the Grassmannian $\mathrm{Gr}(k,n)$ of k-planes in $\mathbb{C}^n$ is of particular importance in numerous situations. Postnikov [**Pos**] gave an independent definition of the totally nonnegative part $\mathrm{Gr}(k,n)_{\geq 0}$, as the subspace represented by $k \times n$ matrices all of whose $k \times k$ minors are nonnegative. For $k = 1$, we have $\mathrm{Gr}(1,n)_{\geq 0} = \mathbb{P}^{n-1}_{\geq 0}$ is a $n-1$ dimensional simplex.

THEOREM 9. *$(\mathrm{Gr}(k,n), \mathrm{Gr}(k,n)_{\geq 0})$ is a normal, simplex-like positive geometry.*

The faces of $(\mathrm{Gr}(k,n), \mathrm{Gr}(k,n)_{\geq 0})$ are positive geometries called *positroid cells*, and their Zariski closures are called *positroid varieties* [**Pos, KLS13, Lam16b**].

REMARK 4. Conjecturally, the space of planar Ising models [**GP**], also called the positive orthogonal Grassmannian [**HW**], and the space of electrical networks [**Lam18**], also called the positive Lagrangian Grassmannian [**CGS, BGKT**] are positive geometries.

We expect that other spaces appearing in total positivity, such as wonderful compactifications or Kac-Moody flag varieties [**BH, He**], are positive geometries.

3.5. Moduli space. Let $X = \mathcal{M}_{0,n}$ be the Deligne-Knudsen-Mumford compactification [**DM**] of the moduli space of n points on $\mathbb{P}^1$. This is a smooth complex projective variety of dimension $n-3$. The open subset $\mathcal{M}_{0,n} \subset X$ is the moduli-space of n distinct points on $\mathbb{P}^1$. It is isomorphic to a hyperplane arrangement complement. The real part $\mathcal{M}_{0,n}(\mathbb{R})$ is a $(n-3)$-dimensional smooth manifold with $(n-1)!/2$ connected components. We define $(\mathcal{M}_{0,n})_{\geq 0} \subset X$ to be the closure in X of one of these connected components. (The group S_n acts on X by permuting the n points, and this action acts transitively on the connected components of $\mathcal{M}_{0,n}(\mathbb{R})$.)

THEOREM 10 ([**AHL21b**, Proposition 8.2]). *$(\mathcal{M}_{0,n}, (\mathcal{M}_{0,n})_{\geq 0})$ is a positive geometry.*

For a discussion of differential forms on $\mathcal{M}_{0,n}$, see for example [**BCS**].

The positive geometry $(\mathcal{M}_{0,n})_{\geq 0}$ is diffeomorphic to the associahedron polytope as a stratified space. This example is not a simplex-like geometry. However, it is birationally simplex-like. Indeed, it is birationally isomorphic to the toric variety $(X_P, X_{P,\geq 0})$ associated to the associahedron; see [**AHL21a**, Section 10]. For closely related geometries, see the cluster configuration spaces of [**AHL21b**] and the positive Chow cells of [**ALS**].

3.6. Cluster varieties. Let Y be a d-dimensional cluster variety, that is, the spectrum $Y = \mathrm{Spec}(A)$ of a cluster algebra of geometric type. For cluster algebras, we follow the convention that frozen variables are inverted [**LS**].

We will assume that Y is a reasonable cluster variety, for example, locally acyclic and of full rank [**Mul**]. Then Y has a natural positive part $Y_{>0}$, the positive part of any cluster torus in Y. Furthermore, Y has a natural top form $\Omega(Y_{>0}) := \prod_i dx_i/x_i$, defined up to sign, where $(x_1, \ldots, x_d)$ is a seed of the cluster algebra.

CONJECTURE 1. *Let Y be a cluster variety, locally acyclic and of full rank. Then there is a compactification X of Y such that $(X, X_{\geq 0})$ is a simplex-like positive geometry, where:*

(1) *$X_{>0} = Y_{>0}$ and $Y = X \setminus \partial X$,*
(2) *$\Omega(X_{\geq 0}) = \Omega(Y_{>0})$,*
(3) *each face positive geometry $(C, C_{\geq 0})$ of $(X, X_{\geq 0})$ is also a compactification of a cluster variety (and its positive part).*

Theorem 9 is the prototypical example of this conjecture. Indeed, the stratification of the Grassmannian induced by positroid cells are the open positroid varieties, which are shown to be cluster varieties in [**GL19+**].

3.7. Grassmann polytopes and amplituhedra. Let $Z : \mathbb{R}^n \to \mathbb{R}^{k+m}$ be a linear map and assume that $n \geq k+m$. Then generically, for a k-dimensional subspace $V \subset \mathbb{R}^n$, we have that $Z(V) \subset \mathbb{R}^{k+m}$ is again k-dimensional. This induces a rational map $Z : \mathrm{Gr}(k,n) \dashrightarrow \mathrm{Gr}(k,k+m)$. Assuming that Z is well-defined on $\mathrm{Gr}(k,n)_{\geq 0}$, the image $P = Z(\mathrm{Gr}(k,n)_{\geq 0})$ is a *Grassmann polytope* [**Lam16b**, **Kar**][1]. Viewing Z as a $n \times (k+m)$ matrix, we call Z positive when all the $(k+m) \times (k+m)$ minors of Z are positive. In this case, $A_{n,k,m} = Z(\mathrm{Gr}(k,n)_{\geq 0})$ is known as the *amplituhedron.*

In the case $k = 1$, Grassmann polytopes are projective polytopes, and the amplituhedron is a cyclic polytope. Thus the following conjecture holds for $k = 1$.

CONJECTURE 2 ([**ABL**]). *Grassmann polytopes and amplituhedra are positive geometries.*

The (conjectural) canonical form of the amplituhedron $A_{n,k,m}$ is very well-studied due to its relation to super Yang-Mills amplitudes [**AT14a**]. Among the remarkable properties the amplituhedron canonical form satisfies, let us mention the conjectural positivity [**AHT**] and the behavior under parity duality [**GL20**].

Grassmann polytopes give a wealth of examples of positive geometries that are not simplex-like. Some other (potential) examples are naive positive parts of flag varieties [**BK, BHL**], higher-loop amplituhedra [**ABL, AT14b**], and momentum amplituhedra [**DFLP**]. These examples, as well as Conjecture 2, may require extending the notion of positive geometry to allow for boundary components to be formal sums of positive geometries; see the recent paper [**DHS**].

4. Combinatorics, topology, etc.

We pose a number of questions on various features positive geometries.

4.1. Positive topology. A positive geometry $X_{\geq 0}$ is *connected* if $X_{>0}$ is a connected manifold and all boundary components $(C, C_{\geq 0})$ are connected positive geometries. Note that this is stronger than the condition that $X_{>0}$ is connected as a topological space. For simplicity, we restrict our discussion to connected positive geometries.

PROBLEM 1. What can we say about the topology of $X_{\geq 0}$ and $X_{>0}$ for connected positive geometries?

Polytopes, positive parts of toric varieties, totally positive parts of flag varieties, and the positive part of the moduli space of n points are all examples of connected positive geometries. In all these cases,

(1) $X_{>0}$ is homeomorphic to an open ball, and $X_{\geq 0}$ is homeomorphic to a closed ball of the same dimension;
(2) the face poset of $X_{\geq 0}$ is a Eulerian and shellable;
(3) the face stratification of $X_{\geq 0}$ endows it with the structure of a regular CW-complex.

For polytopes, these statements are well-known. The toric variety case and the moduli space case are both diffeomorphic to polytopes. For the totally positive flag varieties, see [**GKL22, GKL19**] for (1), [**Wil**] for (2), and [**GKL21**] for (3). It would be interesting to establish (1),(2),(3) for connected Grassmann polytopes. See also [**BGPZ**].

[1] More general Grassmann polytopes are considered in [**Lam16b**].

4.2. Complex topology. Let $(X, X_{\geq 0})$ be a positive geometry, and let $\mathring{X} := X \setminus \partial X$ be the complex algebraic "open stratum" of the positive geometry. For simplicity, we assume that $\mathring{X}$ is a smooth complex algebraic variety.

Since $\Omega(P)$ has poles only along ∂X, it is holomorphic on $\mathring{X}$. Furthermore, $\Omega(P)$ is a holomorphic top-form and thus closed. It therefore defines a class $[\Omega(P)] \in H^d(\mathring{X}) = H^d(\mathring{X}, \mathbb{C})$ in the deRham (and therefore singular) cohomology of $\mathring{X}$.

PROBLEM 2. What can we say about the topology of $\mathring{X}$ and what can we say about the class $[\Omega(P)] \in H^d(\mathring{X})$?

(a) For $X_{\geq 0} = P$ a polytope, $\mathring{X}$ is a hyperplane arrangement complement.
(b) For $X_{\geq 0} = X_{P,\geq 0}$ the positive part of a toric variety, $\mathring{X}$ is a complex algebraic torus.
(c) For $X_{\geq 0} = (G/P)_{\geq 0}$ a totally nonnegative flag variety, $\mathring{X}$ is the top open projected Richardson stratum.

The cohomology of hyperplane arrangement complements are well-understood from both geometric and combinatorial perspectives [**OT**]. The cohomologies of open Richardson varieties were studied in [**GL20+**]. In the case of an open positroid variety, or more generally a Grassmann polytope or amplituhedron, the space $\mathring{X}$ could be considered a Grassmannian variant of a hyperplane arrangement complement. These cohomologies are especially interesting in the positroid case, where they are related to q,t-Catalan numbers.

Curiously, in the simplex-like positive geometry discussed in Section 3, we have $\dim H^d(\mathring{X}) = 1$, and the cohomology group is spanned by $[\Omega(P)]$.

(a) For $X_{\geq 0} = P$ a simplex, we have $\mathring{X} = (\mathbb{C}^*)^d$ which deformation retracts to a torus $(S^1)^d$, whose cohomology is well-known. For Fig. 3(a), the formula $H^d(\mathring{X}) = \mathbb{C} \cdot [\Omega(P)]$ can be proven, for example, using the Gysin long exact sequence.
(b) For the toric variety case, we again have $\mathring{X} = (\mathbb{C}^*)^d \simeq (S^1)^d$.
(c) For $\mathring{X}$ an open projected Richardson variety, the dimension $\dim(H^d(\mathring{X})) = 1$ is given in [**HLZ**, Proposition 8.6]. Let us show that $[\Omega(X_{\geq 0})] \neq 0$ inside $H^d(\mathring{X})$. In the notation of Appendix A, for $[v', w']_P \lessdot [v, w]_P$, we have a Gysin exact sequence

$$\cdots \to H^d(\mathring{\Pi}_{[v,w]_P} \cup \mathring{\Pi}_{[v',w']_P}) \to H^d(\mathring{\Pi}_{[v,w]_P}) \xrightarrow{\text{Res}} H^{d-1}(\mathring{\Pi}_{[v',w']_P}) \to \cdots.$$

By induction on dimension we may assume that $[\Omega((\Pi_{[v',w']_P})_{\geq 0})]$ spans $H^{d-1}(\mathring{\Pi}_{[v',w']_P})$; then $(\star\star)$ in Appendix A shows that $[\Omega((\Pi_{[v,w]_P})_{\geq 0})]$ spans $H^{d-1}(\mathring{\Pi}_{[v,w]_P})$.

The fact that $H^d(\mathring{X})$ is spanned by $[\Omega(P)]$ also suggests a relation to mirror symmetry [**LT, HLZ**]. See also [**LS**] for a discussion of the cluster variety case.

On the other hand, for a general polytope, or a positive geometry that is not simplex-like, $H^d(\mathring{X})$ can be large.

4.3. Triangulations. In Section 1, we give many formulae for the canonical form of a polytope. A fundamental problem in positive geometries is to give (similar or otherwise) formulae for canonical forms of positive geometries.

A collection $(X_1, X_{1,\geq 0}), (X_2, X_{2,\geq 0}), \ldots, (X_r, X_{r,\geq 0})$ of positive geometries is a *subdivision* of a positive geometry $(X, X_{\geq 0})$ if the interiors $X_{i,>0}$ are pairwise disjoint and $\bigcup_i X_{i,\geq 0} = X_{\geq 0}$. (This is weaker than the usual notion of subdivion for polytopes.) Eq. (4) generalizes to positive geometries.

THEOREM 11 ([**ABL**]). *If* $(X_1, X_{1,\geq 0}), (X_2, X_{2,\geq 0}), \ldots, (X_r, X_{r,\geq 0})$ *subdivides* $(X, X_{\geq 0})$ *then* $\Omega(X_{\geq 0}) = \sum_i \Omega(X_{i,\geq 0})$.

In the case of the amplituhedron, Theorem 11 is the main tool used to construct the canonical form [**AT14a, ATT**]. This has led to much work on the triangulations of amplituhedra; see the recent works [**GL20, PSW, ELT**] and references therein.

CONJECTURE 3. *Every positive geometry has a subdivision into positive geometries that are birationally simplex-like.*

A subdivision as in Conjecture 3 is called a triangulation. We believe the adjective "birationally" is necessary because of the possibility of modifying X by a blowup.

4.4. Adjoint. Motivated by Theorem 5, we make the following definition. See [**KPRRSSST**] for a detailed discussion.

DEFINITION 2. The *adjoint hypersurface* $A_{X_{\geq 0}}$ of a positive geometry $(X, X_{\geq 0})$ is the closure of the zero locus of $\Omega(X_{\geq 0})$.

Up to a scalar, determining the adjoint hypersurface is equivalent to determining the canonical form of a positive geometry. Let $C_1, C_2, \ldots, C_r$ be the irreducible components of ∂X. We produce a finite list of closed irreducible subvarieties of X by repeatedly intersecting and taking irreducible components, starting with $C_1, \ldots, C_r$. Define the *residual arrangement* $\mathcal{R}_{X_{\geq 0}}$ to be the collection of such subvarieties that do not intersect $X_{\geq 0}$.

PROPOSITION 1. *Let* $(X, X_{\geq 0})$ *be a positive geometry and assume that all the varieties in the face stratification of* $X_{\geq 0}$ *are smooth. Then the adjoint hypersurface* $A_{X_{\geq 0}}$ *of* $X_{\geq 0}$ *contains* $\mathcal{R}_{X_{\geq 0}}$.

PROOF. Let C be a boundary component of $(X, X_{\geq 0})$. We first show that $A_{C_{\geq 0}}$ is contained in $A_{X_{\geq 0}} \cap C$. Using the smoothness assumption, we let $f_1, f_2, \ldots, f_d$ be local coordinates for X such that C is cut out by $f_1 = 0$ to order one. Locally, $df_1 \wedge df_2 \wedge \cdots \wedge df_d$ is a non-vanishing top form, so we have

$$\Omega(X_{\geq 0}) = \frac{A_{X_{\geq 0}}}{f_1 p_2 p_3 \cdots p_m} df_1 \wedge df_2 \wedge \cdots \wedge df_d,$$

where $p_2, p_3, \ldots$ are regular functions whose vanishing sets are the other boundary components of X, and $A_{X_{\geq 0}}$ here denotes a regular function that vanishes along the adjoint hypersurface. Then

$$\Omega_{C_{\geq 0}} = \mathrm{Res}_C \, \Omega(X_{\geq 0}) = \left(\frac{A_{X_{\geq 0}}}{p_2 p_3 \cdots p_m} \right) \bigg|_C df_2 \wedge \cdots \wedge df_d. \tag{10}$$

The restriction to C is obtained by setting $f_1 = 0$. After cancelling out common factors in the numerator and denominator, we deduce that the adjoint hypersurface $A_{C_{\geq 0}}$ is contained in $A_{X_{\geq 0}} \cap C$.

Now let $Z \subset \mathcal{R}_{X_{\geq 0}}$ be an irreducible component and let C be a boundary component of X that contains Z. Let $D_1, D_2, \ldots, D_r$ be the irreducible components of $C_i \cap C_j$ for two boundary components of $(X, X_{\geq 0})$. Since X is smooth, all these subvarieties are codimension two [**Stacks**, Tag 0AZL]. Reindex so that

$D_1, D_2, \ldots, D_p$ are the boundary components of $(C, C_{\geq 0})$. Suppose first that Z belongs to one of the subvarieties $D_{p+1}, \ldots, D_r$. Since $Z \subset C$, we may assume that $Z \subset D \subset C$ where D is an irreducible component of $C \cap C'$ for C' some other boundary component of $(X, X_{\geq 0})$. But D is not an irreducible component of ∂C, so $\Omega(C_{\geq 0})$ does not have a pole along D. On the other hand, $\Omega(X_{\geq 0})$ has a pole along C'. From (10), we see that this is only possible if $A_{X_{\geq 0}}$ vanishes along D.

Next, suppose that Z does not belong to any of $D_{p+1}, \ldots, D_r$. Then Z can be obtained by repeatedly intersecting and taking irreducible components of the subvarieties $D_1, D_2, \ldots, D_p$. On the other hand, $Z \cap X_{\geq 0} = \emptyset$ which implies that $Z \cap C_{\geq 0} = \emptyset$, so by definition we have $Z \subset \mathcal{R}_{C_{\geq 0}}$. By induction on dimension, we conclude that $Z \subset A_{C_{\geq 0}}$, and since $A_{C_{\geq 0}} \subset A_{X_{\geq 0}} \cap C \subset A_{X_{\geq 0}}$, the result follows. □

We expect Proposition 1 to hold without any smoothness assumption.

PROBLEM 3. When is $A_{X_{\geq 0}}$ characterized by the property of containing $\mathcal{R}_{X_{\geq 0}}$?

In the case that $X_{\geq 0}$ is one-dimensional, Problem 3 has an affirmative answer only when $X_{\geq 0}$ is a closed interval. When $X_{\geq 0}$ is a disjoint union of multiple intervals, the canonical form has a non-trivial zero set, but the residual arrangement is empty.

Note that even in the case of polytopes, the adjoint hypersurface is not always uniquely determined by its vanishing on the residual arrangement [**KR**].

It would be especially interesting to study the residual arrangement of the amplituhedron.

4.5. Positive convexity and dual positive geometries. A positive geometry $(X, X_{\geq 0})$ is called *positively convex* [**ABL**] if $\Omega(X, X_{\geq 0})$ has constant sign on $X_{>0}$. In other words, the poles and zeros of $\Omega(X, X_{\geq 0})$ do not intersect $X_{>0}$. For a polytope P, the dual volume formula (Theorem 3) shows that the canonical form $\Omega(P)$ takes constant sign in the interior of the polytope. Thus polytopes are positively convex positive geometries. Fig. 3(a) and Fig. 4(a,b) are positively convex geometries. Fig. 3(b) is not positively convex. It is conjectured in [**ABL, AHT**] that certain amplituhedra are positively convex geometries.

As suggested by Theorem 3, positive convexity indicates the potential existence of a *dual positive geometry.*

4.6. Integral functions. Let us say only a brief word about the relation between positive geometries and scattering amplitudes which typically arises from taking integrals of the canonical form over positive geometries. One typically considers integrals of the form

$$\int_{X_{>0}} (\text{regulator})\Omega(X_{\geq 0}) \tag{11}$$

Since $\Omega(X_{\geq 0})$ has poles along $\partial X_{\geq 0}$, some regulator in the integrand is necessary for the integral to converge. Different regulators appear in different applications.

The motivating examples of (11) are integrals for scattering amplitudes in physics. We refer the reader to [**HT, FL**] for further discussion.

Let us mention two classes of examples that recover classical special functions. The *stringy canonical forms* of [**AHL21a**] are obtained by taking the regulator to be a monomial $f_1^{s_1} \cdots f_r^{s_r}$ in a collection of rational functions $f_1, f_2, \ldots, f_r$, and viewing the integral as a function of the exponents $s_1, s_2, \ldots, s_r$. Stringy canonical forms are so named because in the case $X_{\geq 0} = (\mathcal{M}_{0,n})_{\geq 0}$ one obtains string theory amplitudes. For $n = 4$, one recovers the beta function, studied clasically by Euler and Legendre, as a special case. Curiously, these integrals also appear as marginal likelihood integrals in algebraic statistics [**ST**].

Another class of examples appear in mirror symmetry and the theory of geometric crystals [**Rie12, Lam16a, LT**]. In this case, one takes the regulator to be $\exp(f)$ for a rational function f called the *superpotential*. When $(X, X_{\geq 0}) = (\mathbb{P}^1, \Delta^1)$, this integral recovers the Bessel function, studied classically by Bernoulli, Bessel, and others.

Appendix A. Proof of Theorem 8

Write $[v', w']_P \leq [v, w]_P$ for the partial order on projected Richardson varieties, and let $[v', w']_P \lessdot [v, w]_P$ denote a cover relation. Thus $(\Pi_{v'}^{w'})_{\geq 0} \subseteq (\Pi_v^w)_{\geq 0}$ for $[v', w']_P \leq [v, w]_P$. We generally follow the notations of [**KLS14, GKL21**].

The projected Richardson variety Π_v^w is irreducible, normal, and of dimension $\ell(w) - \ell(v)$ [**KLS14**]. Write $\partial \Pi_v^w$ for the union of all projected Richardson varieties contained in Π_v^w but of smaller dimension. The irreducible components of $\partial \Pi_v^w$ are called the facets of Π_v^w. In [**KLS14**, Section 5], it is shown that $\partial \Pi_v^w$ is an anticanonical divisor in Π_v^w. It follows that there is a rational top-form ω_v^w, unique up to scalar, that is holomorphic and nonzero on $\mathring{\Pi}_v^w$, with simple poles along the facets of Π_v^w. Furthermore, for $[v', w'] \lessdot [v, w]$, we have

$$\mathrm{Res}_{\Pi_{v'}^{w'}} \omega_v^w = \omega_{v'}^{w'} \text{ up to scalar.} \tag{12}$$

To prove Theorem 8, we define a top-form Ω_v^w that is a scalar multiple of ω_v^w and that satisfies the defining recursion for the canonical form of $(\Pi_v^w, (\Pi_v^w)_{\geq 0})$. Each $\mathring{\Pi}_v^w$ is isomorphic to an open Richardson variety in the full flag variety G/B, and it is enough to consider the G/B case. Then the strata are indexed by intervals $[v, w]$, $v \leq w$ in Bruhat order.

By [**GKL21**], $(G/B)_{\geq 0}$ is a stratified closed ball. We may pick an orientation on the closed ball $(G/B)_{\geq 0}$ that is compatible with orientations on each of the strata. Thus it is enough to define Ω_v^w up to sign: the sign can be fixed by requiring that Ω_v^w induces the chosen orientation.

Let $v \leq w$ and $\mathbf{w}$ be a reduced word for w. Let $\mathbf{v}$ be the positive distinguished subexpression of $\mathbf{w}$, that is, $\mathbf{v}$ is the rightmost subword of $\mathbf{w}$ that is a reduced word for v. Let $J \subset [1, \ell(w)]$ denote the indices not belonging to the subword $\mathbf{v}$. For $\mathbf{t} \in (\mathbb{C}^*)^{|J|}$, we have a group element (see [**GKL21**] for notation)

$$g_{\mathbf{v},\mathbf{w}}(\mathbf{t}) = g_1 g_2 \cdots g_n \in G \tag{13}$$

where $g_k = \dot{s}_{i_k}$ or $g_k = y_{i_k}(t_k)$ depending on whether $k \notin J$ or $k \in J$. The map $\mathbf{t} \mapsto g_{\mathbf{v},\mathbf{w}}(\mathbf{t})B \in G/B$ defines a rational parametrization of Π_v^w. Define the rational form

$$\Omega_v^w = \pm \bigwedge_{k \in J} \frac{dt_k}{t_k}$$

on Π_v^w. It is not immediately clear that Ω_v^w is holomorphic on $\mathring{\Pi}_v^w$. By [**Rie12**, Proof of Proposition 7.2] and [**GKL21**, Proof of Proposition 5.4], up to sign the form Ω_v^w does not depend on the choice of $\mathbf{w}$.

We shall show that

$$(\star) \qquad \Omega_v^w \text{ is equal to } \omega_v^w \text{ up to a scalar.}$$

$$(\star\star) \qquad \operatorname{Res}_{\Pi_{v'}^{w'}} \Omega_v^w = \pm\Omega_{v'}^{w'} \text{ for } [v',w'] \lessdot [v,w]$$

which together proves Theorem 8.

By [**Lam16a**, Proposition 2.11], $(\star)$ holds for $(v,w) = (e,w_0)$. Now let $(v,w) = (e,w)$. Then we can choose $\mathbf{w_0} = \mathbf{w}_1\mathbf{w}_2$ where $\mathbf{w}_2$ is a reduced word for w. Then $g_{e,\mathbf{w_0}}(\mathbf{t}) = g_1(t_1,\ldots,t_r)g_2(t_{r+1},\ldots,t_\ell)$ for $\ell = \ell(w_0)$ and $r = \ell(w_0) - \ell(w)$, and $g_2 = g_{e,\mathbf{w}}$ parametrizes Π_e^w. Thus Ω_1^w is obtained from $\Omega_1^{w_0}$ by repeatedly taking residues $\operatorname{Res}_{t_k=0} \circ \cdots \circ \operatorname{Res}_{t_2=0} \circ \operatorname{Res}_{t_1=0}$. By (12), this proves $(\star)$ for the $v = e$ case. Similarly, when $(v,w) = (v,w_0)$, we pick $\mathbf{w_0} = \mathbf{w}_1\mathbf{w}_2$ where $\mathbf{w}_2$ is a reduced word for v. A calculation shows that in the parametrization given by the group element (13), $\lim_{t_k\to\infty}$ sends $y_{i_k}(t_k)$ to $\dot{s}_{i_k}$. Then $\Omega_v^{w_0}$ is obtained from $\Omega_e^{w_0}$ by repeatedly taking residues $\cdots \circ \operatorname{Res}_{t_{\ell-1}=\infty} \circ \operatorname{Res}_{t_\ell=\infty}$. By (12), this proves $(\star)$ for the $w = w_0$ case.

Next, suppose that we have $[v',w'] \lessdot [v,w]$ where v,w are arbitrary. Then either (a) $w' = w$ or (b) $v' = v$. Suppose we are in case (a). If $w' = w = w_0$, we may pick $\mathbf{w_0} = \mathbf{w}_1\mathbf{w}_2$ where $\mathbf{w}_2$ is a reduced word for v'. Then $(\star\star)$ follows as before. Otherwise if $w \neq w_0$, we let $w_0 = uw$ and left multiply the parametrization $g_{\mathbf{v},\mathbf{w}}(\mathbf{t})$ by $y_{i_1}(a_1)\cdots y_{i_r}(a_r)$ where $u = s_{i_1}\cdots s_{i_r}$ is reduced, obtaining a parametrization $g_{\mathbf{v},\mathbf{w_0}}$, and similarly for $g_{\mathbf{v'},\mathbf{w_0}}$. Using the definition of residue, we have

$$\begin{aligned} \bigwedge_i \frac{da_i}{a_i} \wedge \Omega_{v'}^w &= \pm\Omega_{v'}^{w_0} = \pm \operatorname{Res}_{\Pi_{v'}^{w_0}} \Omega_v^{w_0} \\ &= \pm \operatorname{Res}_{\Pi_{v'}^{w_0}} \left(\bigwedge_i \frac{da_i}{a_i} \wedge \Omega_v^w \right) = \pm \bigwedge_i \frac{da_i}{a_i} \wedge \operatorname{Res}_{\Pi_{v'}^w} \Omega_v^w. \end{aligned}$$

Wedging with $\bigwedge_i \frac{da_i}{a_i}$ is injective for forms on $\mathring{\Pi}_{v'}^w$, so we deduce that $(\star\star)$ holds in case (a). By (12), this implies that $(\star)$ holds for $[v',w]$ if it holds for $[v,w]$. Since we have shown $(\star)$ for $(v,w) = (e,w)$, we deduce that $(\star)$ holds for all $[v,w]$.

There is an automorphism $\phi : G/B \to G/B$ that takes $\mathring{\Pi}_v^w$ to $\mathring{\Pi}_{ww_0}^{vw_0}$ for each $[v,w]$; see [**Lus21**]. Since Ω_v^w satisfies $(\star)$, we have that $\phi^*(\Omega_v^w)$ is a scalar multiple of $\Omega_{ww_0}^{vw_0}$. This scalar multiple must be ± 1, since both $\phi^*(\Omega_v^w)$ and $\Omega_{ww_0}^{vw_0}$ have unit residues on some 0-dimensional strataum Π_u^u. Thus the validity of $(\star\star)$ for case (b) follows from that for case (a). We are done.

Acknowledgments

We apologize for the multiple perspectives, especially those from physics, that we do not mention. We thank the OPAC organizers, Christine Berkesch, Ben Brubaker, Gregg Musiker, Pavlo Pylyavskyy, and Vic Reiner for the invitation to speak.

References

[ABCGPT] Nima Arkani-Hamed, Jacob Bourjaily, Freddy Cachazo, Alexander Goncharov, Alexander Postnikov, and Jaroslav Trnka, *Grassmannian geometry of scattering amplitudes*, Cambridge University Press, Cambridge, 2016, DOI 10.1017/CBO9781316091548. MR3467729

[ABL] Nima Arkani-Hamed, Yuntao Bai, and Thomas Lam, *Positive geometries and canonical forms*, J. High Energy Phys. **11** (2017), 039, front matter+121, DOI 10.1007/jhep11(2017)039. MR3747267

[AHL21a] Nima Arkani-Hamed, Song He, and Thomas Lam, *Stringy canonical forms*, J. High Energy Phys. **2** (2021), Paper No. 069, 59, DOI 10.1007/jhep02(2021)069. MR4260349

[AHL21b] Nima Arkani-Hamed, Song He, and Thomas Lam, *Cluster configuration spaces of finite type*, SIGMA Symmetry Integrability Geom. Methods Appl. **17** (2021), Paper No. 092, 41, DOI 10.3842/SIGMA.2021.092. MR4324836

[AHT] Nima Arkani-Hamed, Andrew Hodges, and Jaroslav Trnka, *Positive amplitudes in the amplituhedron*, J. High Energy Phys. **8** (2015), 030, front matter+24, DOI 10.1007/JHEP08(2015)030. MR3402135

[ALS] Nima Arkani-Hamed, Thomas Lam, and Marcus Spradlin, *Positive configuration space*, Comm. Math. Phys. **384** (2021), no. 2, 909–954, DOI 10.1007/s00220-021-04041-x. MR4259378

[ATT] Nima Arkani-Hamed, Hugh Thomas, and Jaroslav Trnka, *Unwinding the amplituhedron in binary*, J. High Energy Phys. **1** (2018), 016, front matter+40, DOI 10.1007/jhep01(2018)016. MR3764259

[AT14a] Nima Arkani-Hamed and Jaroslav Trnka, *The amplituhedron*, J. High Energy Phys. (2014), Article 33.

[AT14b] Nima Arkani-Hamed and Jaroslav Trnka *Into the amplituhedron*, J. High Energy Phys. (2014), Article 182.

[BHL] Yuntao Bai, Song He, and Thomas Lam, *The amplituhedron and the one-loop Grassmannian measure*, J. High Energy Phys. **1** (2016), 112, front matter+41, DOI 10.1007/JHEP01(2016)112. MR3471468

[BH] Huanchen Bao and Xuhua He, *Product structure and regularity theorem for totally nonnegative flag varieties*, `arXiv:2203.02137`, 2022.

[BGPZ] Pavle V. M. Blagojević, Pavel Galashin, Nevena Palić, and Günter M. Ziegler, *Some more amplituhedra are contractible*, Selecta Math. (N.S.) **25** (2019), no. 1, Paper No. 8, 11, DOI 10.1007/s00029-019-0462-2. MR3908855

[BK] Anthony M. Bloch and Steven N. Karp, *On two notions of total positivity for partial flag varieties*, Adv. Math. **414** (2023), Paper No. 108855, 24, DOI 10.1016/j.aim.2022.108855. MR4536122

[BCS] Francis Brown, Sarah Carr, and Leila Schneps, *The algebra of cell-zeta values*, Compos. Math. **146** (2010), no. 3, 731–771, DOI 10.1112/S0010437X09004540. MR2644933

[BGKT] B. Bychkov, V. Gorbounov, A. Kazakov, and D. Talalaev, *Electrical networks, Lagrangian Grassmannians, and symplectic groups*, Mosc. Math. J. **23** (2023), no. 2, 133–167, DOI 10.17323/1609-4514-2023-23-2-133-167. MR4598191

[CGS] Sunita Chepuri, Terrence George, and David E Speyer, *Electrical networks and Lagrangian Grassmannians*, `arXiv:2106.15418`, 2021.

[DFLP] David Damgaard, Livia Ferro, Tomasz Łukowski, and Matteo Parisi, *The momentum amplituhedron*, J. High Energy Phys. **8** (2019), 042, 21, DOI 10.1007/jhep08(2019)042. MR4014511

[DHS] Gabriele Dian, Paul Heslop, and Alastair Stewart, *Internal boundaries of the loop amplituhedron*, SciPost Phys. **15** (2023), no. 3, Paper No. 098, 43, DOI 10.21468/scipostphys.15.3.098. MR4649738

[DM] P. Deligne and D. Mumford, *The irreducibility of the space of curves of given genus*, Inst. Hautes Études Sci. Publ. Math. **36** (1969), 75–109. MR262240

[EH] Henriette Elvang and Yu-tin Huang, *Scattering amplitudes in gauge theory and gravity*, Cambridge University Press, Cambridge, 2015, DOI 10.1017/CBO9781107706620. MR3329658

[ELT] Chaim Even-Zohar, Tsviqa Lakrec, and Ran J. Tessler, *The amplituhedron BCFW triangulation*, `arXiv:2112.02703`, 2021.

[ELT] Chaim Even-Zohar, Tsviqa Lakrec, and Ran J. Tessler, *The BCFW tiling of the amplituhedron*, Sém. Lothar. Combin. **89B** (2023), Art. 47, 12. MR4659555

[FL] Livia Ferro and Tomasz Łukowski, *Amplituhedra, and beyond*, J. Phys. A **54** (2021), no. 3, Paper No. 033001, 42, DOI 10.1088/1751-8121/abd21d. MR4209128

[Fil] P. Filliman, *The volume of duals and sections of polytopes*, Mathematika **39** (1992), no. 1, 67–80, DOI 10.1112/S0025579300006860. MR1176472

[Gae] Christian Gaetz, *Positive geometries learning seminar. Lecture 3: canonical forms of polytopes from adjoints*, `http://www.math.lsa.umich.edu/~tfylam/posgeom/gaetz_notes.pdf`, 2020.

[GKL22] Pavel Galashin, Steven N. Karp, and Thomas Lam, *The totally nonnegative Grassmannian is a ball*, Adv. Math. **397** (2022), Paper No. 108123, 23, DOI 10.1016/j.aim.2021.108123. MR4377938

[GKL19] Pavel Galashin, Steven N. Karp, and Thomas Lam, *The totally nonnegative part of G/P is a ball*, Adv. Math. **351** (2019), 614–620, DOI 10.1016/j.aim.2019.05.009. MR3954040

[GKL21] Pavel Galashin, Steven N. Karp, and Thomas Lam, *Regularity theorem for totally nonnegative flag varieties*, J. Amer. Math. Soc. **35** (2022), no. 2, 513–579, DOI 10.1090/jams/983. MR4374956

[GL20] Pavel Galashin and Thomas Lam, *Parity duality for the amplituhedron*, Compos. Math. **156** (2020), no. 11, 2207–2262, DOI 10.1112/S0010437X20007411. MR4190044

[GL19+] Pavel Galashin and Thomas Lam, *Positroid varieties and cluster algebras* (English, with English and French summaries), Ann. Sci. Éc. Norm. Supér. (4) **56** (2023), no. 3, 859–884, DOI 10.24033/asens.2545. MR4650160

[GL20+] Pavel Galashin and Thomas Lam, *Positroids, knots, and q,t-Catalan numbers*, Sém. Lothar. Combin. **85B** (2021), Art. 54, 12. MR4311935

[GP] Pavel Galashin and Pavlo Pylyavskyy, *Ising model and the positive orthogonal Grassmannian*, Duke Math. J. **169** (2020), no. 10, 1877–1942, DOI 10.1215/00127094-2019-0086. MR4118643

[He] Xuhua He, *Total positivity in the De Concini-Procesi compactification*, Represent. Theory **8** (2004), 52–71, DOI 10.1090/S1088-4165-04-00213-4. MR2048587

[HP] Johannes M. Henn and Jan C. Plefka, *Scattering amplitudes in gauge theories*, Lecture Notes in Physics, vol. 883, Springer, Heidelberg, 2014, DOI 10.1007/978-3-642-54022-6. MR3184495

[HT] Enrico Herrmann and Jaroslav Trnka, *The SAGEX Review on Scattering Amplitudes. Chapter 7: Positive geometry of scattering amplitudes*, J. Phys. A **55** (2022), no. 44, Paper No. 443008, 34. MR4516661

[HLZ] An Huang, Bong H. Lian, and Xinwen Zhu, *Period integrals and the Riemann-Hilbert correspondence*, J. Differential Geom. **104** (2016), no. 2, 325–369. MR3557307

[HW] Yu-tin Huang and CongKao Wen. *ABJM amplitudes and the positive orthogonal Grassmannian*, J. High Energy Phys. (2014), Article 104.

[Kar] Steven N. Karp, *Defining amplituhedra and Grassmann polytopes* (English, with English and French summaries), 28th International Conference on Formal Power Series and Algebraic Combinatorics (FPSAC 2016), Discrete Math. Theor. Comput. Sci. Proc., BC, Assoc. Discrete Math. Theor. Comput. Sci., Nancy, [2016] ©2016, pp. 683–693. MR4111289

[KPRRSSST] K. Kohn, R. Piene, K. Ranestad, F. Rydell, B. Shapiro, R. Sionn, M.-S. Sorea, and S. Telen, *Adjoints and canonical forms of polypols*, `arXiv:2108.11747`, 2021.

[KR] Kathlén Kohn and Kristian Ranestad, *Projective geometry of Wachspress coordinates*, Found. Comput. Math. **20** (2020), no. 5, 1135–1173, DOI 10.1007/s10208-019-09441-z. MR4156995

[KLS13] Allen Knutson, Thomas Lam, and David E. Speyer, *Positroid varieties: juggling and geometry*, Compos. Math. **149** (2013), no. 10, 1710–1752, DOI 10.1112/S0010437X13007240. MR3123307

[KLS14] Allen Knutson, Thomas Lam, and David E. Speyer, *Projections of Richardson varieties*, J. Reine Angew. Math. **687** (2014), 133–157, DOI 10.1515/crelle-2012-0045. MR3176610

[Lam16a] Thomas Lam, *Whittaker functions, geometric crystals, and quantum Schubert calculus*, Schubert calculus—Osaka 2012, Adv. Stud. Pure Math., vol. 71, Math. Soc. Japan, [Tokyo], 2016, pp. 211–250, DOI 10.2969/aspm/07110211. MR3644825

[Lam16b] Thomas Lam, *Totally nonnegative Grassmannian and Grassmann polytopes*, Current developments in mathematics 2014, Int. Press, Somerville, MA, 2016, pp. 51–152. MR3468251

[Lam18] Thomas Lam, *Electroid varieties and a compactification of the space of electrical networks*, Adv. Math. **338** (2018), 549–600, DOI 10.1016/j.aim.2018.09.014. MR3861712

[LS] Thomas Lam and David E. Speyer, *Cohomology of cluster varieties I: Locally acyclic case*, Algebra Number Theory **16** (2022), no. 1, 179–230, DOI 10.2140/ant.2022.16.179. MR4384567

[LT] Thomas Lam and Nicolas Templier, *The mirror conjecture for minuscule flag varieties*, `arXiv:1705.00758`, 2017.

[CGA] Thomas Lam, *Combinatorics and geometry of amplitudes*, `http://www.math.lsa.umich.edu/~tfylam/Math669.html`, 2024.

[Lus98] G. Lusztig, *Total positivity in partial flag manifolds*, Represent. Theory **2** (1998), 70–78, DOI 10.1090/S1088-4165-98-00046-6. MR1606402

[Lus21] G. Lusztig, *Total positivity in Springer fibres*, Q. J. Math. **72** (2021), no. 1-2, 31–49, DOI 10.1093/qmathj/haaa021. MR4271379

[Mul] Greg Muller and David E. Speyer, *Cluster algebras of Grassmannians are locally acyclic*, Proc. Amer. Math. Soc. **144** (2016), no. 8, 3267–3281, DOI 10.1090/proc/13023. MR3503695

[OT] Peter Orlik and Hiroaki Terao, *Arrangements of hyperplanes*, Grundlehren der mathematischen Wissenschaften [Fundamental Principles of Mathematical Sciences], vol. 300, Springer-Verlag, Berlin, 1992, DOI 10.1007/978-3-662-02772-1. MR1217488

[Pos] Alexander Postnikov, *Total positivity, Grassmannians, and networks*, `http://math.mit.edu/~apost/papers/tpgrass.pdf`, 2007.

[Rie99] Konstanze Rietsch, *An algebraic cell decomposition of the nonnegative part of a flag variety*, J. Algebra **213** (1999), no. 1, 144–154, DOI 10.1006/jabr.1998.7665. MR1674668

[Rie12] Konstanze Rietsch, *A mirror symmetric solution to the quantum Toda lattice*, Comm. Math. Phys. **309** (2012), no. 1, 23–49, DOI 10.1007/s00220-011-1308-8. MR2864786

[PSW] Matteo Parisi, Melissa Sherman-Bennett, and Lauren K. Williams, *The $m = 2$ amplituhedron and the hypersimplex: signs, clusters, tilings, Eulerian numbers*, Comm. Amer. Math. Soc. **3** (2023), 329–399, DOI 10.1090/cams/23. MR4612722

[Stacks] The Stacks Project Authors. Stacks Project. `https://stacks.math.columbia.edu`, 2018.

[ST] Bernd Sturmfels and Simon Telen, *Likelihood equations and scattering amplitudes*, Algebr. Stat. **12** (2021), no. 2, 167–186, DOI 10.2140/astat.2021.12.167. MR4350875

[War] Joe Warren, *Barycentric coordinates for convex polytopes*, Adv. Comput. Math. **6** (1996), no. 2, 97–108 (1997), DOI 10.1007/BF02127699. MR1431788

[Wil] Lauren K. Williams, *Shelling totally nonnegative flag varieties*, J. Reine Angew. Math. **609** (2007), 1–21, DOI 10.1515/CRELLE.2007.059. MR2350779

Department of Mathematics, University of Michigan, 2074 East Hall, 530 Church Street, Ann Arbor, Michigan 48109-1043

Email address: `tfylam@umich.edu`

Proceedings of Symposia in Pure Mathematics
Volume **110**, 2024
https://doi.org/10.1090/pspum/110/02014

An upper bound problem for triangulations of manifolds

Satoshi Murai

1. Introduction

The study of face numbers of simplicial complexes using Stanley–Reisner rings is one of major research topics in algebraic combinatorics. The starting point of this research is Stanley's Upper Bound Theorem (UBT for short) for spheres, which gives sharp upper bounds on face numbers of triangulated spheres for a given number of vertices. Since then, the idea of Stanley–Reisner rings has been applied to obtain a number of interesting results on face numbers including a complete characterization of face numbers (g-theorem) for simplicial polytopes [**BL, Sta80**] and triangulated spheres [**Ad**].

One interesting research topic on this subject is to extend results on face numbers of triangulations of spheres, such as the UBT and the g-theorem, to more general (closed) manifolds. One possible extension of the UBT would be to find an upper bound which holds for triangulations of all manifolds. See [**No**] for results on this direction. The other possible extension of the UBT is to find a sharp upper bound on face numbers of triangulations for each fixed manifold, but, while the UBT for spheres is a basic result in algebraic combinatorics, we seem to miss a good method to obtain such a result for manifolds which are not spheres. The aim of this article is to pose a problem which will hopefully initiate the study of such upper bounds. We first formulate this problem. We say that a simplicial complex Δ is a **triangulation** of a topological space X if its geometric realization $|\Delta|$ is homeomorphic to X. Let $f_i(\Delta)$ be the number of the i-dimensional faces of Δ. The vector $f(\Delta) = (f_{-1}(\Delta), f_0(\Delta), \dots, f_d(\Delta))$, where d is the dimension of Δ and $f_{-1}(\Delta) = 1$, is called the **f-vector** of Δ. We are interested in the following combinatorial invariant of a topological manifold M:

$$f_i^{\max}(M, n) = \max\{f_i(\Delta) \mid \Delta \text{ is an } n\text{-vertex triangulation of } M\},$$

where n is assumed to be larger than or equal to $f_0^{\min}(M)$ the minimal number of vertices which is required to triangulate M. The UBT for spheres can be considered as a result of determining $f_i^{\max}(\mathbb{S}^{d-1}, n)$. Indeed, it tells that $f_i^{\max}(\mathbb{S}^{d-1}, n)$ equals to the f_i of the cyclic d-polytope with n vertices. Using this notation, our question can be stated as follows:

2020 *Mathematics Subject Classification*. Primary 52B05, 13F55, 57Q15.
The author is partially supported by JSPS KAKENHI 21K03190.

QUESTION 1.1. Can we determine $f_i^{\max}(M, n)$ for all i and $n \geq f_0^{\min}(M)$ for (some classes of) closed manifolds M?

It is probably not feasible to get a complete answer to the above question for an arbitrary manifold. Indeed, to get a complete answer, we need to know the number $f_0^{\min}(M)$ but in general it is a very difficult problem to determine this number (see [**Lu**]). On the other hand, it is still an interesting problem to study the above question for particular classes of manifolds, and probably the first case which we should understand is the case when M is a sphere bundle over the circle.

One reason to consider sphere bundles over the circle is that we have enough good information on lower bounds on f-vectors of their triangulations. There are only two topological types of $\mathbb{S}^k$-bundles over $\mathbb{S}^1$. One of them is the product $\mathbb{S}^k \times \mathbb{S}^1$ and the other one is a non-orientable $\mathbb{S}^k$-bundle $\mathbb{S}^k \underline{\times} \mathbb{S}^1$, called a *generalized Klein bottle* (see [**Ste**]). Every triangulation of an $\mathbb{S}^{d-2}$-bundle over $\mathbb{S}^1$ has at least $2d+1$ vertices [**BK**], and there is only one $(2d+1)$-vertex triangulation whose topological type is $\mathbb{S}^{d-2} \times \mathbb{S}^1$ when d is odd and $\mathbb{S}^{d-2} \underline{\times} S^1$ when d is even [**BD, CSS**]. Moreover, it is known that stacked triangulations give a sharp lower bound on f-vectors of their triangulations for a fixed number of the vertices (see [**Ka**, Theorem 1.1] and [**BD**, Theorem 3.12]). Considering these known results, our problem can be formulated as follows.

PROBLEM 1.2. Determine $f_i^{\max}(\mathbb{S}^k \times \mathbb{S}^1, n)$ and $f_i^{\max}(\mathbb{S}^k \underline{\times} \mathbb{S}^1, n)$ for all i and $n \geq 2d+2$.

Here are some partial known results. There is a generalization of the UBT given by Novik [**No**] which guarantees that $f_i^{\max}(\mathbb{S}^{d-2} \times \mathbb{S}^1, n)$ and $f_i^{\max}(\mathbb{S}^{d-2} \underline{\times} \mathbb{S}^1, n)$ are bounded above by the f-vector of the cyclic d-polytope with n vertices, but this bound is not sharp. Also, Chestnut, Sapir and Swartz [**CSS**, Theorem 4.1] gave a complete characterization of all possible pairs (f_0, f_1) for each $\mathbb{S}^{d-2} \times \mathbb{S}^1$ and $\mathbb{S}^{d-2} \underline{\times} \mathbb{S}^1$, which gives an answer to Problem 1.2 for $i = 1$. However, even a reasonable conjecture for precise values of $f_i^{\max}(\mathbb{S}^{d-2} \times \mathbb{S}^1, n)$ and $f_i^{\max}(\mathbb{S}^{d-2} \underline{\times} \mathbb{S}^1, n)$ is not known when $i \geq 2$.

In this note, we present some recent algebraic results which might help to solve this upper bound problem for manifolds, and, for sphere bundles over the circle, we present in Problem 3.3 conjectural upper bounds on face numbers of triangulations of these manifolds which is expected to be close to an actual value of $f_i^{\max}$.

This note is organized as follows. In section 2, we recall the UBT for spheres and its connection to Stanley–Reisner ring theory. In section 3, we pose our main conjectural upper bounds. These bounds use h''-vectors and $\widetilde{g}$-vectors, which appear in recent studies of face numbers of triangulated manifolds, and we explain algebraic meanings of these vectors in section 4. In section 5, we discuss some additional problems related to the contents of this article.

2. Background: The UBT and Stanley–Reisner rings

Before discussing face numbers of manifolds, we recall the classical UBT for spheres and its connection to Stanley–Reisner rings. An algebraic idea to prove the UBT will be used to formulate our conjectural upper bounds on face numbers of triangulations of sphere bundles over the circle. We refer the readers to [**Sta96, Sta14**] for historical comments and necessary background on the UBT.

The upper bound theorem. We first define necessary notation. Let Δ be a finite abstract simplicial complex. An element F of Δ is called a **face** of Δ, and the **dimension** of a face F is $\#F-1$, where $\#W$ denotes the cardinality of a finite set W. The **dimension** of a simplicial complex Δ is the maximum of the dimension of its faces. Recall that $f(\Delta)=(f_{-1}(\Delta),\dots,f_{d-1}(\Delta))$ is the f-vector of a $(d-1)$-dimensional simplicial complex Δ. We write $C(n,d)$ for the boundary complex of the cyclic d-polytope with n vertices (that is, the convex hull of n distinct points on the moment curve $\{(t,t^2,\dots,t^d)\in\mathbb{R}^d \mid t\in\mathbb{R}\}$). Below is the UBT for spheres.

THEOREM 2.1 (UBT for spheres). *If Δ is an n-vertex triangulation of the $(d-1)$-dimensional sphere $\mathbb{S}^{d-1}$, then $f_i(\Delta)\le f_i(C(n,d))$ for all $i=0,1,\dots,d-1$.*

While we write the upper bound theorem in terms of f-vectors to see that it determines $f_i^{\max}(\mathbb{S}^d,n)$ (indeed it tells $f_i^{\max}(\mathbb{S}^{d-1},n)=f_i(C(n,d))$), it is more convenient to state the UBT using other vectors known as h- and g-vectors. The **h-vector** of a $(d-1)$-dimensional simplicial complex Δ is the vector $h(\Delta)=(h_0(\Delta),h_1(\Delta),\dots,h_d(\Delta))$ defined by

$$h_i(\Delta)=\sum_{k=0}^{i}(-1)^{i-k}\binom{d-k}{d-i}f_{k-1}(\Delta).$$

Also, if Δ is a triangulation of $\mathbb{S}^{d-1}$, then its **g-vector** is the vector $g(\Delta)=(g_0(\Delta),g_1(\Delta),\dots,g_{\lfloor\frac{d}{2}\rfloor}(\Delta))$ defined by $g_i(\Delta)=h_i(\Delta)-h_{i-1}(\Delta)$ for $i=0,1,\dots,\lfloor\frac{d}{2}\rfloor$, where $h_{-1}(\Delta)=0$. Below are some remarks on h- and g-vectors.

- We have $f_{i-1}=\sum_{k=0}^{i}\binom{d-k}{d-i}h_k(\Delta)$. In particular, knowing $f(\Delta)$ is equivalent to knowing $h(\Delta)$ (if we know the dimension of Δ).
- Suppose that Δ is a triangulation of the $(d-1)$-sphere $\mathbb{S}^{d-1}$. Then $h(\Delta)$ is non-negative, unimodal and symmetric. In particular, the symmetry tells that knowing $h(\Delta)$ is equivalent to knowing $g(\Delta)$.
- Bounds on h-vectors induce bounds on f-vectors. Also, bounds on g-vectors induce bounds on h-vectors.

It is known that the h-vectors of cyclic polytopes are given by $h_k(C(n,d))=\binom{n-d+k-1}{k}$ for $k\le\frac{d}{2}$. The UBT is actually proved by proving the following statement.

THEOREM 2.2 (UBT, h-version). *If Δ is an n-vertex triangulation of the $(d-1)$-dimensional sphere $\mathbb{S}^{d-1}$, then*

$$h_k(\Delta)\le\binom{n-d+k-1}{k}\qquad \textit{for } k\le\frac{d}{2}.$$

Also, the following g-vector version of the statement also holds.

THEOREM 2.3 (UBT, g-version). *If Δ is an n-vertex triangulation of the $(d-1)$-dimensional sphere $\mathbb{S}^{d-1}$, then*

$$g_k(\Delta)\le\binom{n-d+k-2}{k}\qquad \textit{for } k\le\frac{d}{2}.$$

We note that the g-version implies the h-version, and the h-version implies the f-version of the UBT.

Algebraic meanings of the h- and g-versions. We explain algebraic meanings of h- and g-version of the UBT using Stanley–Reisner rings. Let Δ be a simplicial complex with the vertex set $[n] = \{1, 2, \dots, n\}$ and $S = \mathbb{F}[x_1, \dots, x_n]$ the polynomial ring over an infinite field $\mathbb{F}$. The ideal $I_\Delta = (x^F \mid F \notin \Delta)$, where $x^F = \prod_{i \in F} x_i$, is called the **Stanley–Reisner ideal** and the ring $\mathbb{F}[\Delta] = S/I_\Delta$ is called the **Stanley–Reisner ring**. It is known that for any $(d-1)$-dimensional simplicial complex Δ, there is a sequence of linear forms $\Theta = \theta_1, \dots, \theta_d \in S$ such that $\mathbb{F}[\Delta]/(\Theta\mathbb{F}[\Delta])$ is Artinian. Such a sequence Θ is called a **linear system of parameters** (l.s.o.p. for short) for $\mathbb{F}[\Delta]$. For a homogenous ideal I, the formal power series $\mathrm{Hilb}(S/I, t) = \sum_{k \geq 0} \dim_{\mathbb{F}}(S/I)_k t^k$ is called the **Hilbert series** of S/I, where M_k denotes the degree k homogeneous component of a graded S-module M.

Now, assume that Δ is a triangulation of $\mathbb{S}^{d-1}$. It is known that $\mathbb{F}[\Delta]$ is Cohen–Macaulay, which means that for any (equivalently, some) l.s.o.p. Θ for $\mathbb{F}[\Delta]$ one has

$$\mathrm{Hilb}\left(\mathbb{F}[\Delta]/(\Theta\mathbb{F}[\Delta]), t\right) = \sum_{k=0}^{d} h_k(\Delta) t^k. \tag{2.1}$$

Also, since $\mathbb{F}[\Delta]/(\Theta\mathbb{F}[\Delta]) \cong S/(I_\Delta + (\Theta))$, we have the inequality

$$\dim_{\mathbb{F}}\left(\mathbb{F}[\Delta]/(\Theta\mathbb{F}[\Delta])\right)_k \leq \dim_{\mathbb{F}}(S/(\theta_1, \dots, \theta_d))_k = \dim_{\mathbb{F}}(\mathbb{F}[x_1, \dots, x_{n-d}])_k. \tag{2.2}$$

Considering (2.1), one can see that the inequality (2.2) for $k = 0, 1, \dots, d$ is nothing but the h-version of the UBT (Theorem 2.2) since $\dim_{\mathbb{F}}(\mathbb{F}[x_1, \dots, x_{n-d}])_k = \binom{n-d+k-1}{k}$.

Similarly, the g-version has the following algebraic meaning. An algebraic g-theorem for spheres [**Ad**] (a shorter proof in characteristic 2 can be found in [**APP, KX, PP**]) tells that, if Δ is a triangulation of $\mathbb{S}^{d-1}$ and if $\Theta = \theta_1, \dots, \theta_{d+1} \in S_1$ is sufficiently general, then $\theta_1, \dots, \theta_d$ is an l.s.o.p. for $\mathbb{F}[\Delta]$ and the multiplication

$$\times \theta_{d+1} : \left(\mathbb{F}[\Delta]/(\theta_1, \dots, \theta_d)\mathbb{F}[\Delta]\right)_{k-1} \to \left(\mathbb{F}[\Delta]/(\theta_1, \dots, \theta_d)\mathbb{F}[\Delta]\right)_k$$

is injective for $k \leq \frac{d+1}{2}$ and is surjective for $k \geq \frac{d+1}{2}$, which in particular tells

$$\mathrm{Hilb}\left(\mathbb{F}[\Delta]/(\Theta\mathbb{F}[\Delta]), t\right) = \sum_{k=0}^{\lfloor \frac{d}{2} \rfloor} g_k(\Delta) t^k. \tag{2.3}$$

Then the inequality

$$\dim_{\mathbb{F}}\left(\mathbb{F}[\Delta]/(\Theta\mathbb{F}[\Delta])\right)_k \leq \dim_{\mathbb{F}}\left(S/(\theta_1, \dots, \theta_{d+1})\right)_k = \dim_{\mathbb{F}}\left(\mathbb{F}[x_1, \dots, x_{n-d-1}]\right)_k \tag{2.4}$$

coincides with the g-version of the UBT (Theorem 2.3)[1].

3. The main problem

Now we present the main problem in this article. Recall that the g-version of the UBT can be expressed as follows.

THEOREM 3.1 (UBT, g-version). *If Δ is an n-vertex triangulation of the $(d-1)$-dimensional sphere $\mathbb{S}^{d-1}$ and $R = \mathbb{F}[x_1, \dots, x_{n-d-1}]$, then*

$$g_k(\Delta) \leq \dim_{\mathbb{F}} R_k \quad \text{for} \quad k \leq \frac{d}{2}.$$

[1]The algebraic g-theorem is not necessary to prove Theorem 2.3 since in the equality (2.3) the left-hand side is larger than or equal to the right-hand side even without the algebraic g-theorem, but (2.3) clarifies an algebraic meaning of g-vectors.

Our formulation of the problem is based on this algebraic experession of the UBT.

To state the problem, we need h''- and $\widetilde{g}$-vectors, which are considered to be manifold versions of h- and g-vectors. To simplify the argument, we assume that $\mathbb{F}$ has characteristic 2 in the rest of this note[2]. For a simplicial complex Δ, we write $\widetilde{H}_i(\Delta;\mathbb{F})$ for the ith reduced homology group of Δ with coefficients in $\mathbb{F}$ and $\beta_i(\Delta) = \dim_{\mathbb{F}} \widetilde{H}_i(\Delta;\mathbb{F})$. For a $(d-1)$-dimensional simplicial complex Δ, we define the vectors $h'(\Delta) = (h'_0(\Delta), h'_1(\Delta), \dots, h'_d(\Delta))$ and $h''(\Delta) = (h''_0(\Delta), h''_1(\Delta), \dots, h''_d(\Delta))$ by

$$h'_i(\Delta) = h_i(\Delta) - \binom{d}{i}\left(\sum_{k=1}^{i-1}(-1)^{i-k}\beta_{k-1}(\Delta)\right)$$

and

$$h''_i(\Delta) = \begin{cases} h'_i(\Delta) - \binom{d}{i}\beta_{i-1}(\Delta), & (i \neq d), \\ h'_d(\Delta), & (i = d). \end{cases}$$

When Δ is a triangulation of a connected closed $(d-1)$-manifold, we also define the vector $\widetilde{g}(\Delta) = (\widetilde{g}_0(\Delta), \widetilde{g}_1(\Delta), \dots, \widetilde{g}_{\lfloor \frac{d}{2} \rfloor}(\Delta))$ by

$$\begin{aligned} \widetilde{g}_i(\Delta) &= h_i(\Delta) - h_{i-1}(\Delta) - \binom{d+1}{i}\left(\sum_{k=1}^{i}(-1)^{i-k}\beta_{k-1}(\Delta)\right) \\ &= h'_i(\Delta) - h''_{i-1}(\Delta) - \binom{d+1}{i}\beta_{i-1}(\Delta). \end{aligned}$$

Similar to h-vectors and g-vectors, the above vectors have nice algebraic meanings in terms of Stanley–Reisner rings, and $\widetilde{g}$-vectors of triangulated closed manifolds are considered to play a role of g-vectors of triangulated spheres. Indeed, the following generalization of Theorem 3.1 is known.

THEOREM 3.2. *Let $d \geq 4$, Δ an n-vertex triangulation of a connected closed $(d-1)$-manifold, and $R = \mathbb{F}[x_1, \dots, x_{n-d-1}]$. Then there is an ideal $I \subset R$ that has $\binom{d+1}{2}\beta_1(\Delta)$ generators of degree 2 such that*

$$\widetilde{g}_k(\Delta) \leq \dim_{\mathbb{F}}(R/I)_k \quad \textit{for} \quad k \leq \tfrac{d}{2}.$$

We will explain meanings of h'-, h''- and $\widetilde{g}$-vectors as well as a more general statement which proves Theorem 3.2 in the next section.

Inspired by Theorem 3.2, we pose the following more explicit conjectural upper bounds for triangulations of sphere bundles over the circle.

PROBLEM 3.3. Let $d \geq 4$, $n \geq 2d+1$ and $R = \mathbb{F}[x_1, \dots, x_{n-d-1}]$. Is it true that if Δ is an n-vertex triangulation of an $\mathbb{S}^{d-2}$-bundle over $\mathbb{S}^1$, then

$$\widetilde{g}_k(\Delta) \leq \dim_{\mathbb{F}}(R/(x_1, \dots, x_d)^2)_k \quad \text{for} \quad k \leq \frac{d}{2}?$$

So the problem asks if we can take I in Theorem 3.2 as the ideal $(x_1, \dots, x_d)^2$. At this moment, it is not so clear if the Hilbert series of $R/(x_1, \dots, x_d)^2$ is really a right choice for the maximum of the $\widetilde{g}$-vectors in Problem 3.3. This choice is based on author's computational experiments that construct triangulations of $\mathbb{S}^{d-2}$-bundle over $\mathbb{S}^1$ with many faces, and the author thinks that if the inequality holds then it is sharp (see secton 5).

[2]This is assumed to ignore an orientability issue.

REMARK 3.4. To study the problem it might be convenient to write the inequalities of Problem 3.3 more explicitly. If Δ is a triangulation of an $\mathbb{S}^{d-2}$-bundle over $\mathbb{S}^1$, then it has one-dimensional homology in dimensions 1 and $d-2$ so $\tilde{g}_0(\Delta)=1$, $\tilde{g}_1(\Delta)=g_1(\Delta)$ and

$$\widetilde{g}_k(\Delta)=g_k(\Delta)-(-1)^k\binom{d+1}{k}\quad \text{for } k\geq 2.$$

Also, we have

$$\dim_{\mathbb{F}}(R/(x_1,\dots,x_d)^2)_k=\binom{n-2d+k-2}{k}+d\binom{n-2d+k-3}{k-1}$$

for $k\geq 2$ since we have an isomorphism of the following vector spaces

$$R/(x_1,\dots,x_d)^2\cong \mathbb{F}[x_{d+1},\dots,x_{n-d-1}]\bigoplus(\oplus_{k=1}^d x_k\mathbb{F}[x_{d+1},\dots,x_{n-d-1}]).$$

4. Algebraic meanings of h''- and $\widetilde{g}$-vectors

In this section, we explain algebraic meanings of h''- and $\widetilde{g}$-vectors, in particular, we explain how Theorem 3.2 follows from known results. For further results and background on these vectors, see the survey article [**KN**].

For a graded S-module M, its **socle** is the submodule $\mathrm{Soc}(M)=\{f\in M\mid \mathfrak{m}f=0\}$, where $\mathfrak{m}=(x_1,\dots,x_n)$ is the graded maximal ideal of S. The following theorem provides an algebraic interpretation of h'- and h''-vectors.

THEOREM 4.1. *Let Δ be a triangulation of a connected closed $(d-1)$-manifold with the vertex set $[n]$, Θ an l.s.o.p. for $\mathbb{F}[\Delta]$ and $R=\mathbb{F}[\Delta]/\Theta\mathbb{F}[\Delta]$. Then*

1. *(Schenzel's formula)* $\mathrm{Hilb}(R,t)=\sum_{k=0}^d h'_k(\Delta)t^k$.
2. *(Novik–Swartz) Let $N=\bigoplus_{k=1}^{d-1}\mathrm{Soc}(R)_k$. Then*
   (a) $\mathrm{Hilb}(N,t)=\sum_{k=1}^{d-1}\binom{d}{k}\beta_{k-1}(\Delta)t^k$,
   (b) $\mathrm{Hilb}(R/N,t)=\sum_{k=0}^d h''_k(\Delta)t^k$, *and*
   (c) *R/N is Gorenstein, in particular, $h''_i(\Delta)=h''_{d-i}(\Delta)$ for all i.*

The statement (1) is due to Schenzel [**Sc**] while (2) appear in [**No, NS1, NS2**]. We next give an algebraic interpretation of $\widetilde{g}$-vectors, which essentially appeared in [**APP**] and in [**MN**].

THEOREM 4.2. *Let Δ be a triangulation of a connected closed $(d-1)$-manifold. Let $\Theta=\theta_1,\dots,\theta_{d+1}\in S_1$ be general linear forms and $R'=\mathbb{F}[\Delta]/\Theta\mathbb{F}[\Delta]$. Then*

1. $\dim_{\mathbb{F}}R'_k=h'_k(\Delta)-h''_{k-1}(\Delta)$ *for* $k\leq\frac{d+1}{2}$.
2. *There is an ideal J of R' satisfying the following conditions*
   (a) $\dim_{\mathbb{F}}J_k=\binom{d+1}{k}\beta_{k-1}(\Delta)$ *for* $k\leq\frac{d}{2}$, *and*
   (b) $\mathrm{Hilb}(R'/J,t)=\sum_{k=0}^{\lfloor\frac{d}{2}\rfloor}\widetilde{g}_k(\Delta)t^k$.

PROOF SKETCH. Let R and N be as in Theorem 4.1. We recall the following known facts.

(I) The multiplication $\times\theta_{d+1}:R_{k-1}\to R_k$ is surjective for $k\geq\frac{d}{2}+1$.
(II) The multiplication $\times\theta_{d+1}:(R/N)_{k-1}\to(R/N)_k$ is injective for $k\leq\frac{d+1}{2}$ and is surjective for $k\geq\frac{d+1}{2}$.

The statement (I) is a consequence of [**Sw**, Theorem 2.6] and the algebraic g-theorem [**Ad**]. Also, the statement (II) appears in [**APP**].

Now the injectivity in (II) tells that, for $k \le (d+1)/2$, the kernel of

$$\times\theta_{d+1} : R_{k-1} \to R_k$$

is N_{k-1}, so $\dim(R')_k = h'_k(\Delta) - h''_{k-1}(\Delta)$ proving (1).

We now consider (2). By (1) it suffices to prove (b). By taking the Matlis dual of the short exact sequence $0 \to N \to R \to R/N \to 0$, we have the exact sequence

$$0 \longrightarrow (R/N)^\vee \xrightarrow{\phi} R^\vee \longrightarrow N^\vee \longrightarrow 0$$

where $M^\vee$ denotes the Matlis dual of a graded S-module M. Since $\theta_{d+1}N = 0$, we have $\theta_{d+1}N^\vee = 0$, which tells that $\theta_{d+1}R^\vee$ is a submodule of $\operatorname{Im}\phi$. Since R/N is Gorenstein, we have $R/N \cong (R/N)^\vee \cong \operatorname{Im}\phi$ if we shift the grading by d. Then, since $\phi(\theta_{d+1}(R/N)^\vee) \subset \theta_{d+1}R^\vee$,

$$\operatorname{Im}\phi/(\theta_{d+1}R^\vee) \cong R'/J$$

for some ideal J of R'. We prove that J satisfies the desired conditions.

The surjectivity (I) tells that $\times\theta_{d+1} : R^\vee_{-k} \to R^\vee_{-k+1}$ is injective for $\frac{d}{2}+1 \le k \le d$, so

$$\dim_{\mathbb{F}}(\theta_{d+1}R^\vee)_{-k+1} = \dim_{\mathbb{F}}(R^\vee)_{-k} = \dim_{\mathbb{F}} R_k \qquad \text{for } k \ge d/2+1.$$

Recall $h''_k(\Delta) = h''_{d-k}(\Delta)$ and $\beta_{k-1}(\Delta) = \beta_{d-k}(\Delta)$ for all k. Then, for $k \le \frac{d}{2}$, since $R/N \cong \operatorname{Im}\phi$, we have

$$\begin{aligned}
\dim(\operatorname{Im}\phi/\theta_{d+1}R^\vee)_{-d+k} &= \dim(R/N)^\vee_{-d+k} - \dim(\theta_{d+1}R^\vee)_{-d+k}\\
&= \dim(R/N)_{d-k} - \dim(R)_{d-k+1}\\
&= h''_{d-k}(\Delta) - h'_{d-k+1}(\Delta)\\
&= h''_{d-k}(\Delta) - h''_{d-k+1}(\Delta) - \binom{d}{d-k+1}\beta_{d-k}(\Delta)\\
&= h''_k(\Delta) - h''_{k-1}(\Delta) - \binom{d}{k-1}\beta_{k-1}(\Delta)\\
&= h'_k(\Delta) - h''_{k-1}(\Delta) - \binom{d+1}{k}\beta_{k-1}(\Delta)\\
&= \widetilde{g}_k(\Delta).
\end{aligned}$$

Thus

$$\dim_{\mathbb{F}}(R'/J)_k = \dim_{\mathbb{F}}(\operatorname{Im}\phi/\theta_{d+1}R^\vee)_{-d+1} = \widetilde{g}_k(\Delta)$$

for $k \le \frac{d}{2}$. Also, since $\operatorname{Im}\phi/\theta_{d+1}R^\vee$ is a quotient of

$$\operatorname{Im}\phi/\theta_{d+1}\operatorname{Im}\phi \cong (R/N)/\theta_{d+1}(R/N),$$

the surjectivity in (II) tells $(R'/J)_k = 0$ for $k > \frac{d}{2}$, so $\operatorname{Hilb}(R'/J, t) = \sum_{k=0}^{\lfloor \frac{d}{2} \rfloor} \widetilde{g}_k(\Delta)t^k$ as desired. □

REMARK 4.3. For the proof, we followed the argument in [**MN**]. It might be possible to prove this theorem using the partition complex, a recent argument given in [**AY**], which in turn may lead to a better understanding of the ideal J.

Theorem 3.2 easily follows from the previous result.

PROOF OF THEOREM 3.2. Let Θ, R' and J be as in Theorem 4.2. Since $R' \cong S/(I_\Delta + (\Theta))$ and $S/(\Theta) \cong R = \mathbb{F}[x_1, \dots, x_{n-d-1}]$, we have $R'/J \cong R/I$ for some ideal I in R. It suffices to show that $\dim_{\mathbb{F}} I_2 \geq \binom{d+1}{2}\beta_1(\Delta)$. But this inequality follows from the trivial inequality $\dim_{\mathbb{F}} I_2 \geq \dim_{\mathbb{F}} J_2$ and Theorem 4.2(2-a). $\square$

5. Related problems

Here we posit a few more problems related to Problem 3.3.

Construction of the equality case. To determine $f_i^{\max}$, it is also necessary to study the following problem.

PROBLEM 5.1. For all integers $d \geq 4$ and $n \geq 2d+1$, find an n-vertex triangulation of $\mathbb{S}^{d-2}$-bundle over $\mathbb{S}^1$ such that

$$\widetilde{g}_k(\Delta) = \dim_{\mathbb{F}}(\mathbb{R}[x_1, \dots, x_{n-d-1}]/(x_1, \dots, x_d)^2)_k \quad \text{for} \quad k = 0, 1, \dots, \frac{d}{2}.$$

The author has a number of examples of triangulations of $\mathbb{S}^{d-2} \underline{\times}\, \mathbb{S}^1$ with the above $\tilde{g}$-vectors when d is odd. Here is an instance.

Consider the case of $d = 7$ and $n = 16$. We present an example of a triangulation Γ of $\mathbb{S}^5 \underline{\times}\, \mathbb{S}^1$ whose $\widetilde{g}$-vector coincides with the Hilbert series of the algebra $\mathbb{F}[x_1, \dots, x_8]/(x_1, \dots, x_7)^2$, that is, $\widetilde{g}(\Gamma) = (1, 8, 8, 8)$. Let B be the simplicial complex on the vertex set $\{0, 1, \dots, 8, \bar{1}, \dots, \bar{7}, v_0, \dots, v_6\}$ generated by the following simplices

01234567	01234578	01235678	01345678
$\bar{7}1234567$	$\bar{7}1234578$	$\bar{7}1235678$	$\bar{7}1345678$
$\bar{6}\bar{7}234567$	$\bar{6}\bar{7}234578$	$\bar{6}\bar{7}235678$	
$\bar{5}\bar{7}\bar{6}34567$	$\bar{5}\bar{7}\bar{6}34578$	$\bar{5}\bar{7}\bar{6}35678$	$\bar{5}\bar{7}345678$
$\bar{4}\bar{7}\bar{6}\bar{5}4567$	$\bar{4}\bar{7}\bar{6}\bar{5}4578$		$\bar{4}\bar{7}\bar{5}45678$
$\bar{3}\bar{7}\bar{6}\bar{5}\bar{4}567$	$\bar{3}\bar{7}\bar{6}\bar{5}\bar{4}578$	$\bar{3}\bar{7}\bar{6}\bar{5}5678$	$\bar{3}\bar{7}\bar{5}\bar{4}5678$
$\bar{2}\bar{7}\bar{6}\bar{5}\bar{4}\bar{3}67$		$\bar{2}\bar{7}\bar{6}\bar{5}\bar{3}678$	$\bar{2}\bar{7}\bar{5}\bar{4}\bar{3}678$
$\bar{1}\bar{7}\bar{6}\bar{5}\bar{4}\bar{3}\bar{2}7$	$\bar{1}\bar{7}\bar{6}\bar{5}\bar{4}\bar{3}78$	$\bar{1}\bar{7}\bar{6}\bar{5}\bar{3}\bar{2}78$	$\bar{1}\bar{7}\bar{5}\bar{4}\bar{3}\bar{2}78$
$\bar{7}\bar{6}\bar{5}\bar{4}\bar{3}\bar{2}\bar{1}v_0$	$\bar{6}\bar{5}\bar{4}\bar{3}\bar{2}\bar{1}v_0v_1$	$\bar{5}\bar{4}\bar{3}\bar{2}\bar{1}v_0v_1v_2$	$\bar{4}\bar{3}\bar{2}\bar{1}v_0v_1v_2v_3$
$\bar{3}\bar{2}\bar{1}v_0v_1v_2v_3v_4$	$\bar{2}\bar{1}v_0v_1v_2v_3v_4v_5$	$\bar{1}v_0v_1v_2v_3v_4v_5v_6$	

The simplicial complex B is a shellable triangulated 7-dim ball with the h-vector

$$h(B) = (1, 15, 8, 8, 4, 0, 0, 0, 0)$$

and its boundary ∂B has the g-vector

$$g(\partial B) = (1, 15, 8, 8).$$

Note that 0123456 and $v_0v_1v_2v_3v_4v_5v_6$ are facets of ∂B. Let Γ be the simplicial complex obtained from $\partial B \setminus \{0123456, v_0v_1v_2v_3v_4v_5v_6\}$ by identifying $v_0 \to 0, \dots, v_6 \to 6$. Note that the vertex set of Γ is $\{0, 1, \dots, 8, \bar{1}, \dots, \bar{7}\}$. The operation $\partial B \to \Gamma$ is a combinatorial handle addition (see [**NS1**, §5]), so $|\Gamma|$ is homeomorphic to either $\mathbb{S}^5 \times \mathbb{S}^1$ or $\mathbb{S}^5 \underline{\times}\, \mathbb{S}^1$. A routine computation tells that Γ is not orientable, so it must triangulate $\mathbb{S}^5 \underline{\times}\, \mathbb{S}^1$. Also, by a simple counting, we can see that

$$\widetilde{g}(\Gamma) = g(\partial B) - (0, 7, 0, 0) = (1, 8, 8, 8).$$

More general upper bounds. In Problem 3.3, we consider $\mathbb{S}^{d-2}$-bundles over $\mathbb{S}^1$ since triangulations of these manifolds are well-studied, but of course a bound which holds for more general manifolds would be desirable. The next problem might be a natural generalization of Problem 3.3.

PROBLEM 5.2. Let Δ be an n-vertex triangulation of a connected closed $(d-1)$-dimensional manifold with $\beta_\ell(\Delta) \neq 0$, where $\ell < \frac{d}{2}$. Is it true that

$$\widetilde{g}_k(\Delta) \leq \dim_{\mathbb{F}}(\mathbb{F}[x_1, \ldots, x_{n-d-1}]/(x_1, \ldots, x_{d+1-\ell})^{\ell+1})_k \quad \text{for } k \leq \frac{d}{2}?$$

Interesting special cases of the above problem would be triangulations of the sphere product $\mathbb{S}^\ell \times \mathbb{S}^{d-1-\ell}$.

It would be also interesting to consider similar guesses for manifolds M such that $\beta_k(M) \neq 0$ for many values of k, like a projective space $\mathbb{RP}^d$ or a torus $\mathbb{S}^1 \times \cdots \times \mathbb{S}^1$. But, at this moment, it is not clear how we can find a concrete guess on upper bounds on $\widetilde{g}$-vectors for triangulations of these manifolds from Theorem 4.2.

Buchsbaum complexes. The h''-vectors work well not only for triangulated manifolds but also a more general class of simplicial complexes called Buchsbaum complexes. A $(d-1)$-dimensional simplicial complex Δ is said to be **Buchsbaum** (over $\mathbb{F}$) if $\widetilde{H}_k(\mathrm{lk}_\Delta(F); \mathbb{F}) \neq 0$ for $k \neq d-1-\#F$ for any non-empty face $F \in \Delta$, where $\mathrm{lk}_\Delta(F) = \{G \setminus F \mid F \subset G \in \Delta\}$ is the link of F in Δ. The following result is due to Novik and Swartz [**NS1**] (an algebraic description of the module N in the theorem can be found in [**MNY**]).

THEOREM 5.3. *Let Δ be a $(d-1)$-dimensional Buchsbaum simplicial complex with the vertex set $[n]$, Θ an l.s.o.p. for $\mathbb{F}[\Delta]$ and $R = \mathbb{F}[\Delta]/(\Theta\mathbb{F}[\Delta])$. There is a submodule $N \subset \mathrm{Soc}(R)$ such that*

(1) $\mathrm{Hilb}(N, t) = \sum_{k=1}^{d-1} \binom{d}{k} \beta_{k-1}(\Delta) t^k$,
(2) $\mathrm{Hilb}(R/N, t) = \sum_{k=0}^{d} h''_k(\Delta) t^k$.

Answers to the following problems may help understanding Problems 3.3, 5.1 and 5.2.

PROBLEM 5.4. Let Δ be a $(d-1)$-dimensional Buchsbaum simplicial complex having n vertices. Is it true that if $\beta_\ell(\Delta) \neq 0$ then

$$h''_k(\Delta) \leq \dim_{\mathbb{F}}(\mathbb{F}[x_1, \ldots, x_{n-d}]/(x_1, \ldots, x_{d-\ell})^{\ell+1})_k \quad \text{for } k = 0, 1, \ldots, d?$$

PROBLEM 5.5. For all integers ℓ, d, n satisfying $0 \leq \ell \leq d-1$ and $n \geq 2d-\ell$, can we find a $(d-1)$-dimensional Buchsbaum simplicial complex Δ such that $\beta_\ell(\Delta) \neq 0$ and

$$h''_k(\Delta) = \dim_{\mathbb{F}}(\mathbb{F}[x_1, \ldots, x_{n-d}]/(x_1, \ldots, x_{d-\ell})^{\ell+1})_k \quad \text{for } k = 0, 1, \ldots, d?$$

Acknowledgments

We thank the reviewer for careful reading and helpful comments.

References

[Ad] K. Adiprasito, *Combinatorial Lefschetz theorems beyond positivity*, arXiv:1812.10454, 2018.

[APP] K. Adiprasito, S.A. Papadakis and V. Petrotou, *Anisotropy, biased pairings, and the Lefschetz property for pseudomanifolds and cycles*, arXiv:2101.07245, 2021.

[AY] Karim Adiprasito and Geva Yashfe, *The partition complex: an invitation to combinatorial commutative algebra*, Surveys in combinatorics 2021, London Math. Soc. Lecture Note Ser., vol. 470, Cambridge Univ. Press, Cambridge, 2021, pp. 1–41. MR4273427

[BD] Bhaskar Bagchi and Basudeb Datta, *On stellated spheres and a tightness criterion for combinatorial manifolds*, European J. Combin. **36** (2014), 294–313, DOI 10.1016/j.ejc.2013.07.018. MR3131896

[BL] Louis J. Billera and Carl W. Lee, *A proof of the sufficiency of McMullen's conditions for f-vectors of simplicial convex polytopes*, J. Combin. Theory Ser. A **31** (1981), no. 3, 237–255, DOI 10.1016/0097-3165(81)90058-3. MR635368

[BK] U. Brehm and W. Kühnel, *Combinatorial manifolds with few vertices*, Topology **26** (1987), no. 4, 465–473, DOI 10.1016/0040-9383(87)90042-5. MR919730

[CSS] Jacob Chestnut, Jenya Sapir, and Ed Swartz, *Enumerative properties of triangulations of spherical bundles over S^1*, European J. Combin. **29** (2008), no. 3, 662–671, DOI 10.1016/j.ejc.2007.03.005. MR2397347

[Ka] Gil Kalai, *Rigidity and the lower bound theorem. I*, Invent. Math. **88** (1987), no. 1, 125–151, DOI 10.1007/BF01405094. MR877009

[KX] Kalle Karu and Elizabeth Xiao, *On the anisotropy theorem of Papadakis and Petrotou*, Algebr. Comb. **6** (2023), no. 5, 1313–1330, DOI 10.5802/alco.298. MR4668930

[KN] Steven Klee and Isabella Novik, *Face enumeration on simplicial complexes*, Recent trends in combinatorics, IMA Vol. Math. Appl., vol. 159, Springer, [Cham], 2016, pp. 653–686, DOI 10.1007/978-3-319-24298-9_26. MR3526427

[Lu] F. Lutz, *Triangulated Manifolds with Few Vertices: Combinatorial Manifolds*, arXiv:math/0506372, 2005.

[MN] Satoshi Murai and Eran Nevo, *On r-stacked triangulated manifolds*, J. Algebraic Combin. **39** (2014), no. 2, 373–388, DOI 10.1007/s10801-013-0451-z. MR3159256

[MNY] Satoshi Murai, Isabella Novik, and Ken-ichi Yoshida, *A duality in Buchsbaum rings and triangulated manifolds*, Algebra Number Theory **11** (2017), no. 3, 635–656, DOI 10.2140/ant.2017.11.635. MR3649363

[No] Isabella Novik, *Upper bound theorems for homology manifolds*, Israel J. Math. **108** (1998), 45–82, DOI 10.1007/BF02783042. MR1669325

[NS1] Isabella Novik and Ed Swartz, *Socles of Buchsbaum modules, complexes and posets*, Adv. Math. **222** (2009), no. 6, 2059–2084, DOI 10.1016/j.aim.2009.07.001. MR2562774

[NS2] Isabella Novik and Ed Swartz, *Gorenstein rings through face rings of manifolds*, Compos. Math. **145** (2009), no. 4, 993–1000, DOI 10.1112/S0010437X09004138. MR2521251

[PP] S.A. Papadakis and V. Petrotou, *The characteristic* 2 *anisotropicity of simplicial spheres*, arXiv:2012.09815, 2020.

[Sc] Peter Schenzel, *On the number of faces of simplicial complexes and the purity of Frobenius*, Math. Z. **178** (1981), no. 1, 125–142, DOI 10.1007/BF01218376. MR627099

[Sta80] Richard P. Stanley, *The number of faces of a simplicial convex polytope*, Adv. in Math. **35** (1980), no. 3, 236–238, DOI 10.1016/0001-8708(80)90050-X. MR563925

[Sta96] Richard P. Stanley, *Combinatorics and commutative algebra*, 2nd ed., Progress in Mathematics, vol. 41, Birkhäuser Boston, Inc., Boston, MA, 1996. MR1453579

[Sta14] Richard P. Stanley, *How the upper bound conjecture was proved*, Ann. Comb. **18** (2014), no. 3, 533–539, DOI 10.1007/s00026-014-0238-5. MR3245897

[Ste] N. E. Steenrod, *The classification of sphere bundles*, Ann. of Math. (2) **45** (1944), 294–311, DOI 10.2307/1969267. MR9857

[Sw] Ed Swartz, *Face enumeration—from spheres to manifolds*, J. Eur. Math. Soc. (JEMS) **11** (2009), no. 3, 449–485, DOI 10.4171/JEMS/156. MR2505437

Department of Mathematics, Faculty of Education, Waseda University, 1-6-1 Nishi-Waseda, Shinjuku, Tokyo 169-8050, Japan

Email address: s-murai@waseda.jp

Proceedings of Symposia in Pure Mathematics
Volume **110**, 2024
https://doi.org/10.1090/pspum/110/02007

What is a combinatorial interpretation?

Igor Pak

ABSTRACT. In this survey we discuss the notion of *combinatorial interpretation* in the context of Algebraic Combinatorics and related areas. We approach the subject from the Computational Complexity perspective. We review many examples, state a workable definition, discuss many open problems, and present recent results on the subject.

1. Introduction

1.1. What numbers? Traditionally, *Combinatorics* works with numbers. Not with structures, relations between the structures, or connections between the relations — just numbers. These numbers tend to be nonnegative integers, presented in the form of some exact formula or disguised as probability. More importantly, they always count the number of some combinatorial objects.

This approach, with its misleading simplicity, led to a long series of amazing discoveries, too long to be recounted here. It turns out that many interesting combinatorial objects satisfy some formal relationships allowing for their numbers to be analyzed. More impressively, the very same combinatorial objects appear in a number of applications across the sciences.

Now, as structures are added to Combinatorics, the nature of the numbers and our relationship to them changes. They no longer count something explicit or tangible, but rather something ephemeral or esoteric, which can only be understood by invoking further results in the area. Even when you think you are counting something combinatorial, it might take a theorem or a even the whole theory to realize that what you are counting is well defined (see e.g. §4.6).

This is especially true in *Algebraic Combinatorics* where the numbers can be, for example, dimensions of invariant spaces, weight multiplicities or Betti numbers. Clearly, all these numbers are nonnegative integers, but as defined they do not count anything per se, at least in the most obvious or natural way.

2020 *Mathematics Subject Classification.* Primary 05A20, 05E10; Secondary 05A19, 05C30, 68Q17, 68R05.

Key words and phrases. Combinatorial interpretation, combinatorial inequalities, counting complexity, correlation inequalities, Hamiltonian cycles, linear extensions, symmetric group, Young tableaux, group characters, Kronecker coefficients, Littlewood–Richardson coefficients, Schubert coefficients.

The author was partially supported by the NSF..

1.2. Why combinatorial interpretations? This brings us to the popular belief that one should always look for a combinatorial interpretation (see §15.1). As we see it, there are two reasons for this.

The first reason is clear: when you know what you are counting you have access to a large toolkit already developed in Enumerative Combinatorics and areas further afield. Essentially, the explicit combinatorial objects serve as a common playground where the areas can meet and be understood (see §3.2).

The second, deeper reason, is largely based on the hope is that a combinatorial interpretation would reveal some structures hidden in the algebraic objects they are working with. One can think of combinatorial interpretations as projections — you gather enough projections and hope the whole structure emerges. Consequently, when a combinatorial interpretation is found it is hard to tell if it points to a new structure without further study.

From our perspective, the first reason is terrific and can bring a lot of activity as new combinatorial objects rise to prominence in the areas spurred by applications. Meanwhile, the second reason is unfortunate and indicates that the area does not have a workable definition of a "combinatorial interpretation". This brings us to the following difficult question.

1.3. What do we mean by a combinatorial interpretation? Well, this is what this survey is about. The short answer is #P, a computational complexity class which we discuss at length.

But before we proceed, let us make a trivial comment. The first step towards building a theory is admitting the need for a formal definition. Without that, the negative results are impossible to state while the positive results ring hollow and cannot be fully appreciated for the miracle that they are.

According to Popper's philosophy, a belief needs to be *disprovable* in order to be *scientific* [**Pop62**]. Those who have unquestionable faith in the existence of combinatorial interpretations for all problems they care about, might want to take this lesson to heart.

1.4. There is no there there. We argue that many (all?) long-standing open problems on combinatorial interpretations in Algebraic Combinatorics can be formalized and resolved. We believe that few if any of them will have a solution of the kind that people in the area are looking for.

We aim to give a negative solution to many of the combinatorial interpretation problems using a formal definition we mentioned above. The goal of this paper is to advance this as part of a larger project. Until recently, this seemed overly ambitious and beyond the reach. Hopefully, this survey will leave you more optimistic.

1.5. Why bother? Given that until recently the notion of "combinatorial interpretation" had been informal, why set on a quixotic journey? Let us frame the question in much broader terms, from the perspective of *Computational Combinatorics.* There are two fundamental questions we want to address in our study:

(1) How do you prove that given numbers *have* a (#P) combinatorial interpretation?
(2) How do you prove that they *do not*?

Note that we are not so much interested in *finding* an explicit combinatorial interpretation, just proving membership in #P suffices for our purposes. Mostly, we are interested in development of new tools coming both from Combinatorics and Computational Complexity, to resolve the questions above.

In the last few decades, the area of Algebraic Combinatorics did a great job proving relevance and applicability beyond its boundaries. From this point of view, the open problems of finding combinatorial interpretations for numbers such as the *Kronecker* and *Schubert coefficients* are key benchmarks. Resolving them in either direction would be an important achievement in the whole Mathematics.

1.6. Why computational complexity? In other words, is there perhaps a more suitable and easier to understand language in which the problem can be phrased? Perhaps, in terms of integer points in convex polytopes or tilings in the plane? Indeed, both type of combinatorial interpretations do frequently appear in Algebraic Combinatorics and may seem like the natural place to start.

This is both the easiest and the hardest question to answer. The short answer is this: "Computational complexity provides the broadest and the most flexible language". In fact, prior to converging to #P, we tried a handful of approaches including the ones above. Since the types of "combinatorial interpretations" they gave were rather constrained, we reasoned that negative results would be easier to obtain. Eventually we discarded all such formulations as unconvincing and resigned to the fundamental difficulty and its ever conditional nature of Computational Complexity.

As we see it now, the Computational Complexity is truly foundational for the whole of Mathematics, and allows one to ask questions on the deeper level. While we constrain ourselves with the problem at hand, we refer to [**Wig19**] for the general picture.

1.7. What to expect from this survey. We assume that the reader is familiar with standard notation, results and ideas in Algebraic Combinatorics, see e.g. [**Mac95**, **Man01**, **Sag01**, **Sta99**]. After some hesitation, we decided to assume the same about Computational Complexity. We realize this might be unreasonable, and we will provide plenty of examples, but in the 21st century, a research survey is probably not the best place to include a long list of basic definitions.

We do include a quick overview of basic complexity notions (Section 2), but stop short of getting technical. For more details, we recommend [**MM11**] which is a fun read, [**AB09**, **Pap94b**] for standard textbooks on the subject, and [**Aar16**] for a very readable survey.

1.8. Structure of the paper. In Section 2, we introduce the computational complexity language, followed by Section 3 with many motivating examples. In Section 4, we discuss many examples from Enumerative Combinatorics, the original motivation for combinatorial interpretations (cf. §15.1). In Sections 5 and 6 we survey inequalities in Probabilistic Combinatorics and Order Theory, respectively. The selection of results is somewhat biased and reflects some of our own interests. The goal is prepare the reader for more difficult problems later on.

In Sections 7–10 is the core part of the survey. Here we discuss many functions in Algebraic Combinatorics centered around four subjects: Young tableaux, S_n characters, Kronecker and Schubert coefficients. These should be read in order, as they build on top of each other. Sections 9 and 10 have polemical portions at the end, which some readers might disagree with.

The last part of the survey presets our effort to organize the earlier material and make digestible conclusions. In Section 11, we discuss results and bijections in Algebraic Combinatorics centered around the *LR rule* and the *RSK correspondence*, which we consider crucially important to the subject. In Section 12, we discuss our recent work [**IP22**] which develops tools to prove nonexistence of combinatorial interpretations.

In Section 13, we give an annotated list of #P-completeness and #P-hardness results that we omitted earlier to avoid the confusion. These last three sections are independent from each other and should be accessible to experts in the area who skipped earlier section. We conclude with proofs postponed from earlier sections (Section 14), and some final remarks (Section 15).

Notation. Let $\{0,1\}^n$ denote the set of sequences of zeros and ones of length n, and let $\{0,1\}^*$ the set of all such sequences of finite length. We use $\mathbb{N} = \{0,1,2,\ldots\}$ and $[n] = \{1,\ldots,n\}$. The rest is pretty self-explanatory.

2. Basic Computational Combinatorics

For the purposes of this survey, we will make numerous shortcuts and imprecise statements, largely sacrificing standard definitions and rigor for the sake of clarity and conciseness. We also make our focus quite a bit more narrow than it could be. We beg the experts for forgiveness.

2.1. Combinatorial objects. The notion of a *combinatorial object* can be viewed as follows. A *word* is a binary sequence $x \in \{0,1\}^*$. The *size* of x is the length $|x|$. In other words, combinatorial objects of size N are encoded by words of length N, which in turn correspond to integers $0 \le a < 2^N$.[1] For example, a simple graph on n vertices can be viewed as a word of length $\binom{n}{2}$.

Note that we view *combinatorial object* as more than an abstract concept because the definition includes the *presentation*. For example, a simple graph on n vertices and m edges can also be presented as a list of edges. The resulting word would be of size $\Theta(m \log n)$. This makes some algorithms faster and other slower, but since $m = O(n^2)$ the change is at most polynomial for general graphs, so we can ignore it.

On the other hand, the presentation can make a lot of difference for problems in Algebraic Combinatorics. For example, a partition $\lambda = (4,3,1) \vdash 8$ can be written in *binary* $(100,11,1)$ or in *unary* $(1111,111,1)$. The binary presentation is more compact, so partitions $\lambda = (\lambda_1,\ldots,\lambda_\ell) \vdash n$ with $\ell = O(1)$ have size $O(\log n)$.

On the other hand, for many problems in Algebraic Combinatorics the unary presentation of size $O(n)$ is more appropriate. For example if your problem involves *self-conjugate* partitions $\lambda = \lambda' \vdash n$, we have $\ell(\lambda) \ge \sqrt{n}$, so the binary presentation still requires $\mathrm{poly}(n)$ space.[2] Similarly, the unary presentation is more natural when one works with *Young diagrams*. In summary, every time the problem involves partitions, one should *always* state whether partitions are given in unary or in binary.

---

[1]We realize this is not how some combinatorialists think of combinatorial objects.

[2]Occasionally, people use *Frobenius coordinates* specifically to avoid this issue for partitions with bounded *Durfee size*.

2.2. Classes and functions. The notion of a *combinatorial class* can be viewed as follows. Define the *language* as a subset of binary words: $L \subseteq \{0,1\}^*$. For example, we can consider the language of words which encode all Hamiltonian graphs. Similarly, we can consider the language of unary encodings of Young tableaux counted by the LR rule, i.e. for every triple of partitions (λ, μ, ν), there are exactly $c^{\lambda}_{\mu\nu}$ words in the corresponding language.

We write $\overline{L} := \{0,1\}^* \setminus L$ to denote the *complement* of L. There is some ambiguity here depending on the presentation, but in principle $\overline{L}$ can be very large. For example, the complement to the language of Hamiltonian graphs includes both non-Hamiltonian graphs and all words corresponding to non-graphs.

We usually think of a *combinatorial function* as a function counting combinatorial objects. This can be viewed as special case of the following general definition. Let $f : \{0,1\}^* \to \mathbb{Z}$ an integer function. Suppose $f(x) = 0$ for $x \notin L$, and $f(x) \geq 1$ for $x \in L$. We then say that f is *supported on* L.

For example, the number of Hamiltonian cycles is a function supported on Hamiltonian graphs. Similarly, the LR coefficients is a function $f : (\lambda, \mu, \nu) \to c^{\lambda}_{\mu\nu}$ supported on triples of partitions where LR coefficients are nonzero.

Note that the language can be defined in a variety of ways: by an explicit function f, by a Turing machine, by a formal grammar by an explicit mathematical definition, or by an abstract logical construction. It is important to keep in mind that the complexity of L can reflect the complexity of its definition, but that does not always hold.

For example, the set of exponents n for which the *Fermat's Last Theorem* (FLT) holds, naturally correspond to a language $L = \{3, 4, 5, \ldots\}$. From the mathematical point of view, proving that $n \in L$ becomes increasingly more difficult as n grows, but as a language L is rather simple now that FLT has been proved.

2.3. Decision, search and counting problems. A *decision problem* is a computational problem defined by the membership in the language. For example, HAMILTONICITY is a problem whether the language of Hamiltonian graphs contains a word corresponding to given graph. Similarly, the non-vanishing of LR coefficients problem $c^{\lambda}_{\mu\nu} >^? 0$ is a problem whether a triple of partitions (λ, μ, ν) written in unary is in the language given as a support of the LR coefficients.

A *search problems* is similar to the decision problem and asks not only to decide $x \in^? L$, but also to *verify* the answer by providing a *witness*. We formalize this notion below, but for now let us think of this qualitatively rather than quantitatively.

For example, the search problem associated with HAMILTONICITY ask to find a Hamiltonian cycle, since that implies that Hamiltonicity by definition. On the other hand, for NON-HAMILTONICITY there is no natural witness as nonexistence of a Hamiltonian cycle cannot be easily characterized (for a very good reason, see below). The verifier can simply list all 2^m subgraphs and supplant a proof that none are a Hamiltonian cycle.

While seemingly harder, in some cases it is possible to use the decision problem as a black box to solve the search problem, by applying it repeatedly for smaller instances. For example, given graph G, if G has a Hamiltonian cycle, check if so does $G \smallsetminus e$. Continue removing edges until eventually some edge cannot be removed. Suppose now $G \smallsetminus e$ does not have Hamiltonian cycles. This means that e is an edge in *every* Hamiltonian cycle in G. This reduces the problem to finding a Hamiltonian cycle in G/e obtained from G as contraction by e. Proceed in this fashion until the whole Hamiltonian cycle is constructed.

Given a search problem, a *counting problem* is a problem of computing a function f given by the number of witnesses the verifier can accept. So if $x \notin L$, i.e. the decision problem has a negative answer, the function $f(x) := 0$. Otherwise, the function $f(x) \geq 1$ for all $x \in L$, i.e. function f is supported on L.

For example, the number of Hamiltonian cycles is a function supported on Hamiltonian graphs which naturally arises that way. Similarly, but less obviously, the LR-coefficients $c^{\lambda}_{\mu\nu}$ is the counting function for the number of LR tableaux.

2.4. Polynomial time problems. Until now we avoided using the words *polynomial time*, since it makes definitions quantitative and unnecessarily complicates the matter. But we need it from this point.

The first truly important class for us is P. It is a class of languages where the decision problem can be solved in polynomial time. There is a wide variety of problems in this class, for example testing whether a graph is connected or bipartite. More involved graph theoretic problems in P include planarity and having a perfect matching.[3]

Historically, there was a variety of ways to formalize the definition of P, all of which turn out to be equivalent. We will use a *Turing machine* (TM) mostly out of habit and because it is best known (compared to *RAM* and other equivalent models of computation). From our point of view, using the colloquial *polynomial time algorithm* is absolutely fine.

We distinguish class P from the class FP of nonnegative functions which can be computed in polynomial time.[4] To remember the difference, note that the former outputs 0 or 1, while the latter can output larger numbers. Simple examples of functions in FP include the number of connected components of a graph, the number of (proper) 2-colorings, and the number of spanning trees.[5]

2.5. Polynomial time verifier. Now, given a language L, the *polynomial time verifier* is a Turing machine M such that for some fixed polynomials p, q we have:

- for all $x, w \in \{0,1\}^*$, we have $M(x,w) \in \{0,1\}$,
- for all $x \in \{0,1\}^n$, machine M runs in time $\leq p(n)$,
- for all $x \in L \cap \{0,1\}^n$, there exists w, s.t. $|w| \leq q(n)$ and $M(x,w) = 1$,
- for all $x \in \{0,1\}^n \smallsetminus L$ and all w s.t. $|w| \leq q(n)$, we have $M(x,w) = 0$.

In particular, the verifier *accepts*, i.e. outputs 1, only if w is a witness for $x \in L$. Note that the witness w have to have polynomial size to avoid the type of witnesses we had seen in the Non-Hamiltonicity problem. This constraint is also necessary for M to work polynomial time, since otherwise it would take exponential time just to read the exhaustive list of subgraphs in this case.

Continuing with our favorite example, in the HAMILTONIANCYCLE search problem, the verifier checks if a collection of edges (this is w) is a Hamiltonian cycle in graph G (this is x in the notation above). Clearly, this can be done in polynomial

---

[3]These follow from the *Kuratowski theorem* and the *blossom algorithm*, resp., see e.g. [**Schr03**, §3.1, §24.2].

[4]Some experts use a different definition of FP. The one we use is more common in Counting Complexity.

[5]The latter follows from the *matrix-tree theorem*.

time. Similarly, in the LR TABLEAU search problem, a Young tableau can be verified to be a LR tableau corresponding to a triple (λ, μ, ν) in polynomial time. This is done by checking all equalities for the shape and the content, and all inequalities involved in the definition: non-increase in rows, strict increase in columns, and balance conditions for the right-to-left reading word.

2.6. Complexity classes. Complexity class NP is the class of decision problems $x \in^? L$ for which there exists a polynomial time verifier. Similarly, class coNP is the class of decision problems $x \in^? L$, such that there exists a polynomial time verifier for the complementary problem $x \in^? \overline{L}$.

For example, HAMILTONICITY $\in \mathsf{NP}$ and NON-HAMILTONICITY $\in \mathsf{coNP}$. Clearly, $\mathsf{P} \subseteq \mathsf{NP} \cap \mathsf{coNP}$. There are several hard decision problems known to be in $\mathsf{NP}\cap\mathsf{coNP}$, so it is conjectured that $\mathsf{P} \neq \mathsf{NP}\cap\mathsf{coNP}$. It is also conjectured that $\mathsf{NP} \neq \mathsf{coNP}$. It is known that $\mathsf{P} \neq \mathsf{NP}$ would not imply either of these two conjectures (see e.g. [**Aar16**, §2.2.3]).

Next, complexity class $\#\mathsf{P}$ is the class of counting functions for which there exists a polynomial time verifier. Formally, a function $f : \{0,1\}^* \to \mathbb{N}$ is in $\#\mathsf{P}$ if there exists a polynomial time verifier M and polynomial $q : \mathbb{N} \to \mathbb{N}$, such that for all $n \in \mathbb{N}$ we have:

$$f(x) = \big|\{w \in \{0,1\}^{q(n)} : M(x,w) = 1\}\big| \quad \text{for all} \quad x \in \{0,1\}^n.$$

Observe that $\mathsf{FP} \subseteq \#\mathsf{P}$. Indeed, for $f \in \mathsf{FP}$ the witness for x is any integer in $a \in \big\{0, \ldots, f(x) - 1\big\}$, and the verifier first computes $f(x)$ and then checks if $a < f(x)$. It is widely assumed that $\mathsf{FP} \neq \#\mathsf{P}$. In fact, it is hard to overstate how strong is this assumption.

For example, let $f(G)$ be the number of Hamiltonian cycles in G. Then $f \in \#\mathsf{P}$ since it is counting combinatorial objects which can be verified in polynomial time. Similarly, LR coefficients are given as a function $(\lambda, \mu, \nu) \to c^{\lambda}_{\mu\nu}$ is also in $\#\mathsf{P}$, since it is counting LR-tableaux which can be verified in polynomial time by the argument above. We give many more examples in the next section.

Note that we do not discuss problems that are NP-complete or $\#\mathsf{P}$-complete. That's largely because these notions are largely tangential to this survey. Like with other complexity classes and standard computational complexity notation, we will mention them at will when we need them and hope the reader can catch up. Here is a partial list as a mental check for the reader:

$$\mathsf{P} \subseteq \mathsf{UP} \subseteq \mathsf{NP} \subseteq \Sigma^{\mathsf{P}}_2 \subseteq \mathsf{PH} \subseteq \mathsf{PSPACE}.$$

We do want to emphasize the distinction of NP-complete and NP-hard classes – the former is contained in NP, while the latter does not. The same with $\#\mathsf{P}$-complete and $\#\mathsf{P}$-hard classes.

3. Combinatorial interpretation, first steps

3.1. Main definition. We will be brief. Let $f : \{0,1\}^* \to \mathbb{N}$ be a function. We say that f has a *combinatorial interpretation* if

$$f \in \#\mathsf{P}.$$

Note that until now, we used the term "combinatorial interpretation" in both its technical and colloquial meaning, which usually coincide but can also differ in several special case. For the rest of the paper, we will use it only in the technical sense, and use quotation marks for the colloquial meaning.

3.2. Basic examples and non-examples. We begin with some motivating examples, mostly following [**IP22**].

(1) Let $e : P \to \mathbb{N}$ be the number of linear extensions of P, where $P = (X, \prec)$ is a poset with n elements. Clearly, $e(P) \geq 1$, so we can define a nonnegative function $e'(P) := e(P) - 1$. Now observe that $e' \in \#\mathsf{P}$ simply because finding the lex-smallest linear extension L can be done in polynomial time by a greedy algorithm (see e.g. [**CW95**]), so $e'(P)$ counts linear extensions of P that are different from L.

(2) Let $G = (V, E)$ be a simple graph with $n = |V|$ vertices and $m = |E| \geq 1$ edges. Let $c(G)$ be the number of proper 3-colorings of G. Clearly, $c \in \#\mathsf{P}$. Note that $3^n - c(G)$ is also in $\#\mathsf{P}$, since verifying that a 3-coloring is *not* proper is in P.

Now, taking into account permutations of colors, observe that $f(G) := c(G)/6$ is an integer valued function. To see that $f(G) \in \#\mathsf{P}$, note that of the six possible 3-colorings corresponding to a given 3-coloring one can easily choose the lex-smallest. In other words the combinatorial interpretation for $f(G)$ is the set of lex-smallest 3-colorings of G.

The key point here is that starting with a 3-coloring $\chi : V \to \{1, 2, 3\}$, we can compute in polynomial time the lex-smallest 3-coloring χ' from the set of 6 recolorings of χ. If $\chi = \chi'$, we verify that χ is a combinatorial interpretation, and discard χ if otherwise.

(3) Let $G = (V, E)$ be a simple graph with $|V| = n$ vertices and $|E| = m$ edges. Consider the following elegant inequality by Grimmett [**Gri76**]:

$$(*) \qquad \tau(G) \leq \frac{1}{n}\left(\frac{2m}{n-1}\right)^{n-1}$$

for the number $\tau(G)$ of *spanning trees* in G.[6] We can turn this inequality into a nonnegative integer function as follows.

$$f(G) := (2m)^{n-1} - n(n-1)^{n-1}\tau(G).$$

Recall $\tau(G) \in \mathsf{FP}$ by the *matrix-tree theorem*. Then $f \in \mathsf{FP}$, and so $f(G)$ has a combinatorial interpretation according to our definition.

One could argue that a "combinatorial interpretation" should explain *why* the inequality $(*)$ holds in the first place. In fact, there are several schools of thought on this issue (see a discussion in [**Pak18**, §4]). We believe that the computational complexity approach is both the least restrictive and the most formal way to address this. Indeed, the combinatorial interpretations we study depend solely on the functions themselves and not of the difficulty of the proof of the functions being integer or nonnegative.

(4) Let $h(G)$ be the number of Hamiltonian cycles in G, and let $f(G) := \big(h(G)-1\big)^2$. This is our most basic non-example. While we cannot prove unconditionally that $f \notin \#\mathsf{P}$, we can prove it modulo standard complexity assumptions. Intuitively this is relatively straightforward. Clearly, a poly-time verifier that $f(H) \neq 0$ is also a poly-time verifier that $h(G) \neq 1$. A poly-time verifier for $h(G) \geq 2$ is easy:

[6] The original proof is a nice two-line argument using the AM-GM inequality for the product of eigenvalues of the *Laplacian matrix* of G. One could argue whether this proof is "combinatorial", but it definitely does not extend to an explicit injection.

present two distinct Hamiltonian cycles. On the other hand, a poly-time verifier for $h(G) = 0$ is unlikely since that would imply that $\mathsf{NP} = \mathsf{coNP}$.[7]

(5) As above, let $h(G)$ be the number of Hamiltonian cycles in G. Recall *Fermat's little theorem* states that $p|a^p - a$ for all integers a, and prime p.[8] Let

$$f(G) := \tfrac{1}{p}\big(h(G)^p - h(G)\big).$$

It was shown in [**IP22**, Prop. 7.3.1], that $f(G) \in \#\mathsf{P}$. The proof is very short, and a variation on the original proof in [**Pet72**] (see also [**Ges84, Gol56**]). We reproduce it here in full.

PROOF. Consider sequences $(a_1, \ldots, a_p)$ of integers $1 \le a_i \le h(G)$ and partition them into orbits under the natural cyclic action of $\mathbb{Z}/p\mathbb{Z}$. Since p is prime, these orbits have either 1 or p elements. There are exactly p orbits with one elements, where $a_1 = \ldots = a_p$. The remaining orbits of size p have a total of $h(G)^p - h(G)$ elements. Since p is fixed, the lex-smallest orbit representative can be found in poly-time. □

(6) Recall the following *Smith's theorem* [**Tut46**]. Let $e = (v, w)$ be an edge in a cubic graph G. Then the number $N_e(G)$ of Hamiltonian cycles in G containing e is always even. Denote $f(G, e) := N_e(G)/2$. Is $f \in \#\mathsf{P}$? We don't know. This seems unlikely and remains out of reach with existing technology. But let us discuss the context behind this problem.

There are two main proofs of Smith's theorem. Tutte's original proof in [**Tut46**] uses a double counting argument. An algorithmic version of this proof is given by Jensen [**Jen12**]. The algorithm starts with one Hamiltonian cycle in G containing e, and finds another such cycle. Jensen also shows that this algorithm requires an exponential number of steps in the worst case.

The Price–Thomason *lollipop algorithm* [**Pri77, Tho78**] gives a more direct combinatorial proof of Smith's theorem. This algorithm also partitions the set of all Hamiltonian cycles in G containing e into pairs,[9] and is also exponential, see [**Cam01, Kra99b**].[10]

CONJECTURE 3.1. *The function $f(G, e)$ is not in $\#\mathsf{P}$.*

Note that if either Jessen's algorithm or the lollipop algorithm were poly-time, this would imply that $f \in \#\mathsf{P}$. Indeed, by analogy with (2), a poly-time algorithm would allow us to search for Hamiltonian cycles and only count the ones that are lex-smaller than their pairing partner.

3.3. First observations. From the limited number of examples above, here are a few observations. We will develop them further later on.

(i) In combinatorics, nonnegative integer functions don't come from nowhere. They are either already counting something, e.g. orbits under the action of some group as in (2) and (5), or are byproducts of inequalities as in (1), (3) and (4).

[7] This is because NONHAMILTONICITY is coNP-complete [**Pap94b**, Prop. 10.2].

[8] Fermat stated this result in 1640 without proof, and the first published proof was given by Euler in 1736. According to Dickson, "this is one of the fundamental theorems of the theory of numbers" [**Dic52**, p. V].

[9] Finding another Hamiltonian cycle was first raised in [**CP88**] in the context of Smith's theorem. This was a motivational problem for the complexity class PPA, see [**Pap94a**], as well as large part of our work in [**IP22**]. Whether it is PPA-complete remains open.

[10] There are other results similar to Smith's theorem which can be proved by a parity argument by a variation of the lollipop algorithm, see e.g. [**CE99**] and references therein.

(ii) The inequality in (*i*) could be rather trivial. For example, we have the trivial inequality $e(P) \geq 1$ in (1), the *AM-GM inequality* in (3), and $x^2 \geq 0$ in (4). It is the nature of the inequality that determines whether the function is in #P.
(iii) The computational hardness of the functions works only in one direction: if $f \in \mathsf{FP}$ then $f \in \#\mathsf{P}$, see (3), but if $f \in \#\mathsf{P}$-hard then it can go both ways.
(iv) Even for some classical problems like (6), membership in #P can be open.

4. Sequences

The problems in this section come from Enumerative Combinatorics. Although they are not the most interesting questions from the complexity point of view, the problems of finding combinatorial interpretations of integer sequences are much too famous not be addressed. In our notation and problem selection we largely follow [**Pak18**].

Throughout this section we assume that the input n is in unary. We say that an integer sequence $\{a_n\}$ has a combinatorial interpretation if a function $f : n \to a_n$ is in #P. Similarly, we say that $\{a_n\}$ can be computed in poly-time if $f \in \mathsf{FP}$. By abuse of notation, we also say that $\{a_n\}$ is in #P and FP, respectively.

4.1. Catalan numbers. Recall the *Catalan numbers*

$$\mathrm{Cat}(n) \;=\; \frac{1}{n+1}\binom{2n}{n} \;=\; \binom{2n}{n} - \binom{2n}{n-1},$$

see e.g. [**OEIS**, A000108]. The fractional formula implies that $\mathrm{Cat}(n) > 0$, the subtraction formula implies that $\mathrm{Cat}(n) \in \mathbb{Z}$, but a priori it is not immediately obvious that Catalan numbers have *any* combinatorial interpretations. Of course, there are over 200 "combinatorial interpretations" of various types given in [**Sta15**].

Let us show that $\{\mathrm{Cat}(n)\} \in \#\mathsf{P}$. Recall that $\mathrm{Cat}(n)$ is equal to the number of *ballot sequences*, defined as $0-1$ sequences with n zeros and n ones, s.t. every prefix has at least as many zeroes as ones. This can be checked in time poly(n), which proves that Catalan numbers are in #P.[11]

In fact, just about all "combinatorial interpretations" in [**Sta15**] can also be used to show that Catalan numbers are in #P, but some are trickier than others. For example, $\mathrm{Cat}(n)$ is the number of 123-avoiding permutations in S_n, and one would need to observe that there are $\binom{n}{3} = O(n^3)$ possible 3-subsequences. Thus, verifier checking the 123-avoidance is in P, as desired.

On the other hand, *some* "combinatorial interpretations" in [**Sta15**] are not even counting combinatorial objects, and it can take some effort to give them an equivalent presentation which is in #P. Notably, Exc 195 counts certain regions in $\mathbb{R}^n$ in the complement to the *Catalan hyperplane arrangement* $\mathcal{C}_n$. This setting raises some interesting computational questions.

To present the regions as a combinatorial objects one can use a collection of signs, one for each hyperplane. Since the arrangement $\mathcal{C}_n$ has $\Theta(n^2)$ hyperplanes, a trivial consistency check of all $(n+1)$-tuples of hyperplanes would give an exponential time algorithm for testing whether the resulting region is nonempty. This is not good enough for being in #P.

Now, in this specific case of the Catalan arrangement, there is an easy poly-time testing algorithm which uses the simple structures of hyperplanes in $\mathcal{C}_n$ and avoids

[11]This is because n is in unary. Note that if n is binary, the ballot sequences have exponential length.

the redundancy in the exhaustive testing above.[12] This algorithm is the verifier giving the desired combinatorial interpretation.[13]

4.2. Polynomial time computable combinatorial sequences. Note that since n is in unary, the sequence $\{\text{Cat}(n)\}$ is in FP since it can be computed in polynomial time. The same holds for *Fibonacci numbers* [**OEIS**, A000045], *numbers of involutions* [**OEIS**, A000085], *partition numbers* [**OEIS**, A000110], and myriad other sequences which can be computed via recurrence relation.

Formally, we observe in [**Pak18**, Prop. 2.2] that every *D-algebraic sequence* is in FP when the input n is in unary. Thus, in particular, this holds for all *algebraic* and *P-recursive sequences*, see e.g. [**Sta99**, Ch. 6].

On the other hand, there are sequences which likely cannot be computed in poly(n) time. For example, the number SAW(n) of *self-avoiding walks* of length n in $\mathbb{Z}^2$ starting at the origin, is conjectured not to be in FP [**Pak18**, Conj. 2.14]. Clearly, $\{\text{SAW}(n)\} \in \#\mathsf{P}$, so we turn our attention to sequences which are *unlikely* to be in $\#\mathsf{P}$, or are in $\#\mathsf{P}$ for less obvious reasons.

4.3. Unimodality and log-concavity. Both *unimodality* and *log-concavity* properties of combinatorial sequences are heavily studied in the literature, see e.g. [**Brä15**] (see also [**Bre89, Bre94, Sta89**] for more dated surveys). Following [**Pak19**], every time you have an inequality $X \le Y$, we can convert it into a nonnegative integer $(Y - X)$ and ask if it has a combinatorial interpretation. For combinatorial sequences this is especially notable, and the approach above works well again.

To see explicit examples, recall multi-parameter combinatorial sequences such as *binomial coefficients* $\binom{n}{k}$, *Delannoy numbers* $D(i,j)$ [**OEIS**, A008288], *Stirling numbers* of both kinds [**OEIS**, A008275] and [**OEIS**, A008277], *q-binomial coefficients* $\binom{n}{k}_q$ (see e.g. [**Sta99**, §1.7]), etc. All of these satisfy various unimodality and log-concavity properties, e.g.

$$\begin{aligned}
&\tbinom{n}{k-1}\tbinom{n}{k+1} \le \tbinom{n}{k}^2,\\
&D(i,j) \le D(i+1,j-1) \quad \text{for all} \quad i<j, \quad \text{and}\\
&[q^{m-1}]\tbinom{n}{k}_q \le [q^m]\tbinom{n}{k}_q \quad \text{for all} \quad 0<m\le \tfrac{k(n-k)}{2}.
\end{aligned}$$

We refer to [**Sag92, CPP22a**] for the first two of these inequalities both of which have a direct injective proof. The last inequality is due to Sylvester [**Syl78**], see also [**PP13, Pro82, Sta89**] for modern treatment. Clearly, each of these inequalities has a combinatorial interpretation simply because both sides are in FP. For example, $\{\binom{n}{k}^2 - \binom{n}{k-1}\binom{n}{k+1}\} \in \mathsf{FP}$, etc.

4.4. Partitions. Let $p(n)$ denote the number of integer partitions of n. *Ramanujan's congruence* $p(5n-1) \equiv 0 \pmod 5$ has a famous "combinatorial interpretation" by Dyson, who conjectured (among other things) that $\frac{1}{5}p(5n-1)$ is equal to

[12] In fact, the problem of counting the number of regions in the complement of general rational hyperplane arrangements is in $\#\mathsf{P}$. Indeed, one can use standard results in Linear Programming to give a poly-time verifier for all regions encoded by *subsets* of the set of halfspaces defined by the hyperplanes. This implies that counting regions problem is always in $\#\mathsf{P}$. We thank Tim Chow for this observation, see mathoverflow.net/a/428272

[13] The bijection in the solution of Exc 195 in [**Sta15**] (which requires a proof!) is another approach to have these regions being in bijection with combinatorial objects.

the number of partitions $\lambda \vdash (5n-1)$ with *rank* $\lambda_1 - \lambda_1' = 0 \pmod 5$, see [**Dys44**]. This conjecture was proved in [**AS54**] and later extended in a series of remarkable papers, see [**AG88, GKS90, Mah05**].

In the same direction, Ramanujan proved many more congruences such as $p(25n-1) \equiv 0 \pmod{25}$, see e.g. [**Har40**, §6.6], but there seem to be no Dyson-style rank statistics in this case. On the other hand, now that the congruence is known, it follows that $\left\{\frac{1}{25}p(25n-1)\right\} \in \mathsf{FP}$. This is because

$$\sum_{n=0}^{\infty} p(n)t^n = \prod_{i=1}^{\infty} \frac{1}{1-t^i}$$

is D-algebraic, or because $\{p(n)\}$ can be computed in poly(n) time via *Euler's recurrence* (among several other ways), see [**Pak18**, §2.5] and references therein. This implies that $\left\{\frac{1}{25}p(25n-1)\right\}$ *already has* a combinatorial interpretation.

Similarly, recall the curious inequality $p_{14}(n) \ge p_{23}(n)$ for the numbers of partitions of n into parts $\pm 1 \pmod 5$ and $\pm 2 \pmod 5$, respectively. Finding an explicit injection proving the inequality was suggested by Ehrenpreis, see [**AB89, Kad99**]. From the computational complexity point of view, we already have $\{p_{14}(n) - p_{23}(n)\} \in \mathsf{FP}$, which shows that the desired injection can be computed in poly-time.[14]

Finally, the log-concavity of the partition function [**DP15**], implies that the sequence $\left\{p(n)^2 - p(n-1)p(n+1)\,,\, n > 25\right\}$ is in $\#\mathsf{P}$, simply because $\{p(n)\} \in \mathsf{FP}$.

4.5. Unlabeled graphs. Let u_n be the number of non-isomorphic *unlabeled graphs* on n vertices, see [**OEIS**, A000088]. Wilf conjectured in [**Wilf82**], that $\{u_n\}$ *cannot* be computed in poly(n) time, see also [**Pak18**, Conj. 1.1]. Does $\{u_n\}$ have a combinatorial interpretation? This is not so clear. The difficulty is that we are counting orbits rather than combinatorial objects and there is no obvious way to choose orbit representatives:

Open Problem 4.1. *The sequence* $\{u_n\}$ *is in* $\#\mathsf{P}$.

To understand the context of this problem, consider a closely related sequence. Let a_n be the number of nonisomorphic *unlabeled plane triangulations* on n vertices, see [**OEIS**, A000109]. In [**Pak18**, Conj. 1.3], we conjectured that $\{a_n\}$ can be computed in poly(n) time. This would immediately imply that $\{a_n\}$ is in FP and thus in $\#\mathsf{P}$. Since the conjecture remains open, we show the latter directly:

Proposition 4.2. *The sequence* $\{a_n\}$ *is in* $\#\mathsf{P}$.

We postpone the proof until §14.1. The idea is that the group of automorphisms of triangulations has polynomial size and all automorphisms can be computed explicitly via a poly-time algorithm for the isomorphism of planar graph (cf. [**CM92**]). We are able to compute the whole orbit and then use symmetry breaking by taking lex-smallest orbit representative.

Conjecture 4.3 ([**Pak18**, Conj. 1.3]). *The sequence* $\{a_n\}$ *is in* FP.

We believe that this conjecture can be derived using the tools in [**Fusy05, KS18, Wal05**]. Back to the sequence $\{u_n\}$. If GraphIsomorphism was known

[14]This is similar and partially motivated by the discussion of complexity of partition bijections viewed as algorithms, see [**KP09**, §6.1] and [**Pak06**, §8.4.5].

to be in P,[15] one could try to use the symmetry breaking approach in the proof of Proposition 4.2. Babai's recent *quasipolynomial* upper bound $n^{O((\log n)^c)}$ on graph isomorphism [**Bab18**], falls short of what we need towards resolving Open Problem 4.1.

Note that plane triangulations are dual to 3-connected cubic graphs, so the following problem lies in between Proposition 4.2 and Open Problem 4.1.

CONJECTURE 4.4. *Let $R_k(n)$ be the number of k-regular unlabeled graphs on n vertices. Then $\{R_k(n)\}$ is in #P, for all $k \geq 1$.*

We are optimistic about this conjecture since for k-regular graphs the GRAPHISOMORPHISM problem is in P. This was proved by Luks in [**Luks82**], see also [**BL83**, **SW19**].

Finally, there is a curious connection to log-concavity (see §4.3). Denote by $u_n(m)$ the number of nonisomorphic graphs with n vertices and m edges. It follows from [**PR86**] (see also [**Vat18**]), that $u_n(m)^2 \geq u_n(m-1)u_n(m+1)$. If $\{u_n(m)\} \in$ #P (see Open Problem 4.1), it would make sese to ask if we also have $\{u_n(m-1)u_n(m+1) - u_n(m)^2\} \in$ #P. Analogous questions can be asked about nonisomorphic planar graphs, plane triangulations, etc.

4.6. Knots. Denote by k_n the number of distinct knots with bridge number at most n, see e.g. [**OEIS**, A086825]. Here the *bridge number* is a knot invariant defined as the minimal number of bridges required to draw a knot in the plane, see e.g. [**Mur96**, §4.3].

OPEN PROBLEM 4.5. *The sequence $\{k_n\}$ is not in #P.*

To underscore combinatorial nature of the problem, note that knot diagrams are a (subset of) planar 4-regular graphs with signs at the vertices, so n is the bound on the number of vertices. The difficulty starts with the word "distinct" which is formalized as *non-isotopic* and is also combinatorial in nature: two knots are *isotopic* if they are connected by a finite sequence of *Reidemeister moves*. Unfortunately, from computational point of view, the issue with identifying distinct knots is much deeper than with nonisomorphic graphs.

First, note that it is not at all obvious that the isotopy is decidable. Could it be that the number of necessary Reidemeister moves between two isotopic knots with n crossings grows faster than the busy beaver function? The answer turns out to be "No"; the sequence $\{k_n\}$ is computable indeed. The best known upper bound on the number of Reidemeister moves is the tower of twos of height $2^{O(n)}$ and was given by Coward and Lackenby [**CL14**].[16]

We conclude with a simpler problem, or at least the one that has been resolved. Denote by a_n the number of knot diagrams on n labeled crossings which are isotopic to the *unknot*. The fact that $\{a_n\}$ is in #P follows from a famous result by Hass, Lagarias and Pippenger [**HLP99**].[17] Similarly, denote by b_n the number of knot

[15] Formally, we need an effective version of GRAPHISOMORPHISM, which produces generators for $\mathrm{Aut}(G)$ as a subgroup of S_n. This is known in many cases and related to the notion of *canonical labeling*, see [**Bab19, BL83, SW19**].

[16] There are also various hardness results suggesting that such sequence is hard to compute, see e.g. [**dM+21, Lac17, KT21**].

[17] It is also an immediate corollary from [**Lac15**], which shows that unknot can be obtained by a sequence of $O(n^{11})$ Reidemeister moves.

diagrams on n labeled crossings which are *not* isotopic to the *unknot*. The sequence $\{b_n\}$ is also in #P by a recent result of Agol, see [**Lac17**, §3.5].

5. Subgraphs

Discrete Probability is a major source of combinatorial inequalities, most of which can be converted into nonnegative functions. Whether these functions are in #P is then a challenging problem. In this section we concentrate on various counting subgraphs problems.

5.1. Matchings. Let $G = (V, E)$ be a simple graph, and let $p(G, k)$ denote the number of *k-matchings* in a simple graph $G = (V, E)$ defined as the number of k-subsets of E of pairwise nonadjacent edges. Clearly, $p(G, k) \in$ #P. Following [**Pak19**], consider a function

$$f(G, k) := p(G, k)^2 - p(G, k-1)p(G, k+1).$$

Famously, Heilmann and Lieb proved that $f(G, k) \geq 0$ [**HL72**], see also [**God93**, §6.3] and [**MSS15**] for more context on this remarkable result. It was observed in [**Pak19**] that $f \in$ #P follows immediately from Krattenthaler's injective proof of the Heilmann–Lieb theorem [**Kra96**].

Define $q(G)$ to be the number of spanning subgraphs $H = (V, E')$, $E' \subseteq E$, which contain a perfect matching. Observe that the function $q \in$ #P, since testing whether H has perfect matching is in P, see e.g. [**LP86**, §9.1]. The following subsection shows that this is unlikely for other graph properties.

5.2. Hamiltonian subgraphs. Let $f(G)$ denote the number of Hamiltonian spanning subgraphs of a simple graph $G = (V, E)$. Whether $f \in^?$ #P is a difficult question and does not follow directly from the definition since we need a poly-time algorithm to decide Hamiltonicity of G.[18]

Note that this is a close call, since there is an algorithm to *verify* that each H is Hamiltonian by showing a Hamiltonian cycle C in H. Thus, one would think that pairs (H, C) give a combinatorial interpretation of f, but of course one would need to pick only one such cycle C per H. For example, the lex-smallest C would work, but there is no poly-time algorithm to verify that.

OPEN PROBLEM 5.1. *Function f is not in* #P.[19]

Even more difficult is the $\overline{f}(G) := 2^m - f(G)$ function which counts *non-Hamiltonian* spanning subgraphs of G, since there is no efficient verifier in this case. That makes the following problem a little more approachable, perhaps:

CONJECTURE 5.2. *Function $\overline{f}$ is not in* #P.

[18]This is another example where a combinatorialist might disagree, since the definition $f(G)$ already gives a *kind of* "combinatorial interpretation".

[19]Here and all other open problems and conjectures in this paper we implicitly allow the use of any of the standard complexity assumptions. Otherwise, these open problems are both deeper and less approachable.

5.3. Spanning forests. Let $G = (V, E)$ be a simple connected graph with $n = |V|$ vertices, and let $F(G,k)$ denote the number of spanning forests in G with k edges. A special case of the celebrated result by Adiprasito, Huh and Katz [**AHK18**], proves log-concavity of $\{F(G,k)\}$:

$$(\star) \qquad F(G,k)^2 \geq F(G,k-1) \cdot F(G,k+1) \quad \text{for all} \quad 1 \leq k \leq n-2.$$

Following [**Pak19**], define $f(G,k) := F(G,k)^2 - F(G,k-1)F(G,k+1)$.

CONJECTURE 5.3. *Function f is not in $\#\mathsf{P}$.*

We have relatively little evidence in favor of this conjecture other than we tried very hard and failed to show that $f \in \#\mathsf{P}$. The original proof was significantly strengthened and simplified in [**ALOV18, BH20, CP21**] (see also [**CP22a**] for a friendly exposition).[20]

5.4. Perfect matchings. Let $G = (V, E)$ be a k-regular bipartite multigraph on $2n$ vertices, and let $\mathrm{PM}(G)$ be the number of perfect matchings in G. The celebrated *van der Waerden Conjecture*, now proved (see e.g. [**vL82**] and [**LP86**, §8.1]), is equivalent to

$$\mathrm{PM}(G) \geq \frac{k^n n!}{n^n}.$$

Let $f(G) := n^n \mathrm{PM}(G) - k^n n!$ The following result is a variation on [**IP22**, Thm. 7.1.5], and shows that it is unlikely that f has a combinatorial interpretation.

PROPOSITION 5.4. *Assume that edge multiplicities in graph G are given by $\#\mathsf{P}$ functions. If $f \in \#\mathsf{P}$, then $\mathsf{PH} = \Sigma_2^{\mathsf{P}}$.*

PROOF. Let $n = 2$, $V = \{a_1, a_2, b_1, b_2\}$, and let E consists of edges (a_1, b_1) and (a_2, b_2) with multiplicity x_1, edges (a_1, b_2) and (a_2, b_1) with multiplicity x_2. Then $G = (V, E)$ is bipartite and k-regular, where $k = x_1 + x_2$. We have $\mathrm{PM}(G) = x_1^2 + x_2^2$ and

$$f(G) = 4\mathrm{PM}(G) - 2k^2 = 4(x_1^2 + x_2^2) - 2(x_1 + x_2)^2 = 2(x_1 - x_2)^2.$$

The result now follows from Corollary 2.3.2 in [**IP22**]. □

Compare this result with *Schrijver's inequality* [**Schr98**]

$$\mathrm{PM}(G) \geq \left(\frac{(k+1)^{k-1}}{k^{k-2}} \right)^n,$$

for all $n \geq k \geq 3$. An elementary proof of the $k = 3$ case is given in [**Voo79**]. We challenge the reader to give a direct combinatorial proof of this inequality for any fixed $k > 3$.

Finally, let us mention *Bregman's inequality* [**Bre73**] formerly known as *Minc's conjecture* (see also [**Minc78**, §6.2]). In the special case of k-regular bipartite simple graphs, the setting of the former *Ryser's conjecture*, it gives $\mathrm{PM}(G)^k \leq (k!)^n$. Since the proof in this case is relatively short, it would be interesting to see if this inequality is in $\#\mathsf{P}$.

[20] In [**Sta00**, p. 314], Stanley writes about ($\star$): "*Our own feeling is that these questions have negative answers, but that the counterexamples will be huge and difficult to construct.*" We think of this quote as a suggestion that there is no direct combinatorial proof of ($\star$), pointing in favor of Conjecture 5.3.

5.5. Bunkbed conjecture. Let $G = (V, E)$ be a multigraph. Denote by $G \times K_2$ the *bunkbed graph* obtained as a Cartesian product. Formally, two copies of G are connected by parallel edges as follows: each vertex $v \in V$ corresponds to vertices $v_0 = (v, 0)$ and $v_1 = (v, 1)$ which form an edge (v_0, v_1).

For vertices $v, w \in V$, denote by $B_0(v, w)$ and $B_1(v, w)$ the number of spanning subgraphs H of $G \times K_2$, such that $v_0 \leftrightarrow_H w_0$ and $v_0 \leftrightarrow_H w_1$, respectively. In other words, we are counting subgraphs where w_0 or w_1 lie in the same connected component as v_0.

CONJECTURE 5.5 (*Bunkbed conjecture*). *For all $G = (V, E)$ and all $v, w \in V$, we have $B_0(v, w) \geq B_1(v, w)$.*

This conjecture was formulated by Kasteleyn (c. 1985), see [**vdBK01**, Rem. 5], in the context of *percolation*, and has become popular in the past two decades, see e.g. [**Häg03, Lin11**] and most recently [**Gri23, HNK23**].[21] The fact that it is notoriously difficult to establish, combinatorially or otherwise, suggests the following:

CONJECTURE 5.6. *Function $B_0 - B_1$ is not in* #P.

At first glance this might seem contradictory to the bunkbed conjecture, but notice that it only says that if Conjecture 5.5 holds then it holds for "non-combinatorial reasons", like the van der Waerden Conjecture. More precisely, Conjecture 5.2 rules out a simple direct injection establishing $B_1 \leq B_0$.[22] On the other hand, if Conjecture 5.5 is false, then Conjecture 5.2 is trivially true. In other words, Conjecture 5.2 is *complementary* to the bunkbed conjecture and could be easier to resolve.

Let us note that the bunkbed conjecture is known in a few special cases, such as complete graphs [**vHL19**] and complete bipartite graphs [**Ric22**]. It would be interesting to see if the proofs imply that $B_0 - B_1 \in \#\mathsf{P}$ in all these cases.

5.6. Kleitman's inequality. Let $\mathcal{A}$ be a collection of labeled graphs on $[n] = \{1, \ldots, n\}$. We say that $\mathcal{A}$ is *hereditary*, if for every $G \in \mathcal{A}$ and every spanning subgraph H of G, we have $H \in \mathcal{A}$. Examples of hereditary properties include *planarity*, 3*-colorability*, *triangle-free*, *non-connectivity*, *non-Hamiltonicity*, and not containing a perfect matching.

THEOREM 5.7 (Kleitman [**Kle66**]). *Let $\mathcal{A}$ and $\mathcal{B}$ be hereditary collections of labeled graphs on $[n]$. Then:*

$$|\mathcal{A}| \cdot |\mathcal{B}| \leq 2^{\binom{n}{2}} \cdot |\mathcal{A} \cap \mathcal{B}|.$$

This *Kleitman's inequality* is easier to understand in probabilistic terms, as having a positive correlation between uniform random graph events:

$$\mathbb{P}[G \in \mathcal{A}] \leq \mathbb{P}[G \in \mathcal{A} \,|\, G \in \mathcal{B}].$$

It is then natural to ask if Kleitman's inequality is in #P.

[21] The conjecture is usually formulated more generally, as an inequality for p-percolation. Replacing edges with series-parallel graphs simulates p-percolation on $G \times K_2$ for all rational p, and shows that two formulations are equivalent.

[22] Formally, denote by $\mathcal{B}_0(v, w)$ and $\mathcal{B}_1(v, w)$ the sets of subgraphs counted by B_0 and B_1, respectively. Suppose there exists an injection $\varphi : \mathcal{B}_1(v, w) \to \mathcal{B}_0(v, w)$, s.t. both φ and φ^{-1} (where defined) are computable in polynomial time. Then $B_0(v, w) - B_1(v, w)$ has a combinatorial interpretation as the number of elements in $\mathcal{B}_0(v, w) \smallsetminus \varphi(\mathcal{B}_1(v, w))$.

PROPOSITION 5.8. *Let $\mathcal{A}$ and $\mathcal{B}$ be hereditary collections of labeled graphs on $[n]$, such that the membership problems $G \in^? \mathcal{A}$ and $G \in^? \mathcal{B}$ are in* P. *Then:*

$$2^{\binom{n}{2}} \cdot |\mathcal{A} \cap \mathcal{B}| - |\mathcal{A}| \cdot |\mathcal{B}| \in \#\mathsf{P}.$$

The result follows from Kleitman's original proof. In this context, let us mention the *Ahlswede–Daykin* (AD) *inequality*, which is an advanced generalization of Kleitman's inequality, see §12.5. Other classical inequalities such as the *FKG inequality* and the *XYZ inequality* (see e.g. [**AS16**, Ch. 6]) are direct consequences of the AD inequality.

In [**IP22**, Prop. 2.5.1], we prove that assuming the (univariate) *Binomial Basis Conjecture* (BBC), then AD inequality is not in #P. In fact, our proof shows that already *Harris inequality* [**Har60**] is not in #P under BBC. This is in sharp contrast to Proposition 5.7.

5.7. Ising model. In this section, we consider a counting version of the *Ising model*, see e.g. [**Bax82**, §1.7] for the introduction.

Let $G = (V, E)$ be a multigraph with $n = |V|$ vertices and $m = |E|$ edges. For a subset $S \subseteq V$, let

$$\alpha(S) := \{(v, w) \in E \,:\, v, w \in S\} \cup \{(v, w) \in E \,:\, v, w \notin S\}.$$

Define the *correlation function*[23]

$$\mathrm{Cor}(v, w) := \sum_{S \subseteq V \,:\, v,w \in \alpha(S)} 4^{|\alpha(S)|} \;-\; \sum_{S \subseteq V \,:\, v,w \notin \alpha(S)} 4^{|\alpha(S)|}.$$

Note that the statistical sum here is over *induced subgraphs* rather than the spanning subgraphs in the previous two problems.

Griffiths [**Gri67**] showed that $\mathrm{Cor}(v, w) \geq 0$ by an inductive combinatorial argument. When untangled, it can be used to prove the following:

PROPOSITION 5.9. *The correlation function* $\mathrm{Cor} : (G, v, w) \to \mathbb{N}$ *is in* #P.

The *Griffiths–Kelly–Sherman* (GKS) *inequality* [**Gri67, KS68**] is a triangle-type inequality for the correlation functions:

$$2^m \mathrm{Cor}(v, w) \sum_{S \subseteq V} 4^{|\alpha(S)|} \geq \mathrm{Cor}(v, u)\mathrm{Cor}(u, w) \quad \text{for all } \; u, v, w \in V,$$

cf. [**GP20**, Thm. 3.14] for the planar graphs case.

CONJECTURE 5.10. *The GKS inequality is not in* #P.

We refer to [**GHS70**] for an even more curious *Griffiths–Hurst–Sherman* (GHS) *inequality*, and to [**Ell85**] for Statistical Physics context, unified proofs and further references.

[23] Compared to the original version in [**Gri67, KS68**], we modify the definition by fixing the same weight ($\log 2$) on all edges in E, so the correlation functions have integral values. Since our graphs can have multiple edges, both the Griffiths and the GKS inequalities remain equivalent to the original.

6. Linear extensions

Linear extensions of finite posets is another rich source of inequalities, some of which are mysterious and largely misunderstood, see [**CP23c**]. As structures go, linear extensions occupy the middle ground between *easy* combinatorial objects such as standard Young tableaux or spanning trees, and *hard* objects such as 3-colorings of graphs or Hamiltonian cycles.[24]

6.1. Björner–Wachs inequality. Let $P = (X, \prec)$ be a finite poset. A *linear extension* of P is a bijection $\rho : X \to [n]$, such that $\rho(x) < \rho(y)$ for all $x \prec y$. Denote by $\mathcal{E}(P)$ the set of linear extensions of P, and write $e(P) := |\,\mathcal{E}(P)|$.

For each element $x \in X$, let $B(x) := \{y \in X : y \succcurlyeq x\}$ be the *upper order ideal* generated by x, and let $b(x) := |B(x)|$. The *Björner–Wachs inequality* [**BW89**, Thm 6.3] states that

$$e(P) \prod_{x \in X} b(x) \ \geq \ n!$$

Proposition 6.1 ([**CPP23b**, Thm 1.13]). *The Björner–Wachs inequality is in* $\#\mathsf{P}$.

The proof in [**CPP23b**, §3] is essentially the same as the original combinatorial proof by Björner and Wachs. This is in contrast with a probabilistic proof in [**CPP23b**, §4] via (another) Shepp's inequality [**She80**] which in turn uses the FKG inequality. Similarly, this is in contrast with the Hammett–Pittel analytic proof [**HP08**], and Reiner's *q-analogue* based proof given in [**CPP23b**, §5]. Neither of these three proofs seem to imply the proposition.

6.2. Sidorenko's inequality. A *chain* in a poset $P = (X, \prec)$ is a subset $\{x_1, \ldots, x_\ell\} \subseteq X$, such that $x_1 \prec x_2 \prec \ldots \prec x_\ell$. Denote by $\mathcal{C}(P)$ the set of chains in P.

Suppose $P = (X, \prec)$ and $Q = (X, \prec')$ be two posets on the same set with $|X| = n$ elements, such that $|C \cap C'| \leq 1$ for all $C \in \mathcal{C}(P)$ and $C' \in \mathcal{C}(Q)$. *Sidorenko's inequality* states that $e(P)e(Q) \geq n!$ [**Sid91**].

Proposition 6.2 ([**CPP23b**, Thm 1.15] and [**GG20b**, §3]). *Sidorenko's inequality is in* $\#\mathsf{P}$.

Natural examples of posets (P, Q) as above, are the *permutation posets* $(P_\sigma, P_{\overline{\sigma}})$, where $P_\sigma = ([n], \prec)$ is defined as $i \prec j$ if and only if $i < j$ and $\sigma(i) < \sigma(j)$ for all $i, j \in [n]$, and $\overline{\sigma} := (\sigma(n), \ldots, \sigma(1))$. In this case P_σ is a *2-dimensional poset*, and $P_{\overline{\sigma}}$ is its *plane dual*.

The original proof by Sidorenko was based on combinatorial optimization. Other proofs include and earlier direct surjection by Gaetz and Gao [**GG20a**], and the geometric proof by Bollobás, Brightwell and Sidorenko [**BBS99**], via Stanley's theorem on *poset polytopes* [**Sta86**] and Saint-Raymond's proof of *Mahler's conjecture* for convex corners [**StR81**]. Neither of these three proofs imply the proposition, at least not directly.

[24] See [**CP23c**, §12.2] for a complexity theoretic explanation. We refer to survey articles [**BW00, Tro95**] for the notation, standard background on posets, and further references.

6.3. Stanley's inequality. Let $P = (X, \prec)$ be a finite poset. For an element $x \in X$ and integer $k \in [n]$, denote by $\mathrm{N}(k) = \mathrm{N}(P, x, k)$ the number of linear extensions $\rho \in \mathcal{E}(P)$, such that $\rho(x) = k$. *Stanley's inequality* [**Sta81**, Thm 3.1] states that:

$$\mathrm{N}(k)^2 \geq \mathrm{N}(k-1)\,\mathrm{N}(k+1) \quad \text{for all} \quad 1 < k < n.$$

CONJECTURE 6.3. *Stanley's inequality is not in* #P.

In the past few years, we made a considerable effort trying to resolve this problem. The original proof in [**Sta81**] is an easy but ingenuous reduction from the *Alexandrov–Fenchel inequality*, that was used previously to prove the van der Waerden conjecture (see §5.4). Since poset polytopes are rather specialized, initially we hoped that Stanley's inequality is in #P. Another positive evidence is the recent *effective* characterization of the equality conditions in Stanley's inequality, by Shenfeld and van Handel [**SvH23**] (see also [**MS24**]). In our terminology, they showed that the vanishing problem $\big\{\mathrm{N}(k)^2 =^{?} \mathrm{N}(k-1)\mathrm{N}(k+1)\big\}$ can be decided in poly-time.

With Chan [**CP21**], we develop a new *combinatorial atlas* technology (see also [**CP22a**]), which gave a purely linear algebraic proof of Stanley's inequality and its generalization to weighted linear extensions (see also [**CP22b, CP23a**]). This also allowed us to give a new proof of the equality conditions. Unfortunately, the limit argument in our proof does not allow to give a #P description (see [**CP21**, §17.17]). Separately, with Chan and Panova, we employed several combinatorial approaches in [**CPP23a, CPP22b**] to posets of width two, and an algebraic approach in [**CPP23b**, §7 and §9.11]. Unfortunately, Conjecture 6.3 remains elusive.[25]

6.4. XYZ inequality. Let $P = (X, \prec)$ be a finite poset, and let $x, y, z \in X$ be incomparable elements. Let $P_{xy} := P \cup \{x \prec y\}$, $P_{xz} := P \cup \{x \prec z\}$ and $P_{xyz} := P \cup \{x \prec y, x \prec z\}$. Shepp's classic *XYZ inequality* [**She82**], states that

$$e(P)\, e(P_{xyz}) \;\geq\; e(P_{xy})\, e(P_{xz}).$$

As with other correlation inequalities, the XYZ inequality is easier to understand in terms of uniform random linear extensions $\rho \in \mathcal{E}(P)$:

$$\mathbb{P}[\rho(x) < \rho(y) \,|\, \rho(x) < \rho(z)] \;\geq\; \mathbb{P}[\rho(x) < \rho(y)].$$

CONJECTURE 6.4. *The XYZ inequality is not in* #P.

The original proof (see also [**AS16**, §6.4]), uses the FKG inequality, which, like the AD inequality, is not in #P in full generality. A double counting argument proving the XYZ inequality is given in [**BT02**]. Unfortunately, it does not prove that the inequality is in #P, both because of the double counting argument and the use of BIPARTITEUNBALANCE problem, see [**IP22**, §9.2]. This makes the conjecture especially interesting.

---

[25] Most recently, we proved in [**CP23b**] that the *generalized Stanley's inequality* is not in #P unless the polynomial hierarchy collapses. We also showed that the *vanishing problem* in this case is not in PH, see [**CP23c**, §10] for an overview.

7. Young tableaux

We adopt the standard and largely self-explanatory notation from Algebraic Combinatorics. We refer to [**Sag92**, **Sta99**] for both notation and the background. Unless stated otherwise, we use unary encoding for all partition and integer parameters throughout this section. For convenience, we are using notation $f \leq_{\#} g$ to mean that $(g - f) \in \#\mathsf{P}$.

7.1. Standard Young tableaux. Let $f^\lambda = |\,\mathrm{SYT}(\lambda)|$ be the number of *standard Young tableaux* of shape $\lambda \vdash n$. Recall the *hook-length formula*:

$$\text{(HLF)} \qquad f^\lambda = n! \prod_{(i,j)\in\lambda} \frac{1}{h_\lambda(i,j)},$$

where $h_\lambda(i,j) = \lambda_i + \lambda'_j - i - j + 1$ is the *hook-length* in λ. This implies that $f^\lambda \in \mathsf{FP}$. Consequently, we have that $n!/f^\lambda \in \mathsf{FP}$ and $(f^\lambda)^2 \leq_{\#} n!$

Similarly, let $f^{\lambda/\mu} = |\,\mathrm{SYT}(\lambda/\mu)|$ be the number of standard Young tableaux of skew shape λ/μ, where $|\lambda/\mu| = n$. Recall the *Aitken–Feit determinant formula*:

$$\text{(AFDF)} \qquad f^{\lambda/\mu} = n! \det\left(\frac{1}{(\lambda_i - \mu_j - i + j)!}\right)_{i,j=1}^{\ell(\lambda)}.$$

which proves that $f^\lambda \in \mathsf{FP}$. Consequently, we have that $f^{\lambda/\mu} \leq_{\#} n!$

Now, it follows from the *Naruse hook-length formula* (NHLF) that

$$(\otimes) \qquad f^{\lambda/\mu} \prod_{(i,j)\in\lambda/\mu} h_\lambda(i,j) \;\geq\; n!\,,$$

see [**MPP18b**]. Note that this is a much sharper bound than the one given by the Björner–Wachs inequality (see §6.1).

With Morales and Panova, we gave several proofs of the NHLF and its generalizations, both algebraic and combinatorial [**MPP17**, **MPP18a**]. A recursive proof is given by Konvalinka [**Kon20**]. Finding a direct combinatorial proof which allows efficient sampling from $\mathrm{SYT}(\lambda/\mu)$, perhaps generalizing the *NPS bijection* or the *GNW hook walk*, remains an important open problem, see e.g. [**H+23**, §5.6].

CONJECTURE 7.1. *Inequality* $(\otimes)$ *is in* $\#\mathsf{P}$.

REMARK 7.2. Even the simplest special cases of $(\otimes)$ are hard to establish directly. For example, rotate diagram λ by 180° and denote by λ^* be the resulting skew shape. Let $h^*_\lambda(i,j) := i + j - 1$ be the hook-lengths in λ^*. Inequality $(\otimes)$ in this case follows from

$$(\otimes\otimes) \qquad \prod_{(i,j)\in\lambda^*} h_{(a^b)}(i,j) \;=\; \prod_{(i,j)\in\lambda} h^*_\lambda(i,j) \;\geq\; \prod_{(i,j)\in\lambda} h_\lambda(i,j)\,,$$

where $a = \lambda_1$ and $b = \ell(\lambda)$, see [**MPP18b**, Prop. 12.1]. A direct combinatorial proof of $(\otimes\otimes)$ is given in [**PPS20**]. This proof uses *Karamata's inequality* that is not in $\#\mathsf{P}$ in full generality, see [**IP22**, §7.5].

7.2. Semistandard Young tableaux. Let $K_{\lambda\mu} = |\,\mathrm{SSYT}(\lambda,\mu)|$ be the *Kostka numbers*, where $\lambda, \mu \vdash n$. The inequality $K_{\lambda\mu} \leq_{\#} K_{\lambda 1^n} = f^\lambda$ is easy to show directly. Much more interesting is the following generalization.

Let $\boldsymbol{a} = (a_1, \ldots, a_\ell)$ and $\boldsymbol{b} = (b_1, \ldots, b_\ell)$ be two weakly decreasing vectors. We say that $\boldsymbol{a}$ *majorizes* $\boldsymbol{b}$, write $\boldsymbol{a} \trianglerighteq \boldsymbol{b}$, if $a_1 + \ldots + a_i \geq b_1 + \ldots + b_i$ for all $1 \leq i < \ell$, and $a_1 + \ldots + a_\ell = b_1 + \ldots + b_\ell$. Recall that $K_{\lambda\mu} \leq K_{\lambda\nu}$ for all $\mu \trianglerighteq \nu$.

PROPOSITION 7.3. *$K_{\lambda\mu} \leq_\# K_{\lambda\nu}$ for all $\mu \trianglerighteq \nu$.*

Although not stated in this language, the proof follows easily from the combinatorial proof in [**Whi80**]. Here we are using the following trivial observation: if $f_1, \ldots, f_k \in \#\mathsf{P}$ for $k = \text{poly}(n)$, then $f_1 + \ldots + f_k \in \#\mathsf{P}$.

Finally, let $\lambda, \alpha, \beta, \gamma \vdash n$ be such that $\alpha \trianglerighteq \beta \trianglerighteq \gamma$ and $\alpha + \gamma = 2\beta$. Kostka numbers then satisfy log-concavity property given by the *HMMS inequality*: $K_{\lambda\beta}^2 \geq K_{\lambda\alpha} K_{\lambda\gamma}$ [**H+22**, Thm 2].

OPEN PROBLEM 7.4. *The HMMS inequality is not in $\#\mathsf{P}$.*

The original proof of the HMMS inequality is based on the *Lorentzian property* of *Schur polynomials*. Solving the open problem would give an early indication in favor of Conjecture 5.3, since the log-concavity of forest numbers $\{F(G,k)\}$ is proved in [**BH20**] based on the same general approach.

REMARK 7.5. The *Dominance order* "$\triangleright$" is motivated by a technical part in the proof of the *Young symmetrizer* construction, see e.g. [**Weyl39**, §IV.2] and [**Sag01**, §2.4], and reflects the inherent planarity of Young diagrams. The algebraic proof of Proposition 7.3 given in [**LV73**] is based on iterative calculations.[26] We note that [**Ver06**] emphasizes the importance of the identity

$$(\triangle) \qquad 1 \uparrow_{S_{\mu_1} \times S_{\mu_2} \times \cdots}^{S_n} \bigcap \text{sign} \uparrow_{S_{\mu_1} \times S_{\mu_2} \times \cdots}^{S_n} = \mathbb{S}^\mu$$

(see also [**May75**]). This is a curious byproduct of the dominance order and its reverse.

7.3. Contingency tables. Let $\mathcal{T}(\lambda, \mu)$ be the set of *contingency tables*, defined as nonnegative integer matrices with rows sums λ and column sums μ. Denote by $\mathrm{T}(\lambda, \mu) := |\mathcal{T}(\lambda, \mu)|$ the number of such tables. Note that $K_{\lambda,\mu} \leq_\# \mathrm{T}(\lambda, \mu)$, where the injection sends $A \in \mathrm{SSYT}(\lambda, \mu)$ into a matrix $(m_{ij}) \in \mathcal{T}(\lambda, \mu)$, where m_{ij} is the number of letters j in i-th row of A.

Barvinok's inequality [**Bar07**], states that $\mathrm{T}(\lambda, \mu) \leq \mathrm{T}(\nu, \tau)$ for all $\lambda \trianglerighteq \nu$ and $\mu \trianglerighteq \tau$.

PROPOSITION 7.6. *Barvinok's inequality is in $\#\mathsf{P}$.*

There are two natural proofs of Proposition 7.6. First, following the original proof in [**Bar07**, p. 111], one can use the *RSK correspondence*, which is poly-time by definition. This gives $\mathrm{T}(\lambda, \mu) = \sum_\pi K_{\pi\lambda} K_{\pi\mu}$ and then use Proposition 7.3. We then need to use the *inverse RSK correspondence*, which is also poly-time. A more direct (still rather involved) approach is outlined in [**Pak19**].[27]

REMARK 7.7. We note in passing that the log-concavity property for contingency tables remains open. Formally, let $\lambda, \alpha, \beta, \gamma \vdash n$ be as in the HMMS inequality (see §7.2). Barvinok asks if $\mathrm{T}(\lambda, \beta)^2 \geq^? \mathrm{T}(\lambda, \alpha)\mathrm{T}(\lambda, \gamma)$, see [**Bar07**, p. 110].

[26] Dennis White kindly informed us that he meant [**LV73**] as a missing reference [4] in [**Whi80**].

[27] A crucial part of the injective proof of both White's and Barvinok's inequality is the *parenthesization construction* by Greene and Kleitman, see [**GK76**] and [**GK78**, §3] (see also [**dB+51**]).

7.4. Littlewood–Richardson coefficients. Let $c^{\lambda}_{\mu\nu}$ be the *Littlewood–Richardson* (LR) *coefficients*, where $\lambda \vdash n$, $\mu \vdash k$ and $\nu \vdash n-k$. Standard combinatorial interpretations for LR coefficients imply:

$$c^{\lambda}_{\mu\nu} f^{\mu} f^{\nu} \leq_{\#} f^{\lambda} \quad \text{and} \quad c^{\lambda}_{\mu\nu} f^{\lambda} \leq_{\#} \tbinom{n}{k} f^{\mu} f^{\nu}.$$

We refer to [**Ker84, Whi81, Zel81**] for the motivational explanation on how to derive these inequalities from the RSK correspondence or via the *jeu-de-taquin correspondence*. Taking their product gives $(c^{\lambda}_{\mu\nu})^2 \leq \binom{n}{k}$, as was recently observed in [**PPY19**, §4.1]. We call this the *PPY inequality*.

Open Problem 7.8. *The PPY inequality is not in* #P.

As we discussed earlier, the problem could be resolved either by a direct injection proving the PPY inequality, or by giving an explicit combinatorial interpretation for $\binom{n}{k} - (c^{\lambda}_{\mu\nu})^2$.

The remarkable *Lam–Postnikov–Pylyavskyy inequality* [**LPP07**] states that:[28]

$$\text{(LPP)} \qquad c^{\lambda}_{\mu\nu} \leq c^{\lambda}_{\mu\vee\nu\,\mu\wedge\nu}\,,$$

where $\mu \vee \nu$ and $\mu \wedge \nu$ denote the union and the intersection of Young diagrams. The original proof is heavily algebraic and does not seem to give a clue on how this can be proved injectively, see [**BBR06**] for some special cases.

Open Problem 7.9. *The LPP inequality is not in* #P.[29]

Note a closely related *Björner's inequality*:

$$\tbinom{n}{m} f^{\mu} f^{\nu} \leq \tbinom{n}{r} f^{\mu\vee\nu} f^{\mu\wedge\nu},$$

where $|\mu| = m$, $|\mu| + |\nu| = n$ and $|\mu \wedge \nu| = r$ [**Bjö11**, §6]. It follows by summing over all $\lambda \vdash n$, of (LPP) multiplied by f^{λ}. Of course, Björner's inequality is in FP and follows from the HLF.

7.5. Inverse Kostka numbers. Denote by $\boldsymbol{K} = (K_{\lambda\mu})$ the *Kostka matrix*. Ordering all partitions w.r.t. the size $n = |\lambda| = |\mu|$ and the majorization order "$\trianglerighteq$" and using $K_{\lambda\lambda} = 1$, we conclude that the matrix is upper triangular and thus has an integer inverse. Denote by $\boldsymbol{K}^{-1} = (K^{-1}_{\lambda\mu})$ the *inverse Kostka matrix*, and by $K^{-1}_{\lambda\mu}$ the *inverse Kostka numbers*.

Eğecioğlu and Remmel [**ER90**] showed that $K^{-1}_{\lambda\mu}$ has a signed combinatorial interpretation as a sum over certain *rim-hook tableaux* (RHT) of shape λ and weight μ. It follows from the construction that K^{-1} is in GapP = #P − #P. Direct involutions proving validity $\boldsymbol{K} \cdot \boldsymbol{K}^{-1} = \boldsymbol{K}^{-1} \cdot \boldsymbol{K} = I$ were given in [**ER90, LM06**]. Other signed combinatorial interpretations are given in [**Duan03, PR17**].

Conjecture 7.10. *The function* $|K^{-1}_{\lambda\mu}|$ *is not in* #P.

In other words, the conjecture claims that the absolute value of the inverse Kostka numbers does not have a combinatorial interpretation. Thus, a signed sum over combinatorial objects is the best one can have. Further motivation behind the conjecture will become clear in the next section.

---

[28] The actual LPP inequality in [**LPP07**] is more general and written in terms of skew Schur functions.

[29] We state both open problems in the negative largely because we would much rather see negative solutions than positive. Unfortunately, at the moment there is very little evidence in favor of either direction.

8. Characters

In this section we discuss complexity problems related to S_n characters. As before, n and all partitions are given in unary.

8.1. The values. Let $\chi^\lambda(\mu)$ denote the character value of the irreducible S_n module $\mathbb{S}^\lambda$ on the conjugacy class $[\mu]$, where $\lambda, \mu \vdash n$. The *Murnaghan–Nakayama* (MN) *rule*, see e.g. [**Sag92**, §4.10] and [**Sta99**, §7.17], gives a signed combinatorial interpretation for $\chi^\lambda(\mu)$ as a sum of signs over *rim-hook tableaux* $\mathrm{RHT}(\lambda, \mu)$ of shape λ and weight μ. Similar to the inverse Kostka numbers, it follows from the construction that function $\chi : (\lambda, \mu) \to \mathbb{Z}$ is in $\mathsf{GapP} = \#\mathsf{P} - \#\mathsf{P}$.

Does there exist a combinatorial interpretation of the character square $(\chi^\lambda(\mu))^2$? This is an interesting question, and the answer is even more interesting. On the one hand, the answer is yes when $\mu = (k^{n/k})$ is a rectangle. In this case, all rim-hook tableaux in $\mathrm{RHT}(\lambda, \mu)$ have the same sign, see e.g. [**JK81**, §2.7] and [**SW85**], so $\big(\chi^\lambda(\mu)\big)^2 = |\,\mathrm{RHT}(\lambda, \mu)|^2$ has a combinatorial interpretation as the number of ordered pairs of rim-hook tableaux.[30]

On the other hand, in full generality we have:

THEOREM 8.1 ([**IPP22**]). *If* $(\chi)^2 \in \#\mathsf{P}$, *then* $\mathsf{coNP} = \mathsf{C_{=}P}$ *and* $\mathsf{PH} = \Sigma_2^{\mathsf{P}}$.

In other words, it is *very unlikely* that character square has a combinatorial interpretation, assuming the polynomial hierarchy does not collapse to the second level.

REMARK 8.2. Following [**IPP22**], one way to understand the implications of the theorem is to compare two identities:

$$n! = \sum_{\lambda \vdash n} \big(\chi^\lambda(1)\big)^2 \quad \text{and} \quad n! = \sum_{\pi \in S_n} \big(\chi^\lambda(\pi)\big)^2, \quad \text{for all } \lambda \vdash n.$$

The former is the *Burnside identity* since $\chi^\lambda(1) = f^\lambda$, and follows from the RSK correspondence. The latter is the character orthogonality formula, and the theorem explains why there is no natural analogue of the RSK correspondence in this case. Simply put, the terms on the right are not actually counting any combinatorial objects.[31]

8.2. Row and column sums. In [**Sta00**, §3], Stanley defines

$$a_\lambda := \sum_{\mu \vdash n} \chi^\lambda(\mu) \quad \text{and} \quad b_\lambda := \sum_{\mu \vdash n} \chi^\mu(\lambda), \quad \text{where } \lambda \vdash n.$$

which he calls *row sums* and *column sums* in the character table, respectively. It is known that $b_\lambda = \big|\{\omega \in S_n : \omega^2 = \sigma\}\big|$, where $\sigma \in [\lambda]$ is a fixed permutation of type λ, see e.g. [**Sta99**, Exc 7.69] and [**Mac95**, Ex. 11, p. 120].[32] Thus, the column sums are in $\#\mathsf{P}$. One can ask a similar question about the row sums.

[30] In fact, it follows from [**FS98**, **SW85**] that $\big\{|\,\mathrm{RHT}(\lambda, \mu)|\big\} \in \mathsf{FP}$ in this case.

[31] There is, however, an involutive proof of both character orthogonality relations based on the MN rule [**Whi83**, **Whi85**].

[32] It follows form here that $b_\lambda > 0$ if and only if every even part of λ has even multiplicity, see e.g. [**BO04**].

First, we note that $a_\lambda \geq 0$ for all λ.[33] This follows from the fact that $a_\lambda = \langle \rho_n, \chi^\lambda \rangle$, where ρ_n is the character of the *conjugation representation*, see e.g. [**Sta99**, Exc 7.71].[34]

CONJECTURE 8.3 (cf. Problem 12 in [**Sta00**]). *Row sums $\{a_\lambda\}$ are not in* $\#\mathsf{P}$.

In fact, there are very few cases when a combinatorial interpretation of a_λ is known. For example, $a_{(n)} = p(n)$ and $a_{(1^n)} = |\{\lambda \vdash n : \lambda = \lambda'\}|$, see [**BE16**, Prop. 1]. Proving the conjecture would represent a major advance in the area, as we explain below.

REMARK 8.4. The total sum of the entries in the character tables of S_n is a sequence of interest in its own right, see [**OEIS**, A082733]. So is the determinant of the character table, see [**Jam78**, Cor. 6.5] and [**OEIS**, A007870], and even the permanent [**SS84**].

8.3. Refinements. Denote by $\rho^\mu(\sigma) := |\{\omega \in [\mu] : \omega\sigma = \sigma\omega\}|$, where $[\mu] \subset S_n$ is the conjugacy class of permutations of type $\mu \vdash n$. In other words, ρ^μ is the character of the conjugation action on the conjugacy class $[\mu]$. From above, we have $\rho_n = \sum_{\mu \vdash n} \rho^\mu$. Define *refined row sums* $a_{\lambda\mu} := \langle \rho^\mu, \chi^\lambda \rangle \in \mathbb{N}$, and note that $a_\lambda = \sum_{\mu \vdash n} a_{\lambda\mu}$.

CONJECTURE 8.5. *Refined row sums $\{a_{\lambda\mu}\}$ are not in* $\#\mathsf{P}$.

Clearly, the proof of Conjecture 8.3 implies the same for Conjecture 8.5. We warn the reader that it does not follow from definition that $\{a_{\lambda\mu}\} \in \mathsf{GapP}$. Indeed, the definition states

$$a_{\lambda\mu} := \langle \rho^\mu, \chi^\lambda \rangle = \frac{1}{n!} \sum_{\omega \in S_n} \rho^\mu(\sigma)\chi^\lambda(\omega).$$

Even though the terms on the RHS are in $\#\mathsf{P}$, there is no obvious way to divide the sum by $n!$ The claim is true nonetheless, as we explain later in this section.

Toward the open problem, it was proved by Kraśkiewicz and Weyman [**KW01**] (see also [**Sta99**, Exc 7.88b] and [**RW20**, §3]), that

$$\text{(KW)} \qquad a_{\lambda(n)} = \big|\{A \in \mathrm{SYT}(\lambda) \ : \ \mathrm{maj}(A) = 0 \mod n\}\big|,$$

where $\mathrm{maj}(A)$ denotes *major index* of A. This implies that $\{a_{\lambda(n)}\} \in \#\mathsf{P}$. The following result show how close (KW) gets us to resolving the open problem.

PROPOSITION 8.6 (folklore). *Denote by $\mathcal{D}$ the set of partitions into distinct parts. Then*
$\{a_{\lambda\mu} : \mu \in \mathcal{D}\} \in \#\mathsf{P}$. *Furthermore, if $\{a_{\lambda\mu} : \mu = (r^{n/r})\} \in \#\mathsf{P}$, then $\{a_{\lambda\mu}\} \in \#\mathsf{P}$.*

Versions of this result are well-known. For completeness, we include a short proof in §14.2. The proposition implies that to disprove Conjecture 8.3 it suffices to give a combinatorial interpretation for $\{a_{\lambda r^{n/r}}\}$.

[33] It is known that $a_\lambda \geq 1$ for all $\lambda \vdash n \geq 2$ [**Fru86**]. For the alternating group A_n, this *strict positivity* is proved in [**HZ06**] by a non-combinatorial (and much shorter) argument. For other finite simple groups, strict positivity is proved in [**HSTZ13**].

[34] In fact, this approach can be used to show that row sums of characters are nonnegative integers for all finite groups [**Sol61**].

REMARK 8.7. Curiously, refined row sums can be defined and generalized using Pólya's theory for general permutation groups, see [**Whi19, RW20**]. This approach leads to a plethora of numbers in search of combinatorial interpretations, including some of those in §4.5. From our point of view, this is a good starting place to prove for non-existence of such combinatorial interpretations in full generality.

8.4. Plethysm coefficients. Denote by $p_\lambda(\mu,\nu)$ the *plethysm coefficient*, which can defined in terms of Schur functions as $p_\lambda(\mu,\nu) = \langle s_\mu[s_\nu], s_\lambda\rangle$, see e.g. [**Sta99**, §7.A2] and [**Mac95**, §1.8]. Note that the bracket product $s_\mu[s_\nu]$ is non-commutative and equal to the trace of the composition of irreducible GL-modules corresponding to μ and ν: $S^\mu(S^\nu V)$.

CONJECTURE 8.8 (cf. Problem 9 in [**Sta00**]). *Plethysm coefficients* $\{p_\lambda(\mu,\nu)\}$ *are not in* $\#\mathsf{P}$.

Computing plethysm coefficients is so exceedingly difficult, there are very few special cases when they are known to have a combinatorial interpretation. The formalism in [**LR11**, §4] implies that $\{p_\lambda(\mu,\nu)\} \in \mathsf{GapP}$.[35]

The refined character sums can be expressed in terms of plethysm coefficients:

$$a_{\lambda(m^k)} = p_\lambda\big((k),\rho^{(m)}\big) = \sum_{\nu\vdash m} a_{\nu(m)}\, p_\lambda\big((k),\nu\big),$$

where $\lambda \vdash n = mk$, see e.g. [**AS18, Sun18**]. By (KW), a combinatorial interpretation for plethysm coefficients $p_\lambda(\mu,\nu)$ when μ is a row shape, suffices to disprove Conjectures 8.3 and 8.5. In summary, we have:

COROLLARY 8.9. *Conjecture* 8.3 $\Longrightarrow$ *Conjecture* 8.5. $\Longrightarrow$ *Conjecture* 8.8.

REMARK 8.10. There is a closely related study of multiplicities $\widetilde{a}_{\lambda\mu}$ in the *higher Lie modules*, see [**AS18, Kly74**]. The special case $\mu=(n)$ is given by

$$\widetilde{a}_{\lambda(n)} = \big|\{A\in \mathrm{SYT}(\lambda) \ : \ \mathrm{maj}(A) = 1 \mod n\}\big|,$$

see [**KW01**]. The results are completely parallel here: it is not known whether $\{\widetilde{a}_{\lambda\mu}\}$ are in $\#\mathsf{P}$, and it suffices to resolve this problem in the rectangular case $\widetilde{a}_{\lambda(m^k)}$, which in turn reduces to plethysm coefficients $p_\lambda\big((k),\nu\big)$. We refer to [**AS18, Sun18**] for details and further references.

8.5. Hurwitz numbers. Let $\mu = (\mu_1\ldots,\mu_\ell) \vdash n$, and let $g\le 0$ be a fixed constant. Denote by $H_{g\mu}$ the number of products of transpositions $(i_1,j_1)\cdots(i_m,j_m) = w$ in S_n, such that

- $m = 2g-2+n+\ell(\mu)$,
- $w\in S_n$ has cycle type μ, and
- $\langle (i_1,j_1),\ldots,(i_m,j_m)\rangle \subseteq S_n$ is a transitive subgroup.

The (single) *Hurwitz numbers* are defined as $h_{g\mu} := H_{g\mu}/(n-1)!$, see e.g. [**GJ97, LZ04**]. Although one can use the Frobenius character formulas to prove that they are integers, the following result comes as a surprise.

PROPOSITION 8.11 (Folklore). *Hurwitz numbers* $\{h_{g\mu}\}$ *are in* $\#\mathsf{P}$.

[35] See also [**FI20**] for the binary case.

The proof idea goes back to Hurwitz [**Hur91**], although it takes some effort to obtain it in this form. We refer to [**DPS14**, Prop. 1] for an explicit combinatorial interpretation via bijection with certain *Hurwitz galaxies* defined by the authors. The result extends to *double Hurwitz numbers* which we leave undefined, see e.g. [**GJV05**].

In the minimal case $g = 0$, Hurwitz numbers have a product formula, and thus is in FP. The proposition is remarkable since most proofs of this formula are rather technical, see e.g. an elegant proof in [**BS00**] which also uses both *double counting* and *inclusion-exclusion*. Notably, we recall Dénes's beautiful proof of $h_{0(n)} = n^{n-2}$ formula for the number of minimal factorizations of a long cycle. See [**GY02**] for a bijection in this case.

REMARK 8.12. The subject of Hurwitz numbers is quite extensive and ever growing, so giving comprehensive references is a challenge. Hurwitz numbers naturally arise in the study of branched covers of $\mathbb{CP}^1$ in Enumerative Algebraic Geometry, see [**CM16, ELSV99, Oko00, OP09**]. We refer to [**BS00, DPS14, GJV05, PS02**] for a more combinatorial treatment.

9. Kronecker coefficients

9.1. Reaching for the stars. Let $g(\lambda, \mu, \nu)$, where $\lambda, \mu, \nu \vdash n$, denote the *Kronecker coefficients*:

$$g(\lambda, \mu, \nu) := \langle \chi^\lambda \chi^\mu, \chi^\nu \rangle = \frac{1}{n!} \sum_{\sigma \in S_n} \chi^\lambda(\sigma)\chi^\mu(\sigma)\chi^\nu(\sigma).$$

By definition, $g(\lambda, \mu, \nu) \in \mathbb{N}$. Whether it has a combinatorial interpretation remains a major open problem first posed by Murnaghan [**Mur38, Mur56**].

CONJECTURE 9.1 (cf. Problem 10 in [**Sta00**]). *Kronecker coefficients $\{g(\lambda, \mu, \nu)\}$ are not in* #P.

It is known that $\{g(\lambda, \mu, \nu)\} \in$ GapP, see [**BI08**]. This follows, for example, from

$$g(\lambda, \mu, \nu) = \sum_{\omega \in S_\ell} \sum_{\pi \in S_m} \sum_{\tau \in S_r} \operatorname{sign}(\omega\pi\tau) \cdot \mathrm{T}(\lambda + (1^\ell) - \omega, \mu + (1^m) - \pi, \lambda + (1^r) - \tau),$$

where $\ell(\lambda) = \ell$, $\ell(\mu) = m$, $\ell(\nu) = r$, and $\mathrm{T}(\alpha, \beta, \gamma)$ is the number of *3-dim contingency arrays* with 2-dim marginals given by (α, β, γ), see [**PP17**, Eq. (8)]. The same equation implies that $\{g(\lambda, \mu, \nu)\} \in$ FP for partitions with bounded number of rows: $\ell, m, r = O(1)$, see [**CDW12, PP17**].[36]

9.2. Where to look. There are several families of examples when Kronecker coefficients are known.[37] These include Blasiak's remarkable combinatorial interpretation of $\{g(\lambda, \mu, \nu)\}$, where ν is a *hook* [**Bla17**] (see also [**Liu17**]), and an NP-complete combinatorial interpretation for *simplex-like triples* (λ, μ, ν) in [**IMW17**, §3]. That contrasts with the following:

CONJECTURE 9.2. *Kronecker coefficients $\{g(\lambda, \lambda, \lambda) : \lambda = \lambda'\}$ are not in* #P.

[36]When the encoding is in binary, both GapP and FP claims remain true, but the argument now requires *Barvinok's algorithm* for counting integer points in polytopes of bounded dimension in poly-time [**Bar93**].

[37]For a quick guide to the literature, see e.g. the MathSciNet review of [**Bla17**] by Christopher Bowman.

This conjecture is motivated by our inability to improve upon basic bounds in this case: $1 \le g(\lambda,\lambda,\lambda) \le f^\lambda$ for all $\lambda = \lambda'$. Here the lower bound is proved in [**BB04**], while the upper bound follows from an observation $g(\lambda,\mu,\nu) \le f^\lambda$ in [**PPY19**, §3.1]. Even for the *staircase shape* $\lambda = (k, k-1, \ldots, 1)$ these remain the best known bounds. For the *square shape*, we recently showed a lower bound $g(k^k,k^k,k^k) = e^{\Omega(\sqrt{k})}$ [**PP23**], which is very far from the upper bound $g(k^k,k^k,k^k) = e^{O(k^2 \log k)}$ that is conjecturally tight.

REMARK 9.3. When $\lambda = \mu$ and ν is a *two-row partition*, we have the following formula for Kronecker coefficients:

$$(\ddagger) \qquad g\big(\lambda,\lambda,(n-k,k)\big) = \sum_{\alpha\vdash k}\sum_{\beta\vdash n-k}\big(c^\lambda_{\alpha\beta}\big)^2 - \sum_{\alpha\vdash k-1}\sum_{\beta\vdash n-k+1}\big(c^\lambda_{\alpha\beta}\big)^2,$$

see [**PP14**, Lem. 3.1]. Consider the inequality "RHS of (‡) ≥ 0". While different from the LPP and PPY inequalities in §7.4, it has the same flavor and is sufficiently similar to be out of reach by direct combinatorial tools in full generality.[38]

In a special case when $\lambda = (m^\ell)$ is a rectangle, the equality (‡) gives:

$$g\big(m^\ell, m^\ell, (n-k,k)\big) = [q^k]\tbinom{m+\ell}{\ell}_q - [q^{k-1}]\tbinom{m+\ell}{\ell}_q,$$

see [**PP13, PP14**] (see also [**Val14**, Lem. 7.5]). In one direction, this proves unimodality of *q-binomial coefficients* (see §4.3). In the other direction, this highlights the obstacle towards a natural "combinatorial interpretation" of Kronecker coefficients, since proving this unimodality by an explicit injection is famously difficult.[39]

9.3. Taking a step back. Let α, β, γ be fixed integer partitions, not necessarily of the same size. The *reduced Kronecker coefficients* $\overline{g}(\alpha,\beta,\gamma)$ are defined as *stable limits* of Kronecker coefficients when a long first row is added:

$$(\circ) \qquad \overline{g}(\alpha,\beta,\gamma) := \lim_{n\to\infty} g\big(\alpha[n], \beta[n], \gamma[n]\big),$$

where $\alpha[n] := (n - |\alpha|, \alpha_1, \alpha_2, \ldots)$ and $n \ge |\alpha| + \alpha_1$, see [**Mur38, Mur56**]. Here the sequence $\{g(\alpha[n],\beta[n],\gamma[n])\}$ is weakly increasing and stabilizes already at $n \ge |\alpha| + |\beta| + |\gamma|$, see [**BOR11, Val99**]. In other words, the reduced Kronecker coefficients are a special case of Kronecker coefficients for triples of shapes with a long first row.

The problem of finding a combinatorial interpretation of the reduced Kronecker coefficients goes back to Murnaghan and Littlewood, and has been repeatedly asked over the past decades, see e.g. [**Kir04, Man15**]. Part of the reason is that they generalize the LR coefficients $\overline{g}(\alpha,\beta,\gamma) = c^\alpha_{\beta\gamma}$ for $|\alpha| = |\beta| + |\gamma|$, see [**Lit58**], and thus play an intermediate role.

CONJECTURE 9.4. *The reduced Kronecker coefficients* $\{\overline{g}(\alpha,\beta,\gamma)\}$ *are not in* $\#\mathsf{P}$.

Depending on your point of view, this conjecture is either the harder to prove or the easier to disprove, compared to Conjecture 9.1.

[38] When $\lambda_1 \ge 2k-1$ or $\ell(\lambda) \ge 2k-1$, a combinatorial description of $g\big(\lambda,\mu,(n-k,k)\big)$ is given in [**BO05**].

[39] With Greta Panova, we gave a cumbersome "combinatorial interpretation" for $g\big(m^\ell, m^\ell, (n-k,k)\big)$ in terms of certain trees, see these slides, p. 9. The proof is obtained by recursing O'Hara's q-binomial identity [**O'H90**].

9.4. Questioning the motivation. There are several traditional reasons why one should continue pursuing the multidecade project of finding a "combinatorial interpretation" for the Kronecker coefficients. Let us refute the most important of these, as we see them, one by one.

(1) Estimating the Kronecker coefficients is enormously difficult, especially getting the lower bounds. One might argue:

> Having a "combinatorial interpretation" would be a bonanza for getting good lower bounds on $g(\lambda, \mu, \nu)$.

Sure, quite possibly so. But given the poor state of affairs where in most cases we do not have *any* nontrivial lower bounds obtained by *any method* (cf. [**BBS21**, **PP20a**]), shouldn't that be a reason to *not believe* in the existence of a "combinatorial interpretation"?

(2) Recall the *saturation property* for LR coefficients states that $c^{k\lambda}_{k\mu k\nu} > 0 \iff c^{\lambda}_{\mu\nu} > 0$, for all integer $k \geq 1$. The original proof by Knutson and Tao [**KT99**] crucially relies on a variation of the LR rule. One might argue:[40]

> Having a "combinatorial interpretation" could help proving some sort of saturation property for the Kronecker coefficients.

No, it will not. First, the saturation property fails: $g(1^2, 1^2, 1^2) = 0$ while $g(2^2, 2^2, 2^2) = 1$. Second, Mulmuley's natural weakening of the saturation property in [**Mul11**] also fails, already for partitions with at most two rows [**BOR09**]. Third, even for the reduced Kronecker coefficients, the saturation property fails: $\overline{g}(1^5, 1^5, 3^2) = 0$ while $\overline{g}(2^5, 2^5, 6^2) = 12$ [**PP20b**].[41]

(3) The saturation property for the LR coefficients easily implies that their *vanishing* can be decided in poly-time using Linear Programming [**DM06**, **MNS12**] (see also [**BI13**] for a faster algorithm). One might argue:

> Even without the saturation property, perhaps having a "combinatorial interpretation" could give a complete description or possibly even an efficient algorithm for the vanishing of the Kronecker coefficients.

No, it will not (most likely). Here we are assuming that a "complete description" includes both necessary and sufficient conditions verifiable in poly-time, which puts this problem in $\mathsf{NP} \cap \mathsf{coNP}$. We are also assuming that an "efficient algorithm" is being in P.

Now, it is *already known* that the vanishing problem for the Kronecker coefficients is NP-hard [**IMW17**], so an efficient algorithm implies $\mathsf{P} = \mathsf{NP}$. Similarly, if an NP-hard problem is in $\mathsf{NP} \cap \mathsf{coNP}$, then $\mathsf{NP} = \mathsf{coNP}$. So unless one expects a major breakthrough in Computational Complexity, this approach will not work.

(4) There are obvious social aspects to problem solving. This is an old open problem, perhaps the oldest in the area. Famous people worked on it and reiterated its importance. One might proclaim:

> The victor gets the spoils.

---

[40] This argument appears frequently throughout the literature in different contexts. See e.g. [**Kir04**, **Mul11**] for many conjectured variations and generalizations of the saturation property.

[41] This was independently conjectured by Kirillov [**Kir04**, Conj. 2.33] and Klyachko [**Kly04**, Conj. 6.2.6]. We disprove the conjecture in [**PP20b**], by providing an infinite family of counterexamples. It is, however, concerning how little computational effort was made to check the conjecture which fails for relatively small partitions, yet first disproved by a theoretical argument. Could there be more conjectures which are not sufficiently tested? Perhaps, the "minor but interesting" *Foulkes plethysm conjecture* [**Sta00**, §3] is worth another round of computer testing, see [**CIM17**], as its fate may be similar to that of a stronger *Stanley's conjecture* disproved in [**Pyl04**].

Absolutely! But shouldn't then proving *nonexistence* of a combinatorial interpretation be just as much a "victory" as finding one?[42]

(5) Finally, the *intellectual curiosity* is not to be discounted. The problem is clearly attractive and has led to a lot of nice results even in small special cases. One might reasonably argue:

> While we may never be able to resolve the problem completely, many interesting results might get established along the way.

Sure, of course. But again, why limit yourselves to working only in the positive direction?[43]

10. Schubert coefficients

For notation and standard results on Schubert polynomials, see [**Mac91**] and [**Man01**]. An accessible introduction to combinatorics of reduced factorizations is given in [**Gar02**], and the geometric background is given in [**Ful97**, §10]. A friendly modern introduction is given in [**Gil19**]. The presentation below is self-contained but omits the background.

10.1. RC-graphs. For a permutation $w \in S_n$, denote by $\mathrm{RC}(w)$ the set of *RC-graphs* (also called *pipe dreams*), defined as tilings of a staircase shape with *crosses* and *elbows* as in Figure 10.1 which satisfy two conditions:

(i) curves start in row k on the left and end in column $w(k)$ on top, for all $1 \le k \le n$, and
(ii) no two curves intersect twice.

It follows from these conditions that every $G \in \mathrm{RC}(w)$ has exactly $\mathrm{inv}(w)$ crosses.

For $G \in \mathrm{RC}(w)$, denote by $\boldsymbol{x}^G$ the product of x_i's over all crosses $(i,j) \in G$, see Figure 10.1. Define the *Schubert polynomial* $\mathfrak{S}_w \in \mathbb{N}[x_1, x_2, \ldots]$ as[44]

$$\mathfrak{S}_w(\boldsymbol{x}) := \sum_{G \in \mathrm{RC}(w)} \boldsymbol{x}^G.$$

For example, $\mathfrak{S}_{1432} = x_1x_2x_3 + x_1^2x_3 + x_1x_2^2 + x_2^2x_3 + x_1^2x_2$ as in the figure. Note that Schubert polynomials stabilize when fixed points are added at the end, e.g. $\mathfrak{S}_{1432} = \mathfrak{S}_{14325}$. Thus we can pass to the limit $\mathfrak{S}_w$, where $w \in S_\infty$ is a permutation $\mathbb{N} \to \mathbb{N}$ with finitely many nonfixed points.

Polynomials $\{\mathfrak{S}_w : w \in S_\infty\}$ are known to form a basis in the ring $\mathbb{Z}[x_1, x_2, \ldots]$. *Schubert coefficients* are defined as structure constants:

$$\mathfrak{S}_u \cdot \mathfrak{S}_v = \sum_{w \in S_\infty} c^w_{uv} \, \mathfrak{S}_w \, .$$

It is known that $c^w_{uv} \in \mathbb{N}$ for all $u, v, w \in S_\infty$.

CONJECTURE 10.1 (cf. Problem 11 in [**Sta00**]). *Schubert coefficients $\{c^w_{uv} : u, v, w \in S_\infty\}$ are not in* $\#\mathsf{P}$.

---

[42]For more on this argument, see our blog post "*What if they are all wrong?*" (Dec. 10, 2020), available at wp.me/p211iQ-uT

[43]As the renowned 19th century British philosopher the Cheshire Cat once said, "*it doesn't much matter which way you go*", you will definitely get <u>somewhere</u> "*if only you walk long enough*" [**Car65**, Ch. VI].

[44]The usual definition of Schubert polynomials is algebraic, making this definition a crucial result in the area, see [**BB93**]. Let us mention other combinatorial models of Schubert polynomials: *compatible sequences* [**BJS93**] and *bumpless pipe dreams* [**LLS21**]. See also [**GH23**] for the bijections between them.

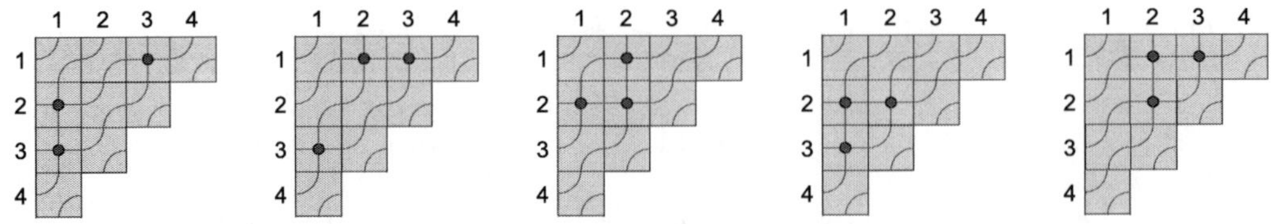

FIGURE 10.1. Graphs in RC(1432) corresponding to the monomials $x_1x_2x_3$, $x_1^2x_3$, $x_1x_2^2$, $x_2^2x_3$ and $x_1^2x_2$, in this order.

For *Grassmannian permutations* (permutations with one descent), Schubert polynomial coincide with Schur polynomials, so Schubert coefficients generalize LR coefficients. This is a starting point of a number of further generalizations, see [**Knu16**, **Kog01**, **MPP14**].

10.2. Schubert–Kostka numbers. For a permutation $w \in S_n$ and an integer vector $a \in \mathbb{N}^n$, the *Schubert–Kostka number* $K_{wa} := [\boldsymbol{x}^a]\,\mathfrak{S}_w$ is the coefficient of a monomial in the Schubert polynomial. By definition, $\{K_{wa}\} \in \#\mathsf{P}$.

PROPOSITION 10.2 (Morales[45]). *Schubert coefficients* $\{c_{uv}^w : u, v, w \in S_\infty\}$ *are in* GapP.

PROOF. Let $\sigma \in S_n$ and let $\rho_n := (n-1, \ldots, 1, 0) \in \mathbb{N}^n$. Define

$$\Omega(\sigma) := \big\{(a,b,c) \in (\mathbb{N}^n)^3 : a + b + c = \sigma\rho_n\big\}.$$

It was shown by Postnikov and Stanley in [**PS09**, Cor. 17.13], that

$$c_{uv}^w = \sum_{\sigma \in S_n} \ \sum_{(a,b,c)\in\Omega(\sigma)} \operatorname{sign}(\sigma) K_{ua} K_{vb} K_{wc}.$$

Separating positive and negative signs shows that $\{c_{uv}^w\} \in \#\mathsf{P} - \#\mathsf{P}$, as desired.[46] □

Following [**PS09**, §17], define the *Schubert–Kostka matrix* $\widehat{\boldsymbol{K}} = (K_{ua})$, which naturally generalizes the Kostka matrix $\boldsymbol{K} = (K_{\lambda\mu})$. Similarly, define the inverse $\widehat{\boldsymbol{K}}^{-1} = (K_{au}^{-1})$ which generalizes the inverse Kostka matrix $\boldsymbol{K}^{-1} = (K_{\lambda\mu}^{-1})$.

PROPOSITION 10.3. *The inverse Schubert–Kostka numbers* $\{K_{au}^{-1} : a \in \mathbb{N}^\infty, u \in S_\infty\}$ *are in* GapP.

Similarly to the previous proposition, the result follows directly from the identity in [**PS09**, Prop. 17.3]. One can ask if the absolute values $\{|K_{au}^{-1}|\}$ are in #P. Conjecture 7.10 implies that the answer is negative, but perhaps this more general problem is an easier place to start.

---

[45] Alejandro H. Morales, personal communication (2016).

[46] One can also use *Monk's rule* (see e.g. [**Gil19**]), to obtain the same result. We thank Alex Yong for this observation.

10.3. Has the problem been resolved? There are two issues around Conjecture 10.1 worth mentioning, as both, in different ways, suggest that the conjecture has already been resolved in the negative (i.e. a combinatorial interpretation has already been found).

First, Izzet Coskun in [**Cos+**] claimed to have completely resolved the problem of finding combinatorial interpretations for Schubert coefficients[47] using the technology of *Mondrian tableaux.*[48] Earlier, he used Mondrian tableaux to give a combinatorial interpretation for *step-two Schubert coefficients* (corresponding to permutations with at most two descents) in [**Cos09**] extending Vakil's earlier work [**Vak06**], see a discussion in [**CV09**].

Unfortunately, paper [**Cos+**] has not been peer reviewed and has been largely ignored by the community (see [**Bil21**] for a notable exception).[49] We should mention that the state of art recent work [**KZ17**] gives a tiling combinatorial interpretation for the *step-three Schubert coefficients.* It seems, we are nowhere close to resolving Conjecture 10.1 in full generality.

Second, Sara Billey suggested in [**Bil21**], that Schubert coefficients *already have* a "combinatorial interpretation", since by definition they are equal to the number of irreducible components in certain intersections of three Schubert varieties,[50] and thus "they already count something". Can one create a #P function out of this definition?

While it is true that Schubert coefficients count the number of certain points in $\mathbb{C}^n$, these points are not necessarily rational. In fact, they are usually roots of a large system of rational polynomials. On the other hand, Billey and Vakil prove in [**BV08**, §4], that there are some remarkable pathologies for these intersections related to realizability and stretchability of pseudoline arrangements. It follows from the *Mnëv universality theorem* [**Mnëv88**] (see also [**Shor91**]), that these problems are *$\exists\mathbb{R}$-complete.*[51]

The $\exists\mathbb{R}$ complexity class is in PSPACE and not expected to have polynomial size verifiers. This suggests that in the worst case, Billey's approach needs a superexponential precision with which one would want to compute the intersection points (i.e. the floating point computation needs superpolynomially many digits), implying that it is unlikely that there exists a poly-time verifier in this case.

The issue is not too different from that of Kronecker coefficients, which can be written as $g(\lambda,\mu,\nu) = \dim\left(\mathbb{S}^\lambda \otimes \mathbb{S}^\mu \otimes \mathbb{S}^\nu\right)^{S_n}$. Thus, one can argue that they *count basis vectors* in this space of invariants. But since the Kronecker coefficients can be exponential in n, see e.g. [**PPY19**], how would one present such a basis so it can be verified (or at least *read*) in poly-time?[52]

10.4. Elements of style. There are both cultural and mathematical reasons why the type of issues we discuss in §9.4 do not apply to the study of Schubert coefficients,[53] so let us quickly address the differences.

Schubert polynomials were originally introduced by Lascoux and Schützenberger in 1980s in the geometric context, see [**Las95**]. They remain deeply entrenched in the area

[47]This paper is undated, but cited already in [**CV09**].

[48]These are aptly named after a Dutch painter Piet Mondrian (1872–1944), who developed his signature style in the "tableau" series in 1920s, and did not live to see his work's influence in Schubert calculus.

[49]The author has maintained that the paper is "currently under revision". Meanwhile, no public statements have been made by experts in the field regarding the accuracy of the work or the provision of a counterexample. We find this state of affairs extremely unfortunate.

[50]Equivalently, this is the number of points in a generic intersection of three Schubert varieties.

[51]See e.g. [**Scha10**] for a computational complexity overview of the *existential theory of the reals* ($\exists\mathbb{R}$), and connections to Mnëv's theorem.

[52]We thank Greta Panova for suggesting this comparison (personal communication).

[53]For the generalized saturation problem in this case, see [**ARY13**] (see also [**Buch02**, §7]).

which offers a melange of tools nonexistent on the combinatorial and representation theoretic side. Arguably, this resulted in a deeper study with a long series of achievements, too long to outline in this survey (see e.g. [**Knu22**]).

By now, the Schubert theory outgrew the "combinatorial interpretation issue" as its raison d'être. Arguably, the area was always about developing the theory to understand the geometry of flag varieties from a combinatorial point of view (cf. [**LS85**]).

This brings us to the curious case of *puzzles*, a type of tilings on a triangular lattice with labeled edges (see e.g. [**Knu16**]).[54] Aesthetically pleasing for sure, is there some additional value such combinatorial interpretation bring to the study?

To us, the answer is a mixture. On the one hand, when the set of tiles is finite they clearly define a combinatorial interpretation.[55] Having a nice or highly symmetric set of tiles can be convenient to prove structural results about vanishing and other properties of the numbers, cf. [**KT99**].

On the other hand, there is nothing surprising in the labeled tiles from the computational point of view. By adding new labels, one can always break them into single triangles, which can then be viewed as a variation on *Wang tiles* [**Wang61**]. The latter can also be broken into triangles leading to the same model of computation. Wang's celebrated insight is that such tilings are just as powerful as Turing machines. In fact, the example of Wang tilings with the boundary was one of Levin's six original NP-complete problems [**Lev73**].

And yet, there is a clear computational difference between various types of increasing tableaux (such as *set-valued tableaux*) vs. puzzles. While the former need some memory to check which numbers appear, the latter can be verified by checking only local label conditions (think of *finite state automata*), making the puzzles (slightly) less powerful as a computational model.[56] Part of the weakness here comes from the fixed triangular region space requirement, further constraining the puzzle tiling model.[57]

11. The magic of the symmetric group

Before we can explain why, in our view, so many numbers in Algebraic Combinatorics are not in #P, we need to deconstruct the wonderful world of the symmetric group. Only after reducing all the great features to just one, we can fully appreciate its power as well as its limitation.

11.1. Pruning down the list. We start with representation theoretic point of view, demystifying some of the magic. All these results are routine and well known, so we restrict ourselves to quick pointers to the literature.

(a) $f^\lambda | n!$ This is non-specific to S_n, since the dimension $\chi(1)$ of an irreducible character of a finite group G always divides the order $|G|$, see e.g. [**Ser77**, §6.5].
(b) $\{f^\lambda\} \in$ #P. This is a consequence of $\chi^\lambda \downarrow^{S_n}_{S_{n-1}}$ having a *simple spectrum*, i.e. multiplicities at most one. Irreducibles in the restriction (given by the *branching*

---

[54] For other models and combinatorial interpretations in this context, see an overview in [**RYY22**, §1.3].

[55] Occasional infinite sets tiles can be reduced to a finite set by breaking them apart and adding colors.

[56] Note that giving a combinatorial interpretation in a weaker computational model is a *stronger* result.

[57] In [**GP14**], we exploited this model by constructing sets of Wang tiling which simulated several classical sequences such as the Catalan numbers. Our model uses a large number of square tiles, just like [**KZ17**] which uses 151 triangular pieces.

rule) correspond to removing a corner from Young diagram λ. Iterating the restrictions along the subgroup chain $S_n \supset S_{n-1} \supset \ldots \supset S_1$ implies that $f^\lambda = |\operatorname{SYT}(\lambda)|$, cf. [**OV96**].
(c) $\sum_{\lambda \vdash n}(f^\lambda)^2 = n!$ This is non-specific to S_n. *Burnside's identity* $\sum \chi(1)^2 = |G|$ holds for all finite groups, cf. Remark 8.2.
(d) $\chi^\lambda(\mu) \in \mathbb{Z}$. This follows from a property of conjugacy classes of S_n, that $\sigma^k \in [\mu]$ for every $\sigma \in [\mu]$ and $(k, \operatorname{ord}(\sigma)) = 1$, see e.g. [**Ser77**, §13.1]. See also §8.2 for references to row and column sums of character tables of general finite groups.
(e) $\{$*Irreps of* $S_n\} \longleftrightarrow \{$*Irreps of* $\operatorname{GL}(N)\}$. This is a direct consequence of the *Schur–Weyl duality*: $S_n \times \operatorname{GL}(N)$ act on $(\mathbb{C}^N)^{\otimes n}$ and the action has simple spectrum $\mathbb{S}^\lambda \otimes V_\lambda$ of tensor products of corresponding irreps, see e.g. [**FH99**, §6.2].
(f) $\{K_{\lambda\mu}\} \in \#\mathsf{P}$. This is a direct consequence of the *highest weight theory* for $\operatorname{GL}(N)$. Indeed, the induced product for S_n reps corresponds to the tensor product of $\operatorname{GL}(N)$ reps, via (*e*). Now use the definition $K_{\lambda\mu} = \langle \chi^\lambda, 1\uparrow_{S_{\mu_1}\times S_{\mu_2}\times\cdots}^{S_n} \rangle$.
(g) $\{f^\lambda\} \in \mathsf{FP}$. This extension of (*b*) is a consequence of dimensions $\dim(V_\lambda)$ of $\operatorname{GL}(N)$ irreps given by a product. The latter is computed as a ratio of two *Vandermonde determinants*. The product formula for $f^\lambda = \dim(\mathbb{S}^\lambda)$ follows by taking the limit, see e.g. [**Sta99**, §7.21].[58]

11.2. Hook-length formula. The results above are both fundamental in the area and have conceptual proofs explaining away some of the magic. The *hook-length formula* (HLF) (see §7.1) may also seem fundamental at first, until one realizes that all proofs are a byproduct of calculations highly specific to symmetric groups. Here is a quick overview of the proofs:

∘ Direct cancelation proof via the product formula in §11.1(*g*), see the original proof in [**FRT54**], see also [**Sta99**, §7.21].
∘ *NPS bijection* [**NPS97**], an intricate argument seemingly based on the *jeu-de-taquin*, but neither using its properties nor originally motivated by it (see e.g. [**Sag01**, §3.10]).
∘ *Hillman–Grassl bijection* [**HG76**], see also [**Kra99a**] (plus a limit argument, see e.g. [**Pak01**]), an elegant bijection that is equivalent to RSK, see [**Gan81**], [**PV10**, §6.5] and [**VW83**, §5].
∘ *Geometric bijection* (plus a limit argument) in [**Pak01**], obtained as a deconstruction of RSK via local transformations (cf. [**Hop14**]).
∘ *Lagrange interpolation* inductive arguments, see [**Ban08, GN04, Kir92, Ver92**]. These are elementary analytic proofs which require the most calculation.
∘ *GNW hook walk* [**GNW79**], a ingenuous argument which embeds the Lagrange interpolation inductive argument in probabilistic disguise as shown by Kerov [**Ker93**].
∘ Inductive proof via bijection of the *branching rule* in [**CKP11**], obtained as a combinatorial deconstruction of the GNW hook walk argument, cf. [**Zei84**].
∘ *Naruse HLF* [**MPP17, MPP18a**], an advanced generalization of the HLF which easily implies it (see also [**Kon20**]). All currently known proofs are technically involved.

[58] If you are unsurprised by this result, try answering if dimensions of $\operatorname{Sp}(2n, q)$ irreps are in FP. What about other simple groups of Lie type? Let me know what you figure out.

Neither of these proofs explains the HLF on a deeper level, since each of them either has a substantive computational part, or outsources the computation to RSK and its relatives.

11.3. Robinson–Schensted–Knuth correspondence. We argue that the *RSK* is the one true miracle in the area. Let us count some of the ways it emerges, in historical order:

- The (original) *Robinson–Schensted algorithm*, later extended by Knuth [**Knu70**].
- The *Burge correspondence* [**Bur74**] (see also [**Ful97**, §A.4.1]).
- The *Hillman–Grassl correspondence* [**HG76**] (see above).
- Schützenberger's *jeu-de-taquin* [**Schü77**] (see also [**Sag01**, §3.7]).
- Viennot's *geometric construction* [**Vie77**] (see also [**Sag01**, §3.6]).
- Quantum version of Désarménien's *straightening algorithm* [**Dés80**], see [**LT96**].
- Steinberg's *unipotent variety* approach [**Ste88**] (see also [**vL00**]).
- Fomin's *growth diagrams* approach [**Fom95**] (see also [**Sta99**, §7.13]).
- Benkart–Sottile–Stroomer *tableau switching* [**BSS96**].
- Our *geometric bijection* [**Pak01**] (see above).
- Lascoux's *double crystal graphs* version [**Las03**].
- The *octahedral map* [**KTW04, DK05b**] (cf. [**DK08, HK06, PV10, Spe07**]).

There are many more versions of RSK — these are just the ones we find most interesting. There are also numerous extensions, generalizations and applications both in and outside of the area. Let us now clarify some items on this list.

Some of the bijections above, such as the geometric construction by Viennot and the double crystal graphs by Lascoux, are restatements of RSK into a different language. Several others require a serious proof that they coincide with RSK, e.g. the Hillman–Grassl correspondence and the jeu-de-taquin. In a difficult case, Désarménien's straightening algorithm is similar but not equal to RSK, but the quantum versions do coincide.[59]

In [**PV10**], we set up a technology which allow one to prove that various maps in the area are in fact linear time reducible to and from the RSK. This allows us to put the *Schützenberger involution*, the tableau switching and the octahedral recurrence on the list.

Now is the time to ask a key question: *Given so many different approaches to so many different problems, why do they result in the same bijection?* There is a formal mathematical reason to explain coincidences between any two RSK appearances. The explanations could be algebraic or combinatorial, but neither would resolve a question given the multitude of instances.

The answer is simple and unambiguous. It does not really matter what is the deep reason behind the RSK, whether it is the *highest weight theory*, the straightening algorithm or something else. What matters is that each RSK appearance is a shadow of one fundamental result that is yet to be formalized. This suggests we treat RSK as *the one true miracle* which causes much of what we consider magical about the symmetric group.

---

[59] This similarity puzzled Gian-Carlo Rota, who asked for years to find a connection, see e.g. [**BT00**]. The mystery was eventually resolved by Leclerc and Thibon in [**LT96**], who noticed that sometimes the leading coefficient in the straightening is "incorrect" because of the cancellation which disappear when $q \neq 1$. We recall Rota's great joy upon learning of this discovery. See also [**GL21**] for another closely related connection.

11.4. LR rule. Consider the *Littlewood–Richardson coefficients* and its many combinatorial interpretations (see e.g. [**vL01**] for an extensive albeit dated survey):

- The *original LR rule*: $c^\lambda_{\mu\nu} = |\operatorname{LR}(\lambda/\mu,\nu)|$, see [**LR34**].
- The *LR variation*: $c^\lambda_{\mu\nu} = |\operatorname{LR}(\mu\circ\nu,\lambda)|$, see e.g. [**RW84**].
- James–Peel *pictures* [**JP79**], see also [**CS84, Zel81**].
- Gelfand–Zelevinsky interpretation using *Gelfand–Tsetlin patterns* [**GZ85**].
- Leaves of the *Lascoux–Schützenberger tree* [**LS85**].
- Kirillov–Reshetikhin *rigged configurations* [**KR88**] (see also [**KSS02**]).
- Berenstein–Zelevinsky *triangles* [**BZ92**].
- Fomin–Greene *good maps* [**FG93**].
- Nakashima's interpretation using *crystal graphs* [**Nak93**] (cf. [**BS17**, §9]).
- Littelmann's *paths* [**Lit94**].
- Knutson–Tao *hives* [**KT99**], see also [**GP00**].
- Kogan's interpretation using *RC-graphs* [**Kog01**].
- Buch's *set-valued tableaux* [**Buch02**].
- Knutson–Tao–Woodward *puzzles* [**KTW04**].
- Danilov–Koshevoy *arrays* [**DK05a**].
- Vakil's *chessgames* [**Vak06**].
- Thomas–Yong *S_3-symmetric LR rule* [**TY08**].
- Purbhoo's *mosaics* [**Pur08**] (see also [**Zin09**]).
- Coskun's *Mondrian tableaux* [**Cos09**].
- Nadeau's *fully packed loop configurations in a triangle* [**Nad13**] (cf. [**FN15**]).

The list above is so lengthy, it is worth examining carefully. Most of these LR rules are byproducts of (often but not always, successful) efforts to find a combinatorial interpretation of more general numbers. Some of these are closely related to each other, while others seem quite different, both visually and mathematically.

Now, on the surface the RSK is nowhere in the picture. We already mentioned [**Ker84, Whi81, Zel81**] which make the connection explicit, but here is a quick outline of how to get the original LR rule.

Start with a skew shape λ/μ, and run the jeu-de-taquin on $\operatorname{SYT}(\lambda/\mu)$. The number of times each $A \in \operatorname{SYT}(\nu)$ is the image $\operatorname{jdt}(\operatorname{SYT}(\lambda/\mu))$ is exactly $c^\lambda_{\mu\nu}$, for all A and ν.[60] Taking the lex-smallest such A (obtained by reading squares of ν left-to-right as commonly done), gives a combinatorial interpretation $c^\lambda_{\mu\nu} = \big|\{B \in \operatorname{SYT}(\lambda/\mu) : \operatorname{jdt}(B) = A\}\big|$.

While the above combinatorial interpretation suffices to show that $\{c^\lambda_{\mu\nu}\} \in$ #P, minor adjustments can be made to beautify the resulting rule. First, note that preimage of squares in every row of A cannot be in the same column of $B = \operatorname{jdt}^{-1}(A)$, so we can relabel A by placing i in the squares of i-th row. We get a unique $A_0 \in \operatorname{SSYT}(\nu,\nu)$. Now, the preimage in $\operatorname{LR}(\lambda/\mu,\nu) := \operatorname{jdt}^{-1}(A_0) \subseteq \operatorname{SSYT}(\lambda/\mu,\nu)$ can be described using the *ballot condition*,[61] giving the usual description of $\operatorname{LR}(\lambda/\mu,\nu)$.[62]

---

[60]Nothing in this claim is obvious: from the fact that jeu-de-taquin is well defined (independent on the order of moves), to the fact that preimage sizes are equinumerous and equal to $c^\lambda_{\mu\nu}$, see e.g. [**Sta99**, App. 1 to Ch. 7].

[61]The ballot condition is often called "lattice" or "Yamanouchi" depending on the context and how the tableau is being read. The differences between these are inconsequential.

[62]The LR variation can be obtained in the same way, by doing jeu-de-taquin to $\operatorname{SYT}(\mu\circ\nu)$ and looking at a preimage of $A \in \operatorname{SYT}(\lambda)$. The difference is that after relabeling, all tableaux

We emphasize that RSK is omnipresent in the LR study [**vL01**]. It helps to approach this historically (we will try to be brief). The LR rule was introduced in [**LR34**]. Soon after, Robinson introduced the first version of RSK in his (incomplete) effort to prove the original LR rule [**Rob38**]. As James describes in [**Jam87**], the LR rule "is much harder to prove than was at first suspected."[63]

Macdonald [**Mac95**, §I.9] credits Schützenberger [**Schü77**] and Thomas's thesis [**Tho74**] with first complete proofs. Both were obtained in the context of RSK and its relatives. Since then, many proofs of the LR rule were discovered, too many to be cited here, all related to RSK relatives (ibid.) We single out the proof in [**KTW04**] based on the associativity property given by the octahedral map, and a geometry inspired proof in [**BKT04**] based on the jeu-de-taquin. Note that both use properties of RSK relatives.[64]

Given that the role of RSK is often invisible without carefully examining the proofs of the LR interpretations above, this brings us to the following question: *Are these "combinatorial interpretations" of LR coefficients equivalent in some formal sense?*

11.5. Little boxes all the same.[65] There are several ways to formalize the question above. First, note that all "combinatorial interpretations" above are naturally in #P, with the exception of crystal graphs which can have exponential size;[66] the issue can be fixed if one follows the bijection in [**NS11**].

Second, all of these combinatorial interpretations are related to the original LR rule via a sequence of explicit poly-time bijections. For example, the LR tableaux are in bijection with: hives and BZ triangles [**PV05**], crystal graphs [**NS11**], Mondrian tableaux [**Liu10**], etc. This is unsurprising, perhaps, compared with parsimonious reductions between many #P-complete problems such as the number of 3-colorings and the number of Hamiltonian cycles, see e.g. [**Pap94b**, §18].

More surprising is that with few notable exceptions these bijections have linear time complexity. For example, when there is a natural presentation as integer points in polytopes, these polytopes are essentially the same and the bijection is given by a special linear map, see [**PV05**]. This holds for LR tableaux interpreted as GT patterns, as well as for hives, BZ triangles and DK arrays, where the natural presentation is binary, see [**DP16, PV05**].

It is thus most surprising, that RSK is behind so many other combinatorial interpretations, the bijection $\mathrm{LR}(\lambda/\mu,\nu) \to \mathrm{LR}(\mu\circ\nu,\lambda)$ being the most natural such example. Recall the (algebraically obvious) *fundamental symmetry* $c^\lambda_{\mu\nu} = c^\lambda_{\nu\mu}$, which is not transparent on the LR tableaux or the highly symmetric BZ triangles. Nor does it follow from integer points in polytopes of GT patterns since the GZ polytopes are asymmetric. In other words, proving $|\,\mathrm{LR}(\lambda/\mu,\nu)| = |\,\mathrm{LR}(\lambda/\nu,\mu)|$ bijectively is rather nontrivial, and indeed is linear time equivalent to computing the RSK map as proved in [**PV10**].

---

in $\mathrm{jdt}^{-1}(A_0) \subseteq \mathrm{SSYT}(\mu\circ\nu,\lambda)$ have the same filling of ν, which can then be omitted from the description. The result if the subset of $\mathrm{SSYT}(\mu,\lambda/\nu)$ with a ν-ballot condition.

[63]See also his famous "get men on the moon" sentence [**Jam87**, p. 117] (also quoted in [**Mac95**, p. 147]).

[64]Let us also mention the *involutive proofs* of variations on the LR rule: [**BZ88, Gas98, RW98, Ste02**]. As far as we can tell, these are essentially the same "verification type proofs" stated in different languages.

[65]See *Malvina Reynolds Sings the Truth*, Columbia Records, 1967, CS9414.

[66]Arguably, moving from the LR tableaux to crystal graphs trades conciseness for elegance, in roughly the same way as moving from standard Young tableaux to vectors in the *Young basis*. While crystal graphs can be inspirational and amenable to generalizations, ultimately all such results can be obtained in the language of LR tableaux (cf. [**Gal17**]).

Applying the fundamental symmetry can double the number of combinatorial interpretations, each time leading to a nontrivial bijections, see e.g. [**ACM09**] for puzzles and [**TKA18**] for hives.[67] It follows from [**PV10**], that the RSK is behind them all. One can argue that the S_3-symmetric LR rule already has the fundamental symmetry "embedded" into the rule. But the way the rule is constructed, to *verify* that the combinatorial interpretation is valid one needs to *perform* the RSK.

Let us mention an experiment we made in [**PV10**, §7], where we used ingredients from different bijections to cook up four(!) bijections proving the fundamental symmetry for LR tableaux. In the scientific method style, we conjectured all four to be identical without much of any checking [**PV10**, Conj. 1]. This conjecture is now completely proved by a combination of results in [**DK05a**] and [**TKA18**]. We followed the same approach to conjecture that two versions of the octahedral map coincide [**PV10**, Conj. 3]; this was later proved in [**HK06**].

Finally, we note a negative sort of evidence: as soon as one needs a different kind of combinatorial interpretation for the LR coefficients which does not involve the RSK, nothing emerges. This is why both Open Problems 7.8 and 7.9 are so challenging, cf. §9.2(‡).

11.6. What gives? Combinatorics of the symmetric group is so vast, it is easy to get lost. There are thousands of papers, hundreds of bijections, and dozens of combinatorial interpretations which we cannot possibly mention here for the lack of space and limited lifespan. And yet, we claim that there is a unifying principle for a large part of the field.

Fundamentally, the Algebraic Combinatorics is the study of bases in symmetric spaces via combinatorics of transition matrices between them. The dimensions of these symmetric spaces tends to be exponential (think S_n irreps, tensor powers, cohomology ring of Grassmannians, etc.) Thus, we need to be able to handle not only the exponential size bases, but also the exponential size of vectors in these bases.

In fact, most *natural bases* of these spaces do have vectors with exponential size support (think Young bases, Schur functions, Schubert polynomial, etc.) Fortunately, the components of the whole vector can often be computed from *name of the vector* (think of semistandard Young tableaux from the Young diagrams for Schur functions, or RC-graphs from permutations). The coefficients are not necessarily positive (e.g. tabloids in the Young basis of S_n irreps can have alternating signs), which is why given a choice it is best to use positive bases, e.g. work with Schur functions rather than Young bases.

Now, when applying *operations* to our symmetric spaces, one needs to be able to extract the standard symmetric bases out of these new large spaces. Since one cannot work with exponential size vectors, the symmetry must be traded for a concise presentation of the lex-smallest vector (in principle, any orbit representative of the underlying group of symmetries). For example, this easily leads to standard Young tableaux as lex-smallest vectors among tabloids of a given shape, see e.g. [**Sag01**, §1.6]. We refer to [**KR84**] for more on this philosophy from the Invariant Theory point of view.

To see a more interesting example, consider the left-right action of $G \times G$ on $\mathbb{C}[G]$, which has a simple spectrum as the sum of $\pi \otimes \pi$ over all irreducibles. For $G = S_n$ one is then tempted to look for how lex-smallest components of vectors in the invariant subspaces of $\mathbb{C}[S_n]$, leading to the *straightening* that Rota liked to much. The details behind the RSK correspondence and the LR rule are more technical, but the underlying story is similar and not much more complicated. And as we emphasize earlier, this all comes down to the miracle of RSK and its relatives.

What is amazing here is not that the resulting algorithm is nice and interrelated, but that it is correct. The former is a property of the underlying algebra, while the latter

[67] Another way to double the number of combinatorial interpretations is to use the *conjugation symmetry* $c^{\lambda}_{\mu\nu} = c^{\lambda'}_{\mu'\nu'}$. This symmetry is easy to prove algebraically. It was proved bijectively in [**HS92**], and the authors attributed to Dennis White a connection to the jeu-de-taquin. It makes sense only for the unary encoding.

is a combinatorial miracle behind RSK. We don't need to work hard to imagine a world where RSK does not exist – the (original) straightening is not in poly-time, or at least not obviously so (this is due to unexpected sign cancellations). If not for the Schur–Weyl duality we could be stuck there for a long time, at least until a quantum version was discovered.

For a more recent example, consider a interesting story of three papers by Thomas–Yong [**TY18**] and Pechenik–Yong [**PY17a**, **PY17b**], where the authors first introduce a generalization of jeu-de-taquin to obtain *some* combinatorial interpretation for the theory they wanted, and then developed *another* generalization of jeu-de-taquin to prove *the desired* combinatorial interpretation (conjectured earlier by Knutson and Vakil, see [**CV09**, §5]).

We finish this section on a positive note. If one *wants* to find a combinatorial interpretation for Kronecker coefficients, in my opinion one would need to find an appropriate generalization of RSK. It is even clear how to start — the map should be from 3-dim contingency arrays into triples of semistandard Young tableaux, so that the number of elements in preimage is always $g(\lambda, \mu, \nu)$, which would then have a combinatorial interpretation as the number of certain contingency arrays.

The idea would be to exploit the *generalized Cauchy identity*

$$\prod_{i,j,k} \frac{1}{1 - x_i y_j z_k} = \sum_{\lambda,\mu,\nu} g(\lambda, \mu, \nu) s_\lambda(\boldsymbol{x}) s_\mu(\boldsymbol{y}) s_\nu(\boldsymbol{z})$$

(see e.g. [**Mac95**, Exc. I.7.10] and [**Sta99**, Exc. 7.78]). Unfortunately, the structure of such arrays is more complicated than in the 2-dim case of contingency tables, so finding lex-smallest (under the action of the triple products of symmetric groups) does not appear to be feasible in full generality.[68] Thus, unsurprisingly, until now this approach is worked out only in a few special cases where lex-smallest contingency arrays are easy to distinguish, see e.g. [**Val00**] and [**IMW17**, §2].[69]

12. How to prove a negative?

By the title of this section we mean: *How to prove that a given function does not have a combinatorial interpretation?* Unfortunately, we really can't, at least not unconditionally. For all we know, it could be that $\mathsf{P} = \mathsf{NP} = \mathsf{PSPACE}$ and $\mathsf{FP} = \#\mathsf{P}$. In that case, poly-time verifiers can do magic, not just give combinatorial interpretations.[70] Now that we accepted the need for *some* complexity assumptions, we can proceed to discussing special cases.

12.1. Nothing comes from nothing. Let $f : \{0,1\}^* \to \mathbb{N}$ be a function. There is always a *mathematical reason* why we have $f(x) \in \mathbb{N}$ for all $x \in \{0,1\}^*$. This reason could be an easy consequence from the definition, an observation, a routine calculation, a standard result in the area, or a technically difficult theorem. Whether this reason can be replaced by a *counting argument* is exactly the same as asking if $f \in \#\mathsf{P}$.

Consider some examples. For $f(x) = 2^{|x|}$, the combinatorial interpretation is "all subsets of $[n]$," where $n = |x|$ is the length of the word x. For $f(x) = x(x-1)(x-2)/3$, the combinatorial interpretation is "all 3-subsets of $[x]$ counted twice". Here the (easy) number theoretic result "$3|m(m-1)(m-2)$ for all $m \in \mathbb{N}$" is proved by a counting argument.

---

[68] A closely related issue is well known in Algebraic Statistics, see e.g. [**Sul18**, §10].

[69] Although we personally don't expect this can be done in full generality (otherwise this would have been done by now), we believe in telling both sides of the story.

[70] See e.g. [**For97**] for a quick review of Counting Complexity.

In the other direction, for $f(x) = 2^x$, there is no combinatorial interpretation since the function is doubly exponential in the size of the input. There is still a (trivial) *counting argument* here, placing this function in a counting class #EXP, but that goes outside the scope of this survey.

Next, consider a polynomial $\varphi(x) = (x-1)^2$ and let $f \in$ #P. The function $\varphi(f)$ is trivially nonnegative and in GapP = #P − #P. However, there is no natural combinatorial interpretation in this case, as we already discussed in §3.2(4). There, we used a substitution[71] $x \leftarrow h(G) :=$ #Hamiltonian cycles in graph G. Note that if we used a different substitution $x \leftarrow e(P) :=$ #linear extensions of a poset P, then $(x-1)^2 \in$ #P, see §3.2(1). This mean that the inequality $(x-1)^2 \geq 0$ may be trivial analytically, but cannot be proved by a counting argument in the worst case even if it can be proved in special cases, as this example shows.

Back to the LR and Kronecker coefficients, there is a very clear algebraic reason why we have $c^{\lambda}_{\mu\nu} \geq 0$ and $g(\lambda, \mu, \nu) \geq 0$. Fundamentally, it is reduced to the (easy) inequality $\langle \zeta, \xi \rangle \geq 0$ for every two characters ζ, ξ of a finite group G. It seems unlikely that this general inequality would have a proof based on a counting argument.

Now note that we have a really good understanding why LR coefficients are in #P, and a very poor understanding why the Kronecker coefficients are not (thus, Conjecture 9.1). As we mentioned above, it is neither unusual nor surprising that an inequality can be in #P is some special cases and not in others. Unfortunately, at the moment, the Kronecker coefficients are much too unapproachable to admit a resolution of the problem.

In summary, we are concerned not so much with whether any particular function is in #P, although we do care about that. Instead, we examine what proof ingredients (of general results) can be shown to be in #P, and refute those which are not. This is the subject of our long and technical paper [**IP22**]. In this section we present a brief and non-technical introduction.

12.2. Polynomials. A rational polynomial $\varphi \in \mathbb{Q}[x]$ is called *integer-valued* if $\varphi(x) \in \mathbb{Z}$ for all $x \in \mathbb{Z}$. It is well known and easy to see that φ is integer-valued if and only if φ is an integer linear combination of binomials: $\varphi \in \mathbb{Z}\langle 1, x, \binom{x}{2}, \binom{x}{3}, \ldots \rangle$. See e.g. [**CC16**] for the background and many related results.

Polynomials φ in the semigroup $\mathbb{N}\langle 1, x, \binom{x}{2}, \binom{x}{3}, \ldots \rangle$ are called *binomial-good* (*binomial-bad*, otherwise). Binomial-good polynomials have a combinatorial interpretation for $\varphi(x)$. Formally, for every function $f \in$ #P, the function $\varphi(f)$ is also in #P. For example, let $f \leftarrow h(G)$ be the number of Hamiltonian cycles in G, and let $\varphi(x) = \frac{1}{3}x(x-1)(x-2)$. Then we have $\varphi(f) = 2\binom{f}{3} \in$ #P as discussed above, since $\varphi(f)$ counts twice the number of triples of distinct Hamiltonian cycles in G.

Let us emphasize that the combinatorial interpretation of $\varphi(f)$ is *oblivious*, i.e. it works with the verifier for the function f as a *black box* without ever looking at the graph G or the notion of Hamiltonicity. Note that the black box is not allowed to *compute* f, and all it can do is give a combinatorial interpretation for f. In other words, the black box verifier for f looks at $x \in \{0,1\}^*$ and says Yes/No in poly-time, and the verifier for $\varphi(f)$ works the same way by calling on the verifier for f.[72]

[71] We elaborate below on the meaning of this substitution.

[72] This notion is completely formal. In computational complexity terminology, this says that φ *relativizes* with respect to *oracle* f. We find the algorithmic notions more transparent.

Denote by $\mathcal{B}$ the set of binomially-good polynomials, and by $\mathcal{O}$ the set of polynomials which have an oblivious combinatorial interpretation. From above, $\mathcal{B} \subseteq \mathcal{O}$. It was proved in [**HVW95**, Thm 3.13] (see also [**IP22**, §4.3]), that $\mathcal{B} = \mathcal{O}$, i.e. we show that binomial-bad polynomials φ cannot have oblivious combinatorial interpretation.[73] For example, $\varphi = (x-1)^2 = 2\binom{x}{2} - x + 1$ is binomial-bad, and for $f \leftarrow h(G)$ we do not expect a combinatorial interpretation.

We also generalize this result to multivariate polynomials, where $\varphi \in \mathbb{Q}[x_1,\ldots,x_k]$ is integer-valued if $\varphi \in \mathbb{Z}\langle\binom{x_1}{d_1}\cdots\binom{x_k}{d_k}\rangle$, see [**Nag19**]. By analogy with the univariate case, we say that φ is *binomial-good* if $\varphi \in \mathbb{N}\langle\binom{x_1}{d_1}\cdots\binom{x_k}{d_k}\rangle$, and prove $\mathcal{B} = \mathcal{O}$ for this generalization [**IP22**, §4.3].

Denote by $\mathcal{C}$ the set of polynomials $\varphi \in \mathbb{Q}[x_1,\ldots,x_k]$ such that $\varphi(f_1,\ldots,f_k) \in$ #P for all $f_1,\ldots,f_k \in$ #P. Clearly, $\mathcal{O} \subseteq \mathcal{C}$. Our *Binomial Basis Conjecture* (BBC) states that $\mathcal{O} = \mathcal{C}$, see [**IP22**, §4.4]. Since this conjecture implies P $\neq$ NP (ibid.), there is little hope to resolve it even in the univariate case. Thus, from the computational complexity point of view, the *"combinatorial interpretation problem"* for polynomials is completely resolved.

EXAMPLE 12.1. Let $\varphi := (x-y)^2$. Clearly, $\varphi(x,y) = 2\binom{x}{2} + 2\binom{y}{2} - 2xy + x + y \notin \mathcal{B}$, which implies that $\varphi \notin \mathcal{C}$. Consider the classical geometric proof of $a^2 + b^2 \geq 2ab$ obtained by reflection of triangles as in Figure 12.1. Since reflected triangles cover both $a \times b$ rectangles, the inequality follows. The reader might want to ponder (before reading the footnote), why does this construction not give an oblivious combinatorial interpretation?[74]

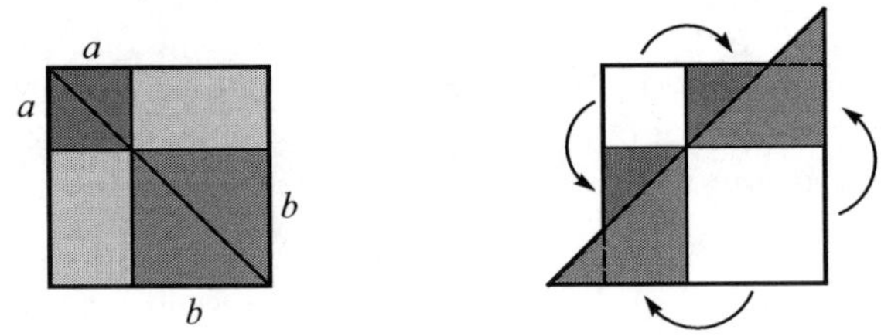

FIGURE 12.1. Geometric proof of the inequality $a^2 + b^2 \geq 2ab$. Here the blue and red triangles are reflected to completely cover yellow rectangles.

12.3. Making weaker assumptions. For some polynomials and some #P functions, the BBC can be replaced by a weaker complexity theoretic assumption. To explain what is going on, we need a few definitions.

Let $f, g : \{0,1\}^* \to \mathbb{Z}$ be two #P functions. We say that f has a *parsimonious reduction* to g, if there is a polynomial reduction $\{0,1\}^* \to \{0,1\}^*$ which maps f into g. Typically, if f and g count the number of solutions to problem A and B, the parsimonious reduction preserves the number of solutions.

[73] To repeat ourselves, we use "oblivious" to restrict combinatorial interpretations of $\varphi(f)$ to only those whose verifier ignores the nature of f. If you have never seen *oblivious algorithms* and this notion seem confusing, just think of "oblivious"="nice".

[74] The difference $a^2 + b^2 - 2ab$ is the area of two triangles on the sides of the square. To decide whether point (i,j) is in one of these triangles, you would need to determine the sign and absolute value of $(a-b)$, which oblivious algorithms are not allowed to do.

It is known, for example, that #SAT has a parsimonious reduction to #HAMILTONIANCYCLES (and vise versa). This means that for every SAT formula Φ, there is a polynomially constructed graph G, such that the number of satisfying assignments of Φ is equal to $h(G)$, see e.g. [**Pap94b**, §18]. The same holds for #3COLORINGS, #WANGTILINGS, and many other #P-complete functions. For the lack of a better name, we call such #P functions *counting-complete*, and use CCF to denote the set of such functions.[75]

On the other hand, unless P = NP, the function #PERFECTMATCHINGS is not in CCF, since the corresponding vanishing problem $\text{PM}(G) =^? 0$ is in P. The same holds for #LINEAREXTENSIONS. So while these two functions are #P-complete, they are not counting-complete.[76]

A map $\varphi : \mathbb{N}^k \to \mathbb{Q}$ is called *monotone* if $\varphi(a_1, \ldots, a_k) \le \varphi(a_1', \ldots, a_k')$ for all integer $a_1 \le a_1', \ldots, a_k \le a_k'$. For example, polynomials $x/2$, $x - 1$ and $x + y$ are monotone, but $x^2 - 2x$ and $(x - y)^2$ are not. Denote by $\mathcal{M}$ the set of monotone polynomials, and note that $\mathcal{B} \subset \mathcal{M}$.

We can now state two results which allow weakening of the "oblivious" assumption:

(1) Let $\varphi(x, y) = (x - y)^2 \notin \mathcal{O}$. We prove a stronger claim in this case, that $\varphi \notin \mathcal{C}$ unless a standard complexity assumption fails. Formally, we prove in [**IP22**, §2.3], that for every two (independent) functions $f, g \in \mathsf{CCF}$, we have $\varphi(f, g) \notin$ #P unless $\Sigma_2^{\mathsf{P}} = \mathsf{PH}$.
This is the strategy used in [**IPP22**] to prove Theorem 8.1. Recall that by MN rule, the character $\chi^\lambda(\mu) \in \mathsf{GapP}$, so it can be written as $(f - g)$ for some (usually, not independent) functions $f, g \in$ #P. We found some instances of (λ, μ), for which the corresponding (f, g) are both independent and counting-complete. This implied the result. We believe the same approach could potentially work for Conjecture 7.10.
(2) Let $\varphi \in \mathbb{Q}[x_1, \ldots, x_k]$ be a non-monotone polynomial, so $\varphi \notin \mathcal{B}$. We prove a stronger claim in this case, that $\varphi \notin \mathcal{C}$ unless a standard complexity assumption fails. Formally, let $f_1, \ldots, f_k \in \mathsf{CCF}$ be independent counting-complete #P functions. We prove in [**IP22**, §2.3], that $\varphi(f_1, \ldots, f_k) \notin$ #P unless UP = coUP, see [**IP22**, §2.4]. In particular, $(f - 1)^2 \notin$ #P unless UP = coUP, cf. §3.2(4).

For another example, recall the *Motzkin polynomial* $M(x, y) := x^2y^4 + x^4y^2 - 3x^2y^2 + 1$. It follows from the AM-GM inequality applied to positive terms, that $M(x, y) \ge 0$ for all $x, y \in \mathbb{R}$. On the other hand, this polynomial is famously not a *sum of squares*, and is a fundamental example in *Semidefinite Optimization*, see e.g. [**Ble13, Mar08**]. Now, observe that $M(x, y)$ is not monotone: $M(0, 1) = 1$ and $M(1, 1) = 0$. This gives $M \notin \mathcal{C}$ unless UP = coUP.

Note that non-monotone polynomial $(x - y)^2$ has a stronger property: $(x - y)^2\psi \notin \mathcal{M}$, for all $\psi \in \mathbb{Z}[x, y]$. By contrast, the multiple $(x-1)^2 \cdot x = 6\binom{x}{3} + 2\binom{x}{2} \in \mathcal{B}$ is binomial-good, and thus monotone.

12.4. Algebraic inequalities. Let $\varphi, \psi \in \mathbb{Q}[t_1, \ldots, t_k]$, so that $\varphi \le \psi$ for all $(t_1, \ldots, t_k) \in \mathbb{N}^k$. Suppose that polynomial $(\psi - \varphi) \in$ #P and has an oblivious combinatorial interpretation. From above, we have $(\psi - \varphi)$ is binomial-good. Since

---

[75] If you like the NSF, this name is both appropriate and rewarding.

[76] This is also why these two problems are more interesting, and the proof of their #P-completeness is more challenging as they cannot have a parsimonious reduction to #SAT.

inequalities are routine building blocks across mathematics, it is worth examining which of them are binomial good.

First, note that the inequality $a^2+b^2 \geq 2ab$ is equivalent to the complete square case (1) above. In [**IP22**, §7.1], we show that a number of standard inequalities are also not in #P (in a sense of substitutions), including the *Cauchy inequality*, the *Minkowski inequality*, and the *Alexandrov–Fenchel inequality*. All these proofs are routine and similar to our proof of Proposition 5.4.

Let us single out the *Hadamard inequality* for real $d \times d$ matrices:

$$\det \begin{pmatrix} a_{11} & \cdots & a_{1d} \\ \vdots & \ddots & \vdots \\ a_{d1} & \cdots & a_{dd} \end{pmatrix}^2 \leq \prod_{i=1}^{d} \left(a_{i1}^2 + \ldots + a_{id}^2\right).$$

Geometrically, it says that the volume of a parallelepiped in $\mathbb{R}^d$ is at most the product of its basis edge lengths, with equality when these edges are orthogonal. Note that the standard proof use eigenvalues, see e.g. [**HLP52**, §2.13] and [**BB61**, §2.11], suggesting that translation into combinatorial language would be difficult.

Denote by $H_d(a_{11}, \ldots, a_{dd})$ the polynomial defined by the Hadamard inequality. Note that we have $H_2(a,b,c,d) \in \mathcal{C}$, since

$$H_2(a,b,c,d) = (a^2+b^2)(c^2+d^2) - \det \begin{pmatrix} a & b \\ c & d \end{pmatrix}^2 = (ac+bd)^2.$$

On the other hand, $H_2(a,b,c,-d) \geq 0$ has different properties from our point of view. Indeed, we have $H_2(a,b,c,-d) = (ac-bd)^2 \notin \mathcal{C}$ unless $\Sigma_2^{\mathsf{P}} = \mathsf{PH}$. Finally, observe that

$$H_3 \begin{pmatrix} x & \binom{x}{3} & 0 \\ 0 & 1 & 1 \\ 1 & 0 & 1 \end{pmatrix} = 3\tbinom{x}{1} + 6\tbinom{x}{2} - 3\tbinom{x}{3} + 28\tbinom{x}{4} + 90\tbinom{x}{5} + 60\tbinom{x}{6} \notin \mathcal{B}.$$

By the BBC, we have $H_3 \notin \mathcal{C}$. Since $H_3 \in \mathcal{M}$, this is the best we can prove with our tools.

12.5. Algebraic inequalities restricted to semialgebraic sets. In many cases, one is interested in polynomial inequalities where the variables are themselves constrained by a system of polynomial inequalities and equations. Our *Diagonalization Theorem* [**IP22**, Thm 6.2.1] gives a complete algebraic characterization of such systems with an oblivious combinatorial interpretation.

The results in [**IP22**] are both technical to state, difficult to prove and hard to apply. Instead of presenting or even outlining them here, we discuss two examples (one easy and one difficult) which give a glimpse at our arguments.

(1) The *Ahlswede–Daykin* (AD) *inequality*, see e.g. [**AS16**, §6.1], for $n=1$ states:

$$\begin{cases} x_0 y_0 \leq u_0 v_0 \ , \ x_0 y_1 \leq u_0 v_1 \\ x_1 y_0 \leq u_0 v_1 \ , \ x_1 y_1 \leq u_1 v_1 \end{cases} \Longrightarrow \quad (x_0+x_1)(y_0+y_1) \leq (u_0+u_1)(v_0+v_1),$$

for all $x_i, y_i, u_i, v_i \geq 0$, where $i \in \{0,1\}$. We prove in [**IP22**, Prop. 2.5.1], that the AD inequality does not have an oblivious combinatorial interpretation. In other words, we prove that $\varphi \notin \mathcal{O}$, where $\varphi := (u_0+u_1)(v_0+v_1) - (x_0+x_1)(y_0+y_1)$ restricted to the above system of four inequalities. Below is the outline of the proof.

Following [**IP22**, §7.4], we rewrite the inequality in terms of #P functions as follows:

$$\begin{cases} \alpha_0\beta_0 + \xi_1 = \gamma_0\delta_0\,,\ \ \alpha_0\beta_1 + \xi_2 = \gamma_0\delta_1 \\ \alpha_1\beta_0 + \xi_3 = \gamma_0\delta_1\,,\ \ \alpha_1\beta_1 + \xi_4 = \gamma_1\delta_1 \end{cases} \Longrightarrow\ (\gamma_0+\gamma_1)(\delta_0+\delta_1)-(\alpha_0+\alpha_1)(\beta_0+\beta_1) \in \#\mathsf{P},$$

for all $(\alpha_0, \alpha_1, \beta_0, \beta_1, \gamma_0, \gamma_1, \delta_0, \delta_1, \xi_1, \xi_2, \xi_3, \xi_4) \in (\#\mathsf{P})^{12}$. Second, make the following substitution into these 12 functions: $\bigl(1, 1, x, x, x, 1, 1, x, 0, 2\binom{x}{2}, 2\binom{x}{2}, 0\bigr)$ and check that the above system of inequalities is satisfied. Now let
$x \leftarrow f$, where $f \in \#\mathsf{P}$. This substitution gives $\varphi(f) = (f-1)^2 \notin \mathcal{B}$, and thus $\varphi \notin \mathcal{O}$.
(2) Let $\boldsymbol{a}, \boldsymbol{b} \in \mathbb{R}^n$ be weakly decreasing, such that $\boldsymbol{a} \trianglerighteq \boldsymbol{b}$, see §7.2. The *Karamata inequality* says for every *convex function* $f : \mathbb{R} \to \mathbb{R}$, we have

$$f(a_1) + \ldots + f(a_n) \ \geq\ f(b_1) + \ldots + f(b_n),$$

see e.g. [**HLP52**, §3.17] and [**BB61**, §28, §30]. We refer to [**BP21**, **PPS20**] for some recent applications to linear extensions and Young tableaux, and to [**MOA11**] for numerous generalizations and further references.
Following [**IP22**, §7.5], we rewrite the inequality in the language of #P functions. Suppose

$$f_1 + \ldots + f_i = g_1 + \ldots + g_i + h_i \quad \text{for all} \quad 1 \leq i \leq n,$$

where $f_i, g_i, h_i \in \#\mathsf{P}$, and $h_n = 0$. Suppose also that

$$f_i = f_{i+1} + d_i, \qquad g_i = g_{i+1} + e_i, \qquad \text{for all} \quad 1 \leq i < n,$$

where $d_i, e_i \in \#\mathsf{P}$. Finally, let $\gamma : \mathbb{Z} \to \mathbb{N}$ be a nonnegative convex function. The *Karamata function* K_γ^n is defined as

$$\mathrm{K}_\gamma^{\langle n\rangle}(f_1, \ldots, f_n, g_1, \ldots, g_n) := \sum_{i=1}^n \gamma(f_i) \ -\ \sum_{i=1}^n \gamma(g_i).$$

Clearly, $\mathrm{K}_\gamma^{\langle n\rangle} \in \mathsf{GapP}$ and nonnegative by the Karamata theorem. It is thus an interesting question if $\mathrm{K}_\gamma^{\langle n\rangle} \in \mathcal{O}$, i.e. if $\mathrm{K}_\gamma^{\langle n\rangle}(f_1, \ldots, g_n) \in \mathcal{O}$, i.e. in #P for all $f_i, g_i, h_i, d_i, e_i \in \#\mathsf{P}$.
For example, for $\gamma(t) = \alpha t + \beta$, we have $\mathrm{K}_\gamma^{\langle n\rangle} = 0$. Similarly,

$$\mathrm{K}_\gamma^{\langle 2\rangle}(f_1, f_2, g_1, g_2) \ =\ (e_1 + h_1)h_1 \in \#\mathsf{P} \quad \text{for} \ \ \gamma(t) = \tbinom{t}{2}, \quad \text{and}$$
$$\mathrm{K}_\gamma^{\langle 3\rangle}(f_1, f_2, f_3, g_1, g_2, g_3) \ =\ (d_1 + e_1)h_1 + (d_2 + e_2)h_2 \in \#\mathsf{P} \quad \text{for} \ \ \gamma(t) = t^2.$$

It follows from here that $2\mathrm{K}_\gamma^{\langle 3\rangle} \in \mathcal{O}$ where $\gamma(t) = \binom{t}{2}$, since linear terms cancel out. We prove in [**IP22**, Prop. 7.5.5], that $\mathrm{K}_\gamma^{\langle 3\rangle} \notin \mathcal{O}$ for $\gamma(t) = \binom{t}{2}$. The proof requires a computation of lattice points in a 12-dimensional polytope defined by linear equations and inequalities corresponding to constraints on f_i, g_i, h_i, d_i, e_i.

12.6. What's next? By now the (exhausted) reader knows what kind of results we want to prove — the many "not in #P" conjectures throughout the paper. And they know how we imagine the plan of attack — by simulating the proofs of positivity and integrality of these combinatorial functions with polynomial equations.

There are two main obstacles on the way. First, the proof of positivity and integrality can be rather involved, so distilling a single reason and expressing it as a polynomial inequality can be difficult.[77] Naturally, one would want to start with counting graphs and linear extensions rather than Kronecker and Schubert coefficients, as the former seem much more generic and less involved. It may take a long time before this project can reach the latter.

Second, the family of polynomials for which we know that they don't have an oblivious combinatorial interpretations is rather large and seems satisfactory for applications. But to make the final results more accessible and convincing to the general audience, it is important to weaken the assumptions (see §12.3). This direction is certainly worth exploring in the context of Computational Complexity.

13. Counting complexity addendum

13.1. We need a list. Throughout the whole survey we tried to mention #P-*completeness* and #P-*hardness* as little as possible. There are two reasons for that: we wanted not to distract the reader from the main problem (membership in #P), and we wanted to minimize the confusion that invariably arises.

There is, however, a direct connection to these complexity classes. As we mention in §2.6, we have FP $\subseteq$ #P, which makes combinatorial interpretation of problems in FP trivial. Naturally, we are thus interested in problems that are not in FP. The best evidence that a function is not in FP yet potentially in #P, is if a function is #P-hard. This is why it is worth checking #P-hardness of functions in every conjecture and open problem that we pose (cf. §15.2).

If this was about NP-completeness, we would stop here and refer to [**GJ79**] along with some recent comprehensive list of NP-complete problems (such as this one on `Wikipedia`). Unfortunately, there does not seem to be such comprehensive sources about #P-completeness. Thus we present an annotated short list of such results, restricted only to functions which we consider relevant to the survey.

13.2. Graph theory problems. We proceed roughly according to the sections in this survey.

(1) #MONOTONE 2SAT is #P-complete but $\notin$ CCF. This implies that #VERTEX-COVER is #P-complete [**Val79a**].

(2) #HAMILTONIANCYCLES is #P-complete [**Val79a**]. Moreover #HC is in CCF, since the proof is based on a parsimonious bijection. In the context of Smiths's theorem (see §3.2), this remains true for #HC in cubic planar graphs [**LOT03**].

(3) #3COLORINGS is #P-complete [**Val79a**]. Moreover #3C is in CCF, since the proof is based on a parsimonious bijection.

(4) #PERFECTMATCHINGS is #P-complete via reduction to PERMANENT [**Val79c**]. This implies that #k-MATCHINGS $\{p(G,k)\}$ is #P-complete (see §5.1). Thus, $\{f(G,k)\}$ is #P-hard by telescoping. The problem remains #P-complete for subgraphs in $\mathbb{Z}^3$ [**Val79b**], and even for the number of 3-dim domino tilings [**PY13**]. For planar graphs #PERFECTMATHINGS is in FP by the *Kasteleyn formula*, see e.g. [**LP86**, §8.3]. The *vanishing problem* $\{\mathrm{PM}(G) =^? 0\}$ is in P, see e.g. [**LP86**, §9.1].

[77] It is now well understood how to translate general mathematical proofs in a formal language of low degree polynomials which can then be "checked" with few queries, see [**A+98**]. The connection is somewhat superficial as the latter uses polynomials over finite field, while in [**IP22**] we work over $\mathbb{C}$. Still, this suggests commonality of the ideas, keeping alive the hope that such translation can be made in special cases.

(5) #SPANNINGFORESTS $\{F(G,k)\}$ is #P-complete (see §5.3). This holds because the total number of spanning forests $1+F(G,1)+\ldots+F(G,n-1) = T(G;2,1)$ is an evaluation of the *Tutte polynomial* known to be #P-complete, see e.g. [**Wel93**, Thm 6.2.9]. This implies that $\{f(G,k)\}$ is #P-hard. For a fixed $k \geq 1$, both $\{F(G,k)\}$ and $\{F(G,n-k)\}$ are in FP [**Myr92**].

(6) #SPANNINGSUBGRAPHS is #P-complete [**PB83**]. This is an evaluation of the Tutte polynomial (see e.g. [**Bol98**, Ch. X]). We conjecture that the function in Conjecture 5.6 is #P-hard.

(7) ISINGMODELSTATISTICALSUM is #P-complete [**JS93**, Thm 15]. In notation of §5.7, this and Proposition 5.9 implies that $\{\mathrm{Cor}(v,w)\}$ is #P-complete. For planar graphs, the problem is in FP by the *Kasteleyn–Fisher determinant formula* [**Kas63, Fis66**].

(8) For general rational hyperplane arrangements, the problem of counting the number of regions in the complement is #P-complete. For membership in #P, see §4.1. The hardness follows from graphical arrangements where the number of regions equal to the evaluation $|\chi_G(-1)|$ of the *chromatic polynomial* (see e.g. [**Sta99**, §3.11, Exc. 94-95]). This evaluation is equal to the *number of acyclic orientations* of G, known to be #P-complete, see e.g. [**Wel93**, Thm 6.2.9].

(9) #LINEAREXTENSIONS $\{e(P)\}$ are #P-complete [**BW91**]. Thus, the function defined by the Björner–Wachs inequality is #P-complete (see §6.1). By telescoping, the function defined by the Stanley inequality is #P-hard (see §6.3). Computing $e(P)$ remains #P-complete for P of height two, and of width two [**DP18**]. The function defined by the Sidorenko inequality (see §6.2) is conjectured to be #P-complete in [**CPP23b**, §9.6].

(10) #WANGTILINGS of a square is #P-complete; this follows e.g. from the proof of Thm 3 in [**DD07**] that the decision problem in NP-complete.

(11) VOLUME is #P-hard via reduction to #LINEAREXTENSIONS of *order polytope* [**BW91**], remains #P-hard for *zonotopes* [**DGH98**]. MIXEDVOLUME is #P-hard for boxes via reduction to PERMANENT (ibid.)

13.3. Algebraic combinatorics problems with binary input. The type of input makes so much difference for problems in Algebraic Combinatorics, we decided to separate them altogether and make a clear indication in the name, so they would be impossible to confuse.

Note that if the problem is in FP in the binary input, then this is also true in the unary input. Vice versa, if the problem is #P-complete or #P-hard in the unary input, then this is also true in the binary input. Similar claims hold for the decision problems as well.

(1) CONTINGENCYTABLESBINARYINPUT $\{\mathrm{T}(\lambda,\mu)\}$ is #P-complete even for $\ell(\lambda)=2$ [**DKM97**]. The *vanishing problem* $\{\mathrm{T}(\lambda,\mu) >^? 0\} \in \mathsf{P}$, since it is equivalent to $\{|\lambda| =^? |\mu|\}$. When $\ell(\lambda)$ is fixed, the problem has FPTAS [**G+11**]. When $\ell(\lambda), \ell(\mu)$ are fixed, the problem is in FP by [**Bar93**].

(2) KOSTKABINARYINPUT $\{K_{\lambda\mu}\}$ is #P-complete [**Nar06**]. The *vanishing problem* $\{K_{\lambda\mu} >^? 0\} \in \mathsf{P}$, since it is equivalent to $\{\lambda \trianglerighteq^? \mu\}$, see §7.2. The *uniqueness problem* $\{K_{\lambda\mu} =^? 1\} \in \mathsf{P}$, see a complete characterization in [**BZ90**].

(3) LITTLEWOODRICHARDSONBINARYINPUT $\{c^\lambda_{\mu\nu}\}$ is #P-complete [**Nar06**]. The *vanishing problem* $\{c^\lambda_{\mu\nu} >^? 0\} \in \mathsf{P}$, see [**DM06, MNS12**].[78]

[78]This is based on the *saturation property*, see §9.4(2), which fails for other root systems. Notably, for the B–C–D Lie types, it holds up to a factor of two [**Sam12**]. It is open whether the vanishing problem is in P in these cases; this would follow from [**DM06**, Conj. 4.7] (cf. [**GOY21**, §5.2] and [**RYY22**, §7.4]).

The *uniqueness problem* $\{c^\lambda_{\mu\nu} =^? 1\} \in \mathsf{P}$. More generally, $\{c^\lambda_{\mu\nu} =^? t\} \in \mathsf{P}$ for every fixed $t \geq 0$, see [**Ike12**, Thm 11.3.2] and [**Ike16**].

(4) KroneckerBinaryInput $g(\lambda, \mu, \nu)$ is #P-hard. This follows easily from the result that Littlewood-RichardsonBinaryInput is #P-complete, and trivially from the unary case. The *vanishing problem* $\{g(\lambda, \mu, \nu) >^? 0\}$ is NP-hard; this follows trivially from the unary case.

(5) ReducedKroneckerBinaryInput $\{\overline{g}(\alpha, \beta, \gamma)\}$ is #P-hard, same reason as above. We conjecture that the *vanishing problem* $\{\overline{g}(\alpha, \beta, \gamma) >^? 0\}$ is NP-hard.

(6) ExcitedDiagrams $\{|\mathcal{E}(\lambda/\mu)|\} \in \mathsf{FP}$ via reduction to *flagged tableaux* (see [**MPP18a**, Cor. 3.7]). This is the number of terms of the summation in the NHLF. The *vanishing problem* is trivial.

(7) CharacterSquaredBinaryInput $\{(\chi^\lambda(\mu))^2\}$ is doubly exponential and thus not in PSPACE. For example, $\chi^\lambda(1) = \mathrm{Cat}(m) = e^{\Omega(n)}$, where $\lambda = (m, m)$ and $n = 2m$. The *vanishing problem* $\{\chi^\lambda(\mu) \neq^? 0\}$ is NP-hard [**PP17**, §7]; a stronger result follows from the unary case.

(8) HurwitzBinaryInput $\{h_{g\mu}\}$ is doubly exponential and thus not in PSPACE. For example, $h_{0(n)} = n^{n-2}$.

13.4. Algebraic combinatorics problems with unary input. This is the most interesting case, and the one we discuss throughout the paper.[79] To minimize the overlap, we don't include here some of the poly-time result which hold already for the binary case.

(1) ContingencyTablesUnaryInput $\{\mathrm{T}(\lambda, \mu)\}$ is not known to be #P-complete, see [**DO04**, §1.1] and [**PP17**, §8.1]. We conjecture this to be true. When $\ell(\lambda)$ is fixed the problem is not #P-complete unless P = NP.[80]

(2) KostkaUnaryInput $\{K_{\lambda\mu}\}$ is not known to be #P-complete [**PP17**, §8.1]. We conjecture this to be true. This would follow from the conjecture that #ContingencyTablesUnaryInput is #P-complete, via the reduction in [**Nar06**]. Moreover, we conjecture that $\{K_{\lambda(2^a 1^b)}\}$ is #P-complete.

(3) LittlewoodRichardsonUnaryInput $\{c^\lambda_{\mu\nu}\}$ is conjectured to be #P-complete [**PP17**, §8.1]. This would follow from the conjecture that KostkaUnaryInput is #P-complete, via the reduction in [**Nar06**]. The *vanishing problem* has a *recursive description*, see [**Zel99**, Prop. 9]. The *vanishing* of *generalized LR coefficients* (tileability using Knutson–Tao puzzles regions in the triangular lattice with given boundary), is NP-complete [**PY14**].

(4) KroneckerUnaryInput $g(\lambda, \mu, \nu)$ is #P-hard. This follows from [**IMW17**]. The *vanishing problem* $\{g(\lambda, \mu, \nu) >^? 0\}$ is NP-hard, ibid. We conjecture that $\{g(\lambda, \lambda, \lambda) : \lambda = \lambda'\}$ is #P-hard, cf. Conjecture 9.2 and [**PP23**], and that $\{g(\lambda, \lambda, \mu) : \mu = (n-k, k)\}$ is also #P-hard, cf. Remark 9.3 and [**PP14**].[81]

(5) ReducedKroneckerUnaryInput $\overline{g}(\alpha, \beta, \gamma)$ is #P-hard [**PP20b**]. The *vanishing problem* $\{\overline{g}(\alpha, \beta, \gamma) >^? 0\}$ is conjectured to be NP-hard in [**PP20b**, §4.4].

(6) CharacterSquaredUnaryInput $\{(\chi^\lambda(\mu))^2\}$ is #P-hard [**Hep94**], and $(\chi^\lambda(\mu))^2 \notin$ #P unless $\Sigma_2^{\mathsf{P}} =$ PH, see Theorem 8.1. The *vanishing problem* $\{\chi^\lambda(\mu) =^? 0\}$ is $\mathsf{C_{=}P}$-complete, and thus NP-hard [**IPP22**]. The *positivity problem* $\{\chi^\lambda(\mu) \geq^? 0\}$ is PP-complete, and thus PH-hard, ibid.

[79] To indicate unary input, the literature often refers to "strong" NP- and #P-completeness, see [**GJ78**] and [**Vaz01**, §8]. We find this terminology misleading and best to be avoided, as some results become weaker while others stronger when the input size changes.

[80] This follows from having FPTAS in the binary input, see e.g. [**Vaz01**, §8.3].

[81] The last conjecture was suggested by Greta Panova (personal communication, Sep. 2022).

(7) InverseKostkaUnaryInput $\{K^{-1}_{\lambda\mu}\}$ has not been studied. We conjecture it is #P-hard. For the *vanishing problem* $\{K^{-1}_{\lambda\mu} =^? 0\}$, we conjecture it is $\mathsf{C_{=}P}$-complete (cf. Conjecture 7.10).

(8) BruhatOrderIdeal $\{B(\sigma)\}$, where $B(\sigma) := |\{\omega \in S_n : \omega \preccurlyeq \sigma\}|$, is #P-complete [**DP18**]. This problem is equivalent to #LinearExtensions for permutation posets: $B(\sigma) = e(P_\sigma)$.

(9) ReducedFactorizations $\{r(w) : w \in S_n\}$ [**Sta84**], are defined as

$$r(w) := \#\{(i_1, \ldots, i_\ell) \, : \, (i_1, i_1 + 1) \cdots (i_\ell, i_\ell + 1) = w, 1 \le i_j < n, \ell = \mathrm{inv}(\omega)\}.$$

The problem is conjectured to be #P-complete [**DP18**, §8.5], cf. [**MPY22**, §6].[82] When $w \in S_n$ is *vexillary* (2143-avoiding), we have $\{r(w)\} \in \mathsf{FP}$ from [**Man01**, Cor. 2.8.2] and the HLF.

(10) SchubertCoefficient c^w_{uv} is not known to be #P-hard.[83] We conjecture this to be true. The *vanishing problem* $\{c^w_{uv} >^? 0\}$ is not known to be NP-hard [**ARY19**, §4].

(11) SchubertKostka $\{K_{ua}\}$ and #RC-graphs $\{\mathfrak{S}_u(1) = \sum_a K_{ua}\}$ have not been studied. We conjecture that both are #P-complete.

(12) IncreasingTableaux $\{g^\lambda := |\,\mathrm{IT}(\lambda)|\}$, where an *increasing tableau* $A \in \mathrm{IT}(\lambda)$ is a plane partition strictly increasing in rows and columns, and no gaps between 1 and maximal entry of A, see [**TY09**]. Note that $g^\lambda \ge f^\lambda$. This problem $\{g^\lambda\}$ was proposed in [**TY11**, §1.3], and conjectured to be #P-complete in [**MPP22**, §7.10].

(13) SetValuedTableaux $\{s(\lambda, k) := |\,\mathrm{SVT}(\lambda, k)|\}$, where a *set-valued tableau* $A \in \mathrm{SVT}(\lambda, k)$ is a surjection from $[k]$ to squares of Young diagram λ, s.t. the numbers increase in rows and columns of the image, see [**Buch02**]. Complexity of $\{s(\lambda, k)\}$ was asked in [**MPY22**, §5.2] and [**H+22**, §5.7]. We conjecture that $\{s(\lambda, k)\}$ is #P-complete.

(14) HurwitzNumberUnaryInput $\{h_{g\mu}\}$ is #P-complete [**PP24+**].

13.5. Related problems. Here are a few additional problems we think are worth solving (all in unary).

(1) Let $\Lambda \subset \mathbb{N}^3$ be a *3-dim Young diagram*. Denote by P_Λ the corresponding poset. We conjecture that $\{e(P_\Lambda)\}$ is #P-complete. In fact, we conjecture this holds for Λ of height two.

(2) Let $\Lambda \subset \mathbb{N}^3$ be a *3-dim Young diagram*. We conjecture that the number of domino tilings of Λ is #P-complete. Again, we conjecture this holds for Λ of height two.

(3) Let $\mathcal{A} = (A_1, \ldots, A_n)$ and $\mathcal{B} = (B_1, \ldots, B_n)$ be two collections of points in $\mathbb{N}^3$. Denote by $f(\mathcal{A}, \mathcal{B})$ the number of collections $(\gamma_1, \ldots, \gamma_n)$ of nonintersecting shortest paths $\gamma_i : A_i \to B_i$, $1 \le i \le n$. We conjecture that $\{f(\mathcal{A}, \mathcal{B})\}$ is #P-complete.

(4) Let $\mathrm{F}(d_1, \ldots, d_n)$ be the number of simple graphs with given degree sequence. We conjecture that $\{\mathrm{F}(d_1, \ldots, d_n)\}$ is #P-complete. For many related results and further references, see e.g. [**Wor18**]. The *vanishing problem* $\{\mathrm{F}(d_1, \ldots, d_n) >^? 0\}$ is in P even in binary, by the *Erdős–Gallai theorem* (see e.g. [**AL05**]).

---

[82] [**MPY22**, §6] observes that $\{r(w)\} \notin$ #P when permutations are presented in binary via the *Lehmer code*.

[83] The argument in [**MQ17**, p. 885] claiming that $\{c^w_{uv}\}$ is #P-complete via reduction to $\{c^\lambda_{\mu\nu}\}$ is erroneous as it conflates the input sizes. The authors acknowledge the mistake (personal communication, 2022).

(5) Let $G \subset S_n$ be a permutation group given by its generators. Recall that the size $|G|$, the size of the commutator $|[G,G]|$ and many other functions are in FP, see e.g. [**Ser03**]. What about the number $c(G)$ of *conjugacy classes*?

(6) We conjecture that tileability of simply-connected regions in $\mathbb{R}^2$ with the unit square and the unit edge equilateral triangle is NP-complete (rotations are allowed, see [**Zin09**, App. 1]), and that the number of such tilings is #P-complete.

(7) Let $Q \subset \mathbb{R}^2$ be a centrally-symmetric polygon with integer side lengths, and let T is a fixed set of rhombi tiles with unit sides. We conjecture that the number of tilings of Q with T is #P-complete.[84] Note that the existence of such tilings is in P, see e.g. [**KS92**]. We refer to [**Ken93**, §5] and [**Pak03**, §5.3, §7] for some background.

(8) We conjecture that for *binary matroids* represented by vectors in $\mathbb{F}_2^n$, the number of bases is #P-complete.[85] Note that for *paving matroids* represented by cycles, and for bicircular matroids, the number of bases is #P-complete, see [**Jer06**, §3] and [**GN06**, §3].

14. Proofs

14.1. Proof of Proposition 4.2. Let $\mathcal{T}_n$ be the set of *rooted plane triangulations* on n vertices, and let $b_n := |\mathcal{T}_n|$. Here the root in a triangulation $G = (V, E)$ is a flag (v, e, F), where $v \in V$, $e = (v, w) \in E$ and F is a face in G containing e. *Tutte's product formula* shows that $\{b_n\}$ can be computed in poly(n) time, see e.g. [**Scha15**]. Moreover, Poulalhon and Schaeffer [**PS06**] gave a bijective proof of Tutte's formula, by constructing a bijection $\Phi : \mathcal{T}_n \to \mathcal{B}_n$ between rooted plane triangulations on n vertices and certain *balanced plane trees* with two additional markings. It follows from the construction that both Φ and Φ^{-1} can be computed in $O(n)$ time.

Let $\Gamma = \mathrm{Aut}(G)$ be the group of automorphisms of graph G. It is easy to see that the stabilizer subgroup $\mathrm{Stab}_G(v, e, F) = 1$, because rooted triangulations have a unique topological embedding into a sphere. In particular, this implies that $|\Gamma| \le 4|E| = O(n)$. Recall that for all planar graphs, the effective graph isomorphism can be done in linear time [**HW74**]. In summary, for each $e' = (v', w') \in E$, it can be decided in poly(n) time whether there is an automorphism $g \in \Gamma$ s.t. $g \cdot v = v'$ and $g \cdot w = w'$. Moreover, when it exists such g can be computed explicitly.

Now, a combinatorial interpretation for $\{a_n\}$ can be constructed as follows. Let $t \in \mathcal{B}_n$. Use the *Depth First Search* (DFS) around t to obtain a unique labeling of vertices of t. Use Φ^{-1} to transferred this labeling onto $\tau := \Phi^{-1}(t) \in \mathcal{T}_n$. Use the argument above to compute all $O(n)$ relabelings τ' of τ. For all such τ', check if they are isomorphic to τ, and if not discard such τ'. Compute $\Phi(\tau')$ for all such rooted triangulations. We obtain exactly $|\Gamma \cdot \tau| = O(n^2)$ balanced plane trees. Accept t if it is lex-smallest, and reject otherwise. The details are straightforward. □

14.2. Proof of Proposition 8.6. Recall that $\rho^{(n)} = 1\uparrow_{C_n}^{S_n}$, where $C_n \simeq \mathbb{Z}_n$ is the usual cyclic subgroup of S_n. Let $\mu = (\mu_1, \ldots, \mu_\ell) \vdash n$ be a partition into distinct parts: $\mu_1 > \ldots > \mu_\ell > 0$. We have:

$$\begin{aligned}\rho^\mu &= 1\uparrow_{C_{\mu_1}\times\cdots\times C_{\mu_\ell}}^{S_n} = \rho^{(\mu_1)} \otimes \cdots \otimes \rho^{(\mu_\ell)} \uparrow_{S_{\mu_1}\times\cdots\times S_{\mu_\ell}}^{S_n} \\ &= \sum_{\lambda \vdash n} \left[\sum_{\nu^{(1)} \vdash \mu_1} \cdots \sum_{\nu^{(\ell)} \vdash \mu_\ell} c^\lambda_{\nu^{(1)},\ldots,\nu^{(\ell)}}\, a_{\nu^{(1)}(\mu_1)} \cdots a_{\nu^{(\ell)}(\mu_\ell)} \right] \chi^\lambda,\end{aligned}$$

[84]For the connection to reduced factorizations, see [**Eln97**].

[85]Despite claims in [**Ver98**], this problem is unresolved since Vertigan's proof remains unwritten, and even the proof idea is unavailable (Dirk Vertigan, personal communication, April 2010).

where the *generalized LR coefficient*

$$c^{\lambda}_{\nu^{(1)},\ldots,\nu^{(\ell)}} \;:=\; \sum_{\tau^{(1)},\ldots,\tau^{(\ell-2)}} c^{\lambda}_{\nu^{(1)}\tau^{(1)}}\, c^{\tau^{(1)}}_{\nu^{(2)}\tau^{(2)}} \cdots c^{\tau^{(\ell-2)}}_{\nu^{(\ell-1)}\nu^{(\ell)}}$$

are written as sums of products of LR coefficients. Combining (KW) and the LR rule, this gives a (rather cumbersome) combinatorial interpretation of the multiplicity $a_{\lambda\mu}$ of χ^{λ}. This is clearly in #P, which proves the first part. The second part follows from the same argument, with rows (μ_i) replaced by rectangles with distinct lengths. □

15. Final remarks

15.1. The term *combinatorial interpretation* seems to be relatively recent and was first used by Hardy [**Har40**, §6.9], in connection with the *Rogers–Ramanujan identities*. There, the meaning was literal, to say that the RR identities can be restated in a combinatorial language. The first modern usage was by Kaplansky and Riordan [**KR46**, p. 262], to say that the *Stirling numbers of second kind* have a "combinatorial interpretation" as the number of *rook placements* on the staircase shape.[86]

Part of the reason is linguistic. For example, MacMahon used plenty of "interpretations" in his celebrated *Combinatory Analysis* [**Mac15**], so the term "combinatory interpretation" can be found in several papers, e.g. in [**Ing26**].

15.2. There is a reason we are so cavalier with many #P-hardness conjectures in Section 13. Roughly speaking, this is because the universe of interesting combinatorial FP functions is quite small, and is reduced to:

- **dynamic programming** (see e.g. [**CLRS09**, §15]),
- **determinant formulas**, e.g. (AFDF), the *matrix-tree theorem* (MTT), the *Lindström–Gessel–Viennot theorem* (LGV, see e.g. [**GJ83**, §5.4.2]), and the *Kasteleyn formula* (cf. §13.2),[87]
- **explicit formulas**, e.g. (HLF) and *MacMahon's box formula* (see e.g. [**Sta99**, Eq. (7.109)]).[88]

If your function clearly does not belong to either of these (not well-defined) classes, there is a good chance it is not in FP. Although in principle there are are *intermediate* counting classes between FP and #P (unless FP = #P), in practice no one has seen them.[89] Various dichotomy NP-completeness results only reinforce this belief, see e.g. classic paper [**HL90**] and the most recent breakthrough [**Zhuk20**]. In summary: *If you don't see how to compute a function using standard approaches, then most likely it is provably hard to do.*

15.3. We have used a lot of space presenting our arguments in support of our conjectures that many combinatorial functions in Algebraic Combinatorics are (probably) not in #P. Here is a quick summary of our arguments, TLDR style:

- **Are you smarter than everyone?** So many people worked on these problems, if there was a combinatorial interpretation it would have been discovered by now.

---

[86] This paper was extremely influential, and the result can be found e.g. in [**Sta99**, Cor. 2.4.2].

[87] Except for the MTT, the clue to all of these is *planarity*. Note that the (AFDF) follows from the LGV and the limit argument. The bijection in [**KPW00**] shows that the MTT implies the Kasteleyn formula.

[88] As we mentioned earlier (see §11.2), both of these formulas are derived from determinant formulas.

[89] Joshua Grochow proposed the number of graph isomorphisms as an intermediate function; that was before Babai's breakthrough [**Bab18**]. Also, a well-known expert once suggested to us (personal communication), that computing the number $\mathrm{T}(\lambda,\mu)$ of contingency tables might be intermediate (in unary); we disagree.

- **You can't get there if you are going in the wrong direction.** If everyone in a community thinks a problem has a positive solution and they can't even agree what would it mean to have a negative solution, there is a chance the problem never gets solved. At the very least the community should hedge and pursue both directions.
- **There is already one miracle. Why are you expecting another one?** The LR rule and all its variations are magical. But they are caused by one true miracle: the RSK. All the variations on the theme (jeu-de-taquin, octahedral map, etc.), are equivalent in a formal sense, and thus simply the RSK in disguise. Given the scarcity of miracles, it seems unreasonable to hope for a positive solution without using the RSK or its relative in an essential way.
- **Nothing comes from nothing.** The proof that a function is positive or integral is based on a sequence of arguments. For certain type of arguments that are given by algebraic inequalities, we can show they do not have oblivious combinatorial interpretations. So either your favorite function is very special and its proof avoids all such arguments, or you need another proof.

15.4. While writing this survey, we had an unmistakable feeling of mapping the *terra incognita*. This reminded us of Nicolas Sanson's 1650 map of North America, where the author made a logical leap and extend shore lines to create the *Island of California*, see below.[90]

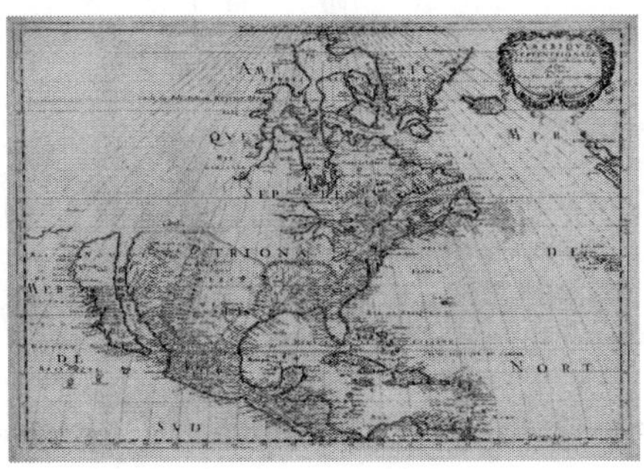

You can imagine why he did it, of course. Unfortunately, it took about fifty years to correct this error. We can't wait to find out if it's us who are making unjustified logical leaps, or it's others who have been using a wrong map. We just hope it will take less than fifty years.

Acknowledgments

We dedicate this survey to the memory of Gian-Carlo Rota whose teaching and philosophy influenced both our approach to the subject and our view of the world. We are embarrassed that it took so long for us to recognize his influence.

This paper would not exist without our recent joint work with Christian Ikenmeyer and our extensive collaboration with Greta Panova. We have also benefited from close collaboration with Swee Hong Chan, Sam Dittmer, Scott Garrabrant, Alejandro Morales, Danny Nguyen, Ernesto Vallejo, Jed Yang and Damir Yeliussizov. We are deeply grateful for the opportunity to work with them all.

We are also thankful for helpful discussions on the subject with Scott Aaronson, Olga Azenhas, Cyril Banderier, Sasha Barvinok, Sara Billey, Artëm Chernikov, Jesús De Loera, Michael Drmota, Éric Fusy, Pavel Galashin, Nikita Gladkov, Darij Grinberg, Josh

[90]Nicolas Sanson (1600–1667)—Raremaps.com, former image transferred from en.wikipedia, Public domain, via Wikimedia Commons.

Grochow, Leonid Gurvits, Zach Hamaker, Sam Hopkins, Mark Jerrum, Allen Knutson, Christian Krattenthaler, Greg Kuperberg, Svante Linusson, Tyrrell McAllister, Fëdor Petrov, Vic Reiner, Gilles Schaeffer, Yair Shenfeld, Richard Stanley, Josh Swanson, Ramon van Handel, Dennis White, Nathan Williams, Alex Yong and Paul Zinn-Justin. We thank Vince Vatter for graciously sharing with us [**Vat18**].

Some of this material was accumulated over the years based on numerous conversations. We sincerely apologize to anyone we forgot to mention.

References

[Aar16] Scott Aaronson, $\mathsf{P} \stackrel{?}{=} \mathsf{NP}$, in *Open problems in mathematics*, Springer, Cham, 2016, 1–122; available at scottaaronson.com/papers/pnp.pdf

[AHK18] Karim Adiprasito, June Huh, and Eric Katz, *Hodge theory for combinatorial geometries*, Ann. of Math. (2) **188** (2018), no. 2, 381–452, DOI 10.4007/annals.2018.188.2.1. MR3862944

[ARY19] Anshul Adve, Colleen Robichaux, and Alexander Yong, *Vanishing of Littlewood-Richardson polynomials is in* P, Comput. Complexity **28** (2019), no. 2, 241–257, DOI 10.1007/s00037-019-00183-6. MR3950166

[AS18] Connor Ahlbach and Joshua P. Swanson, *Cyclic sieving, necklaces, and branching rules related to Thrall's problem*, Electron. J. Combin. **25** (2018), no. 4, Paper No. 4.42, 38, DOI 10.37236/8198. MR3891109

[AS16] Noga Alon and Joel H. Spencer, *The probabilistic method*, 4th ed., Wiley Series in Discrete Mathematics and Optimization, John Wiley & Sons, Inc., Hoboken, NJ, 2016. MR3524748

[ALOV18] Nima Anari, Kuikui Liu, Shayan Oveis Gharan, and Cynthia Vinzant, *Log-concave polynomials II: High-dimensional walks and an FPRAS for counting bases of a matroid*, STOC'19—Proceedings of the 51st Annual ACM SIGACT Symposium on Theory of Computing, ACM, New York, 2019, pp. 1–12, DOI 10.1145/3313276.3316385. MR4003314

[ARY13] David Anderson, Edward Richmond, and Alexander Yong, *Eigenvalues of Hermitian matrices and equivariant cohomology of Grassmannians*, Compos. Math. **149** (2013), no. 9, 1569–1582, DOI 10.1112/S0010437X13007343. MR3109734

[AB89] George E. Andrews and R. J. Baxter, *A motivated proof of the Rogers-Ramanujan identities*, Amer. Math. Monthly **96** (1989), no. 5, 401–409, DOI 10.2307/2325145. MR994031

[AG88] George E. Andrews and F. G. Garvan, *Dyson's crank of a partition*, Bull. Amer. Math. Soc. (N.S.) **18** (1988), no. 2, 167–171, DOI 10.1090/S0273-0979-1988-15637-6. MR929094

[AB09] Sanjeev Arora and Boaz Barak, *Computational complexity: A modern approach*, Cambridge University Press, Cambridge, 2009, DOI 10.1017/CBO9780511804090. MR2500087

[A+98] Sanjeev Arora, Carsten Lund, Rajeev Motwani, Madhu Sudan, and Mario Szegedy, *Proof verification and the hardness of approximation problems*, J. ACM **45** (1998), no. 3, 501–555, DOI 10.1145/278298.278306. MR1639346

[AL05] Richard Arratia and Thomas M. Liggett, *How likely is an i.i.d. degree sequence to be graphical?*, Ann. Appl. Probab. **15** (2005), no. 1B, 652–670, DOI 10.1214/105051604000000693. MR2114985

[AS54] A. O. L. Atkin and P. Swinnerton-Dyer, *Some properties of partitions*, Proc. London Math. Soc. (3) **4** (1954), 84–106, DOI 10.1112/plms/s3-4.1.84. MR60535

[ACM09] Olga Azenhas, Alessandro Conflitti and Ricardo Mamede, On an index two subgroup of puzzle and Littlewood–Richardson tableau $\mathbb{Z}_2 \times S_3$-symmetries, preprint DMUC. 09-51 (2009), 21 pp.; available at hdl.handle.net/10316/13627

[Bab18] László Babai, *Group, graphs, algorithms: the graph isomorphism problem*, Proceedings of the International Congress of Mathematicians—Rio de Janeiro 2018. Vol. IV. Invited lectures, World Sci. Publ., Hackensack, NJ, 2018, pp. 3319–3336. MR3966534

[Bab19] László Babai, *Canonical form for graphs in quasipolynomial time: preliminary report*, STOC'19—Proceedings of the 51st Annual ACM SIGACT Symposium on Theory of

Computing, ACM, New York, 2019, pp. 1237–1246, DOI 10.1145/3313276.3316356. MR4003425

[BL83] László Babai and Eugene M. Luks, Canonical labeling of graphs, in *Proc. 15th STOC* (1983), 171–183.

[BE16] Kirby Baker and Edward Early, Character polynomials and row sums of the symmetric group, preprint (2016), 9 pp.; available at tinyurl.com/yxj5l7fd.

[BO05] Cristina M. Ballantine and Rosa C. Orellana, *A combinatorial interpretation for the coefficients in the Kronecker product* $s_{(n-p,p)} * s_\lambda$, Sém. Lothar. Combin. **54A** (2005/07), Art. B54Af, 29. MR2264933

[Ban08] Jason Bandlow, *An elementary proof of the hook formula*, Electron. J. Combin. **15** (2008), no. 1, Research paper 45, 14, DOI 10.37236/769. MR2398837

[Bar93] Alexander I. Barvinok, *A polynomial time algorithm for counting integral points in polyhedra when the dimension is fixed*, Math. Oper. Res. **19** (1994), no. 4, 769–779, DOI 10.1287/moor.19.4.769. MR1304623

[Bar07] Alexander Barvinok, *Brunn-Minkowski inequalities for contingency tables and integer flows*, Adv. Math. **211** (2007), no. 1, 105–122, DOI 10.1016/j.aim.2006.07.012. MR2313530

[Bax82] Rodney J. Baxter, *Exactly solved models in statistical mechanics*, Academic Press, Inc. [Harcourt Brace Jovanovich, Publishers], London, 1982. MR690578

[BB61] Edwin F. Beckenbach and Richard Bellman, *Inequalities*, Ergebnisse der Mathematik und ihrer Grenzgebiete, (N.F.), Band 30, Springer-Verlag, Berlin-Göttingen-Heidelberg, 1961. MR158038

[BSS96] Georgia Benkart, Frank Sottile, and Jeffrey Stroomer, *Tableau switching: algorithms and applications*, J. Combin. Theory Ser. A **76** (1996), no. 1, 11–43, DOI 10.1006/jcta.1996.0086. MR1405988

[BBR06] François Bergeron, Riccardo Biagioli, and Mercedes H. Rosas, *Inequalities between Littlewood-Richardson coefficients*, J. Combin. Theory Ser. A **113** (2006), no. 4, 567–590, DOI 10.1016/j.jcta.2005.05.002. MR2216453

[BB93] Nantel Bergeron and Sara Billey, *RC-graphs and Schubert polynomials*, Experiment. Math. **2** (1993), no. 4, 257–269. MR1281474

[BZ88] A. D. Berenshteĭn and A. V. Zelevinskiĭ, *Involutions on Gel′fand-Tsetlin schemes and multiplicities in skew* GL_n*-modules* (Russian), Dokl. Akad. Nauk SSSR **300** (1988), no. 6, 1291–1294; English transl., Soviet Math. Dokl. **37** (1988), no. 3, 799–802. MR950493

[BZ90] A. D. Berenshteĭn and A. V. Zelevinskiĭ, *When is the multiplicity of a weight equal to 1?* (Russian), Funktsional. Anal. i Prilozhen. **24** (1990), no. 4, 1–13, 96, DOI 10.1007/BF01077330; English transl., Funct. Anal. Appl. **24** (1990), no. 4, 259–269 (1991). MR1092798

[BZ92] A. D. Berenstein and A. V. Zelevinsky, *Triple multiplicities for* $\mathrm{sl}(r+1)$ *and the spectrum of the exterior algebra of the adjoint representation*, J. Algebraic Combin. **1** (1992), no. 1, 7–22, DOI 10.1023/A:1022429213282. MR1162639

[BB04] Christine Bessenrodt and Christiane Behns, *On the Durfee size of Kronecker products of characters of the symmetric group and its double covers*, J. Algebra **280** (2004), no. 1, 132–144, DOI 10.1016/j.jalgebra.2004.03.028. MR2081924

[BBS21] C. Bessenrodt, C. Bowman, and L. Sutton, *Kronecker positivity and 2-modular representation theory*, Trans. Amer. Math. Soc. Ser. B **8** (2021), 1024–1055, DOI 10.1090/btran/70. MR4350547

[BO04] Jørn B. Olsson, *Christine Bessenrodt, 1958–2022*, Notices Amer. Math. Soc. **70** (2023), no. 7, 1101–1105. MR4615195

[Bil21] Sara Billey, *Basic Schubert Calculus, Part 2*, lecture at *Combinatorial Algebraic Geometry Workshop*, ICERM, Brown University, RI (February 1, 2021); full video and transcript are available at tinyurl.com/yn22dnue, slides are available at tinyurl.com/3vec4hrd

[BJS93] Sara C. Billey, William Jockusch, and Richard P. Stanley, *Some combinatorial properties of Schubert polynomials*, J. Algebraic Combin. **2** (1993), no. 4, 345–374, DOI 10.1023/A:1022419800503. MR1241505

[BV08] Sara Billey and Ravi Vakil, *Intersections of Schubert varieties and other permutation array schemes*, Algorithms in algebraic geometry, IMA Vol. Math. Appl., vol. 146, Springer, New York, 2008, pp. 21–54, DOI 10.1007/978-0-387-75155-9_3. MR2397936

[Bjö11] Anders Björner, *A q-analogue of the FKG inequality and some applications*, Combinatorica **31** (2011), no. 2, 151–164, DOI 10.1007/s00493-011-2644-1. MR2848248

[BW89] Anders Björner and Michelle L. Wachs, *q-hook length formulas for forests*, J. Combin. Theory Ser. A **52** (1989), no. 2, 165–187, DOI 10.1016/0097-3165(89)90028-9. MR1022316

[Bla17] Jonah Blasiak, *Kronecker coefficients for one hook shape*, Sém. Lothar. Combin. **77** ([2016), Art. B77c, 40. MR3650220

[Ble13] Grigoriy Blekherman, *Nonnegative polynomials and sums of squares*, Semidefinite optimization and convex algebraic geometry, MOS-SIAM Ser. Optim., vol. 13, SIAM, Philadelphia, PA, 2013, pp. 159–202. MR3050243

[BP21] I. A. Bochkov and F. V. Petrov, *The bounds for the number of linear extensions via chain and antichain coverings*, Order **38** (2021), no. 2, 323–328, DOI 10.1007/s11083-020-09542-3. MR4274656

[Bol98] Béla Bollobás, *Modern graph theory*, Graduate Texts in Mathematics, vol. 184, Springer-Verlag, New York, 1998, DOI 10.1007/978-1-4612-0619-4. MR1633290

[BBS99] Béla Bollobás, Graham Brightwell, and Alexander Sidorenko, *Geometrical techniques for estimating numbers of linear extensions*, European J. Combin. **20** (1999), no. 5, 329–335, DOI 10.1006/eujc.1999.0299. MR1702372

[BS00] Mireille Bousquet-Mélou and Gilles Schaeffer, *Enumeration of planar constellations*, Adv. in Appl. Math. **24** (2000), no. 4, 337–368, DOI 10.1006/aama.1999.0673. MR1761777

[BDO15] C. Bowman, M. De Visscher, and R. Orellana, *The partition algebra and the Kronecker coefficients*, Trans. Amer. Math. Soc. **367** (2015), no. 5, 3647–3667, DOI 10.1090/S0002-9947-2014-06245-4. MR3314819

[Brä15] Petter Brändén, *Unimodality, log-concavity, real-rootedness and beyond*, Handbook of enumerative combinatorics, Discrete Math. Appl. (Boca Raton), CRC Press, Boca Raton, FL, 2015, pp. 437–483. MR3409348

[BH20] Petter Brändén and June Huh, *Lorentzian polynomials*, Ann. of Math. (2) **192** (2020), no. 3, 821–891, DOI 10.4007/annals.2020.192.3.4. MR4172622

[Bre73] L. M. Brègman, *Certain properties of nonnegative matrices and their permanents* (Russian), Dokl. Akad. Nauk SSSR **211** (1973), 27–30. MR327788

[Bre89] Francesco Brenti, *Unimodal, log-concave and Pólya frequency sequences in combinatorics*, Mem. Amer. Math. Soc. **81** (1989), no. 413, viii+106, DOI 10.1090/memo/0413. MR963833

[Bre94] Francesco Brenti, *Log-concave and unimodal sequences in algebra, combinatorics, and geometry: an update*, Jerusalem combinatorics '93, Contemp. Math., vol. 178, Amer. Math. Soc., Providence, RI, 1994, pp. 71–89, DOI 10.1090/conm/178/01893. MR1310575

[BOR09] Emmanuel Briand, Rosa Orellana, and Mercedes Rosas, *Reduced Kronecker coefficients and counter-examples to Mulmuley's strong saturation conjecture SH*, Comput. Complexity **18** (2009), no. 4, 577–600, DOI 10.1007/s00037-009-0279-z. With an appendix by Ketan Mulmuley. MR2570451

[BOR11] Emmanuel Briand, Rosa Orellana, and Mercedes Rosas, *The stability of the Kronecker product of Schur functions*, J. Algebra **331** (2011), 11–27, DOI 10.1016/j.jalgebra.2010.12.026. MR2774644

[BT02] Graham R. Brightwell and William T. Trotter, *A combinatorial approach to correlation inequalities*, Discrete Math. **257** (2002), no. 2-3, 311–327, DOI 10.1016/S0012-365X(02)00432-6. Kleitman and combinatorics: a celebration (Cambridge, MA, 1999). MR1935730

[BW00] Graham Brightwell and Douglas B. West, Partially ordered sets, Ch. 11 in *Handbook of discrete and combinatorial mathematics*, CRC Press, Boca Raton, FL, 2000, 717–752.

[BW91] Graham Brightwell and Peter Winkler, *Counting linear extensions*, Order **8** (1991), no. 3, 225–242, DOI 10.1007/BF00383444. MR1154926

[Buch02] Anders Skovsted Buch, *A Littlewood-Richardson rule for the K-theory of Grassmannians*, Acta Math. **189** (2002), no. 1, 37–78, DOI 10.1007/BF02392644. MR1946917

[BKT04] Anders Skovsted Buch, Andrew Kresch, and Harry Tamvakis, *Littlewood-Richardson rules for Grassmannians*, Adv. Math. **185** (2004), no. 1, 80–90, DOI 10.1016/S0001-8708(03)00165-8. MR2058780

[BT00] Edwin F. Beschler, David A. Buchsbaum, Jacob T. Schwartz, Richard P. Stanley, Brian D. Taylor, and Michael Waterman, *Gian-Carlo Rota (1932–1999)*, Notices Amer. Math. Soc. **47** (2000), no. 2, 203–216. MR1740352

[BS17] Daniel Bump and Anne Schilling, *Crystal bases: Representations and combinatorics*, World Scientific Publishing Co. Pte. Ltd., Hackensack, NJ, 2017, DOI 10.1142/9876. MR3642318

[Bur74] William H. Burge, *Four correspondences between graphs and generalized Young tableaux*, J. Combinatorial Theory Ser. A **17** (1974), 12–30, DOI 10.1016/0097-3165(74)90024-7. MR351884

[BI08] Peter Bürgisser and Christian Ikenmeyer, *The complexity of computing Kronecker coefficients* (English, with English and French summaries), 20th Annual International Conference on Formal Power Series and Algebraic Combinatorics (FPSAC 2008), Discrete Math. Theor. Comput. Sci. Proc., AJ, Assoc. Discrete Math. Theor. Comput. Sci., Nancy, 2008, pp. 357–368. MR2721467

[BI13] Peter Bürgisser and Christian Ikenmeyer, *Deciding positivity of Littlewood-Richardson coefficients*, SIAM J. Discrete Math. **27** (2013), no. 4, 1639–1681, DOI 10.1137/120892532. MR3108110

[CC16] Paul-Jean Cahen and Jean-Luc Chabert, *What you should know about integer-valued polynomials*, Amer. Math. Monthly **123** (2016), no. 4, 311–337, DOI 10.4169/amer.math.monthly.123.4.311. MR3493376

[Cam01] Kathie Cameron, *Thomason's algorithm for finding a second Hamiltonian circuit through a given edge in a cubic graph is exponential on Krawczyk's graphs*, Discrete Math. **235** (2001), no. 1-3, 69–77, DOI 10.1016/S0012-365X(00)00260-0. Combinatorics (Prague, 1998). MR1829836

[CE99] Kathie Cameron and Jack Edmonds, *Some graphic uses of an even number of odd nodes* (English, with English and French summaries), Ann. Inst. Fourier (Grenoble) **49** (1999), no. 3, 815–827. Symposium à la Mémoire de François Jaeger (Grenoble, 1998). MR1703426

[CW95] E. Rodney Canfield and S. Gill Williamson, *A loop-free algorithm for generating the linear extensions of a poset*, Order **12** (1995), no. 1, 57–75, DOI 10.1007/BF01108590. MR1336537

[Car65] Lewis Carroll, *Alice's adventures in Wonderland*, 150th anniversary edition, Princeton University Press, Princeton, NJ, 2015. Illustrated by Salvador Dalí; With an introduction by Mark Burstein and Thomas Banchoff; Published in association with the National Museum of Mathematics, DOI 10.1515/9781400874262. MR3469831

[CM16] Renzo Cavalieri and Eric Miles, *Riemann surfaces and algebraic curves: A first course in Hurwitz theory*, London Mathematical Society Student Texts, vol. 87, Cambridge University Press, Cambridge, 2016, DOI 10.1017/CBO9781316569252. MR3585685

[CP21] Swee Hong Chan and Igor Pak, *Log-concave poset inequalities*, Journal of Association for Mathematical Research **2** (2024), no. 1, 53–153.

[CP22a] Swee Hong Chan and Igor Pak, *Introduction to the combinatorial atlas*, Expo. Math. **40** (2022), no. 4, 1014–1048, DOI 10.1016/j.exmath.2022.08.003. MR4518642

[CP22b] Swee Hong Chan and Igor Pak, Correlation inequalities for linear extensions, preprint (2022), 23 pp.; `arXiv:2211.16637`.

[CP23a] Swee Hong Chan and Igor Pak, *Multivariate correlation inequalities for P-partitions*, Pacific J. Math. **323** (2023), no. 2, 223–252, DOI 10.2140/pjm.2023.323.223. MR4596256

[CP23b] Swee Hong Chan and Igor Pak, Equality cases of the Alexandrov–Fenchel inequality are not in the polynomial hierarchy, preprint (2023), 35 pp.; extended abstract in *Proc. 56th STOC* (2024), ACM; `arXiv:2309.05764`.

[CP23c] Swee Hong Chan and Igor Pak, Linear extensions of finite posets, preprint (2023), 55 pp.; `arXiv:2311.02743`.

[CPP22a] Swee Hong Chan, Igor Pak, and Greta Panova, *Log-concavity in planar random walks*, Combinatorica **42** (2022), no. suppl. 1, 1011–1026, DOI 10.1007/s00493-021-4860-7. MR4543721

[CPP22b] Swee Hong Chan, Igor Pak, and Greta Panova, *The cross-product conjecture for width two posets*, Trans. Amer. Math. Soc. **375** (2022), no. 8, 5923–5961, DOI 10.1090/tran/8679. MR4469242

[CPP23a] Swee Hong Chan, Igor Pak, and Greta Panova, *Extensions of the Kahn-Saks inequality for posets of width two*, Comb. Theory **3** (2023), no. 1, Paper No. 8, 35, DOI 10.5070/c63160421. MR4565295

[CPP23b] Swee Hong Chan, Igor Pak, and Greta Panova, *Effective poset inequalities*, SIAM J. Discrete Math. **37** (2023), no. 3, 1842–1880, DOI 10.1137/22M1532317. MR4631443

[CIM17] Man-Wai Cheung, Christian Ikenmeyer, and Sevak Mkrtchyan, *Symmetrizing tableaux and the 5th case of the Foulkes conjecture*. part 3, J. Symbolic Comput. **80** (2017), no. part 3, 833–843, DOI 10.1016/j.jsc.2016.09.002. MR3574536

[CDW12] Matthias Christandl, Brent Doran, and Michael Walter, *Computing multiplicities of Lie group representations*, 2012 IEEE 53rd Annual Symposium on Foundations of Computer Science—FOCS 2012, IEEE Computer Soc., Los Alamitos, CA, 2012, pp. 639–648. MR3186652

[CP88] M. Chrobak and S. Poljak, *On common edges in optimal solutions to traveling salesman and other optimization problems*, Discrete Appl. Math. **20** (1988), no. 2, 101–111, DOI 10.1016/0166-218X(88)90057-1. MR943711

[CKP11] Ionuţ Ciocan-Fontanine, Matjaž Konvalinka, and Igor Pak, *The weighted hook length formula*, J. Combin. Theory Ser. A **118** (2011), no. 6, 1703–1717, DOI 10.1016/j.jcta.2011.02.004. MR2793605

[CS84] Michael Clausen and Friedrich Stötzer, *Pictures and Standardtableaux—Grundlagen und Algorithmen* (German), Bayreuth. Math. Schr. **16** (1984), 1–122. MR742686

[CM92] Robert Cori and Antonio Machì, *Maps, hypermaps and their automorphisms: a survey. I, II, III*, Exposition. Math. **10** (1992), no. 5, 403–427, 429–447, 449–467. MR1190182

[CLRS09] Thomas H. Cormen, Charles E. Leiserson, Ronald L. Rivest, and Clifford Stein, *Introduction to algorithms*, 3rd ed., MIT Press, Cambridge, MA, 2009. MR2572804

[Cos09] Izzet Coskun, *A Littlewood-Richardson rule for two-step flag varieties*, Invent. Math. **176** (2009), no. 2, 325–395, DOI 10.1007/s00222-008-0165-3. MR2495766

[Cos+] Izzet Coskun, A Littlewood–Richardson rule for partial flag varieties, undated preprint; available at tinyurl.com/yc4nphjh

[CV09] Izzet Coskun and Ravi Vakil, *Geometric positivity in the cohomology of homogeneous spaces and generalized Schubert calculus*, Algebraic geometry—Seattle 2005. Part 1, Proc. Sympos. Pure Math., vol. 80, Amer. Math. Soc., Providence, RI, 2009, pp. 77–124, DOI 10.1090/pspum/080.1/2483933. MR2483933

[CL14] Alexander Coward and Marc Lackenby, *An upper bound on Reidemeister moves*, Amer. J. Math. **136** (2014), no. 4, 1023–1066, DOI 10.1353/ajm.2014.0027. MR3245186

[DK05a] V. I. Danilov and G. A. Koshevoĭ, *Massifs and the combinatorics of Young tableaux* (Russian, with Russian summary), Uspekhi Mat. Nauk **60** (2005), no. 2(362), 79–142, DOI 10.1070/RM2005v060n02ABEH000824; English transl., Russian Math. Surveys **60** (2005), no. 2, 269–334. MR2152944

[DK05b] V. I. Danilov and G. A. Koshevoy, *The octahedron recurrence and RSK-correspondence*, Sém. Lothar. Combin. **54A** (2005/07), Art. B54An, 16. MR2317686

[DK08] V. I. Danilov and G. A. Koshevoĭ, *The Robinson-Schensted-Knuth correspondence and the bijections of commutativity and associativity* (Russian, with Russian summary), Izv. Ross. Akad. Nauk Ser. Mat. **72** (2008), no. 4, 67–96, DOI 10.1070/IM2008v072n04ABEH002415; English transl., Izv. Math. **72** (2008), no. 4, 689–716. MR2452234

[dB+51] N. G. de Bruijn, Ca. van Ebbenhorst Tengbergen, and D. Kruyswijk, *On the set of divisors of a number*, Nieuw Arch. Wiskunde (2) **23** (1951), 191–193. MR43115

[DM06] Jesús A. De Loera and Tyrrell B. McAllister, *On the computation of Clebsch-Gordan coefficients and the dilation effect*, Experiment. Math. **15** (2006), no. 1, 7–19. MR2229381

[DO04] Jesus De Loera and Shmuel Onn, *The complexity of three-way statistical tables*, SIAM J. Comput. **33** (2004), no. 4, 819–836, DOI 10.1137/S0097539702403803. MR2065334

[dM+21] Arnaud de Mesmay, Yo'av Rieck, Eric Sedgwick, and Martin Tancer, *The unbearable hardness of unknotting*, Adv. Math. **381** (2021), Paper No. 107648, 36, DOI 10.1016/j.aim.2021.107648. MR4215750

[DD07] Erik D. Demaine and Martin L. Demaine, *Jigsaw puzzles, edge matching, and polyomino packing: connections and complexity*, Graphs Combin. **23** (2007), no. suppl. 1, 195–208, DOI 10.1007/s00373-007-0713-4. MR2320628

[DP15] Stephen DeSalvo and Igor Pak, *Log-concavity of the partition function*, Ramanujan J. **38** (2015), no. 1, 61–73, DOI 10.1007/s11139-014-9599-y. MR3396486

[Dés80] J. Désarménien, *An algorithm for the Rota straightening formula*, Discrete Math. **30** (1980), no. 1, 51–68, DOI 10.1016/0012-365X(80)90062-X. MR561764

[DKR78] J. Désarménien, Joseph P. S. Kung, and Gian-Carlo Rota, *Invariant theory, Young bitableaux, and combinatorics*, Advances in Math. **27** (1978), no. 1, 63–92, DOI 10.1016/0001-8708(78)90077-4. MR485944

[Dic52] Leonard Eugene Dickson, *History of the theory of numbers. Vol. I: Divisibility and primality*, Chelsea Publishing Co., New York, 1966. MR245499

[DP18] Samuel Dittmer and Igor Pak, *Counting linear extensions of restricted posets*, Electron. J. Combin. **27** (2020), no. 4, Paper No. 4.48, 13, DOI 10.37236/8552. MR4245223

[DP16] Patrick Doolan and Sangjib Kim, *The Littlewood-Richardson rule and Gelfand-Tsetlin patterns*, Algebra Discrete Math. **22** (2016), no. 1, 21–47. MR3573543

[DRS74] Peter Doubilet, Gian-Carlo Rota, and Joel Stein, *On the foundations of combinatorial theory. IX. Combinatorial methods in invariant theory*, Studies in Appl. Math. **53** (1974), 185–216, DOI 10.1002/sapm1974533185. MR498650

[Duan03] Haibao Duan, *On the inverse Kostka matrix*, J. Combin. Theory Ser. A **103** (2003), no. 2, 363–376, DOI 10.1016/S0097-3165(03)00106-7. MR1996073

[DPS14] Enrica Duchi, Dominique Poulalhon and Gilles Schaeffer, Bijections for simple and double Hurwitz numbers, preprint (2014), 20 pp.; `arXiv:1410.6521`.

[DKM97] Martin Dyer, Ravi Kannan, and John Mount, *Sampling contingency tables*, Random Structures Algorithms **10** (1997), no. 4, 487–506, DOI 10.1002/(SICI)1098-2418(199707)10:4¡487::AID-RSA4¿3.0.CO;2-Q. MR1608222

[DGH98] Martin Dyer, Peter Gritzmann, and Alexander Hufnagel, *On the complexity of computing mixed volumes*, SIAM J. Comput. **27** (1998), no. 2, 356–400, DOI 10.1137/S0097539794278384. MR1616544

[Dys44] F. J. Dyson, *Some guesses in the theory of partitions*, Eureka **8** (1944), 10–15. MR3077150

[ER90] Ömer Eğecioğlu and Jeffrey B. Remmel, *A combinatorial interpretation of the inverse Kostka matrix*, Linear and Multilinear Algebra **26** (1990), no. 1-2, 59–84, DOI 10.1080/03081089008817966. MR1034417

[ELSV99] Torsten Ekedahl, Sergei Lando, Michael Shapiro, and Alek Vainshtein, *On Hurwitz numbers and Hodge integrals* (English, with English and French summaries), C. R. Acad. Sci. Paris Sér. I Math. **328** (1999), no. 12, 1175–1180, DOI 10.1016/S0764-4442(99)80435-2. MR1701381

[Ell85] Richard S. Ellis, *Entropy, large deviations, and statistical mechanics*, Grundlehren der mathematischen Wissenschaften [Fundamental Principles of Mathematical Sciences], vol. 271, Springer-Verlag, New York, 1985, DOI 10.1007/978-1-4613-8533-2. MR793553

[Eln97] Serge Elnitsky, *Rhombic tilings of polygons and classes of reduced words in Coxeter groups*, J. Combin. Theory Ser. A **77** (1997), no. 2, 193–221, DOI 10.1006/jcta.1997.2723. MR1429077

[FN15] Ilse Fischer and Philippe Nadeau, *Fully packed loops in a triangle: matchings, paths and puzzles*, J. Combin. Theory Ser. A **130** (2015), 64–118, DOI 10.1016/j.jcta.2014.10.008. MR3280685

[FI20] Nick Fischer and Christian Ikenmeyer, *The computational complexity of plethysm coefficients*, Comput. Complexity **29** (2020), no. 2, Paper No. 8, 43, DOI 10.1007/s00037-020-00198-4. MR4171310

[Fis66] Michael E. Fisher, On the dimer solution of planar Ising models, *J. Math. Phys.* **7** (1966), 1776–1781.

[Fom95] Sergey Fomin, *Schensted algorithms for dual graded graphs*, J. Algebraic Combin. **4** (1995), no. 1, 5–45, DOI 10.1023/A:1022404807578. MR1314558

[FG93] Sergey Fomin and Curtis Greene, *A Littlewood-Richardson miscellany*, European J. Combin. **14** (1993), no. 3, 191–212, DOI 10.1006/eujc.1993.1024. MR1215331

[FS98] S. Fomin and D. Stanton, *Rim hook lattices*, Algebra i Analiz **9** (1997), no. 5, 140–150; English transl., St. Petersburg Math. J. **9** (1998), no. 5, 1007–1016. MR1604377

[For97] Lance Fortnow, *Counting complexity*, Complexity theory retrospective, II, Springer, New York, 1997, pp. 81–107. MR1478170

[FRT54] J. S. Frame, G. de B. Robinson, and R. M. Thrall, *The hook graphs of the symmetric groups*, Canad. J. Math. **6** (1954), 316–324, DOI 10.4153/cjm-1954-030-1. MR62127

[Fru86] Avital Frumkin, *Theorem about the conjugacy representation of S_n*, Israel J. Math. **55** (1986), no. 1, 121–128, DOI 10.1007/BF02772700. MR858465

[Ful97] William Fulton, *Young tableaux: With applications to representation theory and geometry*, London Mathematical Society Student Texts, vol. 35, Cambridge University Press, Cambridge, 1997. MR1464693

[FH99] William Fulton and Joe Harris, *Representation theory: A first course*, Graduate Texts in Mathematics, vol. 129, Springer-Verlag, New York, 1991, DOI 10.1007/978-1-4612-0979-9. MR1153249

[Fusy05] Éric Fusy, *Counting unrooted maps using tree-decomposition*, Sém. Lothar. Combin. **54A** (2005/07), Art. B54Al, 44. MR2317684

[GG20a] Christian Gaetz and Yibo Gao, *Separable elements and splittings of Weyl groups*, Adv. Math. **374** (2020), 107389, 28, DOI 10.1016/j.aim.2020.107389. MR4157589

[GG20b] Christian Gaetz and Yibo Gao, *The hull metric on Coxeter groups*, Comb. Theory **2** (2022), no. 2, Paper No. 7, 15. MR4449815

[Gal17] Pavel Galashin, *A Littlewood-Richardson rule for dual stable Grothendieck polynomials*, J. Combin. Theory Ser. A **151** (2017), 23–35, DOI 10.1016/j.jcta.2017.04.001. MR3663486

[GP20] Pavel Galashin and Pavlo Pylyavskyy, *Ising model and the positive orthogonal Grassmannian*, Duke Math. J. **169** (2020), no. 10, 1877–1942, DOI 10.1215/00127094-2019-0086. MR4118643

[Gan81] Emden R. Gansner, *Matrix correspondences of plane partitions*, Pacific J. Math. **92** (1981), no. 2, 295–315. MR618067

[GOY21] Shiliang Gao, Gidon Orelowitz, and Alexander Yong, *Newell-Littlewood numbers*, Trans. Amer. Math. Soc. **374** (2021), no. 9, 6331–6366, DOI 10.1090/tran/8375. MR4302162

[GH23] Yibo Gao and Daoji Huang, *The canonical bijection between pipe dreams and bumpless pipe dreams*, Int. Math. Res. Not. IMRN **21** (2023), 18629–18663, DOI 10.1093/imrn/rnad083. MR4665632

[GJ78] M. R. Garey and D. S. Johnson, *"Strong" NP-completeness results: motivation, examples, and implications*, J. Assoc. Comput. Mach. **25** (1978), no. 3, 499–508, DOI 10.1145/322077.322090. MR478747

[GJ79] Michael R. Garey and David S. Johnson, *Computers and intractability: A guide to the theory of NP-completeness*, A Series of Books in the Mathematical Sciences, W. H. Freeman and Co., San Francisco, CA, 1979. MR519066

[GP14] Scott Garrabrant and Igor Pak, Counting with Wang tiles, unpublished (2014); see IMA talk (slides and video) at tinyurl.com/3nc3jry6

[Gar02] Adriano M. Garsia, *The saga of reduced factorizations of elements of the symmetric group*, Publ. LaCIM **29**, Univ. du Québec à Montréal, 2002, 120 pp.

[GM81] A. M. Garsia and S. C. Milne, *A Rogers-Ramanujan bijection*, J. Combin. Theory Ser. A **31** (1981), no. 3, 289–339, DOI 10.1016/0097-3165(81)90062-5. MR635372

[GKS90] Frank Garvan, Dongsu Kim, and Dennis Stanton, *Cranks and t-cores*, Invent. Math. **101** (1990), no. 1, 1–17, DOI 10.1007/BF01231493. MR1055707

[Gas98] Vesselin Gasharov, *A short proof of the Littlewood-Richardson rule*, European J. Combin. **19** (1998), no. 4, 451–453, DOI 10.1006/eujc.1998.0212. MR1630540

[GZ85] I. M. Gel′fand and A. V. Zelevinskiĭ, *Polyhedra in a space of diagrams and the canonical basis in irreducible representations of* $\mathfrak{gl}_3$ (Russian), Funktsional. Anal. i Prilozhen. **19** (1985), no. 2, 72–75. MR800925

[Ges84] Ira M. Gessel, *Combinatorial proofs of congruences*, Enumeration and design (Waterloo, Ont., 1982), Academic Press, Toronto, ON, 1984, pp. 157–197. MR782313

[Gil19] Maria Gillespie, *Variations on a theme of Schubert calculus*, Recent trends in algebraic combinatorics, Assoc. Women Math. Ser., vol. 16, Springer, Cham, 2019, pp. 115–158, DOI 10.1007/978-3-030-05141-9_4. MR3969573

[GN06] Omer Giménez and Marc Noy, *On the complexity of computing the Tutte polynomial of bicircular matroids*, Combin. Probab. Comput. **15** (2006), no. 3, 385–395, DOI 10.1017/S0963548305007327. MR2216475

[GN04] Kenneth Glass and Chi-Keung Ng, *A simple proof of the hook length formula*, Amer. Math. Monthly **111** (2004), no. 8, 700–704, DOI 10.2307/4145043. MR2091544

[GP00] Oleg Gleizer and Alexander Postnikov, *Littlewood-Richardson coefficients via Yang-Baxter equation*, Internat. Math. Res. Notices **14** (2000), 741–774, DOI 10.1155/S1073792800000416. MR1776619

[God93] C. D. Godsil, *Algebraic combinatorics*, Chapman and Hall Mathematics Series, Chapman & Hall, New York, 1993. MR1220704

[Gol56] S. W. Golomb, *Classroom Notes: Combinatorial Proof of Fermat's "Little" Theorem*, Amer. Math. Monthly **63** (1956), no. 10, 718, DOI 10.2307/2309563. MR1529479

[G+11] Parikshit Gopalan, Adam Klivans, Raghu Meka, Daniel Štefankovič, Santosh Vempala, and Eric Vigoda, *An FPTAS for #knapsack and related counting problems*, 2011 IEEE 52nd Annual Symposium on Foundations of Computer Science—FOCS 2011, IEEE Computer Soc., Los Alamitos, CA, 2011, pp. 817–826, DOI 10.1109/FOCS.2011.32. MR2933318

[GJ83] I. P. Goulden and D. M. Jackson, *Combinatorial enumeration*, Wiley-Interscience Series in Discrete Mathematics, John Wiley & Sons, Inc., New York, 1983. With a foreword by Gian-Carlo Rota; A Wiley-Interscience Publication. MR702512

[GJ97] I. P. Goulden and D. M. Jackson, *Transitive factorisations into transpositions and holomorphic mappings on the sphere*, Proc. Amer. Math. Soc. **125** (1997), no. 1, 51–60, DOI 10.1090/S0002-9939-97-03880-X. MR1396978

[GJV05] I. P. Goulden, D. M. Jackson, and R. Vakil, *Towards the geometry of double Hurwitz numbers*, Adv. Math. **198** (2005), no. 1, 43–92, DOI 10.1016/j.aim.2005.01.008. MR2183250

[GY02] Ian Goulden and Alexander Yong, *Tree-like properties of cycle factorizations*, J. Combin. Theory Ser. A **98** (2002), no. 1, 106–117, DOI 10.1006/jcta.2001.3230. MR1897927

[GK76] Curtis Greene and Daniel J. Kleitman, *Strong versions of Sperner's theorem*, J. Combinatorial Theory Ser. A **20** (1976), no. 1, 80–88, DOI 10.1016/0097-3165(76)90079-0. MR389608

[GK78] Curtis Greene and Daniel J. Kleitman, *Proof techniques in the theory of finite sets*, Studies in combinatorics, MAA Stud. Math., vol. 17, Math. Assoc. America, Washington, DC, 1978, pp. 22–79. MR513002

[GNW79] Curtis Greene, Albert Nijenhuis, and Herbert S. Wilf, *A probabilistic proof of a formula for the number of Young tableaux of a given shape*, Adv. in Math. **31** (1979), no. 1, 104–109, DOI 10.1016/0001-8708(79)90023-9. MR521470

[Gri67] Robert B. Griffiths, Correlations in Ising ferromagnets. I, *J. Math. Phys.* **8** (1967), 478–483.

[GHS70] Robert B. Griffiths, C. A. Hurst, and S. Sherman, *Concavity of magnetization of an Ising ferromagnet in a positive external field*, J. Mathematical Phys. **11** (1970), 790–795, DOI 10.1063/1.1665211. MR266507

[Gri76] G. R. Grimmett, *An upper bound for the number of spanning trees of a graph*, Discrete Math. **16** (1976), no. 4, 323–324, DOI 10.1016/S0012-365X(76)80005-2. MR439665

[Gri23] Geoffrey R. Grimmett, *Selected problems in probability theory*, Mathematics going forward—collected mathematical brushstrokes, Lecture Notes in Math., vol. 2313, Springer, Cham, [2023] ©2023, pp. 603–614, DOI 10.1007/978-3-031-12244-6_42. MR4627986

[GL21] Maxim Gurevich and Erez Lapid, *Robinson-Schensted-Knuth correspondence in the representation theory of the general linear group over a non-Archimedean local field*, Represent. Theory **25** (2021), 644–678, DOI 10.1090/ert/578. With an appendix by Mark Shimozono. MR4293086

[Häg03] Olle Häggstöm, Probability on bunkbed graphs, in *Proc. 15th FPSAC* (2003), 9 pp.; available at tinyurl.com/2p8cau7k

[H+23] Zachary Hamaker, Alejandro H. Morales, Igor Pak, Luis Serrano, and Nathan Williams, *Bijecting hidden symmetries for skew staircase shapes*, Algebr. Comb. **6** (2023), no. 4, 1095–1118, DOI 10.5802/alco.285. MR4635091

[HY14] Zachary Hamaker and Benjamin Young, *Relating Edelman-Greene insertion to the Little map*, J. Algebraic Combin. **40** (2014), no. 3, 693–710, DOI 10.1007/s10801-014-0503-z. MR3265229

[HP08] Adam Hammett and Boris Pittel, *How often are two permutations comparable?*, Trans. Amer. Math. Soc. **360** (2008), no. 9, 4541–4568, DOI 10.1090/S0002-9947-08-04478-4. MR2403696

[HS92] Phil Hanlon and Sheila Sundaram, *On a bijection between Littlewood-Richardson fillings of conjugate shape*, J. Combin. Theory Ser. A **60** (1992), no. 1, 1–18, DOI 10.1016/0097-3165(92)90034-R. MR1156644

[Har40] Godfrey H. Hardy, *Ramanujan*, Cambridge Univ. Press, Cambridge, UK, 1940, 236 pp.

[HLP52] G. H. Hardy, J. E. Littlewood, and G. Pólya, *Inequalities*, Cambridge, at the University Press, 1952. 2d ed. MR46395

[Har60] T. E. Harris, *A lower bound for the critical probability in a certain percolation process*, Proc. Cambridge Philos. Soc. **56** (1960), 13–20. MR115221

[HLP99] Joel Hass, Jeffrey C. Lagarias, and Nicholas Pippenger, *The computational complexity of knot and link problems*, J. ACM **46** (1999), no. 2, 185–211, DOI 10.1145/301970.301971. MR1693203

[HSTZ13] Gerhard Heide, Jan Saxl, Pham Huu Tiep, and Alexandre E. Zalesski, *Conjugacy action, induced representations and the Steinberg square for simple groups of Lie type*, Proc. Lond. Math. Soc. (3) **106** (2013), no. 4, 908–930, DOI 10.1112/plms/pds062. MR3056296

[HZ06] G. Heide and A. E. Zalesski, *Passman's problem on adjoint representations*, Groups, rings and algebras, Contemp. Math., vol. 420, Amer. Math. Soc., Providence, RI, 2006, pp. 163–176, DOI 10.1090/conm/420/07975. MR2279239

[HL72] Ole J. Heilmann and Elliott H. Lieb, *Theory of monomer-dimer systems*, Comm. Math. Phys. **25** (1972), 190–232. MR297280

[HL90] Pavol Hell and Jaroslav Ne et il, *On the complexity of H-coloring*, J. Combin. Theory Ser. B **48** (1990), no. 1, 92–110, DOI 10.1016/0095-8956(90)90132-J. MR1047555

[HK06] André Henriques and Joel Kamnitzer, *The octahedron recurrence and $\mathfrak{gl}_n$ crystals*, Adv. Math. **206** (2006), no. 1, 211–249, DOI 10.1016/j.aim.2005.08.007. MR2261754

[Hep94] Charels Hepler, *On the complexity of computing characters of finite groups*, Ph.D. thesis, Univ. of Calgary, 1994, 117 pp.; available at dspace.ucalgary.ca/handle/1880/45530

[HVW95] U. Hertrampf, H. Vollmer and K. Wagner, On the power of number-theoretic operations with respect to counting, in *Proc. 10th CCC* (1995), 299–314.

[HG76] A. P. Hillman and R. M. Grassl, *Reverse plane partitions and tableau hook numbers*, J. Combinatorial Theory Ser. A **21** (1976), no. 2, 216–221, DOI 10.1016/0097-3165(76)90065-0. MR414387

[vHL19] Peter van Hintum and Piet Lammers, *The bunkbed conjecture on the complete graph*, European J. Combin. **76** (2019), 175–177, DOI 10.1016/j.ejc.2018.10.002. MR3886521

[HW74] J. E. Hopcroft and J. K. Wong, *Linear time algorithm for isomorphism of planar graphs: preliminary report*, Sixth Annual ACM Symposium on Theory of Computing (Seattle, Wash., 1974), Association for Computing Machinery, New York, 1974, pp. 172–184. MR433964

[Hop14] Sam Hopkins, RSK via local transformations, unpublished notes (2014), 17 pp.; available at samuelfhopkins.com/docs/rsk.pdf

[Huh18] June Huh, *Combinatorial applications of the Hodge-Riemann relations*, Proceedings of the International Congress of Mathematicians—Rio de Janeiro 2018. Vol. IV. Invited lectures, World Sci. Publ., Hackensack, NJ, 2018, pp. 3093–3111. MR3966524

[Huh22] June Huh, Combinatorics and Hodge theory, in *Proc. ICM* (2022, virtual), Vol. I, 212–239; available at ems.press/books/standalone/273/5393

[H+22] June Huh, Jacob P. Matherne, Karola Mészáros, and Avery St. Dizier, *Logarithmic concavity of Schur and related polynomials*, Trans. Amer. Math. Soc. **375** (2022), no. 6, 4411–4427, DOI 10.1090/tran/8606. MR4419063

[Hur91] A. Hurwitz, *Ueber Riemann'sche Flächen mit gegebenen Verzweigungspunkten* (German), Math. Ann. **39** (1891), no. 1, 1–60, DOI 10.1007/BF01199469. MR1510692

[HNK23] Tom Hutchcroft, Alexander Kent, and Petar Nizić-Nikolac, *The bunkbed conjecture holds in the $p \uparrow 1$ limit*, Combin. Probab. Comput. **32** (2023), no. 3, 363–369, DOI 10.1017/s096354832200027x. MR4574852

[Ike12] Christian Ikenmeyer, *Geometric Complexity Theory, tensor rank, and Littlewood–Richardson coefficients*, Ph.D. thesis, Univ. Paderborn, 2012, 201 pp.

[Ike16] Christian Ikenmeyer, *Erratum to: Small Littlewood-Richardson coefficients [MR3514767]*, J. Algebraic Combin. **44** (2016), no. 1, 31–32, DOI 10.1007/s10801-016-0690-x. MR3514768

[IMW17] Christian Ikenmeyer, Ketan D. Mulmuley, and Michael Walter, *On vanishing of Kronecker coefficients*, Comput. Complexity **26** (2017), no. 4, 949–992, DOI 10.1007/s00037-017-0158-y. MR3723353

[IP22] Christian Ikenmeyer and Igor Pak, *What is in #P and what is not?*, 2022 IEEE 63rd Annual Symposium on Foundations of Computer Science—FOCS 2022, IEEE Computer Soc., Los Alamitos, CA, [2022] ©2022, pp. 860–871. MR4537263

[IPP22] Christian Ikenmeyer, Igor Pak, and Greta Panova, *Positivity of the symmetric group characters is as hard as the polynomial time hierarchy*, Proceedings of the 2023 Annual ACM-SIAM Symposium on Discrete Algorithms (SODA), SIAM, Philadelphia, PA, 2023, pp. 3573–3586, DOI 10.1137/1.9781611977554.ch136. MR4538090

[Ing26] Mark H. Ingraham, *Solution of certain functional equations relative to a general linear set*, Trans. Amer. Math. Soc. **28** (1926), no. 2, 287–300, DOI 10.2307/1989116. MR1501345

[Jam78] G. D. James, *The representation theory of the symmetric groups*, Lecture Notes in Mathematics, vol. 682, Springer, Berlin, 1978. MR513828

[Jam87] Gordon James, *The representation theory of the symmetric groups*, The Arcata Conference on Representations of Finite Groups (Arcata, Calif., 1986), Proc. Sympos. Pure Math., vol. 47, Amer. Math. Soc., Providence, RI, 1987, pp. 111–126, DOI 10.1090/pspum/047.1/933355. MR933355

[JK81] Gordon James and Adalbert Kerber, *The representation theory of the symmetric group*, Encyclopedia of Mathematics and its Applications, vol. 16, Addison-Wesley Publishing Co., Reading, MA, 1981. With a foreword by P. M. Cohn; With an introduction by Gilbert de B. Robinson. MR644144

[JP79] G. D. James and M. H. Peel, *Specht series for skew representations of symmetric groups*, J. Algebra **56** (1979), no. 2, 343–364, DOI 10.1016/0021-8693(79)90342-9. MR528580

[Jen12] Tommy R. Jensen, *Simple algorithm for finding a second Hamilton cycle*, Sib. Èlektron. Mat. Izv. **9** (2012), 151–155. MR2954697

[Jer06] Mark Jerrum, *Two remarks concerning balanced matroids*, Combinatorica **26** (2006), no. 6, 733–742, DOI 10.1007/s00493-006-0039-5. MR2288356

[JS93] Mark Jerrum and Alistair Sinclair, *Polynomial-time approximation algorithms for the Ising model*, SIAM J. Comput. **22** (1993), no. 5, 1087–1116, DOI 10.1137/0222066. MR1237164

[Kad99] Kevin W. J. Kadell, *An injection for the Ehrenpreis Rogers-Ramanujan problem*, J. Combin. Theory Ser. A **86** (1999), no. 2, 390–394, DOI 10.1006/jcta.1998.2929. MR1685539

[KS18] Mihyun Kang and Philipp Sprüssel, *Symmetries of unlabelled planar triangulations*, Electron. J. Combin. **25** (2018), no. 1, Paper No. 1.34, 38, DOI 10.37236/6188. MR3785013

[KS92] Sampath Kannan and Danny Soroker, *Tiling polygons with parallelograms*, Discrete Comput. Geom. **7** (1992), no. 2, 175–188, DOI 10.1007/BF02187834. MR1139079

[KR46] Irving Kaplansky and John Riordan, *The problem of the rooks and its applications*, Duke Math. J. **13** (1946), 259–268. MR16082

[Kas63] P. W. Kasteleyn, *Dimer statistics and phase transitions*, J. Mathematical Phys. **4** (1963), 287–293, DOI 10.1063/1.1703953. MR153427

[KS68] Douglas Gilbert Kelly, *THE GENERAL GRIFFITHS INEQUALITIES ON CORRELATIONS IN ISING FERROMAGNETS*, ProQuest LLC, Ann Arbor, MI, 1967. Thesis (Ph.D.)–Indiana University. MR2616479

[Ken93] Richard Kenyon, *Tiling a polygon with parallelograms*, Algorithmica **9** (1993), no. 4, 382–397, DOI 10.1007/BF01228510. MR1208568

[KPW00] Richard W. Kenyon, James G. Propp, and David B. Wilson, *Trees and matchings*, Electron. J. Combin. **7** (2000), Research Paper 25, 34, DOI 10.37236/1503. MR1756162

[Ker84] S. Kerov, *The Robinson-Schensted-Knuth correspondence and the Littlewood-Richardson rule* (Russian), Uspekhi Mat. Nauk **39** (1984), no. 2(236), 161–162. MR740018

[Ker93] S. V. Kerov, *Transition probabilities of continual Young diagrams and the Markov moment problem* (Russian, with Russian summary), Funktsional. Anal. i Prilozhen. **27** (1993), no. 2, 32–49, 96, DOI 10.1007/BF01085981; English transl., Funct. Anal. Appl. **27** (1993), no. 2, 104–117. MR1251166

[Kir92] A. N. Kirillov, *The Lagrange identity and the hook formula* (Russian, with English summary), Zap. Nauchn. Sem. Leningrad. Otdel. Mat. Inst. Steklov. (LOMI) **172** (1989), no. Differentsial′naya Geom. Gruppy Li i Mekh. Vol. 10, 78–87, 170, DOI 10.1007/BF01480689; English transl., J. Soviet Math. **59** (1992), no. 5, 1078–1084. MR1015700

[Kir04] Anatol N. Kirillov, *An invitation to the generalized saturation conjecture*, Publ. Res. Inst. Math. Sci. **40** (2004), no. 4, 1147–1239. MR2105706

[KR88] A. N. Kirillov and N. Yu. Reshetikhin, *The Bethe ansatz and the combinatorics of Young tableaux* (Russian, with English summary), Zap. Nauchn. Sem. Leningrad. Otdel. Mat. Inst. Steklov. (LOMI) **155** (1986), no. Differentsial′naya Geometriya, Gruppy Li i Mekh. VIII, 65–115, 194, DOI 10.1007/BF01247088; English transl., J. Soviet Math. **41** (1988), no. 2, 925–955. MR869577

[KSS02] Anatol N. Kirillov, Anne Schilling, and Mark Shimozono, *A bijection between Littlewood-Richardson tableaux and rigged configurations*, Selecta Math. (N.S.) **8** (2002), no. 1, 67–135, DOI 10.1007/s00029-002-8102-6. MR1890195

[Kle66] Daniel J. Kleitman, *Families of non-disjoint subsets*, J. Combinatorial Theory **1** (1966), 153–155. MR193020

[Kly74] Aleksandr Anatol′evich Klyachko, *Lie elements in a tensor algebra* (Russian), Sibirsk. Mat. Ž. **15** (1974), 1296–1304, 1430. MR371961

[Kly04] Aleksandr A. Klyachko, Quantum marginal problem and representations of the symmetric group, preprint (2004), 47 pp.; `arXiv:quant-ph/0409113`.

[Knu70] Donald E. Knuth, *Permutations, matrices, and generalized Young tableaux*, Pacific J. Math. **34** (1970), 709–727. MR272654

[Knu16] Allen Knutson, *Schubert calculus and puzzles*, Schubert calculus—Osaka 2012, Adv. Stud. Pure Math., vol. 71, Math. Soc. Japan, [Tokyo], 2016, pp. 185–209, DOI 10.2969/aspm/07110185. MR3644824

[Knu22] Allen Knutson, *Schubert calculus and quiver varieties*, ICM—International Congress of Mathematicians. Vol. VI. Sections 12–14, EMS Press, Berlin, [2023] ©2023, pp. 4582–4605. MR4680415

[KT99] Allen Knutson and Terence Tao, *The honeycomb model of* $\mathrm{GL}_n(\mathbf{C})$ *tensor products. I. Proof of the saturation conjecture*, J. Amer. Math. Soc. **12** (1999), no. 4, 1055–1090, DOI 10.1090/S0894-0347-99-00299-4. MR1671451

[KTW04] Allen Knutson, Terence Tao, and Christopher Woodward, *A positive proof of the Littlewood-Richardson rule using the octahedron recurrence*, Electron. J. Combin. **11** (2004), no. 1, Research Paper 61, 18, DOI 10.37236/1814. MR2097327

[KZ17] Allen Knutson and Paul Zinn-Justin, *Grassmann-Grassmann conormal varieties, integrability, and plane partitions* (English, with English and French summaries), Ann. Inst. Fourier (Grenoble) **69** (2019), no. 3, 1087–1145. MR3986911

[Kog01] Mikhail Kogan, *RC-graphs and a generalized Littlewood-Richardson rule*, Internat. Math. Res. Notices **15** (2001), 765–782, DOI 10.1155/S1073792801000393. MR1849481

[KT21] Dale Koenig and Anastasiia Tsvietkova, *NP-hard problems naturally arising in knot theory*, Trans. Amer. Math. Soc. Ser. B **8** (2021), 420–441, DOI 10.1090/btran/71. MR4273193

[Kon20] Matjaž Konvalinka, *A bijective proof of the hook-length formula for skew shapes*, European J. Combin. **88** (2020), 103104, 14, DOI 10.1016/j.ejc.2020.103104. MR4111716

[KP09] Matjaž Konvalinka and Igor Pak, *Geometry and complexity of O'Hara's algorithm*, Adv. in Appl. Math. **42** (2009), no. 2, 157–175, DOI 10.1016/j.aam.2008.06.005. MR2493975

[KW01] Witold Kraśkiewicz and Jerzy Weyman, *Algebra of coinvariants and the action of a Coxeter element*, Bayreuth. Math. Schr. **63** (2001), 265–284. MR1867283

[Kra96] C. Krattenthaler, *Combinatorial proof of the log-concavity of the sequence of matching numbers*, J. Combin. Theory Ser. A **74** (1996), no. 2, 351–354, DOI 10.1006/jcta.1996.0058. MR1385379

[Kra99a] C. Krattenthaler, *Another involution principle-free bijective proof of Stanley's hook-content formula*, J. Combin. Theory Ser. A **88** (1999), no. 1, 66–92, DOI 10.1006/jcta.1999.2979. MR1713492

[Kra99b] Adam Krawczyk, *The complexity of finding a second Hamiltonian cycle in cubic graphs*, J. Comput. System Sci. **58** (1999), no. 3, 641–647, DOI 10.1006/jcss.1998.1611. MR1705086

[KR84] Joseph P. S. Kung and Gian-Carlo Rota, *The invariant theory of binary forms*, Bull. Amer. Math. Soc. (N.S.) **10** (1984), no. 1, 27–85, DOI 10.1090/S0273-0979-1984-15188-7. MR722856

[Lac15] Marc Lackenby, *A polynomial upper bound on Reidemeister moves*, Ann. of Math. (2) **182** (2015), no. 2, 491–564, DOI 10.4007/annals.2015.182.2.3. MR3418524

[Lac17] Marc Lackenby, *Elementary knot theory*, Lectures on geometry, Clay Lect. Notes, Oxford Univ. Press, Oxford, 2017, pp. 29–64. MR3676592

[LLS21] Thomas Lam, Seung Jin Lee, and Mark Shimozono, *Back stable Schubert calculus*, Compos. Math. **157** (2021), no. 5, 883–962, DOI 10.1112/S0010437X21007028. MR4252201

[LPP07] Thomas Lam, Alexander Postnikov, and Pavlo Pylyavskyy, *Schur positivity and Schur log-concavity*, Amer. J. Math. **129** (2007), no. 6, 1611–1622, DOI 10.1353/ajm.2007.0045. MR2369890

[LZ04] Sergei K. Lando and Alexander K. Zvonkin, *Graphs on surfaces and their applications: Low-Dimensional Topology, II*, Encyclopaedia of Mathematical Sciences, vol. 141, Springer-Verlag, Berlin, 2004. With an appendix by Don B. Zagier, DOI 10.1007/978-3-540-38361-1. MR2036721

[Las95] Alain Lascoux, *Polynômes de Schubert: une approche historique* (French, with English and French summaries), Discrete Math. **139** (1995), no. 1-3, 303–317, DOI 10.1016/0012-365X(95)93984-D. Formal power series and algebraic combinatorics (Montreal, PQ, 1992). MR1336845

[Las03] Alain Lascoux, *Double crystal graphs*, Studies in memory of Issai Schur (Chevaleret/Rehovot, 2000), Progr. Math., vol. 210, Birkhäuser Boston, Boston, MA, 2003, pp. 95–114. MR1985724

[LS82] Alain Lascoux and Marcel-Paul Schützenberger, *Polynômes de Schubert* (French, with English summary), C. R. Acad. Sci. Paris Sér. I Math. **294** (1982), no. 13, 447–450. MR660739

[LS85] Alain Lascoux and Marcel-Paul Schützenberger, *Schubert polynomials and the Littlewood-Richardson rule*, Lett. Math. Phys. **10** (1985), no. 2-3, 111–124, DOI 10.1007/BF00398147. MR815233

[LT96] Bernard Leclerc and Jean-Yves Thibon, *The Robinson-Schensted correspondence, crystal bases, and the quantum straightening at $q = 0$: The Foata Festschrift*, Electron. J. Combin. **3** (1996), no. 2, Research Paper 11, approx. 24, DOI 10.37236/1269. MR1392496

[Lev73] L. A. Levin, *Universal enumeration problems* (Russian), Problemy Peredači Informacii **9** (1973), no. 3, 115–116. MR340042

[LV73] R. A. Liebler and M. R. Vitale, *Ordering the partition characters of the symmetric group*, J. Algebra **25** (1973), 487–489, DOI 10.1016/0021-8693(73)90095-1. MR316544

[Lin11] Svante Linusson, *Erratum to 'On percolation and the bunkbed conjecture'*, Combin. Probab. Comput. **28** (2019), no. 6, 917–918, DOI 10.1017/s0963548319000038. MR4015661

[LOT03] Maciej Liśkiewicz, Mitsunori Ogihara, and Seinosuke Toda, *The complexity of counting self-avoiding walks in subgraphs of two-dimensional grids and hypercubes*, Theoret. Comput. Sci. **304** (2003), no. 1-3, 129–156, DOI 10.1016/S0304-3975(03)00080-X. MR1992330

[Lit03] David P. Little, *Combinatorial aspects of the Lascoux-Schützenberger tree*, Adv. Math. **174** (2003), no. 2, 236–253, DOI 10.1016/S0001-8708(02)00038-5. MR1963694

[Lit94] Peter Littelmann, *A Littlewood-Richardson rule for symmetrizable Kac-Moody algebras*, Invent. Math. **116** (1994), no. 1-3, 329–346, DOI 10.1007/BF01231564. MR1253196

[Lit58] D. E. Littlewood, *Products and plethysms of characters with orthogonal, symplectic and symmetric groups*, Canadian J. Math. **10** (1958), 17–32, DOI 10.4153/CJM-1958-002-7. MR95209

[LR34] Dudley E. Littlewood and Archibald R. Richardson, Group characters and algebra, *Philos. Trans. Roy. Soc. London A* **233** (1934), 99–142.

[Liu10] Ricky Ini Liu, *An algorithmic Littlewood-Richardson rule*, J. Algebraic Combin. **31** (2010), no. 2, 253–266, DOI 10.1007/s10801-009-0184-1. MR2592078

[Liu17] Ricky Ini Liu, *A simplified Kronecker rule for one hook shape*, Proc. Amer. Math. Soc. **145** (2017), no. 9, 3657–3664, DOI 10.1090/proc/13692. MR3665021

[LM06] Nicholas A. Loehr and Anthony Mendes, *Bijective matrix algebra*, Linear Algebra Appl. **416** (2006), no. 2-3, 917–944, DOI 10.1016/j.laa.2006.01.004. MR2242475

[LR11] Nicholas A. Loehr and Jeffrey B. Remmel, *A computational and combinatorial exposé of plethystic calculus*, J. Algebraic Combin. **33** (2011), no. 2, 163–198, DOI 10.1007/s10801-010-0238-4. MR2765321

[LP86] L. Lovász and M. D. Plummer, *Matching theory*, North-Holland Mathematics Studies, vol. 121, North-Holland Publishing Co., Amsterdam; North-Holland Publishing Co., Amsterdam, 1986. Annals of Discrete Mathematics, 29. MR859549

[Luks82] Eugene M. Luks, *Isomorphism of graphs of bounded valence can be tested in polynomial time*, J. Comput. System Sci. **25** (1982), no. 1, 42–65, DOI 10.1016/0022-0000(82)90009-5. MR685360

[MS24] Zhao Yu Ma and Yair Shenfeld, *The extremals of Stanley's inequalities for partially ordered sets*, Adv. Math. **436** (2024), Paper No. 109404, 72, DOI 10.1016/j.aim.2023.109404. MR4669326

[Mac91] Ian G. Macdonald, *Notes on Schubert polynomials*, Publ. LaCIM, UQAM, Montreal, 1991, 116 pp.; available at tinyurl.com/382f7an7

[Mac95] I. G. Macdonald, *Symmetric functions and Hall polynomials*, 2nd ed., Oxford Mathematical Monographs, The Clarendon Press, Oxford University Press, New York, 1995. With contributions by A. Zelevinsky; Oxford Science Publications. MR1354144

[Mac15] Percy A. MacMahon, *Combinatory analysis*, Chelsea Publishing Co., New York, 1960. Two volumes (bound as one). MR141605

[Mah05] Karl Mahlburg, *Partition congruences and the Andrews-Garvan-Dyson crank*, Proc. Natl. Acad. Sci. USA **102** (2005), no. 43, 15373–15376, DOI 10.1073/pnas.0506702102. MR2188922

[Man01] Laurent Manivel, *Symmetric functions, Schubert polynomials and degeneracy loci: Cours Spécialisés [Specialized Courses], 3*, SMF/AMS Texts and Monographs, vol. 6, American Mathematical Society, Providence, RI; Société Mathématique de France, Paris, 2001. Translated from the 1998 French original by John R. Swallow. MR1852463

[Man15] Laurent Manivel, *On the asymptotics of Kronecker coefficients*, J. Algebraic Combin. **42** (2015), no. 4, 999–1025, DOI 10.1007/s10801-015-0614-1. MR3417256

[MSS15] Adam W. Marcus, Daniel A. Spielman, and Nikhil Srivastava, *Interlacing families I: Bipartite Ramanujan graphs of all degrees*, Ann. of Math. (2) **182** (2015), no. 1, 307–325, DOI 10.4007/annals.2015.182.1.7. MR3374962

[MOA11] Albert W. Marshall, Ingram Olkin, and Barry C. Arnold, *Inequalities: theory of majorization and its applications*, 2nd ed., Springer Series in Statistics, Springer, New York, 2011, DOI 10.1007/978-0-387-68276-1. MR2759813

[Mar08] Murray Marshall, *Positive polynomials and sums of squares*, Mathematical Surveys and Monographs, vol. 146, American Mathematical Society, Providence, RI, 2008, DOI 10.1090/surv/146. MR2383959

[May75] S. J. Mayer, *On the irreducible characters of the symmetric group*, Advances in Math. **15** (1975), 127–132, DOI 10.1016/0001-8708(75)90127-9. MR354839

[MPP14] Karola Mészáros, Greta Panova, and Alexander Postnikov, *Schur times Schubert via the Fomin-Kirillov algebra*, Electron. J. Combin. **21** (2014), no. 1, Paper 1.39, 22, DOI 10.37236/3659. MR3177534

[Minc78] Henryk Minc, *Permanents*, Encyclopedia of Mathematics and its Applications, vol. 6, Addison-Wesley Publishing Co., Reading, MA, 1978. With a foreword by Marvin Marcus. MR504978

[Mnëv88] N. E. Mnëv, *The universality theorems on the classification problem of configuration varieties and convex polytopes varieties*, Topology and geometry—Rohlin Seminar, Lecture Notes in Math., vol. 1346, Springer, Berlin, 1988, pp. 527–543, DOI 10.1007/BFb0082792. MR970093

[MPY22] Cara Monical, Benjamin Pankow, and Alexander Yong, *Reduced word enumeration, complexity, and randomization*, Electron. J. Combin. **29** (2022), no. 2, Paper No. 2.46, 28, DOI 10.37236/8560. MR4432979

[MM11] Cristopher Moore and Stephan Mertens, *The nature of computation*, Oxford University Press, Oxford, 2011, DOI 10.1093/acprof:oso/9780199233212.001.0001. MR2849868

[MPP17] Alejandro H. Morales, Igor Pak, and Greta Panova, *Hook formulas for skew shapes II. Combinatorial proofs and enumerative applications*, SIAM J. Discrete Math. **31** (2017), no. 3, 1953–1989, DOI 10.1137/16M1099625. MR3693598

[MPP18a] Alejandro H. Morales, Igor Pak, and Greta Panova, *Hook formulas for skew shapes I. q-analogues and bijections*, J. Combin. Theory Ser. A **154** (2018), 350–405, DOI 10.1016/j.jcta.2017.09.002. MR3718070

[MPP18b] Alejandro H. Morales, Igor Pak, and Greta Panova, *Asymptotics of the number of standard Young tableaux of skew shape*, European J. Combin. **70** (2018), 26–49, DOI 10.1016/j.ejc.2017.11.007. MR3779602

[MPP22] A. H. Morales, I. Pak, and G. Panova, *Hook formulas for skew shapes IV. Increasing tableaux and factorial Grothendieck polynomials*, Zap. Nauchn. Sem. S.-Peterburg. Otdel. Mat. Inst. Steklov. (POMI) **507** (2021), no. Teoriya Predstavleniĭ, Dinamicheski Sistemy, Kombinatornye Methody. XXXIII, 59–98; English transl., J. Math. Sci. (N.Y.) **261** (2022), no. 5, 630–657. MR4355382

[MQ17] Priyanka Mukhopadhyay and Youming Qiao, *Sparse multivariate polynomial interpolation on the basis of Schubert polynomials*, Comput. Complexity **26** (2017), no. 4, 881–909, DOI 10.1007/s00037-016-0142-y. MR3723351

[Mul11] Ketan D. Mulmuley, *Geometric complexity theory V: equivalence between blackbox derandomization of polynomial identity testing and derandomization of Noether's normalization lemma*, 2012 IEEE 53rd Annual Symposium on Foundations of Computer Science—FOCS 2012, IEEE Computer Soc., Los Alamitos, CA, 2012, pp. 629–638. MR3186651

[MNS12] Ketan D. Mulmuley, Hariharan Narayanan, and Milind Sohoni, *Geometric complexity theory III: on deciding nonvanishing of a Littlewood-Richardson coefficient*, J. Algebraic Combin. **36** (2012), no. 1, 103–110, DOI 10.1007/s10801-011-0325-1. MR2927658

[Mur96] Kunio Murasugi, *Knot theory and its applications*, Birkhäuser Boston, Inc., Boston, MA, 1996. Translated from the 1993 Japanese original by Bohdan Kurpita. MR1391727

[Mur38] F. D. Murnaghan, *The Analysis of the Kronecker Product of Irreducible Representations of the Symmetric Group*, Amer. J. Math. **60** (1938), no. 3, 761–784, DOI 10.2307/2371610. MR1507347

[Mur56] Francis D. Murnaghan, *On the Kronecker product of irreducible representations of the symmetric group*, Proc. Nat. Acad. Sci. U.S.A. **42** (1956), 95–98, DOI 10.1073/pnas.42.2.95. MR75213

[Myr92] Wendy Myrvold, *Counting k-component forests of a graph*, Networks **22** (1992), no. 7, 647–652, DOI 10.1002/net.3230220704. MR1189292

[Nad13] Philippe Nadeau, *Fully packed loop configurations in a triangle and Littlewood-Richardson coefficients*, J. Combin. Theory Ser. A **120** (2013), no. 8, 2137–2147, DOI 10.1016/j.jcta.2013.08.006. MR3102178

[Nag19] Trygve Nagell, Über zahlentheoretische Polynome (in German), *Norsk. Mat. Tidsskr* **1** (1919), 14–23; available at tinyurl.com/3cv69sbm

[Nak93] Toshiki Nakashima, *Crystal base and a generalization of the Littlewood-Richardson rule for the classical Lie algebras*, Comm. Math. Phys. **154** (1993), no. 2, 215–243. MR1224078

[NS11] Toshiki Nakashima and Miki Shimojo, *Pictures and Littlewood-Richardson crystals*, Tokyo J. Math. **34** (2011), no. 2, 493–506, DOI 10.3836/tjm/1327931398. MR2918918

[Nar06] Hariharan Narayanan, *On the complexity of computing Kostka numbers and Littlewood-Richardson coefficients*, J. Algebraic Combin. **24** (2006), no. 3, 347–354, DOI 10.1007/s10801-006-0008-5. MR2260022

[NPS97] Jean-Christophe Novelli, Igor Pak, and Alexander V. Stoyanovskii, *A direct bijective proof of the hook-length formula*, Discrete Math. Theor. Comput. Sci. **1** (1997), no. 1, 53–67. MR1605030

[Noy14] Marc Noy, *Random planar graphs and beyond*, Proceedings of the International Congress of Mathematicians—Seoul 2014. Vol. IV, Kyung Moon Sa, Seoul, 2014, pp. 407–430. MR3727618

[O'H90] Kathleen M. O'Hara, *Unimodality of Gaussian coefficients: a constructive proof*, J. Combin. Theory Ser. A **53** (1990), no. 1, 29–52, DOI 10.1016/0097-3165(90)90018-R. MR1031611

[Oko00] Andrei Okounkov, *Toda equations for Hurwitz numbers*, Math. Res. Lett. **7** (2000), no. 4, 447–453, DOI 10.4310/MRL.2000.v7.n4.a10. MR1783622

[OP09] A. Okounkov and R. Pandharipande, *Gromov-Witten theory, Hurwitz numbers, and matrix models*, Algebraic geometry—Seattle 2005. Part 1, Proc. Sympos. Pure Math., vol. 80, Amer. Math. Soc., Providence, RI, 2009, pp. 325–414, DOI 10.1090/pspum/080.1/2483941. MR2483941

[OV96] Andrei Okounkov and Anatoly Vershik, *A new approach to representation theory of symmetric groups*, Selecta Math. (N.S.) **2** (1996), no. 4, 581–605, DOI 10.1007/PL00001384. MR1443185

[Pak01] Igor Pak, *Hook length formula and geometric combinatorics*, Sém. Lothar. Combin. **46** (2001/02), Art. B46f, 13. MR1877632

[Pak03] Igor Pak, *Tile invariants: new horizons*: *Tilings of the plane*, Theoret. Comput. Sci. **303** (2003), no. 2-3, 303–331, DOI 10.1016/S0304-3975(02)00495-4. MR1990769

[Pak06] Igor Pak, *Partition bijections, a survey*, Ramanujan J. **12** (2006), no. 1, 5–75, DOI 10.1007/s11139-006-9576-1. MR2267263

[Pak18] Igor Pak, *Complexity problems in enumerative combinatorics*, Proceedings of the International Congress of Mathematicians—Rio de Janeiro 2018. Vol. IV. Invited lectures, World Sci. Publ., Hackensack, NJ, 2018, pp. 3153–3180. MR3966527

[Pak19] Igor Pak, Combinatorial inequalities, *Notices AMS* **66** (2019), 1109–1112; an expanded version of the paper is available at tinyurl.com/py8sv5v6

[PP13] Igor Pak and Greta Panova, *Strict unimodality of q-binomial coefficients* (English, with English and French summaries), C. R. Math. Acad. Sci. Paris **351** (2013), no. 11-12, 415–418, DOI 10.1016/j.crma.2013.06.008. MR3090120

[PP14] Igor Pak and Greta Panova, *Unimodality via Kronecker products*, J. Algebraic Combin. **40** (2014), no. 4, 1103–1120, DOI 10.1007/s10801-014-0520-y. MR3273403

[PP17] Igor Pak and Greta Panova, *On the complexity of computing Kronecker coefficients*, Comput. Complexity **26** (2017), no. 1, 1–36, DOI 10.1007/s00037-015-0109-4. MR3620776

[PP20a] Igor Pak and Greta Panova, *Bounds on Kronecker coefficients via contingency tables*, Linear Algebra Appl. **602** (2020), 157–178, DOI 10.1016/j.laa.2020.05.005. MR4102189

[PP20b] Igor Pak and Greta Panova, *Breaking down the reduced Kronecker coefficients*, C. R. Math. Acad. Sci. Paris **358** (2020), no. 4, 463–468, DOI 10.5802/crmath.60. MR4134255

[PP23] Igor Pak and Greta Panova, *Durfee squares, symmetric partitions and bounds on Kronecker coefficients*, J. Algebra **629** (2023), 358–380, DOI 10.1016/j.jalgebra.2023.04.006. MR4585439

[PP24+] Igor Pak and Greta Panova, in preparation (2024).

[PPY19] Igor Pak, Greta Panova, and Damir Yeliussizov, *On the largest Kronecker and Littlewood-Richardson coefficients*, J. Combin. Theory Ser. A **165** (2019), 44–77, DOI 10.1016/j.jcta.2019.01.008. MR3906648

[PPS20] Igor Pak, Fedor Petrov, and Viacheslav Sokolov, *Hook inequalities*, Math. Intelligencer **42** (2020), no. 2, 1–8, DOI 10.1007/s00283-019-09941-4. MR4109191

[PV05] Igor Pak and Ernesto Vallejo, *Combinatorics and geometry of Littlewood-Richardson cones*, European J. Combin. **26** (2005), no. 6, 995–1008, DOI 10.1016/j.ejc.2004.06.008. MR2143205

[PV10] Igor Pak and Ernesto Vallejo, *Reductions of Young tableau bijections*, SIAM J. Discrete Math. **24** (2010), no. 1, 113–145, DOI 10.1137/070689784. MR2600656

[PY13] Igor Pak and Jed Yang, *The complexity of generalized domino tilings*, Electron. J. Combin. **20** (2013), no. 4, Paper 12, 23, DOI 10.37236/2554. MR3139397

[PY14] Igor Pak and Jed Yang, Hard tiling problems with triangles and rhombi, unpublished preprint (2014); see IMA talk (slides and video) at tinyurl.com/yck428p9

[Pap94a] Christos H. Papadimitriou, *On the complexity of the parity argument and other inefficient proofs of existence*, J. Comput. System Sci. **48** (1994), no. 3, 498–532, DOI 10.1016/S0022-0000(05)80063-7. 31st Annual Symposium on Foundations of Computer Science (FOCS) (St. Louis, MO, 1990). MR1279412

[Pap94b] Christos H. Papadimitriou, *Computational complexity*, Addison-Wesley Publishing Company, Reading, MA, 1994. MR1251285

[PY17a] Oliver Pechenik and Alexander Yong, *Equivariant K-theory of Grassmannians*, Forum Math. Pi **5** (2017), e3, 128, DOI 10.1017/fmp.2017.4. MR3668467

[PY17b] Oliver Pechenik and Alexander Yong, *Equivariant K-theory of Grassmannians II: the Knutson-Vakil conjecture*, Compos. Math. **153** (2017), no. 4, 667–677, DOI 10.1112/S0010437X16008186. MR3631228

[Pet72] Julius Peterson, Beviser for Wilsons og Fermats Theoremer (in Danish, Proofs of the theorems of Wilson and Fermat), *Tidsskr. Math.* **2** (1872), 64–65.

[PR17] Zoran Z. Petrović and Marko Radovanović, *Recurrence formulas for Kostka and inverse Kostka numbers via quantum cohomology of Grassmannians*, Algebr. Represent. Theory **20** (2017), no. 2, 257–273, DOI 10.1007/s10468-016-9640-5. MR3638348

[Pop62] Karl R. Popper, *Conjectures and refutations: The growth of scientific knowledge*, Basic Books, Inc., Publishers, New York-London, 1962. MR152413

[PS09] Alexander Postnikov and Richard P. Stanley, *Chains in the Bruhat order*, J. Algebraic Combin. **29** (2009), no. 2, 133–174, DOI 10.1007/s10801-008-0125-4. MR2475632

[PS02] Dominique Poulalhon and Gilles Schaeffer, *Factorizations of large cycles in the symmetric group*, Discrete Math. **254** (2002), no. 1-3, 433–458, DOI 10.1016/S0012-365X(01)00361-2. MR1910123

[PS06] Dominique Poulalhon and Gilles Schaeffer, *Optimal coding and sampling of triangulations*, Algorithmica **46** (2006), no. 3-4, 505–527, DOI 10.1007/s00453-006-0114-8. MR2291967

[PR86] M. Pouzet and I. G. Rosenberg, *Sperner properties for groups and relations*, European J. Combin. **7** (1986), no. 4, 349–370, DOI 10.1016/S0195-6698(86)80007-5. MR868553

[Pri77] W. L. Price, *A topological transformation algorithm which relates the Hamiltonian circuits of a cubic planar map*, J. London Math. Soc. (2) **15** (1977), no. 2, 193–196, DOI 10.1112/jlms/s2-15.2.193. MR450112

[Pro82] Robert A. Proctor, *Solution of two difficult combinatorial problems with linear algebra*, Amer. Math. Monthly **89** (1982), no. 10, 721–734, DOI 10.2307/2975833. MR683197

[PB83] J. Scott Provan and Michael O. Ball, *The complexity of counting cuts and of computing the probability that a graph is connected*, SIAM J. Comput. **12** (1983), no. 4, 777–788, DOI 10.1137/0212053. MR721012

[Pur08] Kevin Purbhoo, *Puzzles, tableaux, and mosaics*, J. Algebraic Combin. **28** (2008), no. 4, 461–480, DOI 10.1007/s10801-007-0110-3. MR2443281

[Pyl04] Pavlo Pylyavskyy, *On plethysm conjectures of Stanley and Foulkes: the $2 \times n$ case*, Electron. J. Combin. **11** (2004/06), no. 2, Research Paper 8, 5, DOI 10.37236/1865. MR2120103

[RW20] Victor Reiner and Dennis White, Some notes on Pólya's theorem, Kostka numbers and the RSK correspondence, preprint (2020), 10 pp.; available at tinyurl.com/y58rjzka

[RW98] Jeffrey B. Remmel and Mark Shimozono, *A simple proof of the Littlewood-Richardson rule and applications*, Discrete Math. **193** (1998), no. 1-3, 257–266, DOI 10.1016/S0012-365X(98)00145-9. Selected papers in honor of Adriano Garsia (Taormina, 1994). MR1661373

[RW84] J. B. Remmel and R. Whitney, *Multiplying Schur functions*, J. Algorithms **5** (1984), no. 4, 471–487, DOI 10.1016/0196-6774(84)90002-6. MR769977

[Ric22] Thomas Richthammer, Bunkbed conjecture for complete bipartite graphs and related classes of graphs, preprint (2022), 10 pp.; `arXiv:2204.12931`.

[RYY22] Colleen Robichaux, Harshit Yadav, and Alexander Yong, *Equivariant cohomology, Schubert calculus, and edge labeled tableaux*, Facets of algebraic geometry. Vol. II, London Math. Soc. Lecture Note Ser., vol. 473, Cambridge Univ. Press, Cambridge, 2022, pp. 284–335. MR4381918

[Rob38] G. de B. Robinson, *On the Representations of the Symmetric Group*, Amer. J. Math. **60** (1938), no. 3, 745–760, DOI 10.2307/2371609. MR1507943

[Sag92] Bruce E. Sagan, *Inductive proofs of q-log concavity*, Discrete Math. **99** (1992), no. 1-3, 289–306, DOI 10.1016/0012-365X(92)90377-R. MR1158792

[Sag01] Bruce E. Sagan, *The symmetric group: Representations, combinatorial algorithms, and symmetric functions*, 2nd ed., Graduate Texts in Mathematics, vol. 203, Springer-Verlag, New York, 2001, DOI 10.1007/978-1-4757-6804-6. MR1824028

[StR81] J. Saint-Raymond, *Sur le volume des corps convexes symétriques* (French), Initiation Seminar on Analysis: G. Choquet-M. Rogalski-J. Saint-Raymond, 20th Year: 1980/1981, Publ. Math. Univ. Pierre et Marie Curie, vol. 46, Univ. Paris VI, Paris, 1981, pp. Exp. No. 11, 25. MR670798

[Sam12] Steven V. Sam, *Symmetric quivers, invariant theory, and saturation theorems for the classical groups*, Adv. Math. **229** (2012), no. 2, 1104–1135, DOI 10.1016/j.aim.2011.10.009. MR2855088

[Scha10] Marcus Schaefer, *Complexity of some geometric and topological problems*, Graph drawing, Lecture Notes in Comput. Sci., vol. 5849, Springer, Berlin, 2010, pp. 334–344, DOI 10.1007/978-3-642-11805-0_32. MR2680464

[Scha15] Gilles Schaeffer, *Planar maps*, Handbook of enumerative combinatorics, Discrete Math. Appl. (Boca Raton), CRC Press, Boca Raton, FL, 2015, pp. 335–395. MR3409346

[SS84] Frank W. Schmidt and Rodica Simion, *Addendum to: "On a partition identity"*, J. Combin. Theory Ser. A **40** (1985), no. 2, 456–458, DOI 10.1016/0097-3165(85)90108-6. MR814432

[Schr03] Alexander Schrijver, *Combinatorial optimization. Polyhedra and efficiency. Vol. C: Disjoint paths, hypergraphs; Chapters 70–83*, Algorithms and Combinatorics, vol. 24, Springer-Verlag, Berlin, 2003. MR1956926

[Schr98] Alexander Schrijver, *Counting* 1*-factors in regular bipartite graphs*, J. Combin. Theory Ser. B **72** (1998), no. 1, 122–135, DOI 10.1006/jctb.1997.1798. MR1604705

[Schü72] M. P. Schützenberger, *Promotion des morphismes d'ensembles ordonnés* (French), Discrete Math. **2** (1972), 73–94, DOI 10.1016/0012-365X(72)90062-3. MR299539

[Schü77] M.-P. Schützenberger, *La correspondance de Robinson* (French), Combinatoire et représentation du groupe symétrique (Actes Table Ronde CNRS, Univ. Louis-Pasteur Strasbourg, Strasbourg, 1976), Lecture Notes in Math., Vol. 579, Springer, Berlin-New York, 1977, pp. 59–113. MR498826

[SW19] Pascal Schweitzer and Daniel Wiebking, *A unifying method for the design of algorithms canonizing combinatorial objects*, STOC'19—Proceedings of the 51st Annual ACM SIGACT Symposium on Theory of Computing, ACM, New York, 2019, pp. 1247–1258, DOI 10.1145/3313276.3316338. MR4003426

[Ser03] Ákos Seress, *Permutation group algorithms*, Cambridge Tracts in Mathematics, vol. 152, Cambridge University Press, Cambridge, 2003, DOI 10.1017/CBO9780511546549. MR1970241

[Ser77] Jean-Pierre Serre, *Linear representations of finite groups*, Graduate Texts in Mathematics, Vol. 42, Springer-Verlag, New York-Heidelberg, 1977. Translated from the second French edition by Leonard L. Scott. MR450380

[SvH23] Yair Shenfeld and Ramon van Handel, *The extremals of the Alexandrov-Fenchel inequality for convex polytopes*, Acta Math. **231** (2023), no. 1, 89–204. MR4652411

[She80] L. A. Shepp, *The FKG inequality and some monotonicity properties of partial orders*, SIAM J. Algebraic Discrete Methods **1** (1980), no. 3, 295–299, DOI 10.1137/0601034. MR586157

[She82] L. A. Shepp, *The XYZ conjecture and the FKG inequality*, Ann. Probab. **10** (1982), no. 3, 824–827. MR659563

[Shor91] Peter W. Shor, *Stretchability of pseudolines is NP-hard*, Applied geometry and discrete mathematics, DIMACS Ser. Discrete Math. Theoret. Comput. Sci., vol. 4, Amer. Math. Soc., Providence, RI, 1991, pp. 531–554, DOI 10.1090/dimacs/004/41. MR1116375

[Sid91] A. Sidorenko, *Inequalities for the number of linear extensions*, Order **8** (1991/92), no. 4, 331–340, DOI 10.1007/BF00571183. MR1173139

[OEIS] Neil J. A. Sloane, The Online Encyclopedia of Integer Sequences, `oeis.org`.

[Sol61] Louis Solomon, *On the sum of the elements in the character table of a finite group*, Proc. Amer. Math. Soc. **12** (1961), 962–963, DOI 10.2307/2034401. MR132783

[Spe07] David E. Speyer, *Perfect matchings and the octahedron recurrence*, J. Algebraic Combin. **25** (2007), no. 3, 309–348, DOI 10.1007/s10801-006-0039-y. MR2317336

[Sta81] Richard P. Stanley, *Two combinatorial applications of the Aleksandrov-Fenchel inequalities*, J. Combin. Theory Ser. A **31** (1981), no. 1, 56–65, DOI 10.1016/0097-3165(81)90053-4. MR626441

[Sta84] Richard P. Stanley, *On the number of reduced decompositions of elements of Coxeter groups*, European J. Combin. **5** (1984), no. 4, 359–372, DOI 10.1016/S0195-6698(84)80039-6. MR782057

[Sta86] Richard P. Stanley, *Two poset polytopes*, Discrete Comput. Geom. **1** (1986), no. 1, 9–23, DOI 10.1007/BF02187680. MR824105

[Sta89] Richard P. Stanley, *Log-concave and unimodal sequences in algebra, combinatorics, and geometry*, Graph theory and its applications: East and West (Jinan, 1986), Ann. New York Acad. Sci., vol. 576, New York Acad. Sci., New York, 1989, pp. 500–535, DOI 10.1111/j.1749-6632.1989.tb16434.x. MR1110850

[Sta99] Richard P. Stanley, *Enumerative combinatorics. Vol. 2*, Cambridge Studies in Advanced Mathematics, vol. 62, Cambridge University Press, Cambridge, 1999. With a foreword by Gian-Carlo Rota and appendix 1 by Sergey Fomin, DOI 10.1017/CBO9780511609589. MR1676282

[Sta00] Richard P. Stanley, *Positivity problems and conjectures in algebraic combinatorics*, Mathematics: frontiers and perspectives, Amer. Math. Soc., Providence, RI, 2000, pp. 295–319. MR1754784

[Sta15] Richard P. Stanley, *Catalan numbers*, Cambridge University Press, New York, 2015, DOI 10.1017/CBO9781139871495. MR3467982

[SW85] Dennis W. Stanton and Dennis E. White, *A Schensted algorithm for rim hook tableaux*, J. Combin. Theory Ser. A **40** (1985), no. 2, 211–247, DOI 10.1016/0097-3165(85)90088-3. MR814412

[Ste88] Robert Steinberg, *An occurrence of the Robinson-Schensted correspondence*, J. Algebra **113** (1988), no. 2, 523–528, DOI 10.1016/0021-8693(88)90177-9. MR929778

[Ste02] John R. Stembridge, *A concise proof of the Littlewood-Richardson rule*, Electron. J. Combin. **9** (2002), no. 1, Note 5, 4, DOI 10.37236/1666. MR1912814

[Sul18] Seth Sullivant, *Algebraic statistics*, Graduate Studies in Mathematics, vol. 194, American Mathematical Society, Providence, RI, 2018, DOI 10.1090/gsm/194. MR3838364

[Sun18] Sheila Sundaram, *On conjugacy classes of S_n containing all irreducibles*, Israel J. Math. **225** (2018), no. 1, 321–342, DOI 10.1007/s11856-018-1659-3. MR3805649

[Syl78] James Joseph Sylvester, Proof of the hitherto undemonstrated Fundamental Theorem of Invariants, *Phil. Mag.* **5** (1878), 178–188; reprinted in *Coll. Math. Papers*, vol. 3, Chelsea, New York, 1973, 117–126; available at tinyurl.com/c94pphj

[Tam04] Harry Tamvakis, *The connection between representation theory and Schubert calculus*, Enseign. Math. (2) **50** (2004), no. 3-4, 267–286. MR2116717

[TKA18] I. Terada, R. C. King, and O. Azenhas, *The symmetry of Littlewood-Richardson coefficients: a new hive model involutory bijection*, SIAM J. Discrete Math. **32** (2018), no. 4, 2850–2899, DOI 10.1137/17M1162834. MR3885749

[Tho74] Glânffrwd P. Thomas, *Baxter algebras and Schur functions*, Ph.D. thesis, Swansea, 1974.

[TY08] Hugh Thomas and Alexander Yong, *An S_3-symmetric Littlewood-Richardson rule*, Math. Res. Lett. **15** (2008), no. 5, 1027–1037, DOI 10.4310/MRL.2008.v15.n5.a15. MR2443999

[TY09] Hugh Thomas and Alexander Yong, *A jeu de taquin theory for increasing tableaux, with applications to K-theoretic Schubert calculus*, Algebra Number Theory **3** (2009), no. 2, 121–148, DOI 10.2140/ant.2009.3.121. MR2491941

[TY11] Hugh Thomas and Alexander Yong, *Longest increasing subsequences, Plancherel-type measure and the Hecke insertion algorithm*, Adv. in Appl. Math. **46** (2011), no. 1-4, 610–642, DOI 10.1016/j.aam.2009.07.005. MR2794040

[TY18] Hugh Thomas and Alexander Yong, *Equivariant Schubert calculus and jeu de taquin* (English, with English and French summaries), Ann. Inst. Fourier (Grenoble) **68** (2018), no. 1, 275–318. MR3795480

[Tho78] A. G. Thomason, *Hamiltonian cycles and uniquely edge colourable graphs*, Ann. Discrete Math. **3** (1978), 259–268. MR499124

[Tro95] William T. Trotter, *Partially ordered sets*, Handbook of combinatorics, Vol. 1, 2, Elsevier Sci. B. V., Amsterdam, 1995, pp. 433–480. MR1373664

[Tut46] W. T. Tutte, *On Hamiltonian circuits*, J. London Math. Soc. **21** (1946), 98–101, DOI 10.1112/jlms/s1-21.2.98. MR19300

[Vak06] Ravi Vakil, *A geometric Littlewood-Richardson rule*, Ann. of Math. (2) **164** (2006), no. 2, 371–421, DOI 10.4007/annals.2006.164.371. Appendix A written with A. Knutson. MR2247964

[Val79a] Leslie G. Valiant, *The complexity of enumeration and reliability problems*, SIAM J. Comput. **8** (1979), no. 3, 410–421, DOI 10.1137/0208032. MR539258

[Val79b] L. G. Valiant, *Completeness classes in algebra*, Conference Record of the Eleventh Annual ACM Symposium on Theory of Computing (Atlanta, Ga., 1979), ACM, New York, 1979, pp. 249–261. MR564634

[Val79c] L. G. Valiant, *The complexity of computing the permanent*, Theoret. Comput. Sci. **8** (1979), no. 2, 189–201, DOI 10.1016/0304-3975(79)90044-6. MR526203

[Val99] Ernesto Vallejo, *Stability of Kronecker products of irreducible characters of the symmetric group*, Electron. J. Combin. **6** (1999), Research Paper 39, 7, DOI 10.37236/1471. MR1725703

[Val00] Ernesto Vallejo, *Plane partitions and characters of the symmetric group*, J. Algebraic Combin. **11** (2000), no. 1, 79–88, DOI 10.1023/A:1008795704190. MR1747064

[Val14] Ernesto Vallejo, *A diagrammatic approach to Kronecker squares*, J. Combin. Theory Ser. A **127** (2014), 243–285, DOI 10.1016/j.jcta.2014.06.002. MR3252663

[vdBK01] J. van den Berg and J. Kahn, *A correlation inequality for connection events in percolation*, Ann. Probab. **29** (2001), no. 1, 123–126, DOI 10.1214/aop/1008956324. MR1825144

[vL00] Marc A. A. van Leeuwen, *Flag varieties and interpretations of Young tableau algorithms*, J. Algebra **224** (2000), no. 2, 397–426, DOI 10.1006/jabr.1999.8070. MR1739585

[vL01] Marc A. A. van Leeuwen, *The Littlewood-Richardson rule, and related combinatorics*, Interaction of combinatorics and representation theory, MSJ Mem., vol. 11, Math. Soc. Japan, Tokyo, 2001, pp. 95–145. MR1862150

[vL82] J. H. van Lint, *The van der Waerden conjecture: two proofs in one year*, Math. Intelligencer **4** (1982), no. 2, 72–77, DOI 10.1007/BF03023485. MR672919

[Vat18] Vince Vatter, Notes on the Pouzet–Rosenberg unimodality proof, research notes (March 25, 2018).

[Vaz01] Vijay V. Vazirani, *Approximation algorithms*, Springer-Verlag, Berlin, 2001. MR1851303

[Ver92] A. M. Vershik, *The hook formula and related identities* (Russian, with English summary), Zap. Nauchn. Sem. Leningrad. Otdel. Mat. Inst. Steklov. (LOMI) **172** (1989), no. Differentsial′naya Geom. Gruppy Li i Mekh. Vol. 10, 3–20, 169, DOI 10.1007/BF01480684; English transl., J. Soviet Math. **59** (1992), no. 5, 1029–1040. MR1015695

[Ver06] A. M. Vershik, *A new approach to the representation theory of the symmetric groups. III. Induced representations and the Frobenius-Young correspondence* (English, with English and Russian summaries), Mosc. Math. J. **6** (2006), no. 3, 567–585, 588, DOI 10.17323/1609-4514-2006-6-3-567-585. MR2274866

[Ver98] Dirk Vertigan, *Bicycle dimension and special points of the Tutte polynomial*, J. Combin. Theory Ser. B **74** (1998), no. 2, 378–396, DOI 10.1006/jctb.1998.1860. MR1654172

[Vie77] G. Viennot, *Une forme géométrique de la correspondance de Robinson-Schensted* (French), Combinatoire et représentation du groupe symétrique (Actes Table Ronde CNRS, Univ. Louis-Pasteur Strasbourg, Strasbourg, 1976), Lecture Notes in Math., Vol. 579, Springer, Berlin-New York, 1977, pp. 29–58. MR470059

[VW83] Kiem Phong Vo and Roger Whitney, *Tableaux and matrix correspondences*, J. Combin. Theory Ser. A **35** (1983), no. 3, 328–359, DOI 10.1016/0095-8956(83)90060-6. MR721373

[Voo79] M. Voorhoeve, *A lower bound for the permanents of certain* $(0,1)$*-matrices*, Nederl. Akad. Wetensch. Indag. Math. **41** (1979), no. 1, 83–86. MR528221

[Wal05] Timothy R. Walsh, *Efficient enumeration of sensed planar maps*, Discrete Math. **293** (2005), no. 1-3, 263–289, DOI 10.1016/j.disc.2004.08.036. MR2136069

[Wang61] Hao Wang, Proving theorems by pattern recognition–II, *Bell System Tech. J.* **40** (1961), 1–41.

[Wel93] D. J. A. Welsh, *Complexity: knots, colourings and counting*, London Mathematical Society Lecture Note Series, vol. 186, Cambridge University Press, Cambridge, 1993, DOI 10.1017/CBO9780511752506. MR1245272

[Weyl39] Hermann Weyl, *The Classical Groups. Their Invariants and Representations*, Princeton University Press, Princeton, NJ, 1939. MR255

[Whi80] Dennis E. White, *Monotonicity and unimodality of the pattern inventory*, Adv. in Math. **38** (1980), no. 1, 101–108, DOI 10.1016/0001-8708(80)90059-6. MR594996

[Whi81] Dennis E. White, *Some connections between the Littlewood-Richardson rule and the construction of Schensted*, J. Combin. Theory Ser. A **30** (1981), no. 3, 237–247, DOI 10.1016/0097-3165(81)90020-0. MR618529

[Whi83] Dennis E. White, *A bijection proving orthogonality of the characters of* S_n, Adv. in Math. **50** (1983), no. 2, 160–186, DOI 10.1016/0001-8708(83)90038-5. MR723200

[Whi85] Dennis White, *Orthogonality of the characters of* S_n, J. Combin. Theory Ser. A **40** (1985), no. 2, 265–275, DOI 10.1016/0097-3165(85)90090-1. MR814414

[Whi19] Dennis White, From Schensted to Pólya, in *Open Problems in Algebraic Combinatorics* blog (Dec. 2019), 3 pp.; available at tinyurl.com/anh3czca

[Wig19] Avi Wigderson, *Mathematics and computation: A theory revolutionizing technology and science*, Princeton University Press, Princeton, NJ, [2019] ©2019. MR4205964

[Wilf82] Herbert S. Wilf, *What is an answer?*, Amer. Math. Monthly **89** (1982), no. 5, 289–292, DOI 10.2307/2321713. MR653502

[Wor18] Nicholas Wormald, *Asymptotic enumeration of graphs with given degree sequence*, Proceedings of the International Congress of Mathematicians—Rio de Janeiro 2018. Vol. IV. Invited lectures, World Sci. Publ., Hackensack, NJ, 2018, pp. 3245–3264. MR3966531

[Zhuk20] Dmitriy Zhuk, *A proof of the CSP dichotomy conjecture*, J. ACM **67** (2020), no. 5, Art. 30, 78, DOI 10.1145/3402029. MR4152441

[Zei84] Doron Zeilberger, *A short hook-lengths bijection inspired by the Greene-Nijenhuis-Wilf proof*, Discrete Math. **51** (1984), no. 1, 101–108, DOI 10.1016/0012-365X(84)90027-X. MR755045

[Zel81] A. V. Zelevinsky, *A generalization of the Littlewood-Richardson rule and the Robinson-Schensted-Knuth correspondence*, J. Algebra **69** (1981), no. 1, 82–94, DOI 10.1016/0021-8693(81)90128-9. MR613858

[Zel99] Andrei Zelevinsky, *Littlewood-Richardson semigroups*, New perspectives in algebraic combinatorics (Berkeley, CA, 1996), Math. Sci. Res. Inst. Publ., vol. 38, Cambridge Univ. Press, Cambridge, 1999, pp. 337–345. MR1731821

[Zin09] Paul Zinn-Justin, *Littlewood-Richardson coefficients and integrable tilings*, Electron. J. Combin. **16** (2009), no. 1, Research Paper 12, 33, DOI 10.37236/101. MR2475535

Department of Mathematics, UCLA, Los Angeles, California 90095
Email address: pak@math.ucla.edu

Proceedings of Symposia in Pure Mathematics
Volume **110**, 2024
https://doi.org/10.1090/pspum/110/02008

Trimer covers in the triangular grid: Twenty mostly open problems

James Propp

Abstract. In the past three decades, the study of rhombus tilings and domino tilings of various plane regions has been a thriving subfield of enumerative combinatorics. Physicists classify such work as the study of dimer covers of finite graphs. In this article we move beyond dimer covers to trimer covers, introducing plane regions called *benzels* that play a role analogous to hexagons for rhombus tilings and Aztec diamonds for domino tilings, inasmuch as one finds many (so far mostly conjectural) exact formulas governing the number of tilings.

1. Introduction

If $G = (V, E)$ is a finite graph, a **trimer** on G is a three-element subset of V whose induced subgraph in G is connected, and a **trimer cover** is a partition of V into trimers. Solving the trimer model for G means counting the possible trimer covers. In the physics literature one often assigns weights to the trimers, and solving the weighted model means finding the sum of the weights of all the trimer covers of G, where the weight of an individual trimer cover is the product of the weights of its constituent trimers; this sum is called the *partition function*. As a special case, one can set some of the weights equal to 1 and the rest equal to 0; then the partition function is just the number of trimer covers that only use "permitted" trimers (trimers that have been assigned weight 1).

Physicists have studied the asymptotics of the weighted trimer model on the triangular lattice and obtained formulas for the entropy in various regimes (see [**10**]). These results can be seen as analogous to formulas for the entropy of dimer models on various kinds of finite graphs. However, as far as I am aware there have not been (up to now) any exact enumerative results for trimer models on finite subgraphs of the triangular lattice in the style of the well-known enumerations of rhombus tilings of hexagons and domino tilings of the Aztec diamond [**6**].

In this article I introduce finite subgraphs of the triangular lattice that should interest enumerative combinatorialists inasmuch as the number of trimer covers appears to be given by exact formulas in many cases. See Figure 1 which shows both a trimer cover of a finite subgraph of the triangular lattice and the associated tiling

2020 *Mathematics Subject Classification.* Primary 05C30.

The author is supported by the Simons Foundation Travel Support for Mathematicians program.

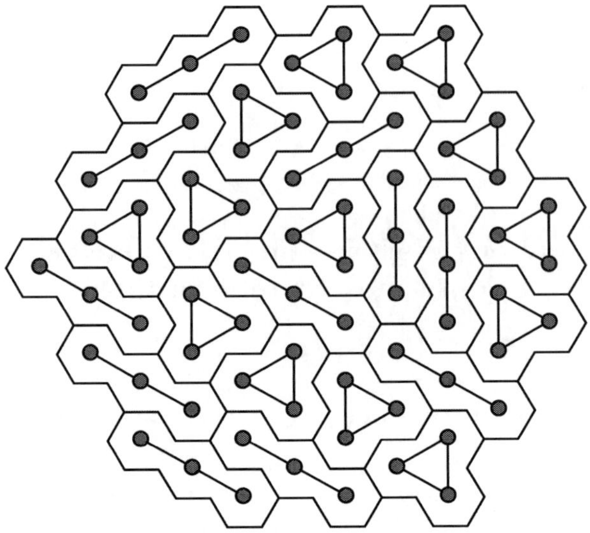

FIGURE 1. The (9,9)-benzel tiled by stones and bones and the associated trimer cover of the (9,9)-benzel graph.

of a dual 2-complex whose faces correspond to vertices of the triangular lattice. In tiling language, the tiled region in the Figure is a *polyhex* and the tiles are *trihexes.* I have dubbed the tiled region a *benzel*, and the allowed tiles *stones* and *bones.* I will switch between the trimer picture and the tiling picture throughout the article. The graph admits an essentially unique coloring with the colors red, blue, and green in which no two adjacent vertices have been assigned the same color; under such a coloring, each stone-trimer and bone-trimer contains exactly one vertex of each color.

This article is a companion to my talk at the Open Problems in Algebraic Combinatorics conference on May 18, 2022; the slides and video are available through the conference website `http://www.samuelfhopkins.com/OPAC/opac.html` and updates on the problems are available at `http://jamespropp.org/benzels.html`.

Consider the complex plane tiled by unit hexagonal cells centered at 1, ω, and ω^2 (here and hereafter ω denotes $e^{2\pi i/3}$); the cell centered at α has corners at $\alpha \pm 1$, $\alpha \pm \omega$, and $\alpha \pm \omega^2$. Given positive integers a, b satisfying $2 \leq a \leq 2b$ and $2 \leq b \leq 2a$, we define the **(a, b)-benzel** as the union of the cells that lie fully inside the hexagon with vertices $a\omega + b$, $-a\omega^2 - b$, $a\omega^2 + b\omega$, $-a - b\omega$, $a + b\omega^2$, and $-a\omega - b\omega^2$ (a hexagon centered at 0 with threefold rotational symmetry whose side-lengths alternate between $2a - b$ and $2b - a$, degenerating to a triangle when $a = 2b$ or $b = 2a$), as shown in Figure 2 for $a = 4$, $b = 6$. (Mnemonically, a is the height of the top edge of the hexagon *above* 0, and b is the depth of the bottom edge of the hexagon *below* 0, using appropriately scaled units.) We lose no generality in assuming $2 \leq a \leq 2b$ and $2 \leq b \leq 2a$, since the (a, b)-benzel as defined above coincides with the $(a, a-b)$-benzel when $a > 2b$ and with the $(b-a, b)$-benzel when $b > 2a$, the former satisfying $a \leq 2(a - b)$ and the latter satisfying $b \leq 2(b - a)$.

Here is an alternative description of benzels in terms of the centers of the cells using barycentric coordinates relative to the triangle with vertices 1, ω, and ω^2. Each cell's center point α belongs to $\mathbb{Z}[\omega]$ and can be represented by the unique

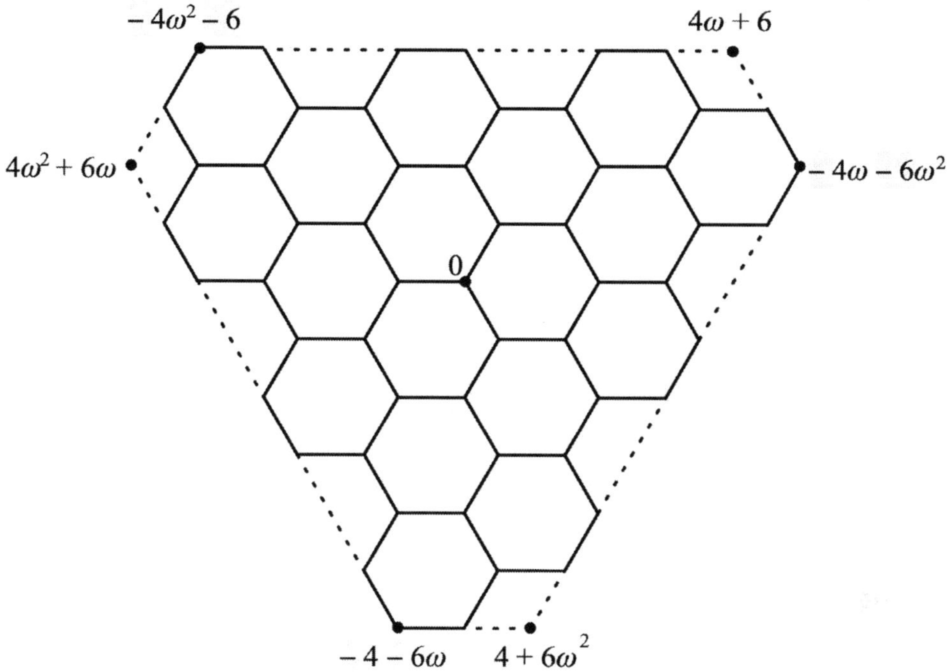

FIGURE 2. The (4,6)-benzel and its enclosing hexagon.

$(i,j,k) \in \mathbb{Z}^3$ satisfying $i+j\omega+k\omega^2 = \alpha$ and $i+j+k=1$ (here and hereafter i denotes a nonnegative integer as opposed to the square root of -1). Then the (a,b)-benzel consists of those cells whose centers (i,j,k) satisfy $-(a-1) \leq j-i, k-j, i-k \leq b-1$. Figure 3 shows the (4,6)-benzel with its cells marked with the barycentric coordinates of their respective center points.

Let $V = \{(i,j,k) \in \mathbb{Z}^3 \,:\, i+j+k=1,\ -(a-1) \leq j-i,\ k-j,\ i-k \leq b-1\}$. Given (i_1,j_1,k_1) and (i_2,j_2,k_2) in V, join (i_1,j_1,k_1) and (i_2,j_2,k_2) by an edge when $|i_1-i_2|+|j_1-j_2|+|k_1-k_2|=2$ (that is, when the unit hexagons centered on those two vertices share an edge). This is the **(a,b)-benzel graph**.

The (a,b)-benzel has threefold rotational symmetry but for most a,b it does not have bilateral symmetry. Exchanging a and b corresponds to reflecting the benzel across a horizontal axis (or if one prefers across an axis making a 60 degree angle with the horizontal axis).

We consider tilings of the (a,b)-benzel by way of five sorts of prototiles. These prototiles (shown in Figure 4) are the **right(-pointing) stone**, the **left(-pointing) stone**, the **vertical bone**, the **rising bone**, and the **falling bone**. The right stone is a benzel (specifically, the (2,2)-benzel) but the left stone is not. As usual, the allowed tiles are the translates of the prototiles. Dually we form spanning subgraphs of the (a,b)-benzel graph whose connected components all consist of three vertices. Figure 1 shows a tiling of the (9,9)-benzel and the associated trimer cover of the (9,9)-benzel graph.

There is a third kind of trihex that consists of three hexagons whose centers form a 120-degree angle. Timothy Chow has dubbed this kind of tile a *phone*.

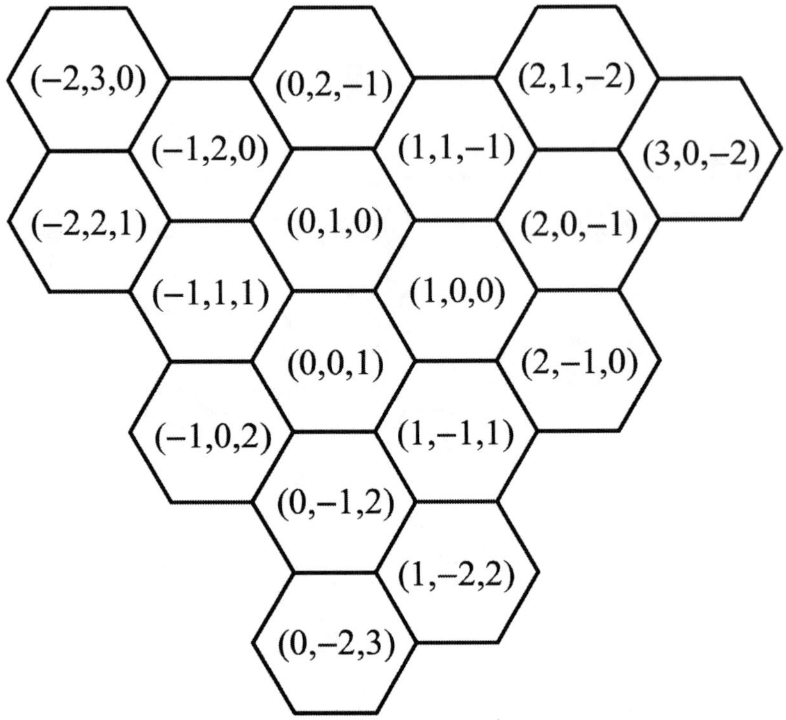

FIGURE 3. Barycentric coordinates for the cells of the (4,6)-benzel.

Since the associated trimers do not respect the tripartite nature of the triangular lattice, I have chosen to exclude those trimers from consideration here. This is not to suggest that such trimers are unworthy of study. Just as enumeration of dimer covers should not be limited to the bipartite case, enumeration of trimer covers should not be limited to the tripartite case. However, it seems best to start with the problems that are likely to be easier. In particular, the Conway-Lagarias invariant discussed below only applies if all our trihexes are stones and bones.

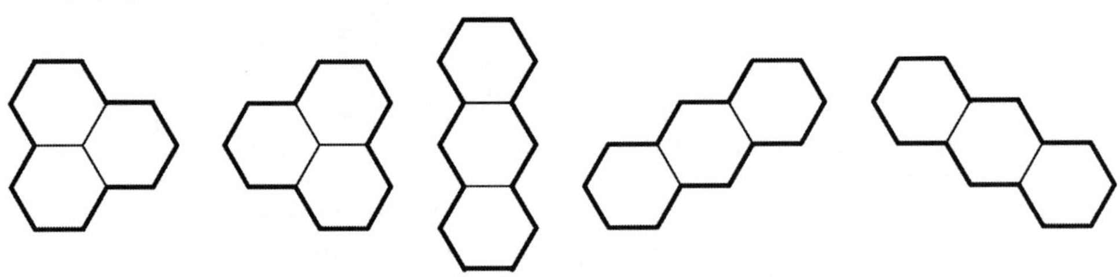

FIGURE 4. The five prototiles: the right stone, the left stone, the vertical bone, the rising bone, and the falling bone.

Conway and Lagarias [2] studied tilings like the ones considered here (calling stones and bones T_2 and L_3 tiles respectively). They showed that for any simply-connected region in the hexagonal grid that can be tiled by stones and bones, the total area of the right stones minus the total area of the left stones does not depend on the specific tiling but only depends on the region being tiled. This is the *Conway-Lagarias invariant* of the region. It can be shown (see the companion article [**4**]) that the area of the (a, b)-benzel (as measured by the number of tiles) is given by

$$\begin{array}{ll} (-a^2 + 4ab - b^2 - a - b)/2 & \text{if } a + b \equiv 0 \text{ or } 2 \pmod 3, \\ (-a^2 + 4ab - b^2 - a - b + 2)/2 & \text{if } a + b \equiv 1 \pmod 3) \end{array}$$

while the value of the Conway-Lagarias invariant is

$$\begin{array}{ll} (3a^2 - 6ab + 3b^2 - a - b)/2 & \text{if } a + b \equiv 0 \pmod 3, \\ (-a^2 + 4ab - b^2 - a - b + 2)/2 & \text{if } a + b \equiv 1 \pmod 3, \\ (3a^2 - 6ab + 3b^2 + a + b - 2)/2 & \text{if } a + b \equiv 2 \pmod 3. \end{array}$$

(The expressions in Theorem 2 of [**4**] have opposite sign because the benzels in that article are mirror images of the benzels considered here.) Note that when $a+b \equiv 1 \pmod 3$, the Conway-Lagarias invariant is equal to the area of the benzel so that the tiling must consist entirely of right-pointing stones; for instance, this is the case with the (4,6)-benzel shown earlier.

The set of 5 prototiles has $2^5 - 1 = 31$ nonempty subsets, and for each, we can ask in how many ways it is possible to tile the (a, b)-benzel, that is, in how many ways it is possible find translates of the prototiles whose interiors are disjoint and whose union is the benzel. There is some redundancy here. Because the benzel has threefold rotational symmetry, and because 120 degree rotations preserve the two stones' orientation (right versus left), the number of tilings depends only on (a) whether right stones are allowed, (b) whether left stones are allowed, and (c) how many of the three kinds of bones are allowed (0, 1, 2, or 3). Thus there are really only $(2)(2)(4) - 1 = 15$ classes of tiling problems to consider. For $0 \leq i, j \leq 1$ and $0 \leq k \leq 3$ we define $T_{ijk}(a, b)$ as the number of ways to tile the (a, b)-benzel if the set of allowed prototiles contains the right stone iff $i = 1$, contains the left stone iff $j = 1$, and contains k of the bones. We say that such a tiling is an i, j, k tiling.

It is not hard to show that for each of the 15 cases, $T_{ijk}(a, b) = T_{ijk}(b, a)$. It is also not hard to show that the $(n, 2n)$-benzel is the same as the $(n, 2n - 1)$-benzel and the $(n, 2n - 2)$-benzel. Consequently, in the tables that follow I provide values for $T_{ijk}(a, b)$ only for $2 \leq a \leq 2b - 2$ and $2 \leq b \leq 2a - 2$. Lastly, it is not hard to show that when $i = 0$ and $a + b \equiv 2 \pmod 3$, $T_{ijk}(a, b) = 0$. That is because in this case the Conway-Lagarias invariant is equal to the positive number $(3(a - b)^2 + (a + b - 2))/2$, implying that every tiling of the (a, b)-benzel must have at least one right-pointing stone.

David desJardins wrote a general purpose program `TilingCount` that I used to enumerate tilings of regions with various sets of allowed prototiles. This led to the questions and conjectures that appear below. I am happy to share the code and the data on which my conjectures are based (much of which can be found in the Online Encyclopedia of Integer Sequences). Table 1 is a map of the first eighteen problems presented in this article as they relate to those fifteen cases. Rows describe which stones are allowed; columns describe how many bones are allowed. (This table omits cases where the number of allowed bone prototiles is zero or one; in such situations at most one tiling exists, even when both of the stone prototiles are allowed.)

TABLE 1. A map of the first eighteen problems.

	Two kinds of bones	Three kinds of bones
No stones	(no tilings exist)	type 003: prob. 1
Left stones	type 012: probs. 2–3	type 013: prob. 4
Right stones	type 102: prob. 5	type 103: probs. 6–7
Both kinds of stones	type 112: probs. 8–13	type 113: probs. 14–18

Benzels behave differently according to whether $a+b$ is 0, 1, or 2 (mod 3), so in what follows we will sometimes subdivide analyses according to the congruence class of $a+b$.

In the cases where only two kinds of bone tiles are permitted, the allowed tilings can be viewed as ribbon tilings, as in [**5**]; indeed, this was the mode of presentation employed in [**2**], with four of the five prototiles being depicted as ribbons. Switching over to the ribbon tilings presentation has already yielded solutions to problems 2 and 3 in [**3**], and is likely to provide leverage on other problems in the first column of the table.

2. No stones, three kinds of bones

Prior to the conference, I was able to show that if an (a,b)-benzel can be tiled by bones alone, then we must have $a = k(3k-1)/2$ and $b = k(3k+1)/2$ (or vice versa) for some $k \geq 2$. Several attending students found a proof that this necessary condition is also sufficient. (Note that such benzels belongs to the case $a+b \equiv 0 \pmod 3$.) Jesse Kim found the most complete solution, providing an explicit proof that the tiling he described works for all k. It appears (see OEIS entry A364134) that the number of such tilings grows exponentially in k^4.

PROBLEM 1. Find an exact formula for $T_{003}(k(3k-1)/2, k(3k+1)/2)$.

Of course, even short of an exact formula, any method of determining the number of tilings that is more efficient than brute-force enumeration (e.g., a recurrence relation) would be of interest.

See [**4**] for more discussion of no-stones tilings of benzels.

3. Left stones, two kinds of bones

Table 2 shows the values of $T_{012}(a,b)$ for $a,b \leq 10$.

TABLE 2. Values of $T_{012}(a,b)$.

$a \backslash b$	2	3	4	5	6	7	8	9	10
2	0								
3		**2**	0						
4		0	0	**2**	0				
5			**2**	0	0	0	0		
6			0	0	**8**	0	0	0	0
7				0	0	0	**8**	0	0
8				0	0	**8**	0	0	0
9					0	0	0	**48**	0
10					0	0	0	0	0

PROBLEM 2. Is it true that $T_{012}(3n, 3n) = 2^n n!$ for $n \geq 1$?

PROBLEM 3. Is it true that $T_{012}(3n+1, 3n+2) = 2^n n!$ for $n \geq 1$?

Comment: Colin Defant, Rupert Li, Benjamin Young and I worked on problem 2 during the conference and later succeeded in solving Problems 2 and 3; see [**3**]. We were also able to verify that the pattern of entries equal to 0 and 1 in Tables 2 through 4 persists for all larger values of a and b.

4. Left stones, three kinds of bones

Table 3 shows the values of $T_{013}(a, b)$ for $a, b \leq 10$.

TABLE 3. Values of $T_{013}(a, b)$.

$a\backslash b$	2	3	4	5	6	7	8	9	10
2	0								
3		**3**	0						
4		0	0	**9**	0				
5			**9**	0	0	**2**	0		
6			0	0	**144**	0	0	0	0
7				**2**	0	0	**1143**	0	0
8				0	0	**1143**	0	0	**825**
9					0	0	0	**73454**	0
10					0	0	**825**	0	0

When a and b are such that the Conway-Lagarias invariant is strictly positive, the (a, b)-benzel cannot be tiled by bones and left stones; the corresponding entries in the table must be zero. On the other hand, when a and b are such that the Conway-Lagarias invariant is negative or zero, the entries in the table are observed to be positive, though I see no reason for concluding that they are.

PROBLEM 4. Is it true that when the Conway-Lagarias invariant associated with the (a, b)-benzel is negative or zero, tilings of type 013 exist?

5. Right stones, two kinds of bones

Table 4 shows the values of $T_{102}(a, b)$ for $a, b \leq 10$.

TABLE 4. Values of $T_{102}(a, b)$.

$a\backslash b$	2	3	4	5	6	7	8	9	10
2	1								
3		0	1						
4		1	**2**	0	1				
5			0	1	**4**	0	1		
6			1	4	0	1	**10**	0	1
7				0	1	**8**	0	1	**28**
8				1	10	0	1	**24**	0
9					0	1	24	0	1
10					1	28	0	1	**48**

PROBLEM 5. Is it true that

$$T_{102}(n+3k, 2n+3k-1) = \prod_{i=1}^{k} \frac{(2i)!(2i+2n-2))!}{(i+n-1)!(i+n+k-1)!}$$

for $k \geq 0$ and $n \geq 1$ (except $(k,n) = (0,1)$)? Equality has been verified for $0 \leq k \leq 5$, $1 \leq n \leq 5$.

Comment: This formula and the a, b symmetry relation together provide a conjectural enumeration of tilings of type 102 of the (a,b)-benzel for all a, b satisfying $a+b \equiv 2 \pmod 3$. (When $a+b \equiv 1$, the number of tilings is 0; when $a+b \equiv 2$, the number of tilings is 1.)

Comment: Three special cases merit special attention. When $k=1$, the right-hand side of the equation is twice the nth Catalan number; when $n=1$, the right-hand side of the equation is $2^k k!$; and when $n=2$, the right-hand side of the equation is $2^k(k+1)!$.

6. Right stones, three kinds of bones

Table 5 shows the values of $T_{103}(a,b)$ for $a, b \leq 10$.

TABLE 5. Values of $T_{103}(a,b)$.

$a\backslash b$	2	3	4	5	6	7	8	9	10
2	1								
3		**0**	1						
4		1	7	**0**	1				
5			0	1	33	**2**	1		
6			1	33	0	1	164	**21**	1
7				2	1	666	0	1	864
8				1	164	0	1	12430	0
9					21	1	12430	0	1
10					1	864	0	1	655721

PROBLEM 6. Is is true that $T_{103}(n, 2n-3) = (3n+3)(3n-7)!/(n-5)!(2n-1)!$ for $n \geq 5$? (The formula also works for $n=3$ and $n=4$ if one treats $1/(-1)!$ and $1/(-2)!$ as 0.) Equality has been verified for $5 \leq n \leq 16$.

Aside from the fact that Problems 2 and 6 involve different prototile sets, the two problems differ in another important way: in problem 2 we have $b/a \to 1$ as $n \to \infty$ while in problem 6 we have $b/a \to 2$ as $n \to \infty$. In the former case we say that the sequence is associated with a **central diagonal** of the table of values while in the latter case we say that the sequence is associated with a **peripheral diagonal** (recall that a/b and b/a cannot exceed 2).

In parallel with the observations that preceded Problem 4, note that when the Conway-Lagarias invariant of a benzel is strictly negative, the benzel cannot be tiled by bones and right-pointing stones; the corresponding entries in the table must be zero. On the other hand, when a and b are such that the Conway-Lagarias invariant is positive or zero, the entries in the table are observed to be positive, though I see no reason for concluding that they are.

PROBLEM 7. Is it true that when the Conway-Lagarias invariant associated with the (a,b)-benzel is positive or zero, tilings of type 103 exist?

7. Both stones, two kinds of bones

Table 6 shows the values of $T_{112}(a,b)$ for $a,b \leq 10$. Table 7 gives the first few values of $T_{112}(3n,3n)$ in factored form.

TABLE 6. Values of $T_{112}(a,b)$.

$a\backslash b$	2	3	4	5	6	7	8	9	10
2	1								
3		**2**	1						
4		1	**4**	**6**	1				
5			6	1	**16**	**22**	1		
6			1	16	**48**	1	**68**	**90**	1
7				22	1	**224**	**512**	1	**304**
8				1	68	512	1	**3360**	6736
9					90	1	3360	**15360**	1
10					1	304	6736	1	**168960**

TABLE 7. Values of $T_{112}(3n,3n)$.

$$\begin{gathered}
2^1,\\
2^4\,3^1,\\
2^{10}\,3^1\,5^1,\\
2^{16}\,7^1\,11^1\,13^1,\\
2^{28}\,3^2\,7^1\,13^1\,17^1,\\
2^{38}\,3^2\,11^1\,17^2\,19^2,\\
2^{50}\,5^1\,11^2\,13^1\,17^1\,19^2\,23^2,\\
2^{64}\,3^3\,5^4\,11^1\,13^2\,19^1\,23^3\,29^1,\\
2^{84}\,3^4\,5^3\,13^2\,17^1\,23^2\,29^3\,31^2,\ \ldots
\end{gathered}$$

PROBLEM 8. With $T(n)$ denoting $T_{112}(3n,3n)$, is it true that the second quotient $T(n)T(n+2)/T(n+1)^2$ is equal to

$$\frac{256(2n+3)^2(4n+1)(4n+3)^2(4n+5)}{27(3n+1)(3n+2)^2(3n+4)^2(3n+5)}$$

for all $n \geq 1$? Equality has been verified for $1 \leq n \leq 7$. See OEIS entry A352207.

Comment: David desJardins found the pattern governing the numbers $T_{112}(n)$, with assistance from Christian Krattenthaler, Greg Kuperberg and other members of the `domino` listserv. The same is true for Problem 10.

Table 8 gives the first few values of $T_{112}(3n+1,3n+1)$ in factored form.

With $T(n) = T_{112}(3n+1,3n+1)$ with $n \geq 1$, it appears that $T(n)$ has no prime factor greater than or equal to $4n$.

PROBLEM 9. Find a formula governing $T_{112}(3n+1,3n+1)$. See OEIS A364481.

Table 9 gives the first few values of $T_{112}(3n+1,3n+2)$ in factored form.

PROBLEM 10. With $T(n)$ denoting $T_{112}(3n+1,3n+2)$, is it true that $T(n)T(n+3)/T(n+1)T(n+2)$ is equal to

$$\frac{65536(2n+3)(2n+5)^2(2n+7)(4n+3)(4n+5)^2(4n+7)^2(4n+9)^2(4n+11)}{729(3n+2)(3n+4)^3(3n+5)^2(3n+7)^2(3n+8)^3(3n+10)}$$

for all $n \geq 1$? Equality has been verified for $1 \leq n \leq 8$. See OEIS entry A364482.

TABLE 8. Values of $T_{112}(3n+1, 3n+1)$.

$$\begin{gathered} 2^2, \\ 2^5\,7^1, \\ 2^{10}\,3^1\,5^1\,11^1, \\ 2^{17}\,7^1\,11^1\,13^2, \\ 2^{30}\,3^1\,13^1\,17^2\,19^1, \\ 2^{38}\,3^1\,11^1\,17^2\,19^3\,23^1, \ldots \end{gathered}$$

TABLE 9. Values of $T_{112}(3n+1, 3n+2)$.

$$\begin{gathered} 2^1\,3^1, \\ 2^9, \\ 2^9\,3^1\,5^1\,7^1\,11^1, \\ 2^{25}\,3^1\,7^1\,13^1, \\ 2^{28}\,3^2\,11^1\,13^1\,17^2\,19^1, \\ 2^{50}\,3^1\,11^1\,17^1\,19^2\,23^1, \\ 2^{49}\,3^2\,5^2\,11^2\,13^2\,17^1\,19^2\,23^3, \\ 2^{81}\,3^3\,5^4\,13^2\,23^2\,29^2\,31^1, \ldots \end{gathered}$$

TABLE 10. Values of $T_{112}(3n-1, 3n)$.

$$\begin{gathered} 1^1, \\ 2^4, \\ 2^5\,3^1\,5^1\,7^1, \\ 2^{16}\,11^1\,13^1, \\ 2^{19}\,3^1\,7^1\,11^1\,13^2\,17^1, \\ 2^{39}\,3^1\,17^2\,19^2, \\ 2^{37}\,5^1\,11^2\,13^1\,17^2\,19^3\,23^2, \ldots \end{gathered}$$

Table 10 gives the first few values of $T_{112}(3n-1, 3n)$ in factored form.

With $T(n) = T_{112}(3n-1, 3n)$ with $n \geq 1$, it appears that $T(n)$ has no prime factor greater than or equal to $4n$.

PROBLEM 11. Find a formula governing $T_{112}(3n-1, 3n)$. See OEIS A364483.

We now switch from central diagonals to peripheral diagonals.

PROBLEM 12. Is it true that $T_{112}(n+2, 2n+1)$ is the "nth large Schröder number" (see OEIS entry A006318) for all $n \geq 1$? Equality has been verified for $1 \leq n \leq 15$.

PROBLEM 13. Is it true that $T_{112}(n+2, 2n)$ is the number of "royal paths in a lattice of order n" (see OEIS entry A006319) for all $n \geq 1$? Equality has been verified for $1 \leq n \leq 15$.

8. Both stones, three kinds of bones

Finally we come to the most permissive situation: all prototiles are allowed. Table 11 shows the values of $T_{113}(a, b)$ for $a, b \leq 10$.

TABLE 11. Values of $T_{113}(a,b)$.

$a\backslash b$	2	3	4	5	6	7	8	9	10
2	1								
3		3	1						
4		1	10	**18**	1				
5			18	1	**84**	**142**	1		
6			1	84	459	1	**724**	**1266**	1
7				142	1	5766	19057	1	**6516**
8				1	724	19057	1	380597	1077681
9					1266	1	380597	3759277	1
10					1	6516	1077681	1	185961668

PROBLEM 14. Is it true that $T_{113} = 1$ when $a+b$ is 1 (mod 3)? (Note that $a+b \equiv 1$ is the situation in which the Conway-Lagarias invariant coincides with the area of the region being tiled, so that all the tiles must be right-pointing stones.)

Comment: This problem is resolved in the affirmative by Theorem 1.1 of [**3**].

It is disappointing that the data for $a+b \not\equiv 1$ (mod 3) do not suggest exact conjectures. On the other hand, it is intriguing that congruence phenomena occur, analogous to Cohn's 2-adic continuity theorem proved in [**1**] and conjectural 2-adic phenomena of a similar kind discussed in [**7**].

PROBLEM 15. Is $T_{113}(n, 2n-4)$ 2-adically continuous as a function of $n \geq 5$?

Comment: The sequence appears to be constant mod 2, constant mod 4, 2-periodic mod 8, and 8-periodic mod 16 (with repeating pattern 4, 4, 4, 4, 4, 12, 4, 12). The case $n = 4$ breaks the pattern. See OEIS entry A364416.

PROBLEM 16. Is $T_{113}(n, 2n-3)$ 2-adically continuous as a function of $n \geq 4$?

Comment: The sequence appears to be constant mod 2, constant mod 4, 2-periodic mod 8, and 8-periodic mod 16 (with repeating pattern 2, 14, 2, 14, 10, 6, 10, 6). The case $n = 3$ breaks the pattern. See OEIS entry A364417.

At the OPAC 2022 meeting, David Speyer suggested that one might use a different definition of a trimer, namely, a path consisting of three vertices and two edges. Thus, each stone would correspond to three different trimers according to which 2 of the 3 possible edges one used. Equivalently, one could use the original definition of a trimer, with the proviso that each stone will count with weight 3. The new numbers do not satisfy any nice patterns, with one exception: in the case where all prototiles are allowed, using stones of weight 3 seems to give rise to the same 2-adic continuity phenomenon we saw in Problems 11 and 12.

Let $T_{ijk}(a, b; 3)$ denote the weighted sum of the i, j, k tilings of the (a, b)-benzel as defined near the end of section 1, where a tiling with m stones has weight 3^m.

PROBLEM 17. Is $T_{113}(n, 2n-4; 3)$ 2-adically continuous as a function of $n \geq 5$?

Comment: The sequence appears to be constant mod 2, constant mod 4, constant mod 8, and 8-periodic mod 16 (with repeating pattern 4, 4, 12, 12, 4, 12, 12, 4). The case $n = 4$ breaks the pattern. See OEIS entry A364418.

PROBLEM 18. Is $T_{113}(n, 2n-3; 3)$ 2-adically continuous as a function of $n \geq 4$?

Comment: The sequence appears to be constant mod 2, constant mod 4, 2-periodic mod 8, and 8-periodic mod 16 (with repeating pattern 14, 6, 10, 2, 6, 14, 2, 10). The case $n = 3$ breaks the pattern. See OEIS entry A364438.

Comment: Given that the number of right stones minus the number of left stones in a tiling of a benzel is the same for all tilings of that benzel, a different way to describe the weighting assigned to tilings in Problems 17 and 18 (equivalent up to a power of 3) is to give weight 1 to left stones and weight 9 to right stones. During the final stages of preparing this article for publication, I realized that one could more generally give weight 1 to left stones and weight m to right stones. The resulting enumerations appear to exhibit p-adic continuity when $m+1$ is a power of the prime p. Other forms of congruential consistency appear in the data as well; for instance, the data for $m = 9$ appear to satisfy 5-adic continuity. I hope to address such patterns in future work.

9. Miscellaneous

The next question is not enumerative; it is an old structural question that has gone unresolved for decades. There are two natural kinds of "2-flips" that can turn one stones-and-bones tiling into another; the first trades two stones of opposite orientation for two bones, and the second trades a stone and a bone for a stone of the same orientation and a bone of a different orientation; see Figure 5.

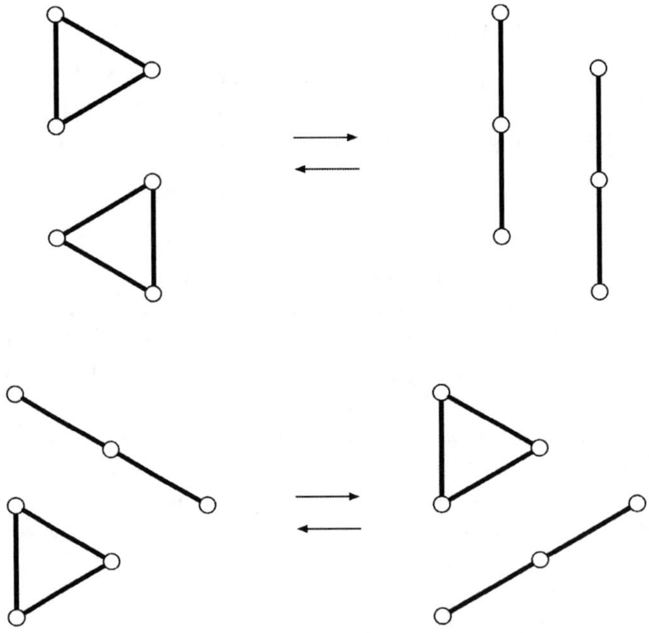

FIGURE 5. 2-flips for changing a trimer cover.

PROBLEM 19. Can every tiling of a finite simply-connected region using stones and bones be mutated into every other such tiling by means of a succession of 2-flips?

Comment: It is known that the hypothesis that the region be simply-connected cannot be dropped. It is also known that if one restricts to tilings of type 112 (that is, if one prohibits one of the three orientations of bones), then the claim is true; Sheffield proved an equivalent claim in the context of ribbon tilings [**8**].

In closing, we turn to the regions that Conway, Lagarias, and Thurston originally studied in papers [**2**, **9**] that inspired much of my work on tilings: triangles of hexagonal cells, with n cells on each side ("T_n regions"). Those three authors showed that if one uses stones alone, T_n can be tiled by T_2's (that is, by stones) precisely when n is congruent to 0, 2, 9, or 11 (mod 12). (In our notation, these are tilings of type 110; such tilings were not discussed above since for benzels they are not interesting from an enumerative perspective.)

The question we ask is, how many such tiling are there? OEIS entry A334875 gives the answers for small values of n. If we look at the prime factorizations of the answers, we notice that the exponent of the prime 2 is creeping upward. Specifically, the multiplicity of the prime 2 in the factorizations of the nonzero terms in this sequence goes 0, 0, 1, 3, 2, 3, 4, 3, 4, 3, 5, 8, 6, 8, ... This is a priori surprising, since the probability that a "random" positive integer is divisible by 2^m decreases exponentially as m increases.

PROBLEM 20. As n goes to infinity within the set of natural numbers congruent to 0, 2, 9, or 11 (mod 12), does the number of tiling of T_n by stones converge 2-adically to 0?

Acknowledgments

I am grateful to the organizers of OPAC 2022 for inviting me to give this presentation. I also thank Colin Defant, Rupert Li, and Ben Young for useful conversations.

References

[1] Henry Cohn, *2-adic behavior of numbers of domino tilings*, Electron. J. Combin. **6** (1999), Research Paper 14, 7, DOI 10.37236/1446. MR1667454

[2] J. H. Conway and J. C. Lagarias, *Tiling with polyominoes and combinatorial group theory*, J. Combin. Theory Ser. A **53** (1990), no. 2, 183–208, DOI 10.1016/0097-3165(90)90057-4. MR1041445

[3] Colin Defant, Rupert Li, James Propp, and Benjamin Young, *Tilings of benzels via the abacus bijection*, Comb. Theory **3** (2023), no. 2, Paper No. 16, 24, DOI 10.5070/c63261995. MR4646097

[4] Jesse Kim and James Propp, *A pentagonal number theorem for tribone tilings*, Electron. J. Combin. **30** (2023), no. 3, Paper No. 3.26, 20, DOI 10.37236/11326. MR4635476

[5] Igor Pak, *Ribbon tile invariants*, Trans. Amer. Math. Soc. **352** (2000), no. 12, 5525–5561, DOI 10.1090/S0002-9947-00-02666-0. MR1781275

[6] James Propp, *Tilings*, Handbook of enumerative combinatorics, Discrete Math. Appl. (Boca Raton), CRC Press, Boca Raton, FL, 2015, pp. 541–588. MR3409350

[7] James Propp, *Some 2-adic conjectures concerning polyomino tilings of Aztec diamonds*, Integers **23** (2023), Paper No. A30, 15, DOI 10.26421/qic23.1-2-1. MR4582847

[8] Scott Sheffield, *Ribbon tilings and multidimensional height functions*, Trans. Amer. Math. Soc. **354** (2002), no. 12, 4789–4813, DOI 10.1090/S0002-9947-02-02981-1. MR1926837

[9] William P. Thurston, *Conway's tiling groups*, Amer. Math. Monthly **97** (1990), no. 8, 757–773, DOI 10.2307/2324578. MR1072815

[10] Alain Verberkmoes and Bernard Nienhuis, *Bethe Ansatz solution of triangular trimers on the triangular lattice*, Phys. Rev. E **63** (2001), 066122. DOI 10.1103/PhysRevE.63.066122

DEPARTMENT OF MATHEMATICS AND STATISTICS, UMASS LOWELL, ONE UNIVERSITY AVENUE, LOWELL, MASSACHUSETTS 01854

Email address: jamespropp@gmail.com

Proceedings of Symposia in Pure Mathematics
Volume **110**, 2024
https://doi.org/10.1090/pspum/110/02018

The mystery of plethysm coefficients

Laura Colmenarejo, Rosa Orellana, Franco Saliola, Anne Schilling, and Mike Zabrocki

ABSTRACT. Composing two representations of the general linear groups gives rise to Littlewood's (outer) plethysm. On the level of characters, this poses the question of finding the Schur expansion of the plethysm of two Schur functions. A combinatorial interpretation for the Schur expansion coefficients of the plethysm of two Schur functions is, in general, still an open problem. We identify a proof technique of combinatorial representation theory, which we call the "s-perp trick", and point out several examples in the literature where this idea is used. We use the s-perp trick to give algorithms for computing monomial and Schur expansions of symmetric functions. In several special cases, these algorithms are more efficient than those currently implemented in SAGEMATH.

1. Introduction

The isomorphism classes of complex irreducible polynomial representations of $GL_n := GL_n(\mathbb{C})$ are indexed by integer partitions λ with at most n parts. We denote such a representation by ρ^λ. Its character is identified with the *Schur polynomial* (see [**39**])

$$s_\lambda(x_1, \ldots, x_n) = \sum_{T \in \mathsf{SSYT}(\lambda)} x^{\mathrm{weight}(T)},$$

where $\mathsf{SSYT}(\lambda)$ is the set of all semistandard Young tableaux of shape λ over the alphabet $\{1, 2, \ldots, n\}$ and $\mathrm{weight}(T)$ is an n-dimensional vector, where the i-th entry contains the number of occurrences of the letter i in T, and where $x^\alpha = x_1^{\alpha_1} x_2^{\alpha_2} \cdots x_n^{\alpha_n}$ for any n-dimensional vector α. The composition of two such representations, say $\rho^\lambda : GL_n \to GL_m$ and $\rho^\mu : GL_m \to GL_\ell$, is also a polynomial representation of GL_n, and its character is denoted by $s_\lambda[s_\mu]$. This operation can be viewed as an operation on symmetric polynomials, which was named *(outer) plethysm* by Littlewood [**29**].

The main objective of this paper is to discuss the following open problem.

2020 *Mathematics Subject Classification.* Primary 05E05, 05E10; Secondary 20C30.

The first author was partially supported by FMQ333 and PID2020-117843GB-I00.
The second author was partially supported by NSF grant DMS-2153998.
The third author was supported by NSERC Discovery Grants.
The fourth author was partially supported by NSF grants DMS–1760329 and DMS–2053350.
The fifth author was supported by NSERC Discovery Grants.

PROBLEM 1.1. Since $s_\lambda[s_\mu]$ is the character of a GL_n-representation, it is an $\mathbb{N}$-linear combination of Schur polynomials. Find a combinatorial interpretation of the coefficients $a^\nu_{\lambda,\mu} \in \mathbb{N}$ in the expansion

$$s_\lambda[s_\mu] = \sum_\nu a^\nu_{\lambda,\mu} s_\nu.$$

In the last century, the problem of understanding the coefficients $a^\nu_{\lambda,\mu}$ has stood as a measure of progress in the field of GL_n-representation theory (see for instance Problem 9 of [**40**]). Here we identify the *s-perp trick* as a possible way to approach this problem by proving known cases in a simple way and finding some new combinatorial descriptions of the $a^\nu_{\lambda,\mu}$. In addition, we demonstrate that the s-perp trick gives an efficient way to compute plethysm coefficients.

The simplest form of the problem occurs when the partitions λ and μ are both of row shape, however even this case is notoriously difficult and explicit formulae are known only in very special circumstances. The following table presents a (non-exhaustive) list of some of the known results in this direction.

Formulae for plethysms of the form $s_m[s_n]$	References
$s_2[s_n]$ in terms of Schur functions	[**28**]
$s_3[s_n]$ in terms of Schur functions	[**43**]
$s_4[s_n]$ in terms of Schur functions	[**16, 21**]
$s_2[s_\lambda]$ in terms of Schur functions	[**8, 45**]
$s_\lambda[s_\mu]$ in terms of fundamental quasisymmetric functions	[**30**]
$s_2[s_b[s_a]]$ and $s_c[s_2[s_a]]$ in terms of Schur functions	[**19**]

Since general formulae for the coefficients $a^\nu_{\lambda,\mu}$ have been elusive, various methods for computing plethysm have been developed [**6, 7, 10, 14, 15, 24, 27, 36, 44, 46, 47**] as well as representation-theoretic approaches [**4, 5, 12, 21, 33, 34, 38**].

The paper is organized as follows. In Section 2, we set up notation in the framework of symmetric functions and recall the definition of plethysm. In Section 3, we describe the s-perp trick to prove symmetric function identities and identify places in the literature where this trick has been used. In Section 4, we state an algorithm for computing the Schur expansion for a symmetric function f given the Schur expansions of $s_r^\perp f$. In Section 5, we apply this algorithm to plethysm expressions to show how it is used to speed up calculations of this type and to prove/derive new combinatorial formulas for the coefficients $a^\nu_{\lambda,\mu}$ in Problem 1.1.

2. Symmetric functions and plethysm

We begin by reviewing some notation in the framework of the ring of symmetric functions. We refer the reader to references like [**32, 37, 39**] for more details.

A *partition* of a positive integer n is a sequence of positive integers $\lambda = (\lambda_1, \lambda_2, \ldots, \lambda_r)$ with $\lambda_1 \geqslant \lambda_2 \geqslant \cdots \geqslant \lambda_r > 0$ whose sum $|\lambda| := \lambda_1 + \lambda_2 + \cdots + \lambda_r$ is n. We use the notation $\lambda \vdash n$ to indicate that λ is a partition of $n \in \mathbb{N}$. The *length* of λ is denoted $\ell(\lambda) := r$. We assume that the empty partition, $\lambda = ()$, is the only partition of $n = 0$ and its length is 0. Also, we use the notation $\overline{\lambda} = (\lambda_2, \lambda_3, \ldots, \lambda_{\ell(\lambda)})$ and λ' to denote conjugate partition.

Let $\mathbb{K}$ be a ring containing $\mathbb{Q}$ as a subfield, such as $\mathbb{C}$ or $\mathbb{Q}(q,t)$. The ring of symmetric functions is defined as

$$\Lambda := \mathbb{K}[p_1, p_2, p_3, \ldots],$$

where the generators p_r are known as the *power sums* and the degree of p_r is equal to r. The subspace of symmetric functions of degree n is denoted $\Lambda_{=n}$ and is spanned by the products $p_\lambda := p_{\lambda_1} p_{\lambda_2} \cdots p_{\lambda_{\ell(\lambda)}}$, where $\lambda \vdash n$.

There are five distinguished bases for $\Lambda_{=n}$ that we will refer to in this exposition: the *power sum* $\{p_\lambda\}_{\lambda \vdash n}$, *complete* (or homogeneous) $\{h_\lambda\}_{\lambda \vdash n}$, *elementary* $\{e_\lambda\}_{\lambda \vdash n}$, *monomial* $\{m_\lambda\}_{\lambda \vdash n}$ and *Schur* $\{s_\lambda\}_{\lambda \vdash n}$ bases. (See [**32, 37, 39**] for the definition of these bases and the basic relations between them.)

The ring of symmetric functions is endowed with a *standard involution* defined by setting $\omega(h_\lambda) = e_\lambda$, $\omega(p_\lambda) = (-1)^{|\lambda| + \ell(\lambda)} p_\lambda$, or $\omega(s_\lambda) = s_{\lambda'}$. It also has a scalar product, known as the *Hall inner product*, which is defined by

$$\langle s_\lambda, s_\mu \rangle = \langle h_\lambda, m_\mu \rangle = \begin{cases} 1 & \text{if } \lambda = \mu, \\ 0 & \text{otherwise.} \end{cases}$$

If multiplication by a symmetric function f is thought of as a linear operator, then the linear operator which is adjoint to multiplication by f with respect to this scalar product is denote $f^\perp$. It satisfies $\langle f^\perp(g), h \rangle = \langle g, fh \rangle$ for all symmetric functions $f, g, h \in \Lambda$ and can be calculated via the formula

$$f^\perp(g) = \sum_\mu \langle g, f s_\mu \rangle \, s_\mu. \tag{2.1}$$

Then for $f \in \Lambda$, we can view $f^\perp$ as a (bi-)linear operator since $(f+g)^\perp = f^\perp + g^\perp$ and $f^\perp(g+h) = f^\perp(g) + f^\perp(h)$. For the Schur basis, $s_\lambda^\perp(s_\mu) = \sum_\nu c^\mu_{\lambda\nu} s_\nu$, where $c^\mu_{\lambda\nu}$ is the Littlewood–Richardson coefficient.

As one of the applications of the s-perp trick we will need the notion of composition of symmetric functions, or *plethysm*, and extend its use to *plethystic notation* on symmetric functions. In [**32**], the operation is denoted $f \circ g$ but it is convenient to express composition using square brackets. If $g \in \Lambda$ with $g = \sum_\lambda c_\lambda p_\lambda$ for some coefficients $c_\lambda \in \mathbb{Q}$,[1] then

$$p_r[g] = \sum_\lambda c_\lambda p_{r\lambda_1} p_{r\lambda_2} \cdots p_{r\lambda_{\ell(\lambda)}}$$

for all positive integers r, and $p_\mu[g] = p_{\mu_1}[g] p_{\mu_2}[g] \cdots p_{\mu_{\ell(\mu)}}[g]$ for all partitions μ. The plethysm of f and g, for $f \in \Lambda$ with $f = \sum_\mu c'_\mu p_\mu$ with $c'_\mu \in \mathbb{K}$, is defined as

$$f[g] = \sum_\mu c'_\mu p_\mu[g].$$

Next, *plethystic notation* extends the operation of plethysm to expressions containing variables from the base ring $\mathbb{K}$. For $E := E(x_1, x_2, x_3, \ldots) \in \mathbb{K}$, we have

$$p_r[E(x_1, x_2, x_3, \ldots)] = E(x_1^r, x_2^r, x_3^r, \ldots)$$

[1]The condition that the coefficients c_λ are in $\mathbb{Q}$ is important because using plethystic notation, if the base ring contains variables (for example, if $\mathbb{K} = \mathbb{Q}(q,t)$), then the variables must be treated differently.

and for $f \in \Lambda$, where $f = \sum_\lambda c_\lambda p_\lambda$ (with $c_\lambda \in \mathbb{K}$),

$$f[E] = \sum_\lambda c_\lambda p_{\lambda_1}[E] p_{\lambda_2}[E] \cdots p_{\lambda_{\ell(\lambda)}}[E] .$$

We will use the symbol X to stand for an arbitrary alphabet of variables $X := x_1 + x_2 + x_3 + \cdots$, but as that expression is sufficiently general it may be replaced with any element of Λ and the identity will hold true. In particular, $f[X]$ is an expression equivalent to f since if $X = p_1$, then by definition, $f[p_1] = f$.

For our purposes, we require the following identities which can be derived from this definition (see [**32**]):

$$f[X+t] = \sum_{r \geqslant 0} (s_r^\perp f)[X] t^r, \qquad f[X-t] = \sum_{r \geqslant 0} (s_{1^r}^\perp f)[X](-t)^r, \tag{2.2}$$

$$f[-X] = (-1)^{\mathsf{degree}(f)} (\omega f)[X], \qquad f[tX] = t^{\mathsf{degree}(f)} f[X], \tag{2.3}$$

where $f \in \Lambda$ is of homogeneous degree equal to $\mathsf{degree}(f)$ and t is a variable in the base ring $\mathbb{K}$. Then for expressions $A_1, A_2, \ldots, A_k \in \Lambda$,

$$f[A_1 + A_2 + \cdots + A_k] = \sum_{\nu^{(*)}} s_{\nu^{(1)}}[A_1] s_{\nu^{(2)}}[A_2] \cdots s_{\nu^{(k)}}[A_k] \left\langle f, s_{\nu^{(1)}} s_{\nu^{(2)}} \cdots s_{\nu^{(k)}} \right\rangle , \tag{2.4}$$

where the sum is over all sequences of partitions $\nu^{(*)} = (\nu^{(1)}, \nu^{(2)}, \ldots, \nu^{(k)})$ with $\sum_{i=1}^k |\nu^{(i)}| = \mathsf{degree}(f)$.

Given two symmetric functions $f, g \in \Lambda$ with known monomial expansion $f = \sum_{i \geqslant 1} x^{a^i}$, where $a^1, a^2, \ldots$ are vectors, the plethysm is also given by

$$g[f] = g(x^{a^1}, x^{a^2}, \ldots).$$

EXAMPLE 2.1. Since $s_1 = x_1 + x_2 + \cdots$, it hence immediately follows that

$$g[s_1] = g(x_1, x_2, \ldots) = g$$

and since $p_n = x_1^n + x_2^n + \cdots$, it follows that if $f = \sum_{i \geqslant 1} x^{a^i}$, then

$$f[p_n] = f(x_1^n, x_2^n, \ldots) = \sum_{i \geqslant 1} x^{a^i n} = p_n[f].$$

EXAMPLE 2.2. For a slightly more advanced example, consider

$$s_2[x_1, x_2] = \underset{\boxed{1\,|\,1}}{x_1^2} + \underset{\boxed{1\,|\,2}}{x_1 x_2} + \underset{\boxed{2\,|\,2}}{x_2^2} \tag{2.5}$$

where we indicated the semistandard Young tableau which contributes to each term. Then the plethysm

$$\begin{aligned}
s_2[s_2[x_1, x_2]] &= s_2[x_1^2, x_1 x_2, x_2^2] \\
&= \underset{\begin{array}{c}\boxed{1\,|\,1}\\ \boxed{\boxed{1\,|\,1}\boxed{1\,|\,1}}\end{array}}{x_1^4} + \underset{\begin{array}{c}\boxed{1\,|\,2}\\ \boxed{\boxed{1\,|\,1}\boxed{1\,|\,2}}\end{array}}{x_1^3 x_2} + \underset{\begin{array}{c}\boxed{1\,|\,3}\\ \boxed{\boxed{1\,|\,1}\boxed{2\,|\,2}}\end{array}}{x_1^2 x_2^2} \\
&\quad + \underset{\begin{array}{c}\boxed{2\,|\,2}\\ \boxed{\boxed{1\,|\,2}\boxed{1\,|\,2}}\end{array}}{x_1^2 x_2^2} + \underset{\begin{array}{c}\boxed{2\,|\,3}\\ \boxed{\boxed{1\,|\,2}\boxed{2\,|\,2}}\end{array}}{x_1 x_2^3} + \underset{\begin{array}{c}\boxed{3\,|\,3}\\ \boxed{\boxed{2\,|\,2}\boxed{2\,|\,2}}\end{array}}{x_2^4} \\
&= s_4[x_1, x_2] + s_{2,2}[x_1, x_2].
\end{aligned}$$

Note that the first row of semistandard Young tableaux consists of the usual tableaux of shape (2) in the alphabet $\{1,2,3\}$ since the Schur polynomial in the example depends on three variables. The second row of tableaux takes into account that by (2.5) each variable itself corresponds to a tableau already, yielding tableaux of tableaux.

3. The s-perp trick

This section identifies a technique in symmetric function theory, which we coin the "*s-perp trick*". It has been used repeatedly in the literature to establish results in combinatorial representation theory. Here we apply it to Macdonald symmetric functions and plethysm to compute the monomial and Schur expansions of these symmetric functions. There are various other algorithms for the computation of plethysm that have appeared in the literature [**1,9,10,27,30,47,48**]. Surprisingly, this method is in many cases more efficient than the current implementation in SageMath; in particular, the computation of the Schur expansion of a single Schur plethysm of the form $s_\lambda[s_m]$ or $s_\lambda[s_{1^m}]$ is usually faster using the s-perp trick.

To explain the s-perp trick, we consider the ring of symmetric functions Λ spanned by the Schur basis s_λ, where λ is a partition of a non-negative integer. The degree of the Schur function indexed by the partition λ is given by the size of the partition. A function $f \in \Lambda$ is of homogeneous degree d if its expansion in the Schur basis only involves Schur functions indexed by partitions of d. Recall that the ring Λ is endowed with a scalar product, where the Schur basis is orthonormal. Also from (2.1), recall the action of the *s-perp operator*.

Definition 3.1. Let λ be a partition and $f \in \Lambda$. The *action of $s_\lambda^\perp$* on f is defined by

$$s_\lambda^\perp f = \sum_\mu \langle f, s_\lambda s_\mu \rangle \, s_\mu \, .$$

Now, we are ready to present the technique studied in this paper.

Proposition 3.2. (*The s-perp trick*) *Let f and g be two symmetric functions of homogeneous degree d. If*

$$s_r^\perp f = s_r^\perp g \qquad \textit{for all } 1 \leqslant r \leqslant d,$$

then $f = g$. The same statement is true if $s_r^\perp$ is replaced by $s_{1^r}^\perp$, mutatis mutandi.

This statement is [**20**, Proposition 6.20.1], where the idea is used to compute the monomial expansion of $\nabla(e_n)$ in the Shuffle Conjecture. This technique occurs relatively frequently in the computation of combinatorial formulas for q,t-symmetric functions with labeled lattice paths (e.g. see the proof of [**23**, Theorem 5.2] in the recent paper by A. Iraci and A. Vanden Wyngaerd).

Proposition 3.2 can be interpreted as a recursive method to determine a symmetric function f by knowing the expressions $s_r^\perp f$ for each r between 1 and the degree of f. In fact, the proof of this proposition follows from the algorithm presented in Section 4, which provides a method for recovering the monomial (Equation (3.1)) and Schur expansions of f from the symmetric functions $s_r^\perp f$.

As we mentioned at the beginning of this section, this technique has been used repeatedly in the literature. For instance, one manifestation of the s-perp trick is to give a monomial expansion of a symmetric polynomial using the recurrence

$$f(x_1, x_2, \ldots, x_n) = \sum_{r \geqslant 0} (s_r^\perp f)(x_1, x_2, \ldots, x_{n-1}) x_n^r \; .$$

This is the method for computing the monomial expansion of the modified Macdonald symmetric functions $\tilde{H}_\lambda[X; q, t]$ (see [**17**]) given by F. Bergeron and M. Haiman in [**3**, Proposition 5]. They provide an explicit formula for the coefficient $d_{\lambda\nu}^{(r)}$ in the expression defined by

$$s_r^\perp \tilde{H}_\lambda[X; q, t] = \sum_\nu d_{\lambda\nu}^{(r)} \tilde{H}_\nu[X; q, t] \; .$$

Denoting $\overline{\mu} = (\mu_2, \mu_3, \ldots, \mu_{\ell(\mu)})$, it follows that if we denote the coefficient of m_μ in $\tilde{H}_\lambda[X; q, t]$ by $L_{\lambda\mu}$, then it can be computed with the recursive formula[2]

$$L_{\lambda\mu} = \sum_\beta d_{\lambda\beta}^{(\mu_1)} L_{\beta\overline{\mu}} \; . \tag{3.1}$$

Another example of the s-perp trick used in combinatorial representation theory appears in a paper by A. Garsia and C. Procesi [**18**], where the authors show that the Frobenius characteristic of a certain quotient module is equal to the Hall–Littlewood symmetric function. Let $H_\lambda[X; q] = \sum_\mu K_{\lambda\mu}(q) s_\mu$ be the Hall–Littlewood symmetric function with $K_{\lambda\mu}(q)$ representing the q-Kostka coefficient (see [**32**, Section III.6]) and let $C_\lambda[X; q]$ be the Frobenius characteristic of a certain module. It is shown in [**18**] that since $s_{1^r}^\perp H_\lambda[X; q]$ and $s_{1^r}^\perp C_\lambda[X; q]$ satisfy exactly the same recursive expression, the identity $H_\lambda[X; q] = C_\lambda[X; q]$ holds.

Iraci, Rhoades and Romero use an almost identical method [**22**, Lemma 3.1] to prove that the diagonal fermionic coinvariants $\bigwedge\{\theta_n, \xi_n\}/I^+$ have Frobenius image of a special case of the Theta conjecture [**11**, Conjecture 9.1].

We will apply this technique to the application of computing the plethysm of symmetric functions.

4. Application: Schur expansions

In the previous section, we saw that the monomial expansion of a symmetric function f can be computed by recursively computing the monomial expansion of $s_r^\perp f$. In this section, we state an algorithm for computing the Schur expansion of a symmetric function f by recursively computing the Schur expansion of $s_r^\perp f$ (respectively $s_{1^r}^\perp f$).

Let addrow($s\_expansion, r$) be a function which takes as input a Schur expansion of an element of Λ and a positive integer r. If each partition indexing the terms in the Schur expansion of f has first part less than or equal to r, then return the expression with each indexing partition having a row of size r appended to it, or 0 otherwise.

REMARK 4.1. The idea for this algorithm occurred to us because we had a symmetric group module and we could combinatorially describe $s_1^\perp f$, where f is the Frobenius image of this module. From this we wanted to deduce a combinatorial formula for f. In general, $s_1^\perp f$ does not characterize f (e.g. consider the symmetric

[2]This algorithm was implemented in 2015 as the default method for computing the $\tilde{H}_\mu[X; q, t]$ symmetric functions in the computer algebra system SAGEMATH [**42**] and there was roughly a 2× speedup against the previous most efficient method for computing these functions.

functions $s_{22}+s_4$ and s_{31} which both satisfy $s_1^\perp(s_{22}+s_4) = s_{21}+s_3 = s_1^\perp(s_{31})$). However the idea seemed quite close. Then we asked what additional information would we need to recover f and realized that there is a general technique that is regularly used in symmetric function theory that could be identified.

Algorithm 1: An algorithm for finding the Schur expansion of f from the Schur expansion of $s_r^\perp f$ for $1 \leqslant r \leqslant \mathsf{degree}(f)$

---

1 function ExpandSchur A;
Input : A - array with $A[r]$ the Schur expansion of $s_r^\perp f$ for $1 \leqslant r \leqslant \mathsf{degree}(f)$
Output: *out* - a Schur expansion of f
2 **begin**
3   *out* $\longleftarrow 0$;
4   **for** $r = length(A)$ *downto* 1 **do**
5   *out* $\longleftarrow$ *out* $+ \,\mathsf{addrow}(A[r] - s_r^\perp(out), r)$;
6   **end**
7   return *out*;
8 **end**

Algorithm 1 works because when $r = k$ at line 4, *out* is equal to the sum of the terms of the Schur expansion of f which have first part of the indexing partition greater than or equal to k and so $f - out$ is in the linear span of Schur functions with parts less than or equal to k. Hence

$$\mathsf{addrow}(A[k] - s_k^\perp(out), k) = \mathsf{addrow}(s_k^\perp(f - out), k)$$

is equal to the sum of terms in the Schur expansion of f with first part of the indexing partition equal to k. When the *for* loop completes, *out* will equal the Schur expansion of f.

This algorithm can be modified to deduce the Schur expansion of f from the Schur expansion of $s_{1^r}^\perp f$, for $1 \leqslant r \leqslant \mathsf{degree}(f)$, by replacing the function addrow with addcol, which adds a column on the partitions indexing the expansions of the Schur functions.

5. Application to plethysm

5.1. s-perp formulae for plethysm. Equation (2.2) can be used to compute $s_r^\perp$ or $s_{1^r}^\perp$ acting on plethysms. The expression that we will specialize for our application is the following identity.

PROPOSITION 5.1. *For partitions λ and μ and a positive integer r,*

(5.1)
$$s_r^\perp s_\lambda[s_\mu] = \sum_{\nu^{(i)}} s_{\nu^{(0)}}[s_\mu] s_{\nu^{(1)}}[s_1^\perp s_\mu] s_{\nu^{(2)}}[s_2^\perp s_\mu] \cdots s_{\nu^{(r')}}[s_{r'}^\perp s_\mu] \left\langle s_\lambda, s_{\nu^{(0)}} s_{\nu^{(1)}} \cdots s_{\nu^{(r')}} \right\rangle$$

where $r' = \min(r, \mu_1)$ and the sum is over all sequences of partitions $(\nu^{(0)}, \nu^{(1)}, \ldots, \nu^{(r')})$ with $\sum_{i=0}^r |\nu^{(i)}| = |\lambda|$ and $\sum_{i=0}^r i|\nu^{(i)}| = r$.

PROOF. If we let $r' = \min(r, \mu_1)$, then by Equation (2.2) we have

$$s_\mu[X+t] = s_\mu + t s_1^\perp s_\mu + t^2 s_2^\perp s_\mu + \cdots + t^{r'} s_{r'}^\perp s_\mu.$$

Now Equation (5.1) follows from an application of Equation (2.4), factoring out a power of t using the right expression of Equation (2.3) and then taking a coefficient of t^r on both sides of the equation. □

A nearly identical proof will derive a similar identity for $s_{1^r}^\perp s_\lambda[s_\mu]$ by applying Equation (2.2) and in addition using Equation (2.3) so that in the product we have terms of the form $(\omega s_{\nu^{(k)}})[s_{1^k}^\perp s_\mu] = s_{(\nu^{(k)})'}[s_{1^k}^\perp s_\mu]$ if k is odd and $s_{\nu^{(k)}}[s_{1^k}^\perp s_\mu]$ if k is even.

PROPOSITION 5.2. *For partitions λ and μ and a positive integer r, $s_{1^r}^\perp s_\lambda[s_\mu]$ is equal to*

(5.2)
$$\sum_{\nu^{(i)}} s_{\nu^{(0)}}[s_\mu](\omega s_{\nu^{(1)}})[s_1^\perp s_\mu] s_{\nu^{(2)}}[s_{1^2}^\perp s_\mu]\cdots(\omega^{r'} s_{\nu^{(r')}})[s_{1^{r'}}^\perp s_\mu]\,\langle s_\lambda, s_{\nu^{(0)}}s_{\nu^{(1)}}\cdots s_{\nu^{(r')}}\rangle$$

where $r' = \min(r, \ell(\mu))$ and the sum is over all sequences of partitions $(\nu^{(0)}, \nu^{(1)}, \ldots, \nu^{(r')})$ with $\sum_{i=0}^r |\nu^{(i)}| = |\lambda|$ and $\sum_{i=0}^r i|\nu^{(i)}| = r$.

Applying Algorithm 1 to either Proposition 5.1 or 5.2 will have too many terms, and so it cannot be considered an efficient method for computing plethysms in general. However, special cases of these identities can be used to recursively compute classes of plethysms much more efficiently than the standard algorithms.

Special cases of Equations (5.1) and (5.2) include

$$s_r^\perp s_\lambda[s_{1^w}] = \sum_{\mu,\gamma} c^\lambda_{\mu\gamma} s_\mu[s_{1^w}]s_\gamma[s_{1^{w-1}}] \qquad s_{1^r}^\perp s_\lambda[s_w] = \sum_{\mu,\gamma} c^\lambda_{\mu\gamma'} s_\mu[s_w]s_\gamma[s_{w-1}],$$

where the sums are over partitions μ and γ with $|\mu| = |\lambda| - r$ and $|\gamma| = r$ and $c^\lambda_{\mu\gamma} := \langle s_\lambda, s_\mu s_\gamma\rangle$ are the Littlewood–Richardson coefficients.

While we saw improvements in recursive computations with these formulas, even smaller classes of functions are the special cases:

$$s_r^\perp s_{1^h}[s_{1^w}] = s_{1^{h-r}}[s_{1^w}]s_{1^r}[s_{1^{w-1}}] \qquad s_{1^r}^\perp s_h[s_w] = s_{h-r}[s_w]s_{1^r}[s_{w-1}] \tag{5.3}$$

$$s_r^\perp s_h[s_{1^w}] = s_{h-r}[s_{1^w}]s_r[s_{1^{w-1}}] \qquad s_{1^r}^\perp s_{1^h}[s_w] = s_{1^{h-r}}[s_w]s_r[s_{w-1}]\,. \tag{5.4}$$

We compared this method with the current method used by the SAGEMATH computer algebra system [**42**] and found it to be significantly faster for computing the plethysms $s_{1^h}[s_{1^w}]$. The current implementation in SAGEMATH uses the definition stated in Section 2, which requires a change of basis from the Schur functions to power sums and back again. An implementation of our method in SAGEMATH was able to compute $s_{1^4}[s_{1^6}]$ in less than a second compared to 36 seconds with the current implementation in SAGEMATH; and to compute $s_{1^6}[s_{1^6}]$ it took 21 seconds compared to well over an hour.

5.2. A formula for $s_\lambda[s_k]$ for $|\lambda| \leqslant 3$. Quasi-polynomial and integer polytope expressions for the coefficients of s_μ in $s_\lambda[s_k]$ with $|\lambda| \leqslant 3$ and k a positive integer have been extensively studied [**2, 10, 13, 21, 24, 35, 41, 43**]. The quasi-polynomial expressions are probably the most efficient means of computation of a single coefficient in this expression. In [**25**], the authors use the expressions for $s_3[s_k]$ and $s_k[s_3]$ to conclude that certain families of these coefficients do not need to be given by Ehrhart functions of rational polytopes.

In the following theorem we use the s-perp trick and some of the properties and analysis that others have derived about these formulae to give a combinatorial interpretation in terms of semi-standard Young tableaux. This formulation gives an idea of what we expect a combinatorial interpretation for the plethysm coefficients should look like in general. The result was obtained independently using different methods by Florence Maas-Gariépy and Étienne Tétreault [**31**], and our proof makes use of one of their earlier results [**41**]. Let $\mathsf{SSYT}_{(k,k,k)}$ represent the set of semi-standard Young tableaux with k 1s, k 2s and k 3s and denote the shape of a tableau by $\mathsf{shape}(S)$.

THEOREM 5.3. *Let* $T \in \left\{ \ytableaushort{123}, \ytableaushort{3,12}, \ytableaushort{2,13}, \ytableaushort{3,2,1} \right\}$ *be a standard tableau with shape a partition of 3. Then for any* $k \geqslant 1$,

$$s_{\mathsf{shape}(T)}[s_k] = \sum_{\substack{S \in \mathsf{SSYT}_{(k,k,k)} \\ \mathsf{type}(S) = T}} s_{\mathsf{shape}(S)}, \tag{5.5}$$

where $\mathsf{type}(S)$ *for* $S \in \mathsf{SSYT}_{(k,k,k)}$ *is given in Definition* 5.4 *below.*

DEFINITION 5.4. For $S \in \mathsf{SSYT}_{(k,k,k)}$, let N_r represent the number of cells with the label r in the second row of S for $1 \leqslant r \leqslant 3$.

- If S is standard and $k = 1$, then let $\mathsf{type}(S) = S$.
- If S has one or two rows, then we define
  (a) $\mathsf{type}(S) = \ytableaushort{123}$ if N_2 is even, $N_3 \geqslant 2N_2$, but $N_3 \neq 2N_2 + 1$.
  (b) $\mathsf{type}(S) = \ytableaushort{3,2,1}$ if N_2 is odd, $N_3 \geqslant 2N_2$, but $N_3 \neq 2N_2 + 1$.
  (c) $\mathsf{type}(S) = \ytableaushort{3,12}$ if N_2 is even, and $N_3 < 2N_2$ or $N_3 = 2N_2 + 1$.
  (d) $\mathsf{type}(S) = \ytableaushort{2,13}$ if N_2 is odd, and $N_3 < 2N_2$ or $N_3 = 2N_2 + 1$.
- If S has three rows and $\overline{S}$ is S with the first column removed, then $\mathsf{type}(S) = \mathsf{type}(\overline{S})^t$.

For T a standard tableau of size 3, let

$$\mathsf{Tab}_{T,k} = \{S \in \mathsf{SSYT}_{(k,k,k)} \mid \mathsf{type}(S) = T\}.$$

Define a linear operator on symmetric functions as

$$s_\mu \downarrow_k = \begin{cases} s_\mu & \text{if } \ell(\mu) \leqslant k, \\ 0 & \text{else.} \end{cases}$$

Also define the following bilinear (and commutative) operator by

$$s_\mu \odot s_\lambda = s_{\mu+\lambda},$$

where the sum of the two partitions is done componentwise, adding zeros if necessary.

LEMMA 5.5. *For* $k \geqslant 1$,

$$s_3[s_k] \downarrow_2 = \sum_S s_{\mathsf{shape}(S)},$$

where the sum is over all $S \in \mathsf{Tab}_{\ytableaushort{123},k}$ *with* $\ell(S) \leqslant 2$.

PROOF. A result of [**41**, Theorem 2.1] says that

$$s_3[s_k]\downarrow_2 = s_{66} \odot s_3[s_{k-4}]\downarrow_2 + \sum_{r=2}^{k} s_{3k-r,r} + s_{3k}.$$

The tableaux $S \in \mathsf{Tab}_{\young(123),k}$ and $\ell(S) \leqslant 2$ also satisfies this recurrence since they either contain $\young(223333)$ in the top row and $\young(111122)$ in the bottom row or not. Those that do are (by induction) enumerated by the expression $s_{66} \odot s_3[s_{k-4}]\downarrow_2$. Those tableaux that do not contain $\young(223333)$ and $\young(111122)$ of type $\young(123)$ are either a single row or have a second row consisting of exactly r 3's for $2 \leqslant r \leqslant k$. $\square$

LEMMA 5.6. *For $k \geqslant 1$,*

$$\sum_{S \in \mathsf{Tab}_{\young(3,12),k}} s_{\mathsf{shape}(S)} = \sum_{S \in \mathsf{Tab}_{\young(2,13),k}} s_{\mathsf{shape}(S)} \ .$$

PROOF. We define a map which we denote

$$\phi\colon \mathsf{Tab}_{\young(3,12),k} \to \mathsf{Tab}_{\young(2,13),k} \ .$$

CASE 1. If the length of S is 3, then let $\overline{S} \in \mathsf{SSYT}_{(k,k,k)}$ be the tableau S with the first column removed. Define $\phi(S)$ to be a column of length 3 added to $\phi^{-1}(\overline{S})$.

CASE 2. If the length of the second row of S is odd, then let S' be the tableau where one takes a 3 from the second row of S and exchanges it with a 2 in the first row.

CASE 3. If the length of the second row of S is even, then let S' be the tableau such that one takes a 3 from the first row of S and exchanges it with a 2 in the second row.

We leave to the reader to check the details that this is a bijection. $\square$

PROOF OF THEOREM 5.3. Assume by induction that (5.5) holds for some fixed k and all $T \in \left\{\young(123), \young(3,12), \young(2,13), \young(3,2,1)\right\}$. By Lemma 5.5 the terms of length less than or equal to 2 in $s_3[s_{k+1}]$ are given by the $S \in \mathsf{Tab}_{\young(123),k+1}$ such that $\ell(S) \leqslant 2$ and by (5.3) we have $s_{111}^{\perp} s_3[s_{k+1}] = s_{111}[s_k]$. Then, it follows by induction that the terms of length 3 are given by those $S \in \mathsf{Tab}_{\young(123),k+1}$ with $\ell(S) = 3$. Therefore,

$$s_3[s_{k+1}] = \sum_{S \in \mathsf{Tab}_{\young(123),k+1}} s_{\mathsf{shape}(S)} \ . \tag{5.6}$$

From [**32**, Section I.8 Example 9],

$$s_2[s_{k+1}] = \sum_{S} s_{\mathsf{shape}(S)},$$

where the sum is over $S \in \mathsf{SSYT}_{(k+1,k+1)}$ with N_2 is even. Therefore,

$$s_{21}[s_{k+1}]+s_3[s_{k+1}] = s_2[s_{k+1}]s_{k+1} = \sum_{S\in \mathsf{Tab}_{\young(123),k+1}} s_{\mathsf{shape}(S)}+ \sum_{S\in \mathsf{Tab}_{\young(3,12),k+1}} s_{\mathsf{shape}(S)},$$

where the last equality holds because $\mathsf{Tab}_{\young(123),k+1} \cup \mathsf{Tab}_{\young(3,12),k+1}$ are all $S \in \mathsf{SSYT}_{(k+1,k+1,k+1)}$ such that N_2 is even. We conclude by subtracting Equation (5.6) from both sides of the equation and Lemma 5.6 that

$$s_{21}[s_{k+1}] = \sum_{S\in \mathsf{Tab}_{\young(3,12),k+1}} s_{\mathsf{shape}(S)} = \sum_{S\in \mathsf{Tab}_{\young(2,13),k+1}} s_{\mathsf{shape}(S)}\ . \tag{5.7}$$

We again use [**32**, Section I.8 Example 9], which says that

$$s_{11}[s_{k+1}] = \sum_S s_{\mathsf{shape}(S)},$$

where the sum is over $S \in \mathsf{SSYT}_{(k+1,k+1)}$ with N_2 is odd so that

$$s_{21}[s_{k+1}]+s_{111}[s_{k+1}] = s_{11}[s_{k+1}]s_{k+1} = \sum_{S\in \mathsf{Tab}_{\young(3,2,1),k+1}} s_{\mathsf{shape}(S)}+ \sum_{S\in \mathsf{Tab}_{\young(2,13),k+1}} s_{\mathsf{shape}(S)}.$$

The last equality holds because $\mathsf{Tab}_{\young(3,2,1),k+1} \cup \mathsf{Tab}_{\young(2,13),k+1}$ are all $S \in \mathsf{SSYT}_{(k+1,k+1,k+1)}$ such that N_2 is odd. It then follows by subtracting Equation (5.7) from both sides of the equation that

$$s_{111}[s_{k+1}] = \sum_{S\in \mathsf{Tab}_{\young(3,2,1),k+1}} s_{\mathsf{shape}(S)}\ .$$

This concludes the proof by induction since Equation (5.5) holds for $k \to k+1$ and all $T \in \left\{ \young(123), \young(3,12), \young(2,13), \young(3,2,1) \right\}$. $\square$

5.3. Explicit formulas for $s_\lambda[s_2]$ and $s_\lambda[s_{1^2}]$ in special cases. A partition λ is called *even* if all columns have even length. A partition λ is called *threshold* if $\lambda'_i = \lambda_i + 1$ for all $1 \leqslant i \leqslant d(\lambda)$ where $d(\lambda)$ is the maximal i such that $(i,i) \in \lambda$.

THEOREM 5.7. *We have*

$$s_h[s_2] = \sum_{\substack{\lambda\vdash 2h\\ \lambda\ even}} s_{\lambda'}, \qquad s_h[s_{1^2}] = \sum_{\substack{\lambda\vdash 2h\\ \lambda\ even}} s_{\lambda}, \tag{5.8}$$

$$s_{1^h}[s_2] = \sum_{\substack{\lambda\vdash 2h\\ \lambda\ threshold}} s_{\lambda'}, \qquad s_{1^h}[s_{1^2}] = \sum_{\substack{\lambda\vdash 2h\\ \lambda\ threshold}} s_{\lambda}. \tag{5.9}$$

REMARK 5.8. The second equation in (5.9) has appeared in [**26**, Theorem 5.2]. The formulas of Theorem 5.7 also appear in [**32**, Page 138] and were originally proven in [**28**, Equations (11.9; 1) through (11.9; 4)], however these proofs are different from the ones we present here.

A *cell* of a partition λ is a pair (r, c), where $1 \leqslant r \leqslant \ell(\lambda)$ and $1 \leqslant c \leqslant \lambda_r$. A *corner* of a partition is a cell of the partition (r, c) such that $(r+1, c)$ and $(r, c+1)$ are not cells of the partition. For the proof of Theorem 5.7, we need the notion of *opposite cell* for each cell (s, t) in a threshold partition λ. The opposite cell $\mathsf{op}(s, t)$ of (s, t) is defined to be $(t+1, s)$ if $s \leqslant t$ and $(t, s-1)$ otherwise.

PROOF OF THEOREM 5.7. We prove the second equation in (5.8) by checking that the right hand side satisfies

$$s_r^{\perp} s_h[s_{1^2}] = s_{h-r}[s_{1^2}] s_r \tag{5.10}$$

from (5.4) holds for all r. We have

$$s_r^{\perp} s_\lambda = \sum_{\nu} s_\nu,$$

where the sum is over all ν such that λ/ν has at most one cell per column and $|\lambda/\nu| = r$. In the columns, where there is a cell in λ/ν, the length of the columns of ν will be odd. This implies that ν contains an even partition μ with ν/μ containing the cells which make those columns odd and so will also have at most one cell per column. Therefore,

$$s_r^{\perp} \sum_{\substack{\lambda \vdash 2h \\ \lambda \text{ even}}} s_\lambda = \sum_{\substack{\nu \vdash 2h-r \\ \nu/\mu \text{ horizontal } r\text{-strip} \\ \text{for some } \mu \vdash 2h-2r \text{ even}}} s_\nu = \sum_{\substack{\mu \vdash 2h-2r \\ \mu \text{ even}}} s_\mu s_r = \Big(\sum_{\substack{\mu \vdash 2h-2r \\ \mu \text{ even}}} s_\mu \Big) s_r$$

where the second to last equality follows from the Pieri rule. This confirms that the right hand side of the second equation in (5.8) satisfies (5.10).

Similarly, we prove the second equation in (5.9) by checking that the right hand side satisfies

$$s_r^{\perp} s_{1^h}[s_{1^2}] = s_{1^{h-r}}[s_{1^2}] s_{1^r}$$

from (5.3) holds for all r. We start with

$$s_r^{\perp} \sum_{\substack{\lambda \vdash 2h \\ \lambda \text{ threshold}}} s_\lambda = \sum_{\substack{\nu \vdash 2h-r \\ \lambda/\nu \text{ vertical } r\text{-strip} \\ \text{for some } \lambda \vdash 2h \text{ threshold}}} s_\nu .$$

For each cell (s, t) in λ/ν in the last sum from rightmost to leftmost (that is largest t to smallest t), check whether $\mathsf{op}(s, t)$ is in ν. If it is, remove $\mathsf{op}(s, t)$ from ν, otherwise leave the partition unchanged. Call the partition with all (possible) opposite cells removed τ. Note that ν/τ is a vertical strip since the cells in λ/ν form a horizontal strip and $\mathsf{op}(s, t)$ involves transposition. For each cell (s, t) in λ/ν from largest to smallest t with $t \geqslant s$, for which $(s', t') = \mathsf{op}(s, t)$ is not in ν, find the cell (x, y) in τ with smallest $y \geqslant t'$ such that $\nu/(\tau \setminus \{(x, y)\})$ is a vertical strip. Remove (x, y) and $\mathsf{op}(x, y)$ from τ. Call the resulting partition μ. Note that μ is threshold and furthermore, ν/μ is a vertical r-strip. Hence

$$s_r^\perp \sum_{\substack{\lambda \vdash 2h \\ \lambda \text{ threshold}}} s_\lambda = \sum_{\substack{\nu \vdash 2h-r \\ \nu/\mu \text{ vertical } r\text{-strip} \\ \text{for some } \mu \vdash 2h-2r \text{ threshold}}} s_\nu = \Big(\sum_{\substack{\mu \vdash 2h-2r \\ \mu \text{ threshold}}} s_\mu \Big) s_{1^r}.$$

The other two formulas follow from the identity

$$s_\lambda[s_{\mu'}] = \sum_\gamma \langle s_\lambda[s_\mu], s_\gamma \rangle s_{\gamma'} \tag{5.11}$$

for $|\mu|$ even (see Equation (2.3)). □

For $h \geqslant 2$ and $1 \leqslant r < h$, we have

$$s_{h-r}[s_{1^2}]s_{1^r}[s_{1^2}] = s_{(h-r,1^r)}[s_{1^2}] + s_{(h-r+1,1^{r-1})}[s_{1^2}]\ , \tag{5.12}$$

since by the dual Pieri rule $s_{h-r}s_{1^r} = s_{(h-r,1^r)} + s_{(h-r+1,1^{r-1})}$.

The next result is new.

COROLLARY 5.9. *We have*

$$s_{(h-1,1)}[s_{1^2}] = \sum_{\mu \in \mathcal{P}_{2h}} s_\mu + \sum_{\nu \vdash 2h \ \textit{even}} (b_\nu - 1)s_\nu, \tag{5.13}$$

$$s_{(h-1,1)}[s_2] = \sum_{\mu \in \mathcal{P}_{2h}} s_{\mu'} + \sum_{\nu \vdash 2h \ \textit{even}} (b_\nu - 1)s_{\nu'}, \tag{5.14}$$

where $\mathcal{P}_{2h}$ is the set of all partitions of $2h$ with columns of even length except two columns of distinct odd length, and b_ν is the number of corners of ν.

PROOF. By (5.12) with $r = 1$, we have $s_{(h-1,1)}[s_{1^2}] = s_{h-1}[s_{1^2}]s_{1^2} - s_h[s_{1^2}]$. On the other hand by Theorem 5.7, the Schur expansion of $s_{h-1}[s_{1^2}]$ contains the sum over s_ν where $\nu \vdash 2h-2$ is even. Multiplication by s_{1^2} adds a vertical strip of size 2 to the partitions in the Schur expansion by the Pieri rule, so that

$$s_{h-1}[s_{1^2}]s_{1^2} = \sum_{\mu \in \mathcal{P}_{2h}} s_\mu + \sum_{\nu \vdash 2h \text{ even}} b_\nu s_\nu.$$

This follows from the fact that adding a vertical strip of length 2 to an even partition gives a partition with two different odd and otherwise even length columns (if the two boxes are added to different columns) or an even partition (if the two boxes are added to the same column). An even partition ν of size $2h$ can be obtained in b_ν ways from an even partition of size $2h-2$ by adding two boxes to a column. Noting that, by Theorem 5.7, the Schur expansion of $s_h[s_{1^2}]$ contains all even partitions of size $2h$ proves (5.13).

Equation (5.14) follows from (5.11). □

REMARK 5.10. Note that there is an involution on the partitions in $\mathcal{S}_{2h} = \mathcal{P}_{2h} \cup \{\nu \vdash 2h \mid \nu \text{ even}\}$ appearing in the expansion in Corollary 5.9. Namely, map the partition $\nu = (\nu_1, \nu_2, \ldots, \nu_\ell)$ with ℓ even (with possibly $\nu_\ell = 0$) to

$$w\colon \nu \mapsto (\nu_1 + \nu_2, \nu_3 + \nu_4, \ldots, \nu_{\ell-1} + \nu_\ell)'.$$

Note that $w(\nu) \in \mathcal{P}_{2h}$ if $\nu \in \mathcal{P}_{2h}$ and $w(\nu)$ is even if ν is even. Also, it is not hard to see that $w^2(\nu) = \nu$ and $b_{w(\nu)} = b_\nu$. The involution w imposes a symmetry on the Schur expansion of $s_{(h-1,1)}[s_{1^2}]$ and $s_h[s_{1^2}]$.

EXAMPLE 5.11. We have

$$s_{(3,1)}[s_{1^2}] = s_{(21^6)} + s_{(2^21^4)} + s_{(2^31^2)} + s_{(321^3)} + s_{(32^21)} + s_{(3^21^2)} + s_{(431)}.$$

Note that $w(431) = (21^6)$, $w(3^21^2) = (2^21^4)$, $w(32^21) = (2^31^2)$, and $w(321^3) = (321^3)$ and indeed the coefficients of s_γ and $s_{w(\gamma)}$ match.

COROLLARY 5.12. *We have*

$$s_{(2,1^{h-2})}[s_{1^2}] = \sum_{\mu\in\mathcal{T}_{2h}} s_\mu + \sum_{\nu\vdash 2h \text{ threshold}} \left\lfloor \frac{b_\nu - 1}{2} \right\rfloor s_\nu, \tag{5.15}$$

$$s_{(2,1^{h-2})}[s_2] = \sum_{\mu\in\mathcal{T}_{2h}} s_{\mu'} + \sum_{\nu\vdash 2h \text{ threshold}} \left\lfloor \frac{b_\nu - 1}{2} \right\rfloor s_{\nu'}, \tag{5.16}$$

where $\mathcal{T}_{2h}$ is the set of all partitions λ of $2h$ with $\lambda'_i = \lambda_i + 1$ for all i except either

(i) *two distinct i in the range $1 \leqslant i \leqslant d(\lambda)$ for which either $\lambda'_i = \lambda_i + 2$ and the cell (λ'_i, i) is a corner or $\lambda'_i = \lambda_i$; or*
(ii) *one i in the range $1 \leqslant i \leqslant d(\lambda)$ with $\lambda'_i = \lambda_i + 3$.*

PROOF. The proof follows the same outline as the proof of Corollary 5.9 using (5.12) with $r = h-1$ (instead of $r = 1$) so that $s_{(2,1^{h-2})}[s_{1^2}] = s_{1^2}s_{1^{h-1}}[s_{1^2}] - s_{1^h}[s_{1^2}]$. □

COROLLARY 5.13. *We have for $h \geqslant 3$*

$$s_{(h-2,1,1)}[s_{1^2}] = \sum_{\mu\vdash 2h} a_\mu s_\mu,$$

$$s_{(h-2,1,1)}[s_2] = \sum_{\mu\vdash 2h} a_\mu s_{\mu'},$$

where

$$a_\mu = \begin{cases} \binom{b_\mu - 1}{2} & \text{if } \mu \text{ is even,} \\ \left(\sum_{\nu\vdash 2(h-2) \text{ even}} c^\mu_{\nu(2,1,1)}\right) - 1 & \text{if } \mu \in \mathcal{P}_{2h}, \\ \sum_{\nu\vdash 2(h-2) \text{ even}} c^\mu_{\nu(2,1,1)} & \text{otherwise.} \end{cases}$$

PROOF. From Equation (5.12) with $r = 2$ we know that $s_{(h-2,1,1)}[s_{1^2}] = s_{h-2}[s_{1^2}]s_{1^2}[s_{1^2}] - s_{(h-1,1)}[s_{1^2}]$. We use the fact that $s_{1^2}[s_{1^2}] = s_{(2,1,1)}$ and the expression for $s_{(h-1,1)}[s_{1^2}]$ from Corollary 5.9. In the Schur expansion of $s_{h-2}[s_{1^2}]$ only even partitions appear. The partitions μ indexing the Schur functions in the product $s_{h-2}[s_{1^2}]s_{1^2}[s_{1^2}] = s_{h-2}[s_{1^2}]s_{(2,1,1)}$ fall into four cases:

- Case 1: μ has exactly 4 odd columns
- Case 2: μ has exactly two equal odd columns
- Case 3: μ has exactly two distinct odd columns
- Case 4: μ is an even partition.

Subtracting the combinatorial formula for $s_{(h-1,1)}[s_{1^2}]$ given by Corollary 5.9 does not subtract anything in Cases 1 and 2, subtracts 1 in Case 3, and $b_\mu - 1$ in Case 4. Using that the Littlewood–Richardson coefficient $c^\mu_{\nu(2,1,1)}$ for all $\nu \subseteq \mu$ with $\nu \vdash 2(h-2)$ even count the coefficients of s_μ in $s_{h-2}[s_{1^2}]s_{(2,1,1)}$ hence proves the last two cases in the formula for a_μ. To prove the first case in a_μ, note that $\sum_{\nu\vdash 2(h-2) \text{ even}} c^\mu_{\nu(2,1,1)}$ for μ even is equal to $\binom{b_\mu}{2}$ since this number is counted by Littlewood–Richardson tableaux of shape μ/ν and weight $(2,1,1)$. Recall that a

Littlewood–Richardson tableau is a semistandard Young tableau such that the row reading of the tableau is a reverse lattice word. For μ and ν even these tableaux have to have 21 in the rightmost column of μ/ν and 31 in the leftmost column of μ/ν. This means we need to pick two corners of μ, which can be done in $\binom{b_\mu}{2}$ ways. Noting $\binom{b_\mu}{2} - (b_\mu - 1) = \binom{b_\mu - 1}{2}$ proves the first equation in the corollary.

The second equation follows again from (5.11). □

For general hooks, we have the following result.

COROLLARY 5.14. *For $h > k \geqslant 0$, we have*

$$s_{(h-k,1^k)}[s_{1^2}] = \sum_{i=0}^{k} \sum_{\substack{\mu \vdash 2h \\ \nu \vdash 2(h-k+i) \text{ even} \\ \rho \vdash 2(k-i) \text{ threshold}}} (-1)^i c^{\mu}_{\nu\rho} s_\mu. \tag{5.17}$$

PROOF. We prove this result by induction on k. For $k = 0$, Equation (5.17) reads

$$s_h[s_{1^2}] = \sum_{\substack{\mu \vdash 2h \\ \nu \vdash 2h \text{ even} \\ \rho \vdash 0 \text{ threshold}}} c^{\mu}_{\nu\rho} s_\mu = \sum_{\mu \vdash 2h \text{ even}} s_\mu,$$

which is true by (5.8).

Now assume by induction that (5.17) holds for $k - 1$. Note that by (5.12)

$$s_{(h-k,1^k)}[s_{1^2}] = s_{h-k}[s_{1^2}] s_{1^k}[s_{1^2}] - s_{(h-k+1,1^{k-1})}[s_{1^2}].$$

By Theorem 5.7 and the Littlewood–Richardson rule, the term $s_{h-k}[s_{1^2}]s_{1^k}[s_{1^2}]$ is the term $i = 0$ in (5.17). By induction, the second term equals

$$\begin{aligned} s_{(h-k+1,1^{k-1})}[s_{1^2}] &= \sum_{i=0}^{k-1} \sum_{\substack{\mu \vdash 2h \\ \nu \vdash 2(h-k+1+i) \text{ even} \\ \rho \vdash 2(k-1-i) \text{ threshold}}} (-1)^i c^{\mu}_{\nu\rho} s_\mu \\ &= -\sum_{i=1}^{k} \sum_{\substack{\mu \vdash 2h \\ \nu \vdash 2(h-k+i) \text{ even} \\ \rho \vdash 2(k-i) \text{ threshold}}} (-1)^i c^{\mu}_{\nu\rho} s_\mu, \end{aligned}$$

which are the remaining terms in (5.17), proving the claim. □

REMARK 5.15. It can be deduced from Corollary 5.14 that the coefficient of s_μ with μ even in $s_{(h-3,1,1,1)}[s_{1^2}]$ is $(b_\mu - 1)(b_\mu - 2)(2b_\mu - 3)/6$. The coefficient of s_μ with μ even in $s_{(h-4,1,1,1,1)}[s_{1^2}]$ is not dependent solely on b_μ since the coefficient of $s_{(3^2 2^2 1^4)}$ in $s_{(3,1,1,1,1)}[s_{1^2}]$ is 1 while the coefficient of $s_{(4^2 2^2 1^2)}$ is 2.

Acknowledgments

The authors would like to thank AIM for the opportunity to collaborate through their SQuaRE program.

References

[1] Y. Agaoka. *An algorithm to calculate plethysms of Schur functions*; Dedicated to Professor Motoyoshi Sakuma on his 70th birthday. Memoirs of the Faculty of Integrated Arts and Sciences, Hiroshima University. IV, Science reports, **21** (1995), 1–17.

[2] Y. Agaoka. *Decomposition formulas of the plethysm* $\{m\}\otimes\{\mu\}$ *with* $|\mu|=3$, Technical report, Hiroshima University, 2002.

[3] François Bergeron and Mark Haiman, *Tableaux formulas for Macdonald polynomials*, Internat. J. Algebra Comput. **23** (2013), no. 4, 833–852, DOI 10.1142/S0218196713400122. MR3078059

[4] C. Bowman and R. Paget, *The partition algebra and the plethysm coefficients I: Stability and Foulkes' conjecture*, Journal of Algebra, 2023.

[5] C. Bowman, R. Paget, and M. Wildon, *The partition algebra and the plethysm coefficients II: ramified plethysm*, `arXiv:2311.02721`, 2023.

[6] J. O. Carbonara, J. B. Remmel, and M. Yang, *A combinatorial rule for the Schur function expansion of the plethysm* $s_{(1^a,b)}[p_k]$, Linear and Multilinear Algebra **39** (1995), no. 4, 341–373, DOI 10.1080/03081089508818407. MR1365453

[7] Luisa Carini and J. B. Remmel, *Formulas for the expansion of the plethysms* $s_2[s_{(a,b)}]$ *and* $s_2[s_{(n^k)}]$, Discrete Math. **193** (1998), no. 1-3, 147–177, DOI 10.1016/S0012-365X(98)00139-3. Selected papers in honor of Adriano Garsia (Taormina, 1994). MR1661367

[8] Christophe Carré and Bernard Leclerc, *Splitting the square of a Schur function into its symmetric and antisymmetric parts*, J. Algebraic Combin. **4** (1995), no. 3, 201–231, DOI 10.1023/A:1022475927626. MR1331743

[9] M. J. Carvalho and S. D'Agostino, *Plethysms of Schur functions and the shell model*, J. Phys. A **34** (2001), no. 7, 1375–1392, DOI 10.1088/0305-4470/34/7/311. MR1819938

[10] Y. M. Chen, A. M. Garsia, and J. Remmel, *Algorithms for plethysm*, Combinatorics and algebra (Boulder, Colo., 1983), Contemp. Math., vol. 34, Amer. Math. Soc., Providence, RI, 1984, pp. 109–153, DOI 10.1090/conm/034/777698. MR777698

[11] Michele D'Adderio, Alessandro Iraci, and Anna Vanden Wyngaerd, *Theta operators, refined delta conjectures, and coinvariants*, Adv. Math. **376** (2021), Paper No. 107447, 59, DOI 10.1016/j.aim.2020.107447. MR4178919

[12] Melanie de Boeck, Rowena Paget, and Mark Wildon, *Plethysms of symmetric functions and highest weight representations*, Trans. Amer. Math. Soc. **374** (2021), no. 11, 8013–8043, DOI 10.1090/tran/8481. MR4328690

[13] Suzie C. Dent and Johannes Siemons, *On a conjecture of Foulkes*, J. Algebra **226** (2000), no. 1, 236–249, DOI 10.1006/jabr.1999.8169. MR1749887

[14] D. G. Duncan, *Note on a formula by Todd*, J. London Math. Soc. **27** (1952), 235–236, DOI 10.1112/jlms/s1-27.2.235. MR47662

[15] D. G. Duncan, *Note on the algebra of S-functions*, Canad. J. Math. **6** (1954), 509–510, DOI 10.4153/cjm-1954-054-3. MR64745

[16] H. O. Foulkes, *Plethysm of S-functions*, Philos. Trans. Roy. Soc. London Ser. A **246** (1954), 555–591, DOI 10.1098/rsta.1954.0008. MR62098

[17] A. M. Garsia and M. Haiman, *Some natural bigraded* S_n*-modules and* q,t*-Kostka coefficients*, Electron. J. Combin. **3** (1996), no. 2, Research Paper 24, approx. 60, DOI 10.37236/1282. The Foata Festschrift. MR1392509

[18] A. M. Garsia and C. Procesi, *On certain graded* S_n*-modules and the* q*-Kostka polynomials*, Adv. Math. **94** (1992), no. 1, 82–138, DOI 10.1016/0001-8708(92)90034-I. MR1168926

[19] Álvaro Gutiérrez and Mercedes H. Rosas, *Partial symmetries of iterated plethysms*, Ann. Comb. **27** (2023), no. 3, 493–518, DOI 10.1007/s00026-023-00652-4. MR4633754

[20] James Haglund, *The q, t-Catalan numbers and the space of diagonal harmonics*, University Lecture Series, vol. 41, American Mathematical Society, Providence, RI, 2008. With an appendix on the combinatorics of Macdonald polynomials, DOI 10.1007/s10711-008-9270-0. MR2371044

[21] Roger Howe, $(\mathrm{GL}_n, \mathrm{GL}_m)$*-duality and symmetric plethysm*, Proc. Indian Acad. Sci. Math. Sci. **97** (1987), no. 1-3, 85–109 (1988), DOI 10.1007/BF02837817. MR983608

[22] Alessandro Iraci, Brendon Rhoades, and Marino Romero, *A proof of the fermionic theta coinvariant conjecture*, Discrete Math. **346** (2023), no. 7, Paper No. 113474, 11, DOI 10.1016/j.disc.2023.113474. MR4579856

[23] Alessandro Iraci and Anna Vanden Wyngaerd, *Pushing our way from the valley Delta to the generalised valley Delta*, Algebr. Comb. **5** (2022), no. 4, 715–725, DOI 10.5802/alco.239. MR4484456

[24] Thomas Kahle and Mateusz Michałek, *Plethysm and lattice point counting*, Found. Comput. Math. **16** (2016), no. 5, 1241–1261, DOI 10.1007/s10208-015-9275-7. MR3552845

[25] Thomas Kahle and Mateusz Michałek, *Obstructions to combinatorial formulas for plethysm*, Electron. J. Combin. **25** (2018), no. 1, Paper No. 1.41, 9, DOI 10.37236/6597. MR3785020

[26] C. Klivans and V. Reiner, *Shifted set families, degree sequences, and plethysm*, Electron. J. Combin. **15** (2008), no. 1, Research Paper 14, 35, DOI 10.37236/738. MR2383434

[27] T. M. Langley and J. B. Remmel, *The plethysm $s_\lambda[s_\mu]$ at hook and near-hook shapes*, Electron. J. Combin. **11** (2004), no. 1, Research Paper 11, 26, DOI 10.37236/1764. MR2035305

[28] Dudley E. Littlewood, *The Theory of Group Characters and Matrix Representations of Groups*, Oxford University Press, New York, 1940. MR2127

[29] D. E. Littlewood, *Invariant theory, tensors and group characters*, Philos. Trans. Roy. Soc. London Ser. A **239** (1944), 305–365, DOI 10.1098/rsta.1944.0001. MR10594

[30] Nicholas A. Loehr and Gregory S. Warrington, *Quasisymmetric expansions of Schur-function plethysms*, Proc. Amer. Math. Soc. **140** (2012), no. 4, 1159–1171, DOI 10.1090/S0002-9939-2011-10999-7. MR2869102

[31] F. Maas-Gariépy and É. Tétreault, *3-Plethysms of homogeneous and elementary symmetric functions*, `arXiv:2209.14942`, 2022.

[32] I. G. Macdonald, *Symmetric functions and Hall polynomials*, 2nd ed., Oxford Mathematical Monographs, The Clarendon Press, Oxford University Press, New York, 1995. With contributions by A. Zelevinsky; Oxford Science Publications. MR1354144

[33] Rosa Orellana, Franco Saliola, Anne Schilling, and Mike Zabrocki, *Plethysm and the algebra of uniform block permutations*, Algebr. Comb. **5** (2022), no. 5, 1165–1203, DOI 10.5802/alco.243. MR4511162

[34] Rowena Paget and Mark Wildon, *Plethysms of symmetric functions and representations of* $\mathrm{SL}_2(\mathbf{C})$, Algebr. Comb. **4** (2021), no. 1, 27–68, DOI 10.5802/alco.150. MR4226555

[35] S. P. O. Plunkett, *On the plethysm of S-functions*, Canadian J. Math. **24** (1972), 541–552, DOI 10.4153/CJM-1972-047-7. MR294526

[36] Jeffrey B. Remmel and Mark Shimozono, *A simple proof of the Littlewood-Richardson rule and applications*, Discrete Math. **193** (1998), no. 1-3, 257–266, DOI 10.1016/S0012-365X(98)00145-9. Selected papers in honor of Adriano Garsia (Taormina, 1994). MR1661373

[37] Bruce E. Sagan, *The symmetric group: Representations, combinatorial algorithms, and symmetric functions*, 2nd ed., Graduate Texts in Mathematics, vol. 203, Springer-Verlag, New York, 2001, DOI 10.1007/978-1-4757-6804-6. MR1824028

[38] Thomas Scharf and Jean-Yves Thibon, *A Hopf-algebra approach to inner plethysm*, Adv. Math. **104** (1994), no. 1, 30–58, DOI 10.1006/aima.1994.1019. MR1272068

[39] Richard P. Stanley, *Enumerative combinatorics. Vol. 2*, Cambridge Studies in Advanced Mathematics, vol. 62, Cambridge University Press, Cambridge, 1999. With a foreword by Gian-Carlo Rota and appendix 1 by Sergey Fomin, DOI 10.1017/CBO9780511609589. MR1676282

[40] Richard P. Stanley, *Positivity problems and conjectures in algebraic combinatorics*, Mathematics: frontiers and perspectives, Amer. Math. Soc., Providence, RI, 2000, pp. 295–319. MR1754784

[41] É. Tétreault, *Recurrence relation for plethysm*, `arXiv:2008.13070`, 2020.

[42] The Sage Developers *Sagemath, the Sage Mathematics Software System (Version 9.3)*, 2021. `https://www.sagemath.org`.

[43] R. M. Thrall, *On symmetrized Kronecker powers and the structure of the free Lie ring*, Amer. J. Math. **64** (1942), 371–388, DOI 10.2307/2371691. MR6149

[44] J. A. Todd, *A note on the algebra of S-functions*, Proc. Cambridge Philos. Soc. **45** (1949), 328–334, DOI 10.1017/s0305004100024932. MR29875

[45] Marc A. A. van Leeuwen, *Some bijective correspondences involving domino tableaux*, Electron. J. Combin. **7** (2000), Research Paper 35, 25, DOI 10.37236/1513. MR1769066

[46] Mark Wildon, *A generalized SXP rule proved by bijections and involutions*, Ann. Comb. **22** (2018), no. 4, 885–905, DOI 10.1007/s00026-018-0409-x. MR3884760

[47] Mei Yang, *An algorithm for computing plethysm coefficients* (English, with English and French summaries), Proceedings of the 7th Conference on Formal Power Series and Algebraic

Combinatorics (Noisy-le-Grand, 1995), Discrete Math. **180** (1998), no. 1-3, 391–402, DOI 10.1016/S0012-365X(97)00127-1. MR1603696

[48] Mei Yang, *The first term in the expansion of plethysm of Schur functions* (English, with English and French summaries), Discrete Math. **246** (2002), no. 1-3, 331–341, DOI 10.1016/S0012-365X(01)00266-7. Formal power series and algebraic combinatorics (Barcelona, 1999). MR1887494

Department of Mathematics, North Carolina State University, 2311 Stinson Drive, Raleigh, North Carolina 27607

Email address: laura.colmenarejo.hernando@gmail.com

URL: https://sites.google.com/view/l-colmenarejo/home

Mathematics Department, Dartmouth College, 6188 Kemeny Hall, Hanover, New Hampshire 03755

Email address: Rosa.C.Orellana@dartmouth.edu

URL: https://math.dartmouth.edu/~orellana/

Département de mathématiques, Université du Québec à Montréal, Canada

Email address: saliola.franco@uqam.ca

URL: http://lacim.uqam.ca/~saliola/

Department of Mathematics, University of California, One Shields Avenue, Davis, California 95616-8633

Email address: anne@math.ucdavis.edu

URL: http://www.math.ucdavis.edu/~anne

Department of Mathematics and Statistics, York University, 4700 Keele Street, Toronto, Ontario M3J 1P3, Canada

Email address: zabrocki@mathstat.yorku.ca

URL: http://garsia.math.yorku.ca/~zabrocki/

Proceedings of Symposia in Pure Mathematics
Volume **110**, 2024
https://doi.org/10.1090/pspum/110/02020

Problems in the enumeration of tilings

Tri Lai

Abstract. Enumeration of tilings is the mathematical study concerning the total number of coverings of regions by similar pieces without gaps or overlaps. Enumeration of tilings has become a vibrant subfield of combinatorics with connections and applications to diverse mathematical areas. In 1999, James Propp published his well-known list of 32 open problems in the field. The list has got much attention from experts around the world. After two decades, most of the problems on the list have been solved. In this paper, we propose a new set of tiling problems. This survey paper contributes to the *Open Problems in Algebraic Combinatorics 2022* conference (OPAC 2022) at the University of Minnesota.

1. Introduction

Enumeration of tilings is a subfield of combinatorics studying the total number of coverings (called "*tilings*") of regions by similar pieces without gaps or overlaps. The first major result of the enumeration of tilings is usually credited to Percy Alexander MacMahon (1854–1929) with his beautiful formula for the number of lozenge tilings of a hexagon in the triangular lattice. However, MacMahon did not study tilings; instead, he worked on the enumeration of plane partitions. More than 100 years ago, he proved his celebrated theorem on the number of plane partitions fitting in a given box [**74**]. Much later (in the 1980s), G. David and C. Tomei showed a simple bijection between lozenge tilings of a centrally symmetric hexagon of side-lengths a, b, c, a, b, c (in cyclic order) and plane partitions fitting in an $a \times b \times c$-box [**25**], as in Figure 2.3. (Strictly speaking, David and Tomei only showed the bijection for the case $a = b = c$; however, their bijection could be easily extended for the general case.) This way, MacMahon's theorem implies a product formula for the tiling number of a hexagon. Since then, MacMahon has been considered as one of the founding fathers of the field.

Tiling-counting problems in the square lattice have already been investigated here and there for decades before the David–Tomei's bijection, say under the form of mathematical puzzles. For instance, no one really knows the author of the folklore puzzle on the number of ways to cover a rectangular stripe of width 2 by domino

2020 *Mathematics Subject Classification.* Primary 05A15, 05B45.
Key words and phrases. perfect matchings, plane partitions, tilings.
This research was supported in part by Simons Collaboration Grant (# 585923).

pieces[1]. Several results of the same flavor appeared in recreational and discrete mathematics, for instance, the work of David Klarner in the 1960s [**47**].

A couple of significant results in the enumeration of tilings come from statistical physics. In the early 1960s, the physicists P.W. Kasteleyn [**46**] and H.N.V. Temperley and M.E. Fisher [**91**] found an explicit formula for the number of dimer configurations of a rectangular grid graph. This result equivalently gives the enumeration of domino tilings of a rectangle. It would be a flaw here if we do not mention the well-known result of Fisher and Stephenson [**30**] concerning interactions of holes in dimer systems on the square lattice. Similarly, we cannot ignore the 1990 classical articles of J.H. Conway and J.C. Lagarias [**22**] and W. Thurston [**92**] investigating the connection between tilings and group theory.

In 1999, James Propp published his well-known article, "*Enumeration of Matchings: Problems and Progress*" [**77**], tracking the progress on a list of 32 open problems in the field. He presented this list in a 1996 lecture as part of the special program on algebraic combinatorics organized at MSRI. In a review on MathSciNet of the American Mathematical Society, Christian Krattenthaler (Professor at the University of Vienna) wrote: "*This list of problems was very influential; it called forth tremendous activity, resulting in the solution of several of these problems (but by no means all), in the development of interesting new techniques, and, very often, in results that move beyond the problems.*"

The enumeration of tilings has become a vibrant subfield of enumerative and algebraic combinatorics with connections and applications to diverse areas of mathematics, including representation theory, linear algebra, cluster algebra, group theory, mathematical physics, graph theory, probability, and discrete dynamical systems, just to name a few. We also refer the reader to the excellent survey (also by Propp) [**79**] for many connections and applications of the enumeration of tilings.

As most of Propp's problems have been solved, we would like to propose a new set of tiling problems. This batch of problems is independent of Propp's list, and we do not include here the still-open problems from the 1999 list. The author does not attempt to collect all open problems in the field of enumeration of tilings. The choice of problems in this paper reflects the author's personal taste.

2. Preliminaries

In this section, we introduce several important definitions that will be used in the paper.

A tiling problem usually starts with a given infinite lattice, e.g., the square lattice or the triangular lattice. The lattice divides the plane into small pieces, called *fundamental regions.* The union of any two fundamental regions sharing an edge is called a *tile.* A covering of a lattice region by tiles so that there are no gaps or over laps is called a *tiling* of the region.

For example, if our lattice is the triangular lattice, then the fundamental regions are unit equilateral triangles, and the tiles are *lozenges.* We orient the triangular lattice so that it accepts horizontal lattice lines. The lozenges in the triangular lattice have three possible orientations: left, vertical, and right, as in Figure 2.1. The corresponding tilings are called *lozenge tilings.* See Figure 2.2 for examples of lozenge tilings.

[1]The answer is the Fibonacci number.

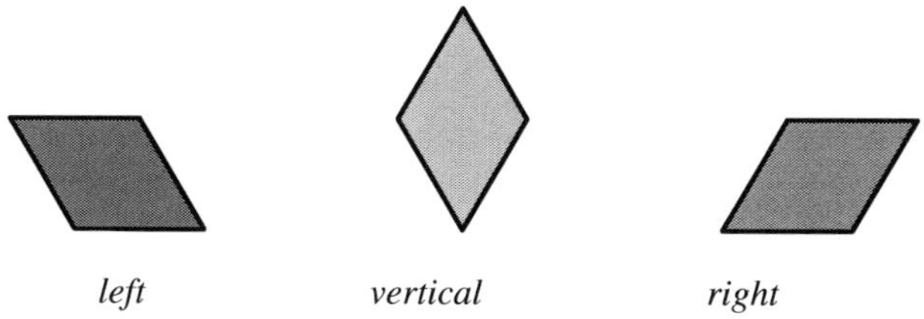

FIGURE 2.1. Three orientations of the lozenges.

If we consider the square lattice, then the fundamental regions are unit squares, the tiles are vertical and horizontal *dominoes*, and the corresponding tilings are called *domino tilings*. See Figures 2.4 and 2.5 for examples of domino tilings.

A tiling can carry *weight*, and we use the notation $\mathrm{T}(R)$ for the sum of weights of all tilings of a region R. We call $\mathrm{T}(R)$ the *tiling generating function* of R. When each tiling is weighted by 1, $\mathrm{T}(R)$ is exactly the tiling number of R.

A *perfect matching* of a graph G is a collection of (vertex-)disjoint edges that covers all vertices of the graph. There is a natural bijection between the tilings of a region R and the perfect matchings of its *planar dual graph* G, the graph whose vertices are fundamental regions in R and whose edges connect exactly two fundamental regions sharing an edge. We use the notation $\mathrm{M}(G)$ for the number of perfect matchings of a graph G. When perfect matchings are weighted, $\mathrm{M}(G)$ denotes the weighted sum of the perfect matchings.

We will frequently mention the following three well-known tiling enumerations of the centrally symmetric hexagon (or *semiregular hexagon*) (see Figure 2.2(a)), the *semi-hexagon* with dents on the base (see Figure 2.2(b)), and the *halved hexagon* (illustrated in Figure 2.2(c)).

The enumeration of tilings of the semiregular hexagon of side-lengths a, b, c, a, b, c (in counter-clockwise order, starting from the north side[2]) is credited to P. A. MacMahon [**74**] in the early 1900s. However, MacMahon did not work on tilings; what he proved is an enumeration of plane partitions fitting in a given box (or '*boxed plane partitions*'). Let $\lambda = (\lambda_1, \lambda_2, \ldots, \lambda_k)$ be a non-decreasing sequence of positive integers. A *plane partition of shape* λ is a left-aligned array of positive integers with k rows and λ_1 columns so that row i has λ_i entries, and that the entries are weekly decreasing across each row and down each column. The entries of the array are called the *parts* of the plane partition. A plane partition with at most a rows, at most b columns, and parts at most c can be viewed as a monotonic stack of unit cubes fitting in an $a \times b \times c$ box, and the latter, in turn, are in bijection with lozenge tilings of a semiregular hexagon [**25**]. For example, we consider the plane partition of shape $(4, 3, 2)$:

$$\pi = \begin{array}{|c|c|c|c|} \hline 6 & 4 & 2 & 1 \\ \hline 3 & 2 & 1 \\ \cline{1-3} 2 & 1 \\ \cline{1-2} \end{array}.$$

By adding zeroes to the end of lower rows in π, we have a 3×4 rectangular array of nonnegative integers, as in the right picture of Figure 2.3. We next embed

[2]From now on, we always list the side-lengths of a region in this order.

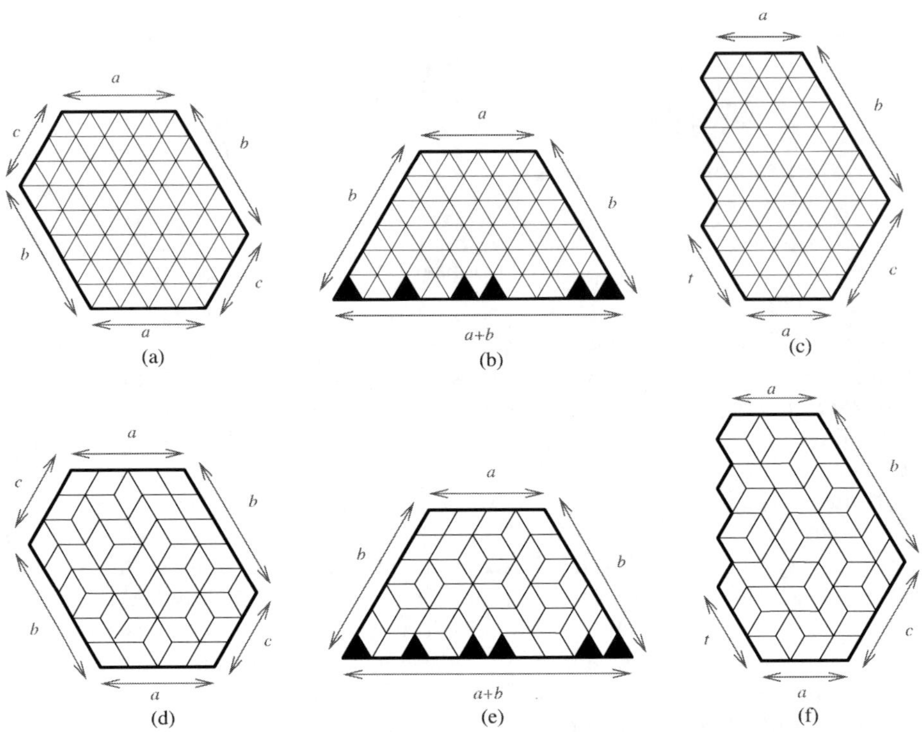

FIGURE 2.2. Three popular regions in the enumeration of lozenge tilings: (a) the semiregular hexagon, (b) the semi-hexagon with dents on the base, and (c) the halved hexagon. Sample tilings are shown in the second row.

this array on the plane Oij and put the corresponding number of unit cubes over each entry of the board. This way, one can interpret the plane partition π as a monotonic stack of unit cubes in the middle picture of Figure 2.3. This stack, in turn, can be projected on the plane $i + j + k = 0$ to obtain a lozenge tiling of a hexagon shown in the left picture. From this point of view, MacMahon's classical theorem [**74**] on boxed plane partitions can be stated in the language of the volume generating function of the stacks as follows.

Let q be an indeterminate. The *q-factorial* is defined as $[n]_q! := \prod_{i=1}^{n} \frac{1-q^n}{1-q}$, where $[0]_q! = 1$; and the *q-hyperfactorial* is $\mathbf{H}_q(n) := [0]_q![1]_q! \dots [n-1]_q!$, where $\mathbf{H}_q(0) = 1$.

THEOREM 1 (MacMahon's Theorem [**74**]). *For non-negative integers a, b, c*

$$\sum_{\pi} q^{|\pi|} = \prod_{i=1}^{a}\prod_{j=1}^{b}\prod_{k=1}^{c} \frac{q^{i+j+k-1}-1}{q^{i+j+k-2}-1}$$

$$= \frac{\mathbf{H}_q(a)\,\mathbf{H}_q(b)\,\mathbf{H}_q(c)\,\mathbf{H}_q(a+b+c)}{\mathbf{H}_q(a+b)\,\mathbf{H}_q(b+c)\,\mathbf{H}_q(c+a)}, \tag{2.1}$$

where the sum is taken over all monotonic stacks π fitting in an $(a \times b \times c)$-box, and where $|\pi|$ denotes the volume *of π.*

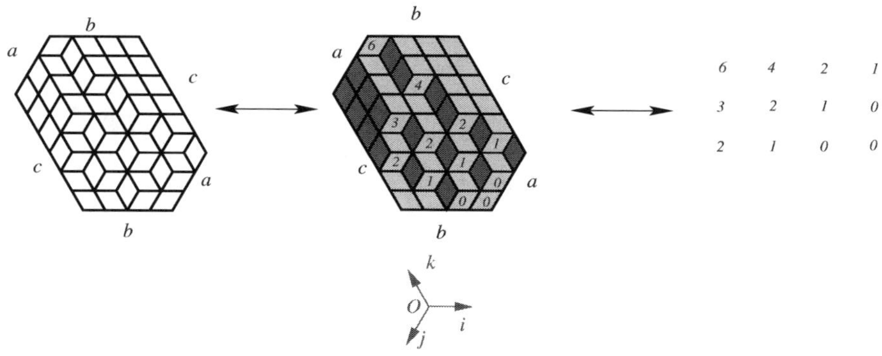

FIGURE 2.3. Correspondence between lozenge tilings of a hexagon, stacks of unit cubes fitting in a rectangular box, and plane partitions. The picture first appeared in [**58**].

Setting q tends to 1 in the above theorem, one obtains the tiling number of the semiregular hexagon. The beauty of this formula has inspired a large body of work, focusing on the enumeration of lozenge tilings of hexagons with defects.

The tiling enumeration of the semi-hexagon with dents on the base is due to H. Cohn, M. Larsen, and J. Propp [**20**, Proposition 2.1] when they give a bijection between lozenge tilings of the region and the semi-strict Gelfand–Tsetlin patterns [**36**]. It is well-known that the lozenge tilings of a dented semi-hexagon also correspond to the *column-strict plane partitions* (or *reverse semi-standard Young tableaux*). The enumeration of the halved hexagon was first found by R. Proctor [**76**, Corollary 4.1] in the form of the number of a particular class of staircase plane partitions. His result also implies the enumeration of the *transpose-complementary plane partitions*, one of the ten symmetry classes of plane partitions [**86**].

In 1961, Kasteleyn [**46**] and Temperley and Fisher [**91**] proved the following formula for the number of domino tilings of a rectangular board. See Figure 2.4 for an example of domino tilings of a rectangular board.

THEOREM 2 (Kasteleyn, Temperley, and Fisher [**46**],[**91**]). *The number of domino tilings of a* $2m \times 2n$ *chessboard is equal to*

$$2^{2mn} \prod_{j=1}^{m} \prod_{k=1}^{n} \left(\cos^2\left(\frac{j\pi}{2m+1}\right) + \cos^2\left(\frac{k\pi}{2n+1}\right)\right). \tag{2.2}$$

The *Aztec diamond* of order n is the region consisting of all unit squares inside the contour $|x| + |y| \leq n+1$ (see Figure 2.5 for examples). Elkies, Kuperberg, Larsen, and Propp [**28, 29**] show that:

THEOREM 3 (Elkies, Kuperberg, Larsen, and Propp). *The number of domino tilings of the Aztec diamond of order* n *is* $2^{n(n+1)/2}$.

This work has ignited a number of papers about domino tilings.

We often use the *q-Pochhammer symbol* in our tiling formulas:

$$(x;q)_n := \begin{cases} 1 & \text{if } n = 0; \\ (1-x)(1-xq)\cdots(1-xq^{n-1}) & \text{if } n > 0; \\ \frac{1}{(1-xq^{-1})(1-xq^{-2})\cdots(1-xq^{n})} & \text{if } n < 0. \end{cases} \tag{2.3}$$

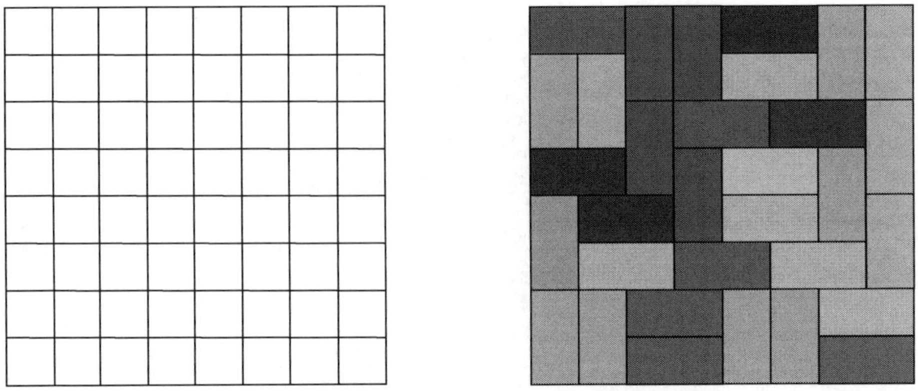

FIGURE 2.4. An 8×8 rectangular board (left) and domino tiling (right).

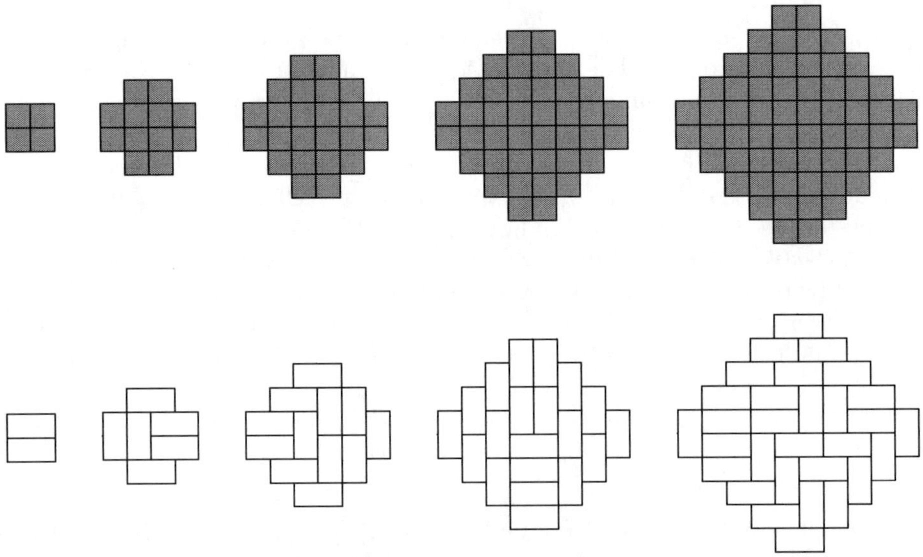

FIGURE 2.5. Top row: the Aztec diamonds of orders $1, 2, 3, 4$ and 5, from left to right. Bottom row: tilings of the Aztec diamonds in the top row.

Strictly speaking, the above q-Pochhammer symbol is not well-defined when n is a negative integer and $a = q^k$ for some $1 \leq k \leq -n$. However, this is not the case in our paper. When $q = 1$, we use notation $(x)_n$ instead of $(x;1)_n$ for the regular Pochhammer symbol.

3. Weighted enumerations of lozenge tilings

Weighted enumeration is usually more challenging and often gives more insights than 'plain' counting. This section is devoted to the weighted enumerations of lozenge tilings of regions in the triangular lattice.

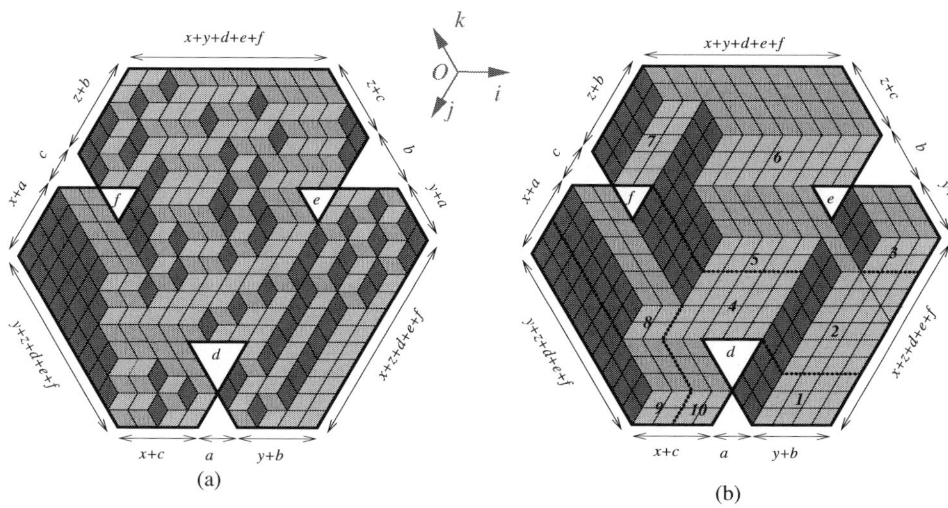

FIGURE 3.1. (a) Viewing a lozenge tiling of a hexagon with three dents as a stack of unit cubes fitting in a compound box. (b) The empty compound box with the floors of the rooms labeled $1, 2, \ldots, 10$. The picture first appeared in [**57**].

3.1. Generalized boxed plane partitions. MacMahon's boxed partition formula (2.1) gives a weighted enumeration of the lozenge tilings of a hexagon. As the enumeration is a function in q, we often call it a "*q-enumeration.*" Unfortunately, such elegant q-enumerations are *very rare* in the domain of enumeration of lozenge tilings. Besides the three new q-enumerations below, only a few are known (see, e.g., [**50, 86, 87**] and the list of references therein).

In [**18, 57, 58**], MacMahon's classical formula in Theorem 1 has been generalized by investigating q-enumerations of several families of hexagons with defects on the boundary. We usually call this type of defects "*dents.*" In order to give an illustrated example, let us focus on only the main result in [**57**].

Our region here is a 9-parameter generalization of the dented hexagon that first appeared in Problem 3 on Propp's well-known list of open problems [**77**][3]. In particular, we consider a certain hexagon with three bowtie-shaped dents on three non-consecutive sides, denoted by $F\begin{pmatrix} x & y & z\\ a & b & c\\ d & e & f\end{pmatrix}$. We view the lozenge tilings of a F-type region as monotonic stacks of unit cubes fitting in a *compound box* $\mathcal{B} = \mathcal{B}\begin{pmatrix} x & y & z\\ a & b & c\\ d & e & f\end{pmatrix}$, which is the union of 10 'rooms' (see Figure 3.1(a)). In Figure 3.1(b), we have a picture of the lozenge tiling corresponding to the empty stack, and this also gives a 3-D picture of the compound box $\mathcal{B}$. The floors of the rooms are labeled $1, 2, \ldots, 10$. One readily sees that the stacks of unit cubes here have the same monotonicity as the ordinary plane partitions (namely, the tops of the

[3]Propp's Problem 3 was first solved by T. Eisenkölbl [**27**].

columns are weakly decreasing along the directions of the i- and j-axes). Like the case of MacMahon's theorem, the volume generating function of these stacks is always given by a simple product formula in terms of q-hyperfactorials.

THEOREM 4 (Theorem 1.2 in [**57**]). *For non-negative integers a, b, c, d, e, f, x, y, z*

$$\sum_{\pi} q^{|\pi|} =$$

$$\frac{\mathbf{H}_q(x)\,\mathbf{H}_q(y)\,\mathbf{H}_q(z)\,\mathbf{H}_q(a)^2\,\mathbf{H}_q(b)^2\,\mathbf{H}_q(c)^2\,\mathbf{H}_q(d)\,\mathbf{H}_q(e)\,\mathbf{H}_q(f)\,\mathbf{H}_q(d+e+f+B)^4}{\mathbf{H}_q(a+d)\,\mathbf{H}_q(b+e)\,\mathbf{H}_q(c+f)\,\mathbf{H}_q(d+e+B)\,\mathbf{H}_q(e+f+B)\,\mathbf{H}_q(f+d+B)}$$
$$\times \frac{\mathbf{H}_q(A+2B)\,\mathbf{H}_q(A+B)^2}{\mathbf{H}_q(A+B+x)\,\mathbf{H}_q(A+B+y)\,\mathbf{H}_q(A+B+z)}$$
$$\times \frac{\mathbf{H}_q(a+b+d+e+B)\,\mathbf{H}_q(a+c+d+f+B)\,\mathbf{H}_q(b+c+e+f+B)}{\mathbf{H}_q(a+d+e+f+B)^2\,\mathbf{H}_q(b+d+e+f+B)^2\,\mathbf{H}_q(c+d+e+f+B)^2}$$
$$\times \frac{\mathbf{H}_q(a+d+x+y)\,\mathbf{H}_q(b+e+y+z)\,\mathbf{H}_q(c+f+z+x)}{\mathbf{H}_q(a+b+y)\,\mathbf{H}_q(b+c+z)\,\mathbf{H}_q(c+a+x)}$$
$$\times \frac{\mathbf{H}_q(A-a+B+z)\,\mathbf{H}_q(A-b+B+x)\,\mathbf{H}_q(A-c+B+y)}{\mathbf{H}_q(b+c+e+f+B+z)\,\mathbf{H}_q(c+a+d+f+B+x)\,\mathbf{H}_q(a+b+d+e+B+y)}, \tag{3.1}$$

where the sum is taken over all monotonic stacks π fitting in the compound box $\mathcal{B} = \mathcal{B}\begin{pmatrix} x & y & z \\ a & b & c \\ d & e & f \end{pmatrix}$, and where $A = a+b+c+d+e+f$, $B = x+y+z$.

We believe that there are more elegant q-enumerations of lozenge tilings waiting for us to explore.

PROBLEM 1. *Find more dented regions whose corresponding volume generating functions are given by simple product formulas.*

We hope that, after collecting enough examples of nice q-enumerations, we can solve the following problem:

PROBLEM 2. *Characterize the compound boxes, which yield nice volume generating functions.*

3.2. Elliptic Weight. A. Borodin, V. Gorin, and E. M. Rains [**4**] provide a different way to define weight for tilings, called "elliptic weight," as follows. We will adapt and specialize Borodin–Gorin–Rains' elliptic weight in this section.

The j-axis of our coordinate runs along a lattice line with the unit equal to 1/2 times the side-length of a lozenge. The i-axis is perpendicular to the j-axis at a lattice vertex (this vertex is the origin of our coordinate system); the unit on the i-axis is equal to $\sqrt{3}/2$ times the side-length of a lozenge. Figure 3.2 shows a particular placement of our coordinate system.

Only one of the three types of lozenges, as shown in Figure 2.1, that have the diagonals parallel to the i- and j-axes. (In Figure 3.2, these lozenges are the right lozenges.) Each such lozenge whose center at the point (i, j) is weighted by

$$\mathrm{wt}_1(i,j) = \frac{Xq^j + Yq^{-j}}{2}, \tag{3.2}$$

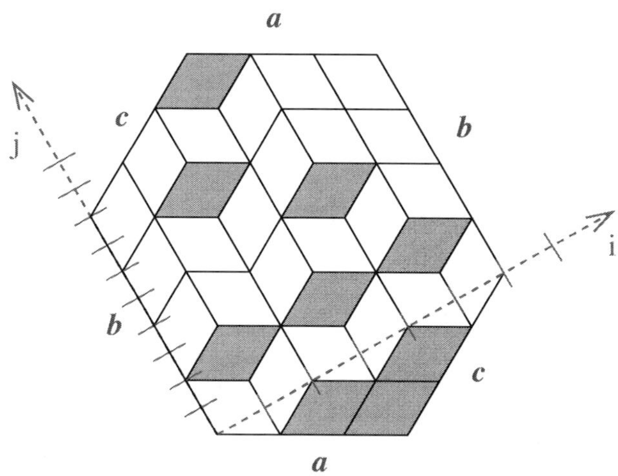

FIGURE 3.2. Assigning weight to lozenges.

where X, Y, q are three indeterminates. We note that the weight does not depend on i. All other lozenges have weight 1. The weight of a tiling is now the product of the weights of its lozenges.

This weight behaves very well and could be considered as the generalization of the "volume weight" investigated in the previous section. Indeed, when $X = 2$ and $Y = 0$, the weight becomes $\mathrm{wt}_2(i, j) = q^j$. With the weighting system as in Figure 3.2, each tiling τ of the hexagon is weighted by $C \cdot q^{2 \cdot \mathrm{Vol}(\tau)}$, where $\mathrm{Vol}(\tau)$ is the volume of the stack of unit cubes corresponding to the tiling τ, and C is a constant independent from the choice of the tiling τ. Sometimes, we call wt_2 the "natural weight" (as it is essentially equivalent to the weight of the boxed plane partitions).

We are especially interested in the following specialization of wt_1:

$$\mathrm{wt}_3(i, j) = \frac{q^j + q^{-j}}{2}. \tag{3.3}$$

One can view wt_3 as a symmetrization of wt_2, and we often call wt_3 the "symmetric weight." As mentioned in the previous section, the natural weight $\mathrm{wt}_2(i, j) = q^j$ does not often give nice tiling generating functions. In contrast, the symmetric weight $\mathrm{wt}_3(i, j)$ behaves much better. For example, it has been shown by M. Ciucu, T. Eisenkölbl, C. Krattenthaler, and D. Zare [**13**] that the "plain" tiling number (unweighted counting of tilings) of a "*cored hexagon*" is always given by a simple product formula (see Figure 3.7). However, there is no such formula for the tiling generating function associated with wt_2. On the other hand, as shown by H. Rosengren [**82**], the symmetric weight wt_3 gives a simple product formula for the tiling generating function of the cored hexagon. We have observed a similar fact for a halved hexagon with defects [**68**]: the natural weight wt_2 does not give a nice q-enumeration, but the symmetric weight wt_3 does.

Despite its very nice behavior, there are only a few results about weight wt_3. It deserves more attention from experts in the field. It would be interesting to see if the weight wt_3 yields nice tiling generating functions for known regions.

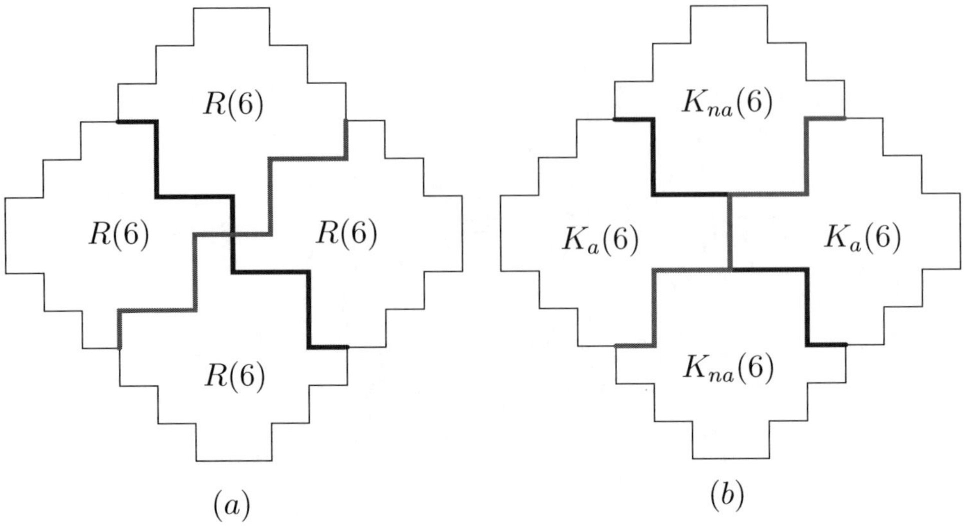

FIGURE 3.3. Three kinds of quartered Aztec diamonds of order 6. The figure first appeared in [**52**].

PROBLEM 3. *Find the tiling generating functions with respect to the symmetric weight* wt_3 *for known families of regions. For instance, one would like to find the tiling generating function for the "S-*cored hexagon*" introduced by Ciucu and Krattenthaler* [**17**] *as a generalization of the cored hexagon in* [**13**] *(see Figure* 6.10 *for an example of S-cored hexagon). This would give a generalization for Rosengren's enumeration in* [**82**].

As most of the known tiling enumerations are unweighted ones, there would be many things to do with this problem.

All the weights wt_1, wt_2, and wt_3 can be viewed as some rational functions in $z = q^j$. We want to find more weights of this type, such that they give nice tiling generating functions, say for several well-known families of regions, like the hexagons, the semi-hexagons, and the halved hexagons.

PROBLEM 4. *Find new "nice" weights that are rational functions in* q^j.

Jockusch and Propp [**41**] introduced the "*quartered Aztec diamonds*" as quarters of an Aztec diamond divided by two zigzag cuts passing the center (see Figure 3.3). These regions have been re-investigated and generalized in [**52–55**]. These papers showed that one could transform a "*quartered Aztec rectangle*" (a natural generalization of the quartered Aztec diamond) into a quartered hexagon using certain local graph transformations. See Figure 3.4 for an example of a quartered Aztec rectangle and a quartered hexagon. It turns out the tiling numbers of the two regions are only different by a multiplicative factor, which is a perfect power of 2. As a nice q-formula for the tiling generating function of the quartered hexagon has been found in [**68**], one would like to find such a q-formula for the quartered Aztec diamond and quartered Aztec rectangles.

PROBLEM 5. *Find a nice q-enumeration for the domino tilings of the quartered Aztec diamond and quartered Aztec rectangle.*

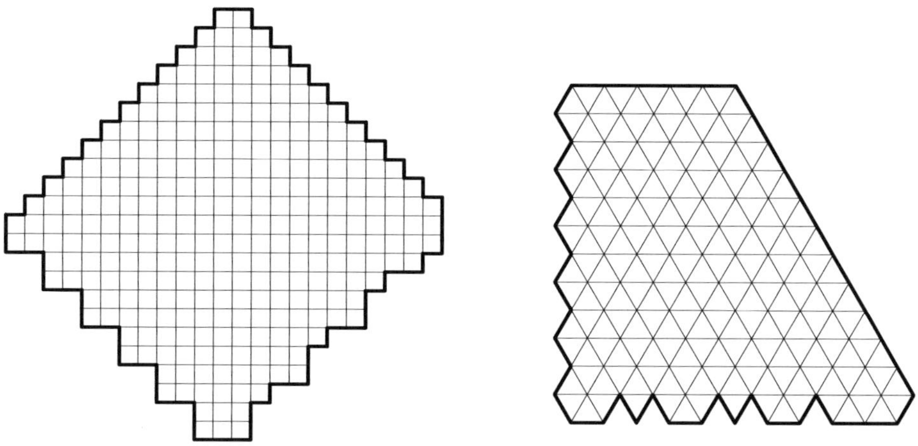

(a) (b)

FIGURE 3.4. (a) A quartered Aztec rectangle and (b) A quartered hexagon.

3.3. Unusual weights. A special weight of lozenge tilings of a hexagon has been inspired by the well-known trace formula of R. Stanley. By letting $c \to \infty$ in MacMahon's classical formula in Theorem 1, we obtain another well-known formula of MacMahon:

$$\sum_{\pi \in \mathcal{P}(a,b)} q^{|\pi|} = \prod_{i=1}^{a} \prod_{j=1}^{b} \frac{1}{1 - q^{i+j-1}}, \tag{3.4}$$

where $\mathcal{P}(a,b)$ denotes the set of plane partitions with at most a rows and b columns. Generalizing MacMahon's formula (3.4), Stanley [**83**] proved the *trace generating function*:

$$\sum_{\pi \in \mathcal{P}(a,b)} q^{|\pi|} t^{\mathrm{tr}(\pi)} = \prod_{i=1}^{a} \prod_{j=1}^{b} \frac{1}{1 - tq^{i+j-1}}, \tag{3.5}$$

where the *trace* is defined by $\mathrm{tr}(\pi) := \sum_i \pi_{i,i}$. E. R. Gansner [**34, 35**] later extended Stanley's work by showing that

$$\sum_{\pi \in \mathcal{P}(a,b)} \prod_{-a<\ell<b} q_\ell^{\mathrm{tr}_\ell(\pi)} = \prod_{i=0}^{a-1} \prod_{j=0}^{b-1} \left(1 - \prod_{\ell=-i}^{j} q_\ell \right)^{-1}, \tag{3.6}$$

where the *ℓ-trace* is defined as $\mathrm{tr}_\ell(\pi) := \sum_{j-i=\ell} \pi_{i,j}$.

Strictly speaking, Stanley's and Gansner's trace formulas above do not give any weighted enumerations for boxed plane partitions (equivalently, lozenge tilings of hexagons), as there is no upper bound for the parts of plane partitions. Recently, S. Kamioka [**42**] provides elegant boxed versions for the above formulas.

For a plane partition π of shape λ and a number $1 \le k \le \pi_{1,1}$, we define the *k-truncation* $\pi^{(k)}$ of π to be the plane partition obtained from π by removing all entries less than k. The shape $\lambda^{(k)}(\pi)$ of $\pi^{(k)}$ is called the *k-cross-section* of π. In particular, we have $\pi^{(1)} = \pi$ and $\lambda^{(1)}(\pi) = \lambda$ (see Figure 3.5 for the interpretation

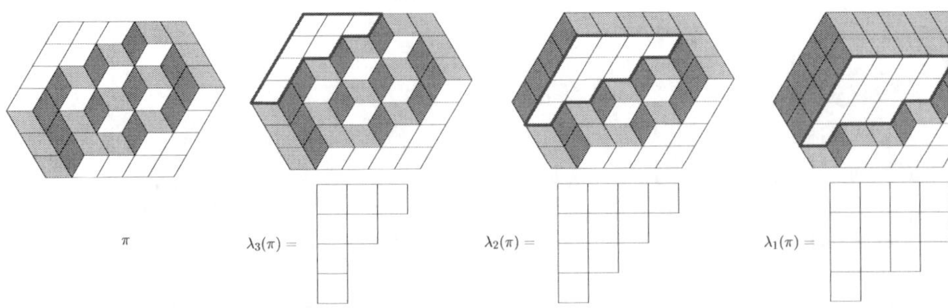

FIGURE 3.5. The plane partition π in the form of a stack of unit cubes and three cross-sections of π.

of the cross-sections). The *Durfee square* of a partition is the largest square fitting in its Ferrers diagram.

THEOREM 5 (Theorem 9 in [**42**]).

$$\sum_{\pi\in\mathcal{P}(a,b,c)} q^{|\pi|}t^{\mathrm{tr}(\pi)}\prod_{k=1}^{\pi_{1,1}}\frac{(q^{n-k+1};q)_{D_k(\pi)}}{(tq^{n-k+1};q)_{D_k(\pi)}}=\prod_{i=1}^{a}\prod_{j=1}^{b}\prod_{k=1}^{c}\frac{1-tq^{i+j+k-1}}{1-tq^{i+j+k-2}}, \tag{3.7}$$

where $D_k(\pi)$ is the side-length of the Durfee square of the k-cross-section $\lambda^{(k)}(\pi)$ of π.

This enumeration can be viewed as a weighted enumeration of the lozenge tilings of a hexagon. In particular, the tiling $\tau=\tau_\pi$ corresponding to the plane partition π is weighted by $\mathrm{wt}_K(\tau)=q^{|\pi|}t^{\mathrm{tr}(\pi)}\prod_{k=1}^{\pi_{1,1}}\frac{(q^{n-k+1};q)_{D_k(\pi)}}{(tq^{n-k+1};q)_{D_k(\pi)}}$. It is worth noticing that Kamioka also proves a finite version of Gansner's formula (see Theorem 17 in the same paper). He later generalizes further the results to arbitrary shapes in [**43**].

Unlike the weights $\mathrm{wt}_1, \mathrm{wt}_2$, and wt_3 in the previous section, Kamioka's theorem does *not* provide the weights for lozenges. However, one could define the corresponding lozenge-weights as follows. Divide the semiregular hexagon $Hex(a,b,c)$ into hooks as in Figure 3.6. Each right lozenge is labeled as in the figure. Now, each right lozenge with label x is weighted by q^{-x}, except for the ones intersecting with the dotted line, which are weighted by $(tq)^{-x}\frac{(q^n;q^{-1})_{n-x}}{(tq^n;q^{-1})_{n-x}}$. All lozenges of different orientations are weighted by 1. It is easy to see that the product of lozenge-weights in tiling τ_π is equal to $C\cdot\mathrm{wt}_K(\tau_\pi)$, where C is a constant independent from the choice of tilings. We get back Kamioka's weight by normalizing the lozenge-weights. See [**56**] for more details.

In his classical paper [**86**], Stanley lists ten symmetry classes of plane partitions. Each of the ten classes is equivalent to a particular type of symmetric tilings of a hexagon. We would expect the existence of Kamioka's trace formula for symmetric tilings.

PROBLEM 6. *Find versions of Kamioka's trace formula for symmetric tilings.*

We conclude this subsection by investigating another unusual weight of tilings inspired by the *(-1)-phenomenon* found by J. Stembridge [**89**], and more generally, the *Cyclic Sieving Phenomenon* by V. Reiner, D. Stanton, and D. White [**81**]. In

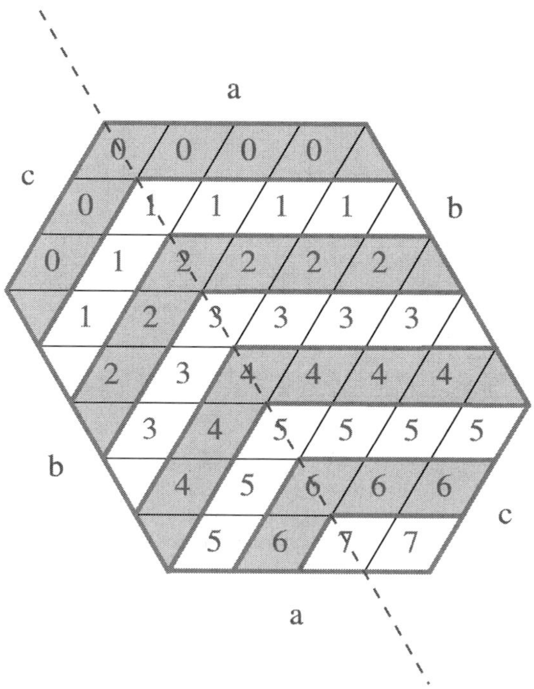

FIGURE 3.6. The weight for lozenge tiling inspired by Kamioka's formula.

[**13**], the authors proved striking formulas for specific (-1)-enumerations of the tilings of the cored hexagons and their cyclically symmetric tilings (see Theorems 4, 5, and 7 therein). In particular, if we extend the base of the triangular hole in the cored hexagon to the right, then each tiling τ is weighted by $(-1)^{n(\tau)}$, where $n(\tau)$ is the number of edges of lozenges of the tiling τ contained in the extended side (see Figure 3.7(a); in this case $n(\tau) = 2$). The critical point of the definition of $n(\tau)$ is that when we encode each tiling as a family of the non-intersecting lattice paths in the spirit of the well-known Lindström–Gessel–Viennot Theorem (see, e.g., [**37**, **71**, **88**]), the sign of the path family is exactly $(-1)^{n(\tau)}$. In the case of cyclically symmetric tilings, the weight is quite different. We consider the fundamental region of the cyclically symmetric tilings (which is illustrated as the parallelogram with bold sides in Figure 3.7(b)), then each tiling is weighted by $(-1)^{n_6(\tau)}$, where $n_6(\tau)$ is the sum of the horizontal distances between the shaded lozenges and the lower-left border of the fundamental region ($n_6(\tau) = 0 + 0 + 1 + 2 + 2 = 5$ in Figure 3.7).

One would ask for similar (-1)-enumerations for the following three generalizations of the cored hexagons:

(1) the "*S-cored hexagons*" (i.e., a hexagon with a cluster of four triangles removed from the center; see Figure 6.9(a)) in [**17**],
(2) the "*F-cored hexagons*" (i.e., a hexagon with an array of alternating triangles removed from the center) in [**11**], and
(3) the hexagons with three arrays of triangles removed in [**61**].

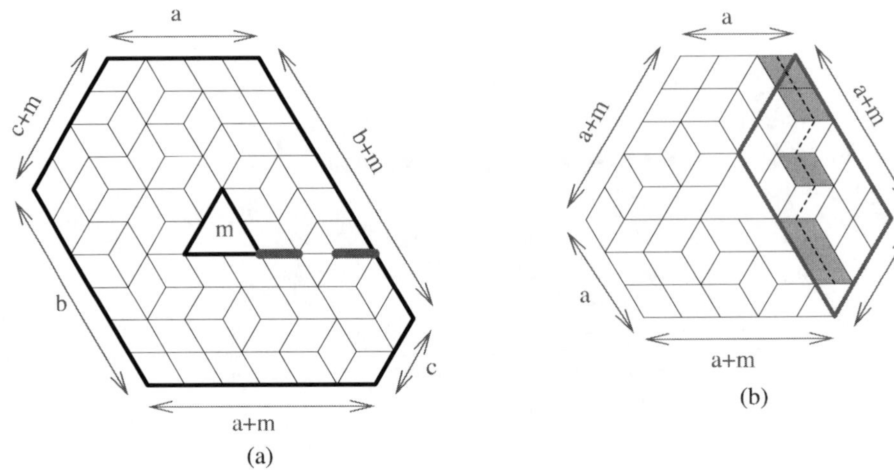

FIGURE 3.7. Weight in the (-1)-enumerations of a cored hexagon.

It is worth noticing that the unweighted enumeration of cyclically symmetric tilings of a S-cored hexagon has been provided by Ciucu in [**12**].

PROBLEM 7. (a) *Find the* (-1)*-enumerations of tilings of the three generalizations of core hexagons listed above.*

(b) *Find the* (-1)*-enumeration of cyclically symmetric tilings for the* S*-cored hexagon.*

Recall that Rosengren proved a simple product formula for the tiling generating of the cored hexagons using the weight $\mathrm{wt}_3 = \frac{q^j+q^{-j}}{2}$. We note that the versions of Rosengren's q-enumeration for the three generalizations of the cored hexagons are not known at this point.

PROBLEM 8. *Find a version of Rosengren's q-enumeration for each of the three generalizations of the cored hexagons above.*

Inspired by the above (-1)-enumeration, we would like to find a signed version of Rosengren's q-enumeration for the cored hexagons, and more generally, for its generalizations.

PROBLEM 9. *(a) Find a signed version of Rosengren's q-enumeration for the cored hexagon.*

(b) Generalize part (a) to the three generalizations of the cored hexagons above.

4. Shuffling phenomenon

This section is devoted to an exciting property of tilings, the "shuffling phenomenon"[4], that was first introduced in [**67**]. A minor modification in a region would lead to an unpredictable change in the tiling number. However, in some particular situations, the tiling number is changed by only a simple multiplicative

[4]The "shuffling phenomenon" here is *not* related to the "domino shuffling" operation in [**28, 29**].

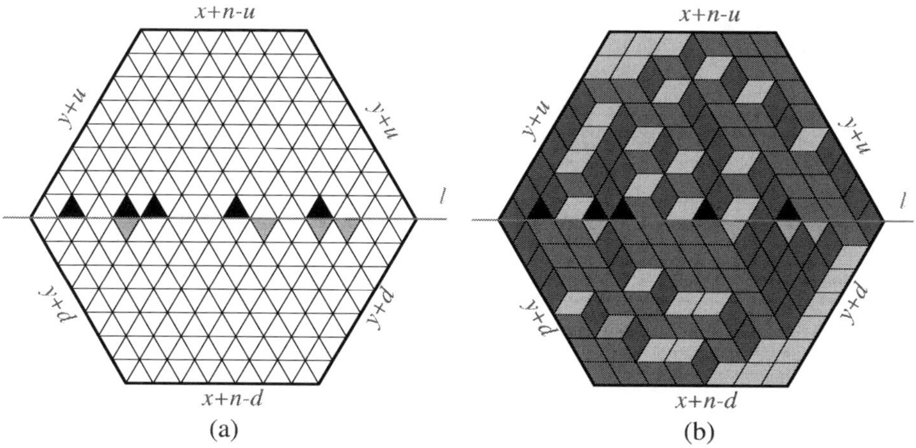

FIGURE 4.1. The region $H_{4,3}(2, 4, 5, 8, 11;\ 4, 9, 11, 12)$ (left) and a lozenge tiling of its (right). The back and shaded triangles indicate the unit triangles removed.

factor. See e.g. [**6, 7, 19, 21, 31, 32, 60, 62, 64, 68**] for the recent development of the phenomenon.

Let x, y, z, u, d be nonnegative integers, such that $u, d \leq n$. Consider a symmetric hexagon of side-lengths $x+n-u, y+u, y+d, x+n-d, y+d, y+u$. We remove u up-pointing and d down-pointing unit triangles along the horizontal lattice line l that contains the west and the east vertices of the hexagon. Let $U = \{s_1, s_2, \dots, s_u\}$ and $D = \{t_1, t_2, \dots, t_d\}$ denote the position sets of the up-pointing and down-pointing removed unit triangles (ordered from left to right), respectively. Assume that n is the size of the union $U \cup D$. Denote by $H_{x,y}(U; D)$ the hexagon with the above setup of removed unit triangles. We call it a *doubly–dented hexagon*. We now allow to 'shuffle' the positions of the up- and down-pointing unit triangles in the symmetric difference $U \Delta D$ to obtain new position sets U' and D', respectively. The following theorem shows that the above shuffling of unit triangles changes the tiling number of the region by only a simple multiplicative factor. Moreover, the factor can be written in a similar form to Cohn–Larsen–Propp's tiling formula of a semi-hexagon [**20**, Proposition 2.1].

THEOREM 6 (Shuffling Theorem; Theorem 2.1 in [**67**]). *For nonnegative integers x, y, u, d, n $(u, d \leq n)$ and four ordered subsets $U = \{s_1, s_2, \dots, s_u\}$, $D = \{t_1, t_2, \dots, t_d\}$, $U' = \{s'_1, s'_2, \dots, s'_u\}$, and $D' = \{t'_1, t'_2, \dots, t'_d\}$ of $[x+y+n]$, such that $U \cup D = U' \cup D'$ and $U \cap D = U' \cap D'$. We have*

$$\frac{\mathrm{T}(H_{x,y}(U; D))}{\mathrm{T}(H_{x,y}(U'; D'))} = \frac{\displaystyle\prod_{1 \leq i < j \leq u} \frac{s_j - s_i}{j - i} \cdot \prod_{1 \leq i < j \leq d} \frac{t_j - t_i}{j - i}}{\displaystyle\prod_{1 \leq i < j \leq u} \frac{s'_j - s'_i}{j - i} \cdot \prod_{1 \leq i < j \leq d} \frac{t'_j - t'_i}{j - i}}, \tag{4.1}$$

recall that we use the notation $\mathrm{T}(R)$ for the number of tilings of the unweighted region R (when R is weighted, $\mathrm{T}(R)$ denotes the tiling generating function of R).

We would like to emphasize that neither $H_{x,y}(U;D)$ nor $H_{x,y}(U;' D')$ has a nice tiling number in general. We also note that several stronger versions of the above Shuffling Theorem were provided in [**67**].

Denote by $S_{a,b}(t_1, t_2, \ldots, t_b)$ the semihexagon of side-lengths $a, b, a+b, b$ (in counter-clock wise order, from the top side) with the dents at the positions $t_1, t_2, \ldots, t_b$ on the base. By, say, [**87**, equation (7.105)] and Cohn–Larsen–Propp's enumeration [**20**, Proposition 2.1], we have[5]

$$\mathrm{T}(S_{a,b}(t_1, t_2, \ldots, t_b)) = \prod_{1 \le i < j \le b} \frac{t_j - t_i}{j - i} = \mathbf{s}_{\lambda(\{t_1,\ldots,t_b\})}(\mathbf{1}^b), \tag{4.2}$$

where $\lambda(\{t_1, \ldots, t_b\})$ is the the partition $(t_b - b + 1, \ldots, t_2 - 1, t_1)$. On the other hand, it is not hard to see that we also have

$$\begin{aligned}\mathrm{T}\left(H_{x,y}(U;D)\right) &= \sum_{|S|=y} \mathrm{T}(S_{x+n-u,u+y}(U \cup S))\,\mathrm{T}(S_{x+n-d,d+y}(D \cup S))\\ &= \sum_{|S|=y} s_{\lambda(U\cup S)}(\mathbf{1}^{u+y}) s_{\lambda(D\cup S)}(\mathbf{1}^{d+y}),\end{aligned} \tag{4.3}$$

where the sum runs over all y-subsets S of the complement of $U \cup D$.

Indeed, each tiling of $H_{x,y}(U;D)$ contains exactly y vertical lozenges along the axis l. Moreover, these vertical lozenges must be at the positions in the complement of $U \cup D$. Grouping those tilings that correspond to the same set of vertical lozenges, one could write the tiling number $\mathrm{T}(H_{x,y}(U;D))$ as the sum of tiling numbers:

$$\mathrm{T}(H_{x,y}(U;D)) = \sum_{|S|=y} \mathrm{T}(H^S_{x,y}(U;D)),$$

where $H^S_{x,y}(U;D)$ denotes the region obtained from $H_{x,y}(U;D)$ by removing y vertical lozenges at the positions in S. It is easy to see that each tiling of $H^S_{x,y}(U;D)$ can be separated into two tilings of two semi-hexagons by the axis l. It means that

$$\mathrm{T}(H^S_{x,y}(U;D)) = \mathrm{T}(S_{x+n-u,u+y}(U \cup S))\,\mathrm{T}(S_{x+n-d,d+y}(D \cup S)).$$

This implies the first identity in (4.3); the second identity follows from (4.2).

Applying (4.3), one can write our Shuffling Theorem in terms of Schur functions as

$$\frac{\sum_{|S|=y} \mathbf{s}_{\lambda(U\cup S)}(\mathbf{1}^{u+y})\mathbf{s}_{\lambda(D\cup S)}(\mathbf{1}^{d+y})}{\sum_{|S|=y} \mathbf{s}_{\lambda(U'\cup S)}(\mathbf{1}^{u+y})\mathbf{s}_{\lambda(D'\cup S)}(\mathbf{1}^{d+y})} = \frac{\mathbf{s}_{\lambda(U)}(\mathbf{1}^{u})\mathbf{s}_{\lambda(D)}(\mathbf{1}^{d})}{\mathbf{s}_{\lambda(U')}(\mathbf{1}^{u})\mathbf{s}_{\lambda(D')}(\mathbf{1}^{d})}, \tag{4.4}$$

where the sums are taken over all y-subsets S of the complement of $U \cup D$.

The q-analog of the Shuffling Theorem in [**67**] implies that (4.4) still holds (up to a q-power) when $\mathbf{1}^n$ is replaced by $(q, q^2, q^3, \ldots, q^n)$. It would be interesting to know if there is a more general Schur function identity behind (4.4).

PROBLEM 10. *Find a general Schur function identity behind the shuffling phenomenon. More particularly, find a* 'simple condition' *under which we have the*

[5]The notation $\mathbf{1}^n$ in the argument of the Schur function $s_{\lambda(\{t_1,\ldots,t_b\})}(\mathbf{1}^b)$ stands for n arguments equal to 1.

following identity

(4.5)
$$\frac{\sum_{|S|=y} \mathbf{s}_{\lambda(U\cup S)}(\mathbf{X}^{m+y})\mathbf{s}_{\lambda(D\cup S)}(\mathbf{X}^{m+(d-u)+y})}{\sum_{|S|=y} \mathbf{s}_{\lambda(U'\cup S)}(\mathbf{X}^{m+y})\mathbf{s}_{\lambda(D'\cup S)}(\mathbf{X}^{m+(d-u)+y})} = C\cdot\frac{\mathbf{s}_{\lambda(U)}(\mathbf{X}^{m})\mathbf{s}_{\lambda(D)}(\mathbf{X}^{m+(d-u)})}{\mathbf{s}_{\lambda(U')}(\mathbf{X}^{m})\mathbf{s}_{\lambda(D')}(\mathbf{X}^{m+(d-u)})},$$

where $\mathbf{X}^m$ *denotes the sequence of variables* $x_1, x_2, \ldots, x_m$*, and where* C *is a monomial in* x_i*'s.*

It is worth noticing that Seok Hyun Byun [**6**] and Markus Fulmek [**32**] independently found a simple alternative proof for the Shuffling Theorem by using the connections between lozenge tilings and Schur function as in (4.2). While their proofs explain the identity (4.4) and its q-analog, they did not solve the problem above.

By Cohn–Larson–Propp's enumeration again, our Shuffling Theorem can be written purely in terms of tiling numbers as

$$\frac{\mathrm{T}(H_{x,y}(U;D))}{\mathrm{T}(H_{x,y}(U';D'))} = \frac{\mathrm{T}(S_{x+y+n-u,u}(U))\,\mathrm{T}(S_{x+y+n-d,d}(D))}{\mathrm{T}(S_{x+y+n-u,u}(U'))\,\mathrm{T}(S_{x+y+n-d,d}(D'))}. \tag{4.6}$$

One readily sees that the two semi-hexagons in the numerator of the right-hand side of (4.6) are obtained by dividing the doubly-dented hexagon $H_{x+y,0}(U;D)$ along the horizontal axis l. More precisely, say by the Region-Splitting Lemma [**57, 58**], we have

$$\mathrm{T}(S_{x+y+n-u,u}(U))\,\mathrm{T}(S_{x+y+n-d,d}(D)) = \mathrm{T}(H_{x+y,0}(U;D))$$

and

$$\mathrm{T}(S_{x+y+n-u,u}(U'))\,\mathrm{T}(S_{x+y+n-d,d}(D')) = \mathrm{T}(H_{x+y,0}(U';D')).$$

This means that identity (4.6) could be rewritten as

$$\mathrm{T}(H_{x,y}(U;D))\,\mathrm{T}(H_{x+y,0}(U';D')) = \mathrm{T}(H_{x,y}(U';D'))\,\mathrm{T}(H_{x+y,0}(U;D)). \tag{4.7}$$

Both sides of (4.7) count pairs of tilings of doubly-dented hexagons. There should be a bijection between these sets of pairs of tilings.

PROBLEM 11. *Find a bijection that proves identity* (4.7).

Several other examples of the shuffling phenomenon have been found. One of them is the shuffling phenomenon for a hexagon with a removed "*triad of bowties*" [**19**]. We consider the hexagon H of side-lengths $x+a+b+c, y+a'+b'+c', z+a+b+c, x+a'+b'+c', y+a+b+c, z+a'+b'+c'$. We remove from the interior of H three bowties shapes located at the vertices of an equilateral triangle Δ of side-length $t+a'+b'+c'$. Denote by $R = R^{\Delta}_{x,y,z}(a,a',b,b',c,c')$ the resulting region (see Figure 4.2(a) for an example; the triangle with red dotted sides indicates the triangle Δ).

We now adjust the side-lengths of the lobes in each bowtie so that the sum of the side-lengths is unchanged. Assume that the (a,a')-, (b,b')-, and (c,c')-bowties are deformed into the (d,d')-, (e,e')-, and (f,f')-bowties, respectively, where $a+a' = d+d'$, $b+b' = e+e'$, and $c+c' = f+f'$. We now create a 'sibling' region R' of R as follows. We start with the hexagon H' of side-lengths $x+d+e+f, y+d'+e'+f', z+d+e+f, x+d'+e'+f', y+d+e+f, z+d'+e'+f'$. We now remove the above three new bowties from H' so that they are located at the vertices of a triangle Δ' of side-length $t+d'+e'+f'$ and that the distances from the top, the left, and the right bowties to the north, the southeast, and the southwest sides

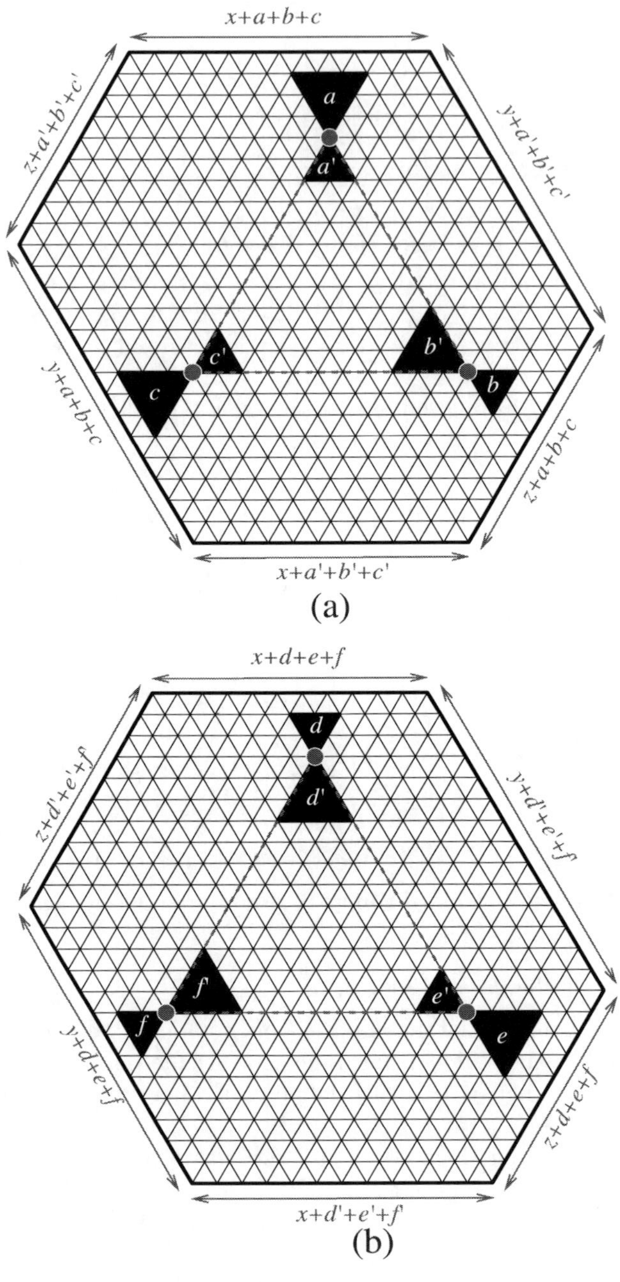

FIGURE 4.2. Two sibling hexagons with a triad of bowties removed.

of H' are equal to the corresponding distances in the original region. Denote by $R' = R^{\Delta'}_{x,y,z}(d, d', e, e', f, f')$ the new region.

It has been shown that the ratio of tiling numbers of the above two regions is given by a simple product formula [**19**]. We would also like to emphasize that, in general, the tiling numbers of the regions

$$R^{\Delta}_{x,y,z}(a,a',b,b',c,c') \quad \text{and} \quad R^{\Delta'}_{x,y,z}(d,d',e,e',f,f')$$

are *not* given by simple product formulas (see Theorem 1 in [**19**]). See Figure 4.2 for an example of $x = 4, y = 6, z = 2, t = 4, a = 3, a' = 2, b = 2, b' = 3, c = 3, c' = 2, d = 2, d' = 3, e = 3, e' = 2, f = 2, f' = 3$; the triangles Δ and Δ' are the ones with red vertices in pictures (a) and (b), respectively. We note that it actually needs three more parameters to define the triangle Δ and that the triangle Δ' is uniquely determined by Δ. It means that each of the regions R and R' depends on 12 parameters.

We have realized that the above instance of the shuffling phenomenon could be extended to the weighted case. In particular, we consider the generating functions with the elliptic weight $\mathrm{wt}_1 = \frac{Xq^j+Yq^{-j}}{2}$ of

$$R^{\Delta}_{x,y,z}(a,a',b,b',c,c') \quad \text{and} \quad R^{\Delta'}_{x,y,z}(d,d',e,e',f,f').$$

Like the unweighted case, these tilling generating functions are not given by any simple product formulas; however, their ratio seems to be a simple product.

PROBLEM 12. *Prove that the ratio of tiling generating functions*

$$\frac{\mathrm{T}_1(R^{\Delta}_{x,y,z}(a,a',b,b',c,c'))}{\mathrm{T}_1(R^{\Delta'}_{x,y,z}(d,d',e,e',f,f'))}$$

is always given by a simple product formula. Here, we use the notations $\mathrm{T}_i(R)$ *for the tiling generating function of* R *using the weight* wt_i, $i = 1, 2, 3$, *as described in the previous section.*

We want to point out that each tiling generating function in the above problem depends on *fifteen* parameters. Solving this problem would be technically challenging. We note that the conjectural formula for the above ratio has been found in the particular case when $X = 2$ and $Y = 0$ (i.e., the case of the natural weight wt_2). However, a formula in the general case is still unknown.

Like its unweighted version, the conjecture would imply a number of results in the weighted enumeration of lozenge tilings. Let us point out one of them. Rosengren in [**82**] found an intriguing consequence of his weighted enumeration of the cored hexagons (see Corollary 2.2 therein). Intuitively, if we reflect the removed triangle through the center of the hexagon, then the tiling generating function of the core hexagon changes by only a simple multiplicative factor. The problem (if solved) would give a conceptual explanation for his observation. Indeed, we consider the special case when $X = Y = 1$, $a = d'$, $a' = b = b' = c = c' = d = e = e' = f = f' = 0$, and q is replaced by $\sqrt{q}$. Then the ratio between the tiling generating functions of $R_{x,y,z}(a,0,0,0,0,0)$ and (the horizontal reflection of) $R_{x,y,z}(0,a,0,0,0,0)$ is exactly the ratio in Rosengren's Corollary 2.2.

It has been shown that the original shuffling phenomenon also holds for the reflectively symmetric tilings and centrally symmetric tilings of the doubly–dented hexagons [**60, 64**]. It suggests we do the same for hexagons with a triad of bowtie holes. In particular, we first focus on the *cyclically symmetric tilings*, i.e., the tilings invariant under 120° rotations. We note that the region $R^{\Delta}_{x,y,z}(a,a',b,b',c,c')$ must be cyclically symmetric itself in order to have a cyclically symmetric tiling. More precisely, we must have $x = y = z$, $a = b = c$, $a' = b' = c'$, and the triangle Δ must

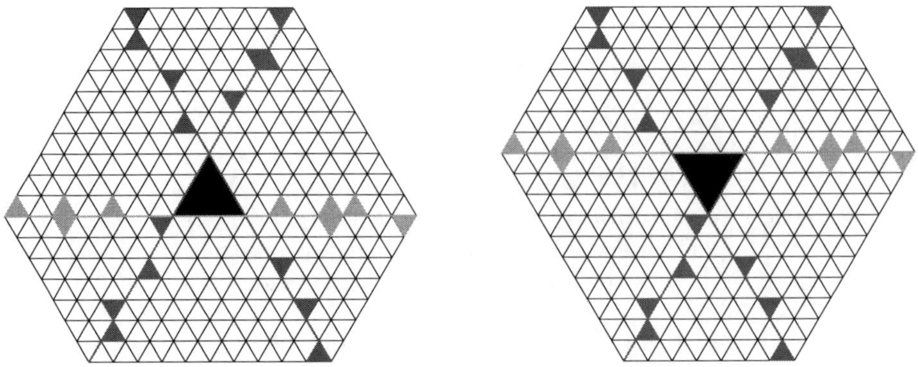

FIGURE 4.3. The hexagons with holes on three crossing lines.

be at the center of the region. To emphasize the symmetric location of the triangle, we use the notation Δ_0 instead of Δ.

PROBLEM 13. *Let x, a, a', d, d' be non-negative integers so that $a + a' = d + d'$. Find a formula for the ratio of tiling generating functions*

$$\frac{\mathrm{T}_c(R^{\Delta_0}_{x,x,x}(a, a', a, a', a, a'))}{\mathrm{T}_c(R^{\Delta'_0}_{x,x,x}(d, d', d, d', d, d'))}.$$

Here we use the notation $\mathrm{T}_c(R)$ for the weighted sum of cyclically symmetric tilings of R using the weight $\mathrm{wt}_1 = \frac{Xq^j+Yq^{-j}}{2}$ (or some specialization).

We are also interested in the shuffling phenomenon for the reflectively symmetric tilings of hexagons with a triad of bowties. The region $R^{\Delta}_{x,y,z}(a, a', b, b', c, c')$ must be reflectively symmetric itself in order to have a reflectively symmetric tiling. More precisely, we must have $y = z$, $b = c$, $b' = c'$, and the triangle Δ must be on the symmetry axis. To emphasize the symmetric location of the triangle, we use the notation Δ_1 instead of Δ.

PROBLEM 14. *Let $x, y, a, a', b, b', d, d', e, e'$ be non-negative integers so that $a + a' = d + d'$ and $b + b' = e + e'$. Find a formula for the ratio of tiling generating functions*

$$\frac{\mathrm{T}_r(R^{\Delta_1}_{x,y,y}(a, a', b, b', b, b'))}{\mathrm{T}_r(R^{\Delta'_1}_{x,y,y}(d, d', e, e', e, e'))}.$$

Here we use the notation $\mathrm{T}_r(R)$ for the weighted sum of reflectively symmetric tilings of R using the weight $\mathrm{wt}_3 = \frac{q^j+q^{-j}}{2}$.

We note that in Problem 14, the tilings are weighted by $\mathrm{wt}_3 = \frac{q^j+q^{-j}}{2}$, a special case of the weight $\mathrm{wt}_1 = \frac{Xq^j+Yq^{-j}}{2}$ in Problem 13. Our data suggest that the general weight wt_1 does not yield a nice product formula for the tiling ratio in this case.

Byun generalized the shuffling phenomenon for a triad of bowties in [**19**]. He showed that the tiling number of a hexagon with holes on three crossing lines only changes by a simple multiplicative factor if we flip the central triangular hole and then translate these lines of holes (see Theorem 2.1 in [**7**]). See Figure 4.3 for an illustration. More precisely, Byun's region is obtained from a cored-hexagon of side-lengths $n, n+x, n, n+x, n, n+x$ (in clockwise order, starting from the north side) with the (up-pointing) triangular hole of side-length x in the center. We next extend the sides of the central hole to three equal segments distinguished by three different colors in the left region. On these segments, we put some additional holes. Byun uses *twelve* sequences $A_1, A_2, \ldots, A_6, B_1, B_2, \ldots, B_6$ to record the position of the additional holes, for brevity, we denote $\mathbf{A} := (A_1, \ldots, A_6)$ and $\mathbf{B} := (B_1, \ldots, B_6)$. The resulting region is denoted by $H_{n,x}(\mathbf{A}, \mathbf{B})$. We now consider the hexagon of side-lengths $n+x, n, n+x, n, n+x, n$ (also in clockwise order, starting from the north side) with a *down-pointing* central hole of size x. We now implant the three segments of holes in the previous region on the extended sides of the new central hole in the new cored-hexagon. Denote the new region by $\overline{H}_{n,x}(\mathbf{A}, \mathbf{B})$ (see the right region in the figure).

One would like to generalize this exciting result to the tiling generating function with respect to the weight wt_1. More precise, we assign weight to lozenges in the hexagon with holes on three crossing lines $H_{n,x}(\mathbf{A}, \mathbf{B})$ (resp., $\overline{H}_{n,x}(\mathbf{A}, \mathbf{B})$) using the weight wt_1 (or some specialization of its), with the horizontal axis running along the base and the vertical axis passing the top (resp., bottom) vertex of the central triangular hole. We conjecture that the ratio of the two resulting generating functions is given by a simple product formula.

PROBLEM 15. *Generalizing Byun's Theorem 2.1(a) in* [**7**] *to tiling generating functions using the weight* wt_1 *(or some specialization).*

Inspired by Problems 13 and 14 above, we are also interested in the symmetric version of Problem 15 above.

PROBLEM 16. *(a) Generalizing Byun's Theorem 2.1(b) in* [**7**] *to generating functions of cyclically symmetric tilings using the weight* wt_1 *(or some specialization).*

(b) Investigate the ratio of generating functions of reflectively symmetric tilings of hexagons with holes on three crossing lines.

As shown in [**51, 52**], in many cases, one can convert a lozenge tiling problem to a domino tiling problem, and vice versa. It suggests the existence of many shuffling theorems for domino tilings. In fact, Propp and Stanley gave an elegant instance in their paper about domino tilings with barriers [**80**].

PROBLEM 17. *Find more examples of the shuffling phenomenon for domino tilings.*

5. Connections to electrical networks

This section is devoted to connections between tilings and other mathematical areas. Among many interesting connections, we focus on the connection to the study of electrical networks. Again, we recommend the reader to the excellent survey paper of James Propp [**79**] for various connections and applications of the enumeration of tilings.

The study of *electrical networks* comes from classical physics with the work of Ohm and Kirchhoff more than 100 years ago. A *circular planar electrical network* (or simply *electrical network*) is a graph $G = (E, V)$ embedded in a disk with a set of distinguished vertices $N \subseteq V$ on the circle, called *nodes*, and a *conductance*

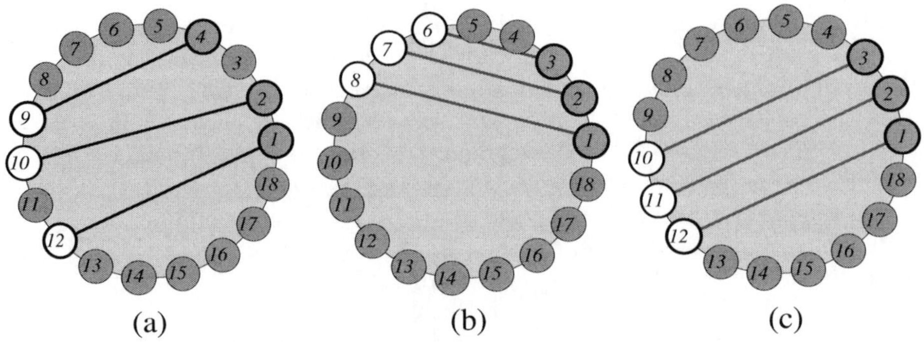

FIGURE 5.1. (a) A non-contiguous minor. (b) A contiguous minor that is not a central minor. (c) A central minor.

function $wt : E \to \mathbb{R}^+$. The electrical networks were first studied systematically by Colin de Verdière [**26**] and Curtis, Ingerman, Moores, and Morrow [**23, 24**]. We refer the reader to, e.g. [**1, 40, 45, 69, 70, 90**], for the recent development of the topic.

We arrange the indices $1, 2, \dots, n$ of a $n \times n$ matrix M in counter-clockwise order around the circle. Let $A = \{a_1, a_2, \dots, a_k\}$ and $B = \{b_1, b_2, \dots, b_k\}$ be two sets of indices so that $a_1, a_2, \dots, a_k, b_k, b_{k-1}, \dots, b_1$ appear in counter-clockwise order around the circle. The *circular minor* $\det M_A^B$ is defined to be the minor of M obtained from the rows $a_1, a_2, \dots, a_k$, and the columns $b_k, b_{k-1}, \dots, b_1$. When A and B are non-interlaced around the circle, we can represent the circle minor $\det M_A^B$ by a disk diagram with k chords connecting notes a_i to b_i. See Figure 5.1 for examples.

A *contiguous minor* of a matrix M is a circular minor whose row indices and whose column indices are contiguous on the circle. A *(small) central minor* is a non-interlaced contiguous minor whose row indices and column indices are opposite (or almost opposite depending on the parity of n) around the circle. There are $\binom{n}{2}$ central minors, whether n is even or odd. See Figures 5.1 and 5.2 for examples of these special types of minors.

Associated with an electrical network is a *response matrix* that measures the response of the network to potential applied at the nodes. It has been shown that a matrix M is the response matrix of an electrical network if and only if it is symmetric with row and column sums equal to zero and each circular minor $\det M_A^B$ is non-negative (see Theorem 4 in [**23**]). An electrical network is called *well-connected* if, for any two non-interlaced sets of nodes A and B, there are k pairwise vertex-disjoint paths in G connecting the nodes in A to the nodes in B, where $|A| = |B| = k$.

R. Kenyon and D. Wilson [**45**] generalize the work of Colin de Verdière [**26**] by showing how to test the well-connectivity of an electrical network by checking the positivity of the $\binom{n}{2}$ central minors of the response matrix. The test is based on their interesting finding of the connection between linear algebra, electrical networks, and tilings.

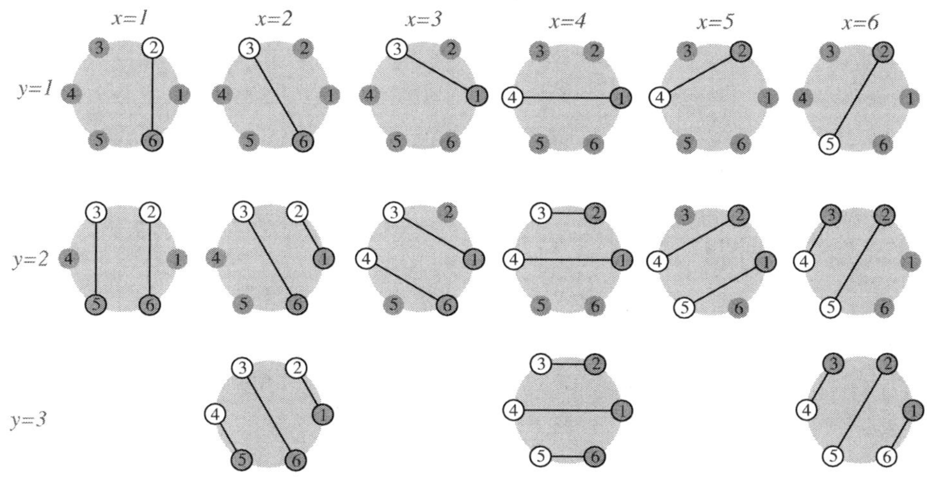

FIGURE 5.2. All 15 small central minors when $n = 6$.

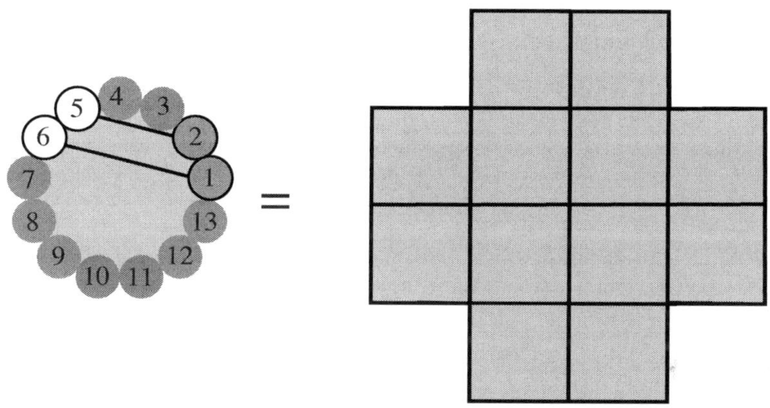

FIGURE 5.3. Encoding a contiguous minor as an Aztec diamond. We refer the reader to the fully detailed version of the picture in [**59**, Fig. 1.6]. In the full version, the lattice points on the right-hand side are replaced by the corresponding central minors. These details are too small for printing, so are omitted here.

THEOREM 7 (Kenyon–Wilson's Theorem [**45**]). *Any contiguous minor can be written as a Laurent polynomial in central minors. Moreover, this Laurent polynomial is the tiling generating function of an Aztec diamond.*

See Figure 5.3 for an illustration of the theorem.

We are also interested in a larger family of minors, called *semicontiguous minors*. A *semicontiguous minor* is a circular minor $\det M_A^B$, where *at least one* of A and B is contiguous. Kenyon and Wilson conjectured that any semicontiguous minor could also be written as the tiling generating function of some region on the square lattice. This conjecture was recently proved in [**59**] by building a special family of regions on the square lattice whose tiling generating functions are given by the semicontiguous minors. In particular, our region is obtained from an *Aztec*

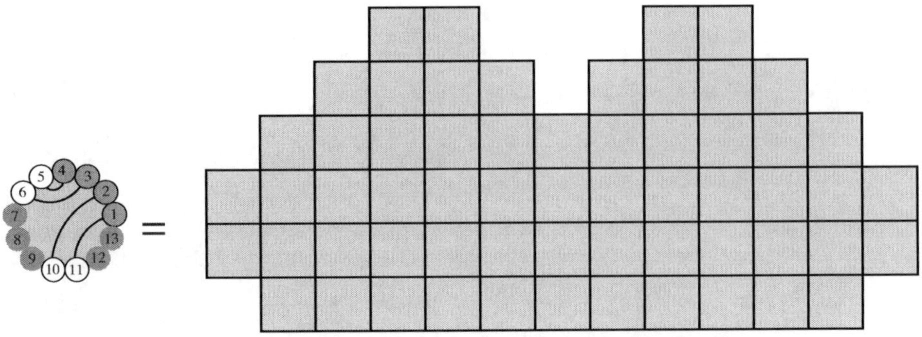

FIGURE 5.4. Illustration of the correspondence between a semi-contiguous minor and a region in the square lattice. We refer the reader to the fully detailed version of the picture in [**59**, Fig.1.7].

rectangle (a natural generalization of the Aztec diamond) or a special union of two Aztec diamonds by trimming the base along a zigzag path determined by the non-contiguous index set. It would be interesting to know if any general circular minor can be encoded as domino tilings of a region.

PROBLEM 18. *Can any circular minor be written as the tiling generating function of a region on a square lattice? If the answer is "NO," characterize all such circular minors.*

6. Other problems

We consider the number of cyclically symmetric tilings of two families of hexagons with four triangles removed as follows.

Let x, y, t, a be non-negative integers. Our first family consists of the hexagons with side-lengths $t+x+3a,\ t,\ t+x+3a,\ t,\ t+x+3a,\ t$, in which an up-pointing triangle of side-length x has been removed from the center. In addition, three up-pointing triangles of side-length a have been removed in a symmetric way along the intervals connecting the center to the midpoints of the southern, northeastern, and northwestern sides of the hexagon. We assume that the distance from the central hole to each of the three satellite holes is $2y$. Denote by $\mathcal{H}_{t,y}(a,x)$ the resulting region (see Figure 6.1 for an example; the black triangles represent the triangles that have been removed).

The second family also consists of hexagons with four similar triangular holes; however, the a-triangles now lie on the other side of the center, as shown in Figure 6.2. The resulting region is denoted by $\overline{\mathcal{H}}_{t,y}(a,x)$. Two simple product formulas for the numbers of cyclically symmetric tilings of $\mathcal{H}_{t,y}(a,x)$ and $\overline{\mathcal{H}}_{t,y}(a,x)$ were provided in [**65**].

Recently, M. Ciucu and I. Fischer [**15**] conjectured a striking connection between the whole number of tilings and the number of cyclically symmetric tilings of $\mathcal{H}_{t,y}(a,x)$.

PROBLEM 19 (Conjecture 1 in [**15**]). *Show that*

$$\frac{\mathrm{T}(\mathcal{H}_{t,y}(a,x))}{\mathrm{T}_c(\mathcal{H}_{t,y}(a,x))^3} = \frac{\mathrm{T}(\mathcal{H}_{t,0}(a,x))}{\mathrm{T}_c(\mathcal{H}_{t,0}(a,x))^3}\left[\prod_{i=1}^{y}\frac{(x+6i-4)(x+3a+6i-2)}{(x+6i-2)(x+3a+6i-4)}\right]^2, \tag{6.1}$$

where $\mathrm{T}_c(R)$ *denotes the number of cyclically symmetric tilings of* R.

We note that when $y = 0$, all four holes in $\mathcal{H}_{t,0}(a, x)$ are glued together, and the region has the same tiling number as a cored hexagon. Recall that the number of tilings of a cored hexagon is given by a simple product formula in [**13**], so the expression on the right-hand side of (6.1) can be expressed as a simple product (see the discussion before Conjecture 2 in [**15**]). As the number of cyclically symmetric

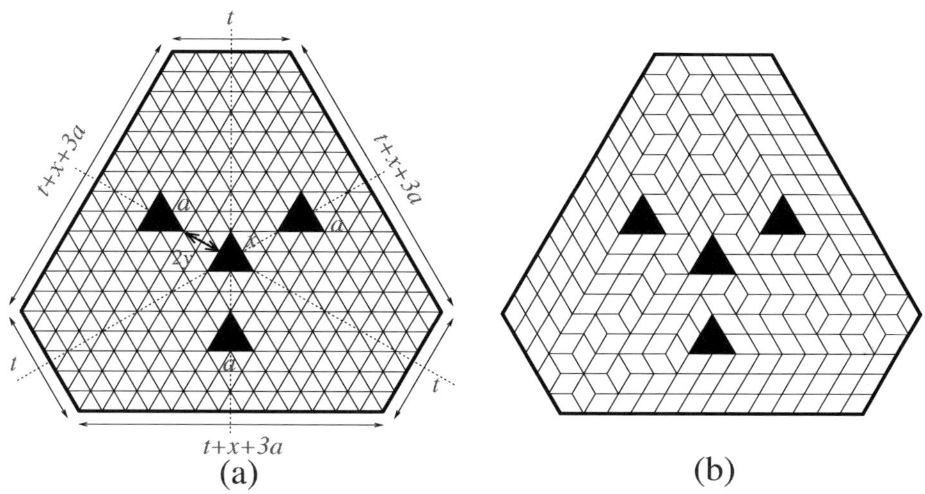

FIGURE 6.1. (a) The hexagon with four holes $\mathcal{H}_{5,1}(2, 2)$. (b) A cyclically symmetric tiling of $\mathcal{H}_{5,1}(2, 2)$. This picture first appeared in [**65**].

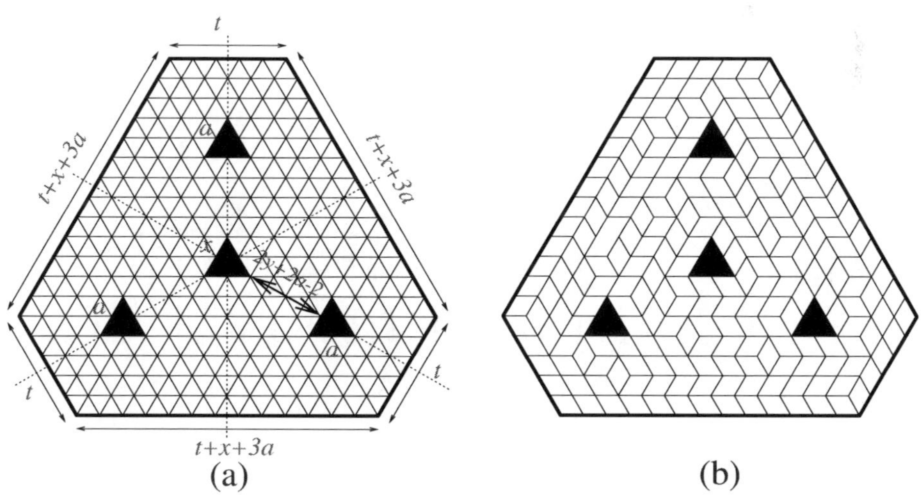

FIGURE 6.2. (a) The hexagon with four holes $\overline{\mathcal{H}}_{5,1}(2, 2)$. (b) A cyclically symmetric tiling of $\overline{\mathcal{H}}_{5,1}(2, 2)$. This picture first appeared in [**65**].

tilings of $\mathcal{H}_{t,y}(a,x)$ has been found, one only needs to find the tiling number of $\mathcal{H}_{t,y}(a,x)$ to prove the conjecture. However, as discussed in [**15**], this task would not be easy.

The similarity of $\mathcal{H}_{t,y}(a,x)$ and $\overline{\mathcal{H}}_{t,y}(a,x)$ suggests the existence of a similar formula for the ratio $\frac{\mathrm{T}(\overline{\mathcal{H}}_{t,y}(a,x))}{\mathrm{T}_c(\overline{\mathcal{H}}_{t,y}(a,x))^3}$. However, such a formula was not provided in [**15**]. It would be interesting to obtain that formula.

PROBLEM 20. *Find a formula for the ratio of tiling numbers*

$$\frac{\mathrm{T}(\overline{\mathcal{H}}_{t,y}(a,x))}{\mathrm{T}_c(\overline{\mathcal{H}}_{t,y}(a,x))^3}. \tag{6.2}$$

It is worth noticing that when $a = x = 0$, i.e., when all four triangular holes are all vanished, the equation (6.1) becomes

$$\frac{\mathrm{T}(Hex(n,n,n))}{\mathrm{T}_c(Hex(n,n,n))^3} = \left[\frac{\left(\frac{1}{3}\right)_n}{\left(\frac{2}{3}\right)_n}\right]^2,$$

which follows from the well-known enumerations of symmetry classes of plane partitions (see, i.e., [**86**]). In particular, the tiling number in the numerator is precisely the number of boxed plane partitions; the tiling number in the denominator is the number of cyclically symmetric plane partitions. Macdonald conjectured the following weighted enumeration of cyclically symmetric plane partitions [**73**, Ex. 18, p. 85]:

$$\sum_{\pi} q^{|\pi|} = \prod_{i=1}^{n} \frac{1-q^{3i-1}}{1-q^{3i-2}} \prod_{1\le i<j\le n} \frac{1-q^{3(2i+j-1)}}{1-q^{3(2i+j-2)}} \prod_{1\le i<j,k\le n} \frac{1-q^{3(i+j+k-1)}}{1-q^{3(i+j+k-2)}}, \tag{6.3}$$

where the sum is over all cyclically symmetric plane partitions π that are contained in a $(n \times n \times n)$-box. The unweighted version of the conjecture was proved by Andrews [**2**]; the full conjecture was proved by Mills, Robbins, and Rumsey [**75**].

At this point, one would think about similar weighted versions of the tiling numbers in (6.1). Unfortunately, there have been no such nice q-enumerations in general. However, in the case when $x = 0$ (i.e., the central hole vanishes) and the three satellite holes are attached to the hexagon's boundary, we have a nice q-enumeration, as claimed in Theorem 4. We want to obtain a formula for the weighted sum of the cyclically symmetric tilings of the region $F\begin{pmatrix} x & x & x \\ a & a & a \\ d & d & d \end{pmatrix}$ (an F-type region in Theorem 4 must have $a = b = c$, $d = e = f$, $x = y = z$ in order to have a cyclically symmetric tiling). See Figure 6.3 for an example of such tiling. Equivalently, we are interested in finding the following volume generating function:

PROBLEM 21. *Find the volume generating function of the cyclically symmetric stacks fitting in the compound room* $\mathcal{B}\begin{pmatrix} x & x & x \\ a & a & a \\ d & d & d \end{pmatrix}$.

Ciucu and Fischer [**14**, Theorems 2.3 and 2.4] proved and generalized two conjectures of Ciucu and Krattenthaler [**16**, Conjectures A.1 and A.2] by enumerating the regions $D_{x,y,z,m}$ and $D'_{x,y,z,m}$ in Figure 6.4. These regions look essentially the same, and the only difference is that the lozenges running along two zigzag cuts in

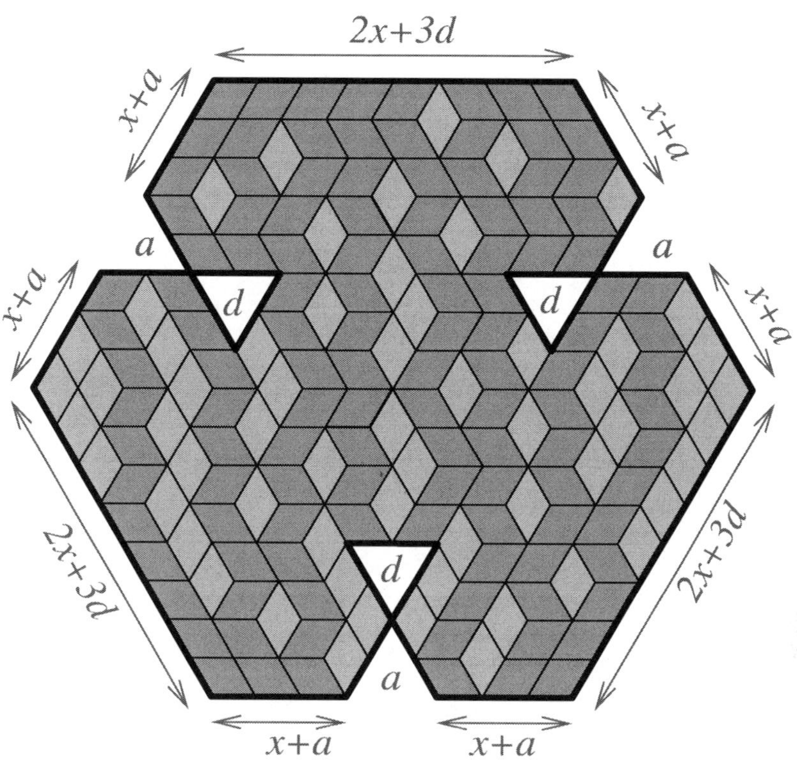

FIGURE 6.3. A cyclically symmetric tiling of the region with three bowties removed.

$D'_{x,y,z,m}$ are weighted by 1/2 (see the lozenges with shaded cores). One can view these regions as one-third of an F-type region in the previous problem. From the tiling formulas, we could realize that the weighted enumeration of tilings of $D'_{x,y,z,m}$

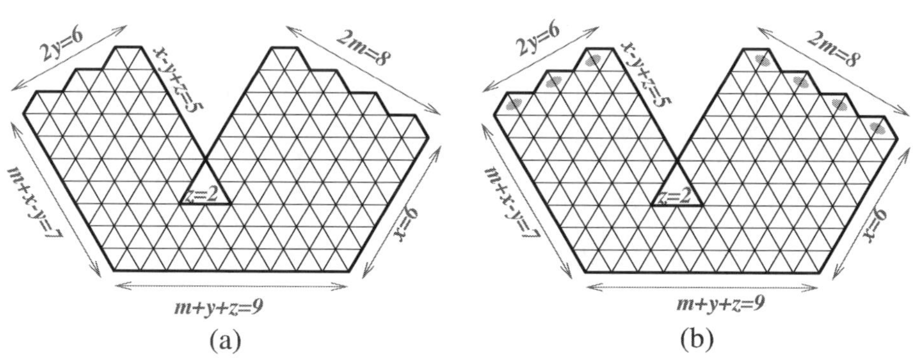

FIGURE 6.4. Two regions enumerated by Ciucu and Fischer: (a) $D_{6,3,2,4}$ and (b) $D'_{6,3,2,4}$. The lozenges with shaded cores are weighted by 1/2.

is obtained from the tiling formula of $D_{x,y,z,m}$ by replacing x by $x-1/2$. This property reminds us of the "*combinatorial reciprocity phenomenon*": even though the region $D_{x,y,z,m}$ is *not* defined when x is a half-integer, its tiling formula gives the number of combinatorial objects of a different sort (here are the tilings of $D'_{x,y,z,m}$) when evaluated at half-integers. We refer the reader to, e.g., [**3, 78, 84**] for more discussions about the phenomenon.

PROBLEM 22. *Explain combinatorially the above $(\frac{1}{2})$-phenomenon for the tiling numbers of the region $D_{x,y,z,m}$ and $D'_{x,y,z,m}$.*

A similar thing happens in [**62**, Theorems 1.3 and 1.4]. In Theorem 1.3, we consider the ratio of tiling generating functions of pairs of (dented) quartered hexagons $Q_x(\mathbf{a})$ and $Q_y(\mathbf{a})$. They have the same dents on the right sides at the positions in $\mathbf{a}=(a_i)_{i=1}^m$; the only difference is in their widths. We assign weights to lozenges of $Q_x(\mathbf{a})$ and $Q_y(\mathbf{a})$ as in Figure 6.5(b): each vertical lozenge with label x is weighted by $(q^x+q^{-x})/2$, for $x\geq 1$ (lozenges of other orientations are all weighted by 1). The pair of quartered hexagons $Q'_x(\mathbf{a})$ and $Q'_y(\mathbf{a})$ in Theorem 1.4 are the same as that in Theorem 1.3; the only difference is the weights of lozenges. Figure 6.5(c) illustrates the weight assignment of lozenges in $Q'_x(\mathbf{a})$ and $Q'_y(\mathbf{a})$ (the lozenges with shaded cores on the left side are weighted by 1/2). We have

$$\frac{\mathrm{T}(Q_x((a_i)_{i=1}^m))}{\mathrm{T}(Q_y((a_i)_{i=1}^m))}=q^{2(y-x)(\sum_{i=1}^m a_i-m^2)}\prod_{i=1}^m\frac{(q^{2(2y+a_i+1)};q^2)_{2i-a_i-1}}{(q^{2(2x+a_i+1)};q^2)_{2i-a_i-1}}\tag{6.4}$$

and

$$\frac{\mathrm{T}(Q'_x((a_i)_{i=1}^m))}{\mathrm{T}(Q'_y((a_i)_{i=1}^m))}=q^{2(y-x)(\sum_{i=1}^m a_i-m^2)}\prod_{i=1}^m\frac{(q^{2(2y+a_i)};q^2)_{2i-a_i-1}}{(q^{2(2x+a_i)};q^2)_{2i-a_i-1}}.\tag{6.5}$$

It is easy to see that $\frac{\mathrm{T}(Q'_x((a_i)_{i=1}^m))}{\mathrm{T}(Q'_y((a_i)_{i=1}^m))}$ is obtained from $\frac{\mathrm{T}(Q_x((a_i)_{i=1}^m))}{\mathrm{T}(Q_y((a_i)_{i=1}^m))}$ by replacing x by $x-1/2$ and y by $y-1/2$. It would be interesting to find a direct explanation for this, i.e., we want an explanation without requiring any calculation of the ratios tiling generating functions.

PROBLEM 23. *Explain the $(\frac{1}{2})$-phenomenon for the ratios of tiling generating functions of the quartered hexagons $Q_x((a_i)_{i=1}^m)$ and $Q'_x((a_i)_{i=1}^m)$.*

In their excellent paper about the tilling enumeration of hexagons with a maximal corner cut off [**16**], Ciucu and Krattenthaler found an unusual pattern for the tiling number of a triangular region denoted by $\mathcal{TT}_n$ (see the shaded region in Figure 6.6 for $\mathcal{TT}_6$). The number of tilings of $\mathcal{TT}_n$ factors as follows for $n\leq 7$:

$$\begin{aligned}
\mathrm{T}(\mathcal{TT}_1)&=2;\\
\mathrm{T}(\mathcal{TT}_2)&=3^2;\\
\mathrm{T}(\mathcal{TT}_3)&=2^2\cdot 13;\\
\mathrm{T}(\mathcal{TT}_4)&=2^2\cdot 5^2\cdot 31;\\
\mathrm{T}(\mathcal{TT}_5)&=2\cdot 3^2\cdot 19^2\cdot 37;\\
\mathrm{T}(\mathcal{TT}_6)&=2\cdot 7^3\cdot 13\cdot 43\cdot 127;\\
\mathrm{T}(\mathcal{TT}_7)&=2^7\cdot 3^5\cdot 5^3\cdot 7\cdot 13\cdot 73.
\end{aligned}$$

They realized that the amount of factorization is remarkable (the authors have computed and factored $\mathrm{T}(\mathcal{TT}_n)$ up to $n = 30$) and comparable to that of the numbers enumerating domino tilings of squares (given by the well-known formula of Kasteleyn [**44**] and Temperley and Fisher [**91**]). Based on this observation, they posed the following problem.

PROBLEM 24 (Problem 1.5 in [**16**]). *Find a formula for the number of lozenge tilings of $\mathcal{TT}_n$ that explains a large amount of prime factorization of these numbers.*

In a recent paper [**63**], a common generalization for three famous families of regions in the enumeration of tilings, namely the hexagons, the semi-hexagons, and the halved hexagons, has been introduced. We consider a hexagon, then cut off a maximal k-staircase whose each step has the width k. When $k = 1$, we have exactly the halved hexagon; when $k = 0$, nothing is cut off, and the region is still the hexagon. When $k \geq 2$, we have new regions that are similar to the halved

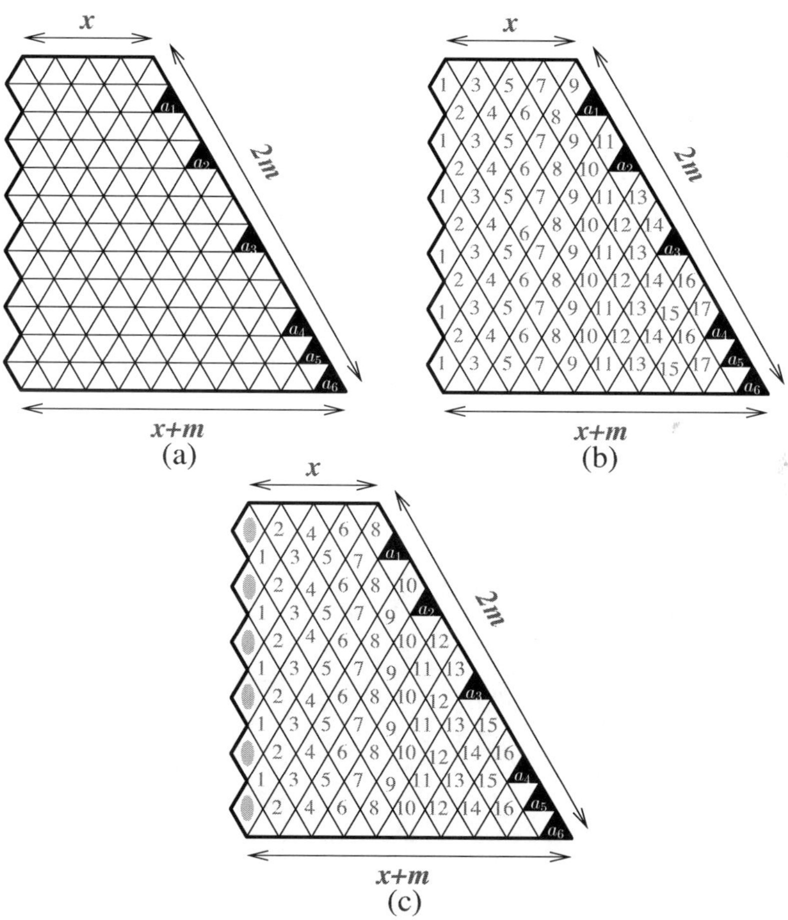

FIGURE 6.5. (a) The quartered hexagon with dents on the right side. (b) How to assign weights to lozenges in $Q_4(2,4,7,10,11,12)$. (c) How to assign weights to lozenges in $Q'_4(2,4,5,10,11,12)$.

hexagons; however, the cut is tilted. We call the new regions the *k-halved hexagons* or the *tilted halved hexagons* (see Figure 6.7).

We actually consider a more general situation when we allow some "dents" on the staircase of the k-halved hexagon. Label the staircase levels from the bottom to the top by $1, 2, \ldots, l+h$. We allow removing the up-pointing unit triangles at h corners. Assume that the remaining steps have labels $a_1, a_2, \ldots, a_l$ as they appear from bottom to the top. Denote by $H_{x,t,h}(a_1, a_2, \ldots, a_l)$ the resulting region (illustrated in Figure 6.8).

As there are nice q-enumerations of the semiregular hexagons, semi-hexagons, and halved hexagons, we would expect a nice q-enumeration of the k-halved hexagons.

PROBLEM 25. *Find a q-enumeration of tilings of a k-halved hexagon.*

In the enumeration of tilings, the symmetric regions often behave better than the asymmetric ones. Many families of regions may not have a nice tiling number in the general case. However, their tilings are enumerated by simple products in the symmetric case. Let us revisit Ciucu–Krattenthaler's S-cored hexagon, a hexagon with a cluster of four triangles (called a "shamrock") removed. The S-cored hexagons have a nice tiling number in only two situations: (1) the "shamrock" is removed from the center [**17**], and (2) the shamrock is removed from the boundary [**57**] (see Figure 6.9 for examples). In general, if we remove the shamrock from a place that is different from the center or the boundary, then the number of tilings is not given by a simple product formula. However, in the case of reflectively symmetric hexagons, it has been shown [**66**] that we can remove the shamrock at *any* position along the symmetry axis and still get a beautiful tiling formula (see Figure 6.10(a)).

We have the same observation for hexagons with a family of vertically aligned triangular holes of side-length 2. In the general case, we do not have a nice tiling number. However, in the case of reflectively symmetric hexagons, Ciucu proved

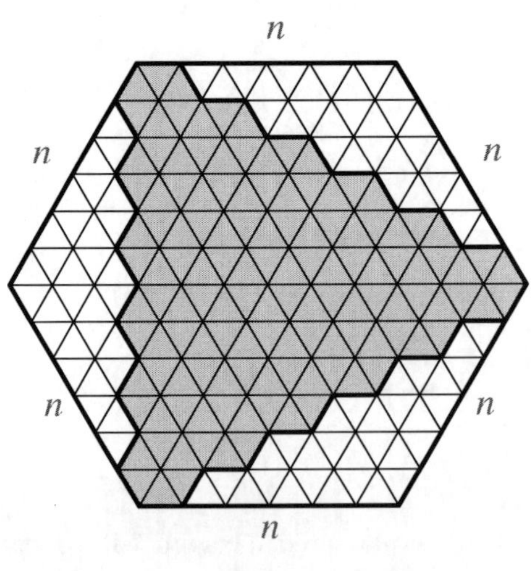

FIGURE 6.6. The region $\mathcal{TT}_6$.

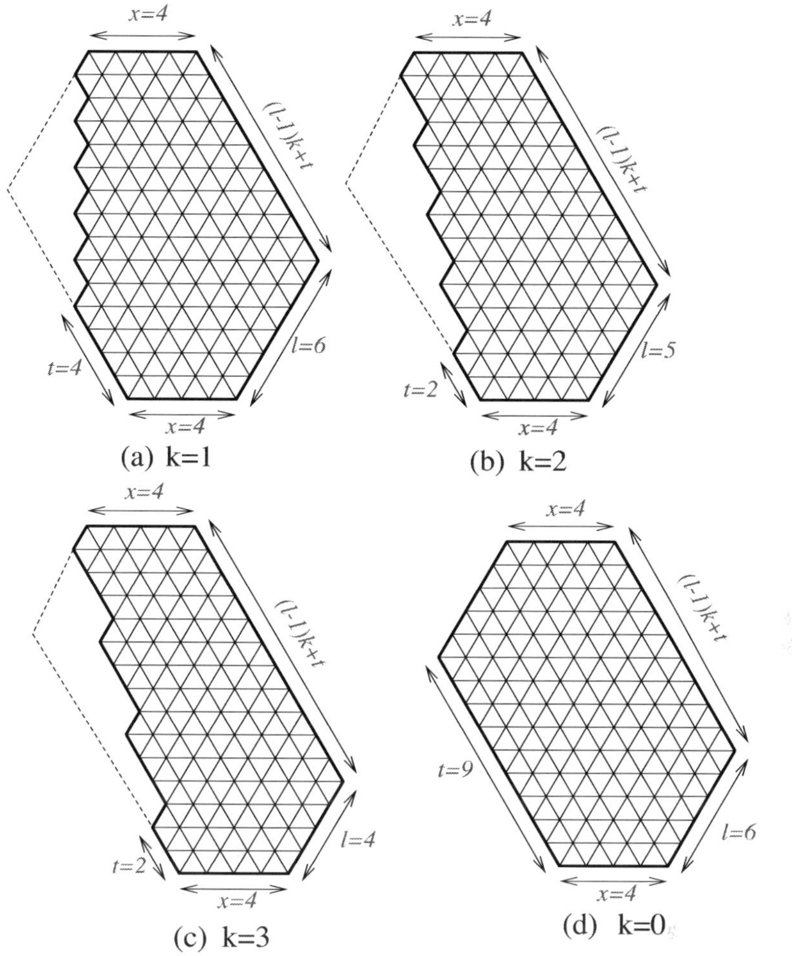

FIGURE 6.7. The k-halved hexagons with no dent.

a simple product formula for the tiling number [**10**]. See Figure 6.11(a) for an illustrated picture.

Our data suggest that if we place the shamrock holes and the family of triangular holes $1/2$ unit off the symmetry axis, we still have nice tiling numbers in the above two families of regions. See Figures 6.10(b) and 6.11(b). In general, we often have the following "*off-center phenomenon*": if we place the hole(s) not on the symmetry axis but $1/2$ unit off the symmetry axis, then the number of tilings seems to be as nice as the tiling number in the symmetric case. This fact is not true anymore if we move the holes just 1 unit away from the symmetric axis. It would be interesting to have a precise explanation for this phenomenon.

PROBLEM 26. *Explain the off-center phenomenon.*

It is worth noticing that the enumeration of tilings of regions with holes[6] is especially challenging. One of the difficulties is that the Lindström–Gessel–Viennot

[6]Several authors used the term "*holey regions*" for "regions with holes".

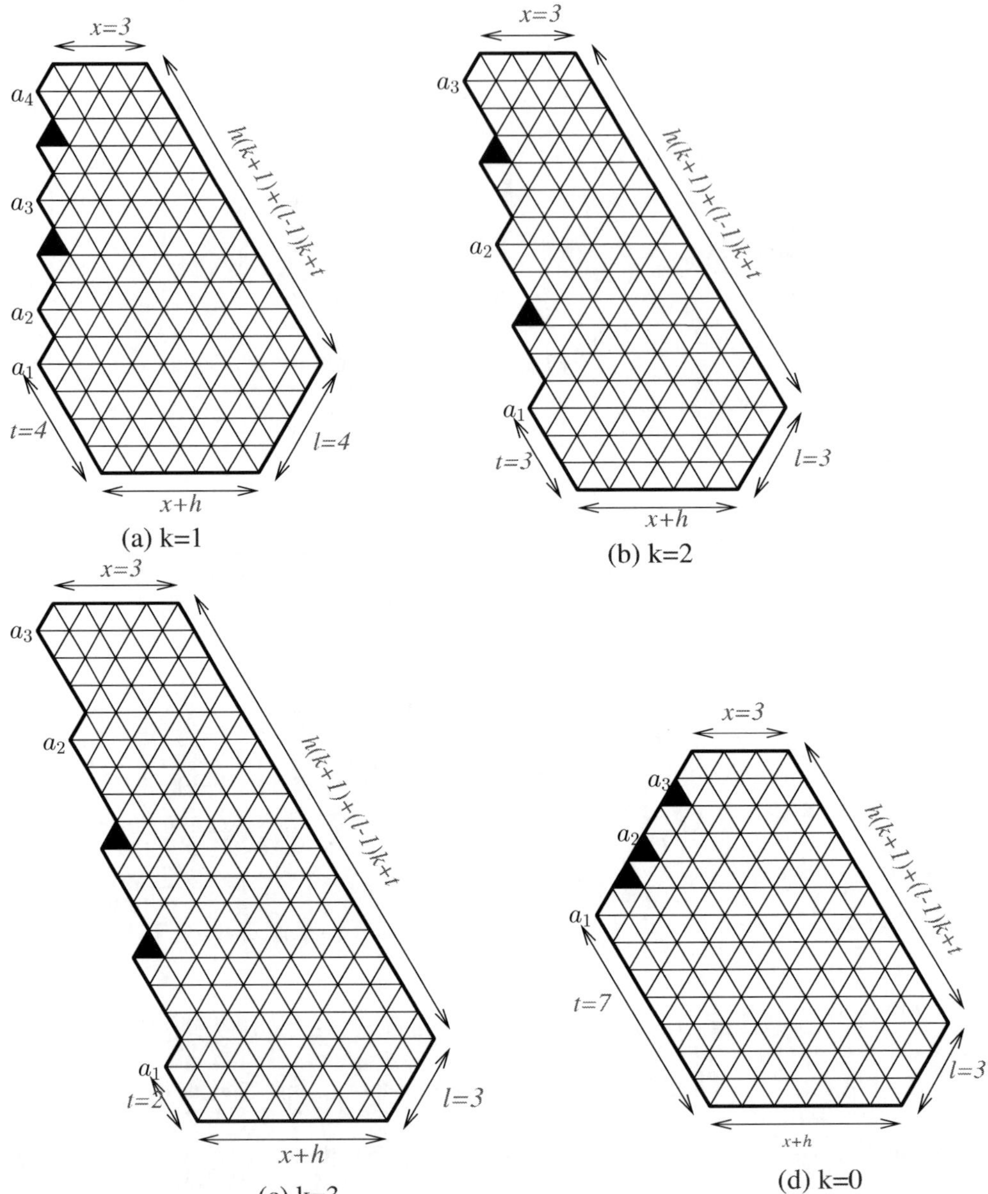

FIGURE 6.8. Several k-halved hexagons with dents.

determinant does *not* give the tiling number; it gives the *signed* tiling number instead. We refer the reader to, say, the proofs in Section 4 of [**13**] for a detailed explanation. In the reflectively symmetric case, one can go around this obstacle by using a powerful tool, usually mentioned as *Ciucu's factorization theorem* (see [**9**, Theorem 1.2]). This theorem allows us to simplify the case of symmetric regions with holes to the case of simply connected regions, say by dividing the region along

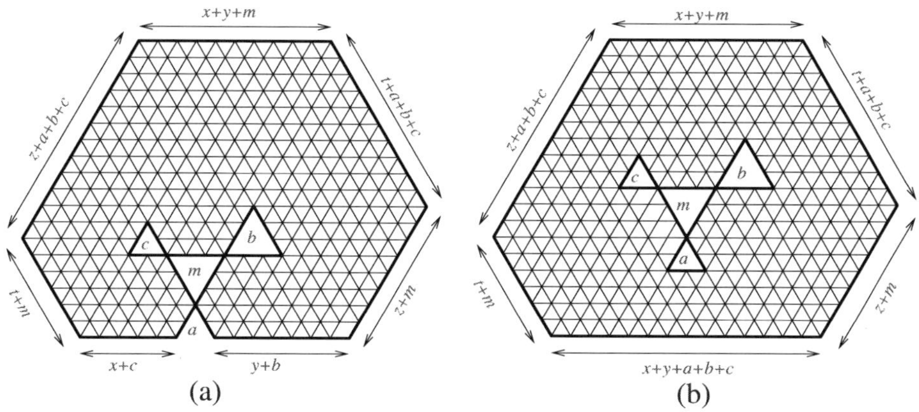

FIGURE 6.9. (a) A hexagon with a shamrock hole in the center in [**17**]. (b) A hexagon with a shamrock hole on the boundary in [**57**]. The picture first appeared in [**66**].

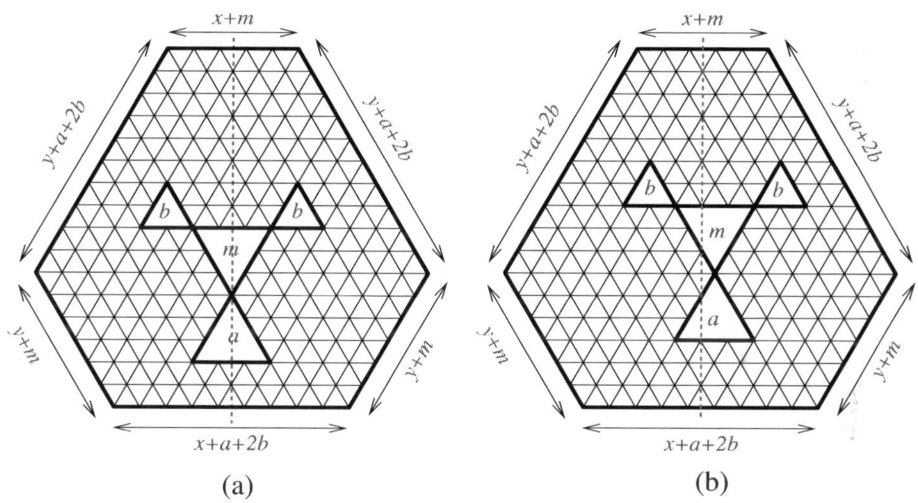

FIGURE 6.10. (a) Placing a shamrock hole on the symmetry axis of the hexagon. (b) Placing a shamrock hole 1/2 unit off the symmetry axis of the hexagon.

the symmetric axis into two smaller regions with no holes. See [**10**, Section 3] for more details of the method. However, if we slide the holes 1/2 unit away from the symmetry axis, then one cannot use Ciucu's theorem (as the new region is not symmetric anymore). It would be very interesting to find explicit formulas for the numbers of tilings of the two regions in Figures 6.10(b) and 6.11(b).

PROBLEM 27. *Find a formula for the number of tilings of a symmetric hexagon with a shamrock hole at* 1/2 *unit off the symmetric axis (as in Figure* 6.10*(b)).*

PROBLEM 28. *Enumerate the tilings of a symmetric hexagon with a family of aligned* 2*-triangles at* 1/2 *unit off the symmetric axis (as in Figure* 6.11*(b)).*

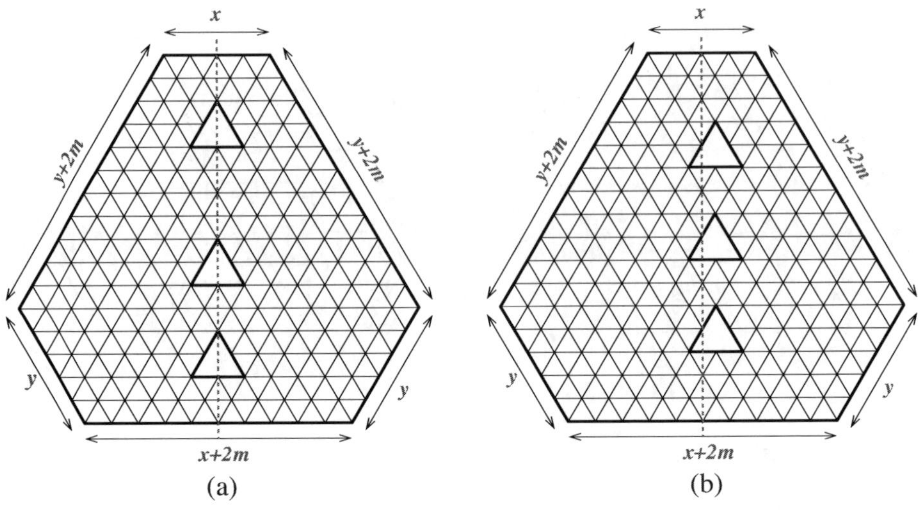

FIGURE 6.11. (a) Placing triangular holes of side-length 2 on the symmetry axis of the hexagon. (b) Placing triangular holes 1/2 unit off the symmetry axis.

The "*Aztec pillow graphs*" were first introduced in [**77**]. The width of an Aztec pillow is always even. When the width of an Aztec pillow is $4k+2$, the upper half of the graph has $k+1$ up-steps (k steps of size 3, and one step of size 1), followed by $k+1$ down-steps of size 1 (as we go from the left to right), and the lower half is simply a $180°$-rotation of the upper half. When the Aztec pillow's width is $4k$, the upper half has k up-steps of size 3, followed by k down-steps of size 1. Denote by AP_n the Aztec pillow of width $2n$. See Figure 6.12 for the graphs AP_n, for $n = 1, 2, \dots, 9$. Forest Tong [**93**] conjectured the following elegant divisibility of the matching numbers of the Aztec pillows.

PROBLEM 29. *Prove that* $\mathrm{M}(AP_m) \mid \mathrm{M}(AP_n)$ *whenever* $(m+3) \mid (n+3)$, *where* $\mathrm{M}(G)$ *denotes the number of perfect matching of graph* G.

Tong has verified this conjecture computationally for $m, n < 77$.

It is worth noticing that the Aztec pillows' definition has been generalized by Christopher Hanusa. The original Aztec pillows above are the *Aztec 3-pillows.* Forest Tong also observed that *none* of the 5-pillows, 7-pillows, and 9-pillows (as defined in [**38**]) seem to share the above divisibility property.

Next, we investigate a nice property of the *matching polynomial* (see, e.g., [**72**, p. xxxii,p. 333ff]).

Let G be a graph with no loops. Denote by $m_k(G)$ the number of (partial) matchings of the graph G with exactly k edges, where $m_0(G) = 1$ by convention. The *matching polynomial* of G is defined to be

$$\mathcal{M}(G) = \sum_{k \geq 0} m_k(G) x^k.$$

A sequence $(a_i)_{i\geq 0}$ is a *Pólya frequency sequence* (PFS) if the infinite Toeplitz matrix $(M_{i,j})_{i,j\geq 0}$ defined by $M_{i,j} = a_{j-i}$ (where $a_k = 0$ if $k < 0$ by convention),

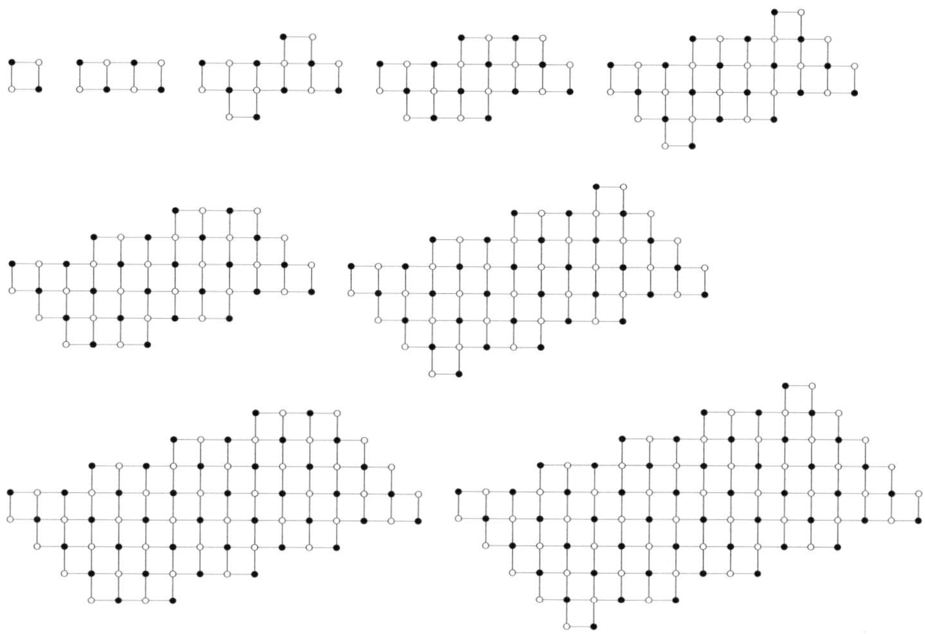

FIGURE 6.12. The Aztec pillows AP_n, for $n = 1, 2, 3, \ldots, 9$ (reading from left to right, top to bottom).

i.e,

$$(M_{i,j})_{i,j\geq 0} = \begin{pmatrix} a_0 & a_1 & a_2 & a_3 & a_4 & \cdots \\ 0 & a_0 & a_1 & a_2 & a_3 & \cdots \\ 0 & 0 & a_0 & a_1 & a_2 & \cdots \\ 0 & 0 & 0 & a_0 & a_1 & \cdots \\ 0 & 0 & 0 & 0 & a_0 & \cdots \\ 0 & 0 & 0 & 0 & 0 & \cdots \\ \vdots & \vdots & \vdots & \vdots & \vdots & \ddots \end{pmatrix},$$

has all nonnegative minors. By definition, a Pólya frequency sequence is log-concave, as the log-concavity is equivalent to the fact that all 2×2 minors of $(M_{i,j})_{i,j\geq 0}$ are nonnegative.

O. J. Heilmann and E. H. Lieb [**39**, Theorem 4.2] proved that the matching polynomial $\mathcal{M}(G)$ has all real roots. This implies that the sequence $(m_k(G))_{k\geq 0}$ is a Pólya frequency sequence (see, e.g., [**5**]). It would be interesting to find a combinatorial proof for this fact.

PROBLEM 30. *Prove combinatorially that the sequence of matching numbers $(m_k(G))_{k\geq 0}$ is a Pólya frequency sequence.*

We note that Krattenthaler [**49**] provided a combinatorial proof for a special case of Problem 30, namely the log-concavity of the sequence $(m_k(G))_{k\geq 0}$.

We can generalize Problem 30 to weighted graphs as follows. To each edge e of G, we assign a weight x_e. Then we define the weighted matching number $m_k(G, \mathbf{x})$

by

$$m_k(G,\mathbf{x}) = \sum_{M} \prod_{e \in M} x_e,$$

where the sum is taken over all k-element matchings M in G. Assume $P_i = P_i(\mathbf{x})$ be a polynomial in $\mathbb{Z}[\mathbf{x}]$, for $i = 1, 2, 3, \ldots$. We now define a sequence $(P_i)_{i=1}^{\infty}$ to be an $\mathbf{x}$-*PFS* if all minors of the Toeplitz matrix $(P_{j-i})_{i,j\geq 0}$ (where $P_k = 0$ if $n < 0$) are polynomials in the x_e's with nonnegative coefficients.

PROBLEM 31. *Prove combinatorially that the sequence of* $\mathbf{x}$-*matching numbers* $(m_k(G,\mathbf{x}))_{k\geq 0}$ *is an* $\mathbf{x}$-*PFS.*

Recently, P. Galashin and P. Pylyavskyy [**33**] consider a similar positivity for a planar bipartite graph $G = (V_1, V_2, E)$ embedded on a cylinder $\mathcal{O}$. Let τ_1 and τ_2 be two perfect matchings on G. We always orient the edges in a perfect matching τ of G from a vertex in V_1 to a vertex in V_2. Define the difference of two perfect matching $\tau_1 - \tau_2$ to be the directed graph on $\mathcal{O}$ with vertices $V_1 \cup V_2$ obtained by superimposing τ_1 and τ_2, and the reversed direction of edges in τ_2. This way $\tau_1 - \tau_2$ is always a disjoint union of directed simple cycles, which can be viewed as singular 1-cycles on $\mathcal{O}$. We define the *relative height* of two perfect matchings $h(\tau_1, \tau_2)$ to be the image of $H_1(\mathcal{O}, \mathbb{Z}) \simeq \mathbb{Z}$ of the sum of these cycles. Fix a minimal-height perfect matching τ_0. Next, we can define the *absolute height* of a perfect matching as $h(\tau) := h(\tau, \tau_0)$. For the reader who is unfamiliar with height function theory, Galashin and Pylyavskyy's above paper would provide needed definitions.

We also assume that the edges of G are weighted by x_e's as above. We now define $H_i(\mathbf{x})$ to be the sum of weights of all perfect matchings with height i, where the weight of a perfect matching is the product of its edge-weights as usual.

PROBLEM 32 (Conjecture 6.1 in [**33**]). *Prove that the sequence* $(H_i(\mathbf{x}))_{i\geq 0}$ *is an* $\mathbf{x}$-*PFS.*

We note that while the PFS properties in Problems 30 and 31 have been proved (and we are asking for a combinatorial proof), the question about the PFS property in Problem 32 is still open. Several special cases of Problem 32 have been proved in [**8**].

We conclude this section with a problem of a rather different flavor. Let k be a fixed positive integer. Denote by $A_n = A_{n,k}$ the number of domino tilings of a $k \times n$ rectangle. Form the generating function

$$F_k(x) = \sum_{n\geq 0} A_n x^n.$$

It has been shown that $F_k(x)$ can be written as a rational function, say $F_k(x) = \frac{P_k(x)}{Q_k(x)}$ with P_k and Q_k polynomials with integer coefficients, and $Q_k(0) = 1$ [**48**]. Stanley proves that all the roots of $Q_k(x)$ are real and nonzero, and exactly half of the roots are positive [**85**]. He also conjectures a special pattern for the roots of the polynomial:

PROBLEM 33. *Prove that* $Q_k(x)$ *has distinct roots.*

Acknowledgments

The author would like to thank Gregg Musiker, James Propp, Pavlo Pylyavskyy, Richard Stanley, Dennis Staton, Seok Hyun Byun, and an anonymous reviewer for helpful comments and fruitful discussions. Problem 6 was suggested by Dennis Stanton, and Problems 30–32 were introduced to the author by Pavlo Pylyavskyy.

References

[1] Joshua Alman, Carl Lian, and Brandon Tran, *Circular planar electrical networks: posets and positivity*, J. Combin. Theory Ser. A **132** (2015), 58–101, DOI 10.1016/j.jcta.2014.11.004. MR3311338

[2] George E. Andrews, *Macdonald's conjecture and descending plane partitions*, Combinatorics, representation theory and statistical methods in groups, Lect. Notes Pure Appl. Math., vol. 57, Dekker, New York, 1980, pp. 91–106. MR588196

[3] Matthias Beck and Raman Sanyal, *Combinatorial reciprocity theorems: An invitation to enumerative geometric combinatorics*, Graduate Studies in Mathematics, vol. 195, American Mathematical Society, Providence, RI, 2018, DOI 10.1090/gsm/195. MR3839322

[4] Alexei Borodin, Vadim Gorin, and Eric M. Rains, *q-distributions on boxed plane partitions*, Selecta Math. (N.S.) **16** (2010), no. 4, 731–789, DOI 10.1007/s00029-010-0034-y. MR2734330

[5] Francesco Brenti, *Unimodal, log-concave and Pólya frequency sequences in combinatorics*, Mem. Amer. Math. Soc. **81** (1989), no. 413, viii+106, DOI 10.1090/memo/0413. MR963833

[6] Seok Hyun Byun, *A short proof of two shuffling theorems for tilings and a weighted generalization*, Discrete Math. **345** (2022), no. 3, Paper No. 112710, 9, DOI 10.1016/j.disc.2021.112710. MR4340746

[7] Seok Hyun Byun, *Lozenge tilings of hexagons with holes on three crossing lines*, Adv. Math. **398** (2022), Paper No. 108230, 22, DOI 10.1016/j.aim.2022.108230. MR4379748

[8] L. Cheng, A. Peng, and V. Srinivasan, *Real-rootedness of polynomials from planar graphs on a cylinder*, University of Minnesota Combinatorics and Algebra REU, Summer 2020. `https://www-users.cse.umn.edu/~reiner/REU/REU2020notes/Problem4_REUreport.pdf`.

[9] Mihai Ciucu, *Enumeration of perfect matchings in graphs with reflective symmetry*, J. Combin. Theory Ser. A **77** (1997), no. 1, 67–97, DOI 10.1006/jcta.1996.2725. MR1426739

[10] M. Ciucu, *Plane partitions I: a generalization of MacMahon's formula* Mem. Amer. Math. Soc., 178(839) (2005), 107–144.

[11] Mihai Ciucu, *The other dual of MacMahon's theorem on plane partitions*, Adv. Math. **306** (2017), 427–450, DOI 10.1016/j.aim.2016.10.008. MR3581307

[12] Mihai Ciucu, *Symmetries of shamrocks, Part I*, J. Combin. Theory Ser. A **155** (2018), 376–397, DOI 10.1016/j.jcta.2017.11.013. MR3741434

[13] M. Ciucu, T. Eisenkölbl, C. Krattenthaler, and D. Zare, *Enumeration of lozenge tilings of hexagons with a central triangular hole*, J. Combin. Theory Ser. A **95** (2001), no. 2, 251–334, DOI 10.1006/jcta.2000.3165. MR1845144

[14] Mihai Ciucu and Ilse Fischer, *Proof of two conjectures of Ciucu and Krattenthaler on the enumeration of lozenge tilings of hexagons with cut off corners*, J. Combin. Theory Ser. A **133** (2015), 228–250, DOI 10.1016/j.jcta.2015.02.008. MR3325633

[15] Mihai Ciucu and Ilse Fischer, *Lozenge tilings of hexagons with removed core and satellites*, Ann. Inst. Henri Poincaré D **10** (2023), no. 3, 407–501, DOI 10.4171/aihpd/131. MR4620362

[16] Mihai Ciucu and Christian Krattenthaler, *Enumeration of Lozenge tilings of hexagons with cut-off corners*, J. Combin. Theory Ser. A **100** (2002), no. 2, 201–231, DOI 10.1006/jcta.2002.3288. MR1940333

[17] Mihai Ciucu and Christian Krattenthaler, *A dual of MacMahon's theorem on plane partitions*, Proc. Natl. Acad. Sci. USA **110** (2013), no. 12, 4518–4523, DOI 10.1073/pnas.1217100110. MR3047654

[18] Mihai Ciucu and Tri Lai, *Lozenge tilings of doubly-intruded hexagons*, J. Combin. Theory Ser. A **167** (2019), 294–339, DOI 10.1016/j.jcta.2019.05.004. MR3952173

[19] Mihai Ciucu, Tri Lai, and Ranjan Rohatgi, *Tilings of hexagons with a removed triad of bowties*, J. Combin. Theory Ser. A **178** (2021), Paper No. 105359, 39, DOI 10.1016/j.jcta.2020.105359. MR4175890

[20] Henry Cohn, Michael Larsen, and James Propp, *The shape of a typical boxed plane partition*, New York J. Math. **4** (1998), 137–165. MR1641839

[21] Daniel Condon, *Lozenge tiling function ratios for hexagons with dents on two sides*, Electron. J. Combin. **27** (2020), no. 3, Paper No. 3.60, 24, DOI 10.37236/9363. MR4245173

[22] J. H. Conway and J. C. Lagarias, *Tiling with polyominoes and combinatorial group theory*, J. Combin. Theory Ser. A **53** (1990), no. 2, 183–208, DOI 10.1016/0097-3165(90)90057-4. MR1041445

[23] E. B. Curtis, D. Ingerman, and J. A. Morrow, *Circular planar graphs and resistor networks*, Linear Algebra Appl. **283** (1998), no. 1-3, 115–150, DOI 10.1016/S0024-3795(98)10087-3. MR1657214

[24] E. Curtis, E. Mooers, and J. Morrow, *Finding the conductors in circular networks from boundary measurements* (English, with English and French summaries), RAIRO Modél. Math. Anal. Numér. **28** (1994), no. 7, 781–814, DOI 10.1051/m2an/1994280707811. MR1309415

[25] Guy David and Carlos Tomei, *The problem of the calissons*, Amer. Math. Monthly **96** (1989), no. 5, 429–431, DOI 10.2307/2325150. MR994034

[26] Yves Colin de Verdière, *Réseaux électriques planaires. I* (French), Comment. Math. Helv. **69** (1994), no. 3, 351–374, DOI 10.1007/bf02564493. MR1289333

[27] Theresia Eisenkölbl, *Rhombus tilings of a hexagon with three fixed border tiles*, J. Combin. Theory Ser. A **88** (1999), no. 2, 368–378, DOI 10.1006/jcta.1999.3000. MR1723802

[28] Noam Elkies, Greg Kuperberg, Michael Larsen, and James Propp, *Alternating-sign matrices and domino tilings. I*, J. Algebraic Combin. **1** (1992), no. 2, 111–132, DOI 10.1023/A:1022420103267. MR1226347

[29] Noam Elkies, Greg Kuperberg, Michael Larsen, and James Propp, *Alternating-sign matrices and domino tilings. II*, J. Algebraic Combin. **1** (1992), no. 3, 219–234, DOI 10.1023/A:1022483817303. MR1194076

[30] Michael E. Fisher and John Stephenson, *Statistical mechanics of dimers on a plane lattice. II. Dimer correlations and monomers*, Phys. Rev. (2) **132** (1963), 1411–1431. MR158705

[31] Markus Fulmek, *Generating functions of lozenge tilings for hexagonal regions via nonintersecting lattice paths*, Enumer. Comb. Appl. **1** (2021), no. 3, Paper No. S2R24, 17, DOI 10.54550/eca2021v1s3r24. MR4409690

[32] Markus Fulmek, *A simple explanation for the "shuffling phenomenon" for lozenge tilings of dented hexagons*, Discrete Math. **344** (2021), no. 7, Paper No. 112396, 6, DOI 10.1016/j.disc.2021.112396. MR4240693

[33] Pavel Galashin and Pavlo Pylyavskyy, *Linear recurrences for cylindrical networks*, Int. Math. Res. Not. IMRN **13** (2019), 4047–4080, DOI 10.1093/imrn/rnx241. MR3978434

[34] Emden R. Gansner, *The enumeration of plane partitions via the Burge correspondence*, Illinois J. Math. **25** (1981), no. 4, 533–554. MR630832

[35] Emden R. Gansner, *The Hillman-Grassl correspondence and the enumeration of reverse plane partitions*, J. Combin. Theory Ser. A **30** (1981), no. 1, 71–89, DOI 10.1016/0097-3165(81)90041-8. MR607040

[36] I. M. Gel′fand and M. L. Tsetlin, *Finite-dimensional representations of the group of unimodular matrices* (Russian), Doklady Akad. Nauk SSSR (N.S.) **71** (1950), 825–828. MR35774

[37] Ira Gessel and Gérard Viennot, *Binomial determinants, paths, and hook length formulae*, Adv. in Math. **58** (1985), no. 3, 300–321, DOI 10.1016/0001-8708(85)90121-5. MR815360

[38] Christopher R. H. Hanusa, *A Gessel-Viennot-type method for cycle systems in a directed graph*, Electron. J. Combin. **13** (2006), no. 1, Research Paper 37, 28, DOI 10.37236/1063. MR2212510

[39] Ole J. Heilmann and Elliott H. Lieb, *Theory of monomer-dimer systems*, Comm. Math. Phys. **25** (1972), 190–232. MR297280

[40] Patricia Hersh and Richard Kenyon, *Shellability of face posets of electrical networks and the CW poset property*, Adv. in Appl. Math. **127** (2021), Paper No. 102178, 37, DOI 10.1016/j.aam.2021.102178. MR4220681

[41] W. Jockusch and J. Propp, *Antisymmetric monotone triangles and domino tilings of quartered Aztec diamonds*, Unpublished work.

[42] S. Kamioka, *Plane partitions with bounded size of parts and biorthogonal polynomials*, `arXiv:1508.01674`, 2015.

[43] Shuhei Kamioka, *Multiplicative partition functions for reverse plane partitions derived from an integrable dynamical system*, Sém. Lothar. Combin. **78B** (2017), Art. 29, 12. MR3678611

[44] P. W. Kasteleyn, *The statistics of dimers on a lattice*, Physica, **27** (1961), 1209–1225.

[45] Richard W. Kenyon and David B. Wilson, *The space of circular planar electrical networks*, SIAM J. Discrete Math. **31** (2017), no. 1, 1–28, DOI 10.1137/140997798. MR3590647

[46] P.W. Kesteleyn, *The statistics of dimers on alttice*, SPhysica., **27** (1961) 1209–1225.

[47] David A. Klarner, *Packing a rectangle with congruent n-ominoes*, J. Combinatorial Theory **7** (1969), 107–115. MR248643

[48] David Klarner and Jordan Pollack, *Domino tilings of rectangles with fixed width*, Discrete Math. **32** (1980), no. 1, 45–52, DOI 10.1016/0012-365X(80)90098-9. MR588907

[49] C. Krattenthaler, *Combinatorial proof of the log-concavity of the sequence of matching numbers*, J. Combin. Theory Ser. A **74** (1996), no. 2, 351–354, DOI 10.1006/jcta.1996.0058. MR1385379

[50] C. Krattenthaler, *Plane partitions in the work of Richard Stanley and his school*, The mathematical legacy of Richard P. Stanley, Amer. Math. Soc., Providence, RI, 2016, pp. 231–261, DOI 10.1090//mbk/100/14. MR3618037

[51] Tri Lai, *Enumeration of hybrid domino-lozenge tilings*, J. Combin. Theory Ser. A **122** (2014), 53–81, DOI 10.1016/j.jcta.2013.10.001. MR3127678

[52] Tri Lai, *Enumeration of tilings of quartered Aztec rectangles*, Electron. J. Combin. **21** (2014), no. 4, Paper 4.46, 27, DOI 10.37236/4246. MR3292283

[53] Tri Lai, *A simple proof for the number of tilings of quartered Aztec diamonds*, Electron. J. Combin. **21** (2014), no. 1, Paper 1.6, 13, DOI 10.37236/3429. MR3177501

[54] Tri Lai, *A new proof for the number of lozenge tilings of quartered hexagons*, Discrete Math. **338** (2015), no. 11, 1866–1872, DOI 10.1016/j.disc.2015.04.024. MR3357773

[55] Tri Lai, *Enumeration of antisymmetric monotone triangles and domino tilings of quartered Aztec rectangles*, Discrete Math. **339** (2016), no. 5, 1512–1518, DOI 10.1016/j.disc.2015.12.027. MR3475563

[56] T. Lai, *A new proof for a triple product formula for plane partition*, `arXiv:1710.02241`, 2017.

[57] Tri Lai, *A q-enumeration of lozenge tilings of a hexagon with four adjacent triangles removed from the boundary*, European J. Combin. **64** (2017), 66–87, DOI 10.1016/j.ejc.2017.04.001. MR3658821

[58] Tri Lai, *A q-enumeration of lozenge tilings of a hexagon with three dents*, Adv. in Appl. Math. **82** (2017), 23–57, DOI 10.1016/j.aam.2016.07.002. MR3566085

[59] Tri Lai, *Proof of a conjecture of Kenyon and Wilson on semicontiguous minors*, J. Combin. Theory Ser. A **161** (2019), 134–163, DOI 10.1016/j.jcta.2018.07.008. MR3861774

[60] Tri Lai, *A shuffling theorem for reflectively symmetric tilings*, Discrete Math. **344** (2021), no. 7, Paper No. 112390, 10, DOI 10.1016/j.disc.2021.112390. MR4240689

[61] Tri Lai, *Lozenge tilings of hexagons with central holes and dents*, Electron. J. Combin. **27** (2020), no. 1, Paper No. 1.61, 63, DOI 10.37236/8716. MR4093162

[62] Tri Lai, *Ratio of tiling generating functions of semi-hexagons and quartered hexagons with dents*, Enumer. Comb. Appl. **2** (2022), no. 1, Paper No. S2R5, 14, DOI 10.54550/eca2022v2s1r5. MR4416829

[63] T. Lai, *Tilted halved hexagons: Hexagons, semi-hexagons, and halved hexagon under one roof*, `arXiv:2006.10826`, 2020.

[64] Tri Lai, *A shuffling theorem for reflectively symmetric tilings*, Discrete Math. **344** (2021), no. 7, Paper No. 112390, 10, DOI 10.1016/j.disc.2021.112390. MR4240689

[65] Tri Lai and Ranjan Rohatgi, *Cyclically symmetric lozenge tilings of a hexagon with four holes*, Adv. in Appl. Math. **96** (2018), 249–285, DOI 10.1016/j.aam.2018.01.003. MR3767510

[66] Tri Lai and Ranjan Rohatgi, *Enumeration of lozenge tilings of a hexagon with a shamrock missing on the symmetry axis*, Discrete Math. **342** (2019), no. 2, 451–472, DOI 10.1016/j.disc.2018.10.024. MR3876181

[67] T. Lai and R. Rohatgi, *A shuffling theorem for lozenge tilings of doubly-dented hexagons*, `arXiv:1905.08311`, 2019.

[68] Tri Lai and Ranjan Rohatgi, *Tiling generating functions of halved hexagons and quartered hexagons*, Ann. Comb. **25** (2021), no. 2, 471–493, DOI 10.1007/s00026-021-00537-4. MR4268297

[69] Thomas Lam, *Electroid varieties and a compactification of the space of electrical networks*, Adv. Math. **338** (2018), 549–600, DOI 10.1016/j.aim.2018.09.014. MR3861712

[70] Thomas Lam and Pavlo Pylyavskyy, *Electrical networks and Lie theory*, Algebra Number Theory **9** (2015), no. 6, 1401–1418, DOI 10.2140/ant.2015.9.1401. MR3397406

[71] Bernt Lindström, *On the vector representations of induced matroids*, Bull. London Math. Soc. **5** (1973), 85–90, DOI 10.1112/blms/5.1.85. MR335313

[72] L. Lovász and M. D. Plummer, *Matching theory*, North-Holland Mathematics Studies, vol. 121, North-Holland Publishing Co., Amsterdam; North-Holland Publishing Co., Amsterdam, 1986. Annals of Discrete Mathematics, 29. MR859549

[73] I. G. Macdonald, *Symmetric functions and Hall polynomials*, 2nd ed., Oxford Mathematical Monographs, The Clarendon Press, Oxford University Press, New York, 1995. With contributions by A. Zelevinsky; Oxford Science Publications. MR1354144

[74] Percy A. MacMahon, *Combinatory analysis. Vol. I, II (bound in one volume)*, Dover Phoenix Editions, Dover Publications, Inc., Mineola, NY, 2004. Reprint of *An introduction to combinatory analysis* (1920) and *Combinatory analysis. Vol. I, II* (1915, 1916). MR2417935

[75] W. H. Mills, David P. Robbins, and Howard Rumsey Jr., *Proof of the Macdonald conjecture*, Invent. Math. **66** (1982), no. 1, 73–87, DOI 10.1007/BF01404757. MR652647

[76] Robert A. Proctor, *Odd symplectic groups*, Invent. Math. **92** (1988), no. 2, 307–332, DOI 10.1007/BF01404455. MR936084

[77] James Propp, *Enumeration of matchings: problems and progress*, New perspectives in algebraic combinatorics (Berkeley, CA, 1996), Math. Sci. Res. Inst. Publ., vol. 38, Cambridge Univ. Press, Cambridge, 1999, pp. 255–291. MR1731819

[78] James Propp, *A reciprocity theorem for domino tilings*, Electron. J. Combin. **8** (2001), no. 1, Research Paper 18, 9, DOI 10.37236/1562. MR1855859

[79] James Propp, *Tilings*, Handbook of enumerative combinatorics, Discrete Math. Appl. (Boca Raton), CRC Press, Boca Raton, FL, 2015, pp. 541–588. MR3409350

[80] James Propp and Richard Stanley, *Domino tilings with barriers*, J. Combin. Theory Ser. A **87** (1999), no. 2, 347–356, DOI 10.1006/jcta.1999.2967. MR1704265

[81] V. Reiner, D. Stanton, and D. White, *The cyclic sieving phenomenon*, J. Combin. Theory Ser. A **108** (2004), no. 1, 17–50, DOI 10.1016/j.jcta.2004.04.009. MR2087303

[82] Hjalmar Rosengren, *Selberg integrals, Askey-Wilson polynomials and lozenge tilings of a hexagon with a triangular hole*, J. Combin. Theory Ser. A **138** (2016), 29–59, DOI 10.1016/j.jcta.2015.09.006. MR3423472

[83] Richard P. Stanley, *The conjugate trace and trace of a plane partition*, J. Combinatorial Theory Ser. A **14** (1973), 53–65, DOI 10.1016/0097-3165(73)90063-0. MR309748

[84] Richard P. Stanley, *Combinatorial reciprocity theorems*, Advances in Math. **14** (1974), 194–253, DOI 10.1016/0001-8708(74)90030-9. MR411982

[85] Richard P. Stanley, *On dimer coverings of rectangles of fixed width*, Discrete Appl. Math. **12** (1985), no. 1, 81–87, DOI 10.1016/0166-218X(85)90042-3. MR798013

[86] Richard P. Stanley, *Symmetries of plane partitions*, J. Combin. Theory Ser. A **43** (1986), no. 1, 103–113, DOI 10.1016/0097-3165(86)90028-2. MR859302

[87] Richard P. Stanley, *Enumerative combinatorics. Vol. 2*, Cambridge Studies in Advanced Mathematics, vol. 62, Cambridge University Press, Cambridge, 1999. With a foreword by Gian-Carlo Rota and appendix 1 by Sergey Fomin, DOI 10.1017/CBO9780511609589. MR1676282

[88] John R. Stembridge, *Nonintersecting paths, Pfaffians, and plane partitions*, Adv. Math. **83** (1990), no. 1, 96–131, DOI 10.1016/0001-8708(90)90070-4. MR1069389

[89] John R. Stembridge, *Some hidden relations involving the ten symmetry classes of plane partitions*, J. Combin. Theory Ser. A **68** (1994), no. 2, 372–409, DOI 10.1016/0097-3165(94)90112-0. MR1297179

[90] Y. Su, *Electrical Lie algebra of classical types*, `arXiv:1410.1188`, 2014.

[91] H. N. V. Temperley and Michael E. Fisher, *Dimer problem in statistical mechanics—an exact result*, Philos. Mag. (8) **6** (1961), 1061–1063. MR136398

[92] William P. Thurston, *Conway's tiling groups*, Amer. Math. Monthly **97** (1990), no. 8, 757–773, DOI 10.2307/2324578. MR1072815

[93] Forest Tong, *Generalizing the divisibility property of rectangle domino tilings*, Electron. J. Combin. **23** (2016), no. 3, Paper 3.5, 21, DOI 10.37236/4520. MR3558042

Department of Mathematics, University of Nebraska–Lincoln, Nebraska 68588
Email address: `tlai3@unl.edu`

Proceedings of Symposia in Pure Mathematics
Volume **110**, 2024
https://doi.org/10.1090/pspum/110/02017

Some questions arising from the study of cohomology on flag varieties

Zhao Gao, Claudiu Raicu, and Keller VandeBogert

ABSTRACT. A fundamental problem at the confluence of algebraic geometry and representation theory is to describe the cohomology of line bundles on flag varieties over a field of characteristic p. When $p = 0$, the solution is given by the celebrated Borel–Weil–Bott Theorem, while for $p > 0$ the problem is widely open. In this note we describe a collection of open questions that arise from the study of particular cases of the general theory, focusing on their combinatorial and commutative algebra aspects.

1. Introduction

Let $\mathbf{k}$ be an algebraically closed field of characteristic p, and consider the **(complete) flag variety** Fl_n parametrizing flags of subspaces

$$V_\bullet : \qquad 0 \subset V_1 \subset \cdots \subset V_{n-1} \subset \mathbf{k}^n, \qquad \text{where } \dim(V_i) = i.$$

Flag varieties have a rich theory that overlaps many important areas in mathematics: algebraic geometry, representation theory, algebraic group theory, singularity theory, combinatorics, commutative algebra etc. This brief note is motivated by the following foundational question about flag varieties.

PROBLEM 1.1. Describe the cohomology of line bundles on Fl_n.

In characteristic $p = 0$, the answer comes from the Borel–Weil–Bott theorem [**9**, **37**] (see also [**25**, Section II.5], [**40**, Chapter 4], [**8**, Section 11.1]), but in positive characteristic the problem remains open, despite important progress over the years. Perhaps a more modest goal, which remains equally elusive, is to determine which cohomology groups are zero, and which ones are not.

PROBLEM 1.2. Characterize the vanishing of cohomology of line bundles on Fl_n.

The best results concerning Problem 1.2 are Kempf's Vanishing theorem [**26**, **27**], and Andersen's characterization of the non-vanishing of H^1 [**6**]. Essentially complete answers to Problems 1.1 and 1.2 are known for $n = 2$ (when $Fl_n = \mathbf{P}^1$) and for $n = 3$ [**17**, **23**, **24**, **29**], but the problems remain open already when $n = 4$.

Problems 1.1 and 1.2 can be formulated within the broad context of **generalized flag varieties** G/B, where G is a reductive algebraic group, and B is a Borel

---

2020 *Mathematics Subject Classification.* Primary 14M15, 20G05, 05E05.
Raicu was supported by the NSF Grant No. 2302341.
The third author was supported by the NSF Grant No. 2202871.

subgroup, and there is a wealth of important results in this setting that we will not touch upon (see the influential book by Jantzen [**25**] and the recent survey by Andersen [**7**]). The varieties Fl_n correspond to the case when $G = \mathrm{GL}_n$ and B is the Borel subgroup of upper triangular matrices.

Our own interest in Problems 1.1 and 1.2 comes from the many applications Borel–Weil–Bott theory has had in describing homological invariants in commutative algebra: syzygies [**28, 40**], Ext modules [**32**], and local cohomology [**30, 34, 35**] for determinantal varieties, syzygies of Segre varieties [**31**], vanishing results for Koszul modules [**4, 5**] etc. The main obstacle to extending such results to positive characteristic comes from the difficulty of computing cohomology in positive characteristic, as illustrated in what follows.

2. Arithmetic complexes

Consider $d \geq 1$ and a collection of non-negative integers $w_0, \cdots, w_d$. Write $[d] = \{1, \cdots, d\}$ and think of $[d]$ as indexing edges, and of w_i as weights on the vertices of the path graph

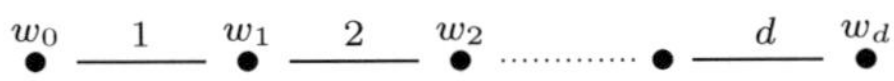

Every subset $J \subseteq [d]$ determines a decomposition of the path graph into disjoint intervals, and we define the **weight** of an interval to be the sum of weights of the vertices it contains. The removal of an element $j \in J$ breaks up exactly one interval, say of weight $w = w(J,j)$, into two intervals of weights $w' = w'(J,j)$ and $w'' = w''(J,j)$, with $w' + w'' = w$. We define the complex of $\mathbf{k}$-vector spaces $C_\bullet = C_\bullet(w_0, \cdots, w_d)$, where

$$C_k = \bigoplus_{|J|=k} \mathbf{k} \cdot e_J$$

and differentials $\partial : C_k \longrightarrow C_{k-1}$ given by

$$\partial(e_J) = \sum_{j \in J} (-1)^{s(J,j)} \binom{w(J,j)}{w'(J,j)} \cdot e_{J \setminus \{j\}}, \tag{2.1}$$

where $s(J,j)$ denotes the cardinality of the set $\{i \in [d] : i < j,\ i \notin J\}$. For an example of a binomial coefficient arising in (2.1), take $J = \{1,2,3\}$, $j = 2$:

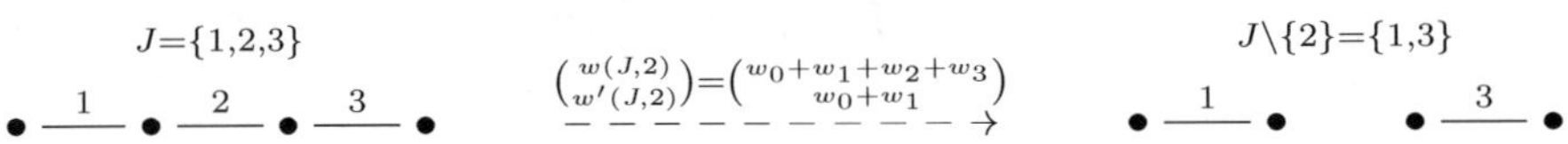

The complex $C_\bullet$ for $d = 3$ is illustrated below in Figure 1. In the special case when $w_0 = w_1 = w_2 = w_3 = 1$, we obtain

$$0 \longrightarrow \mathbf{k} \xrightarrow{\begin{bmatrix} 4 \\ 6 \\ 4 \end{bmatrix}} \mathbf{k}^3 \xrightarrow{\begin{bmatrix} -3 & 2 & 0 \\ -3 & 0 & 3 \\ 0 & -2 & 3 \end{bmatrix}} \mathbf{k}^3 \xrightarrow{\begin{bmatrix} 2 & -2 & 2 \end{bmatrix}} \mathbf{k} \longrightarrow 0 \tag{2.2}$$

which is exact if $p = \mathrm{char}(\mathbf{k})$ satisfies $p = 0$ or $p \geq 5$, but has non-trivial homology when $p = 2, 3$. If we make the convention that $\binom{0}{0} = 1$ and take $w_0 = \cdots = w_d = 0$, then $C_\bullet$ is essentially the complex computing the reduced homology of a simplex, which is exact in all characteristics. If all the binomial coefficients $\binom{w}{w'}$ appearing in the definition of $C_\bullet(w_0, \cdots, w_d)$ are non-zero in $\mathbf{k}$ (for instance if $p = 0$ or $p > w_0 + \cdots + w_d$), then it is also easy to see that the resulting complex is exact. In

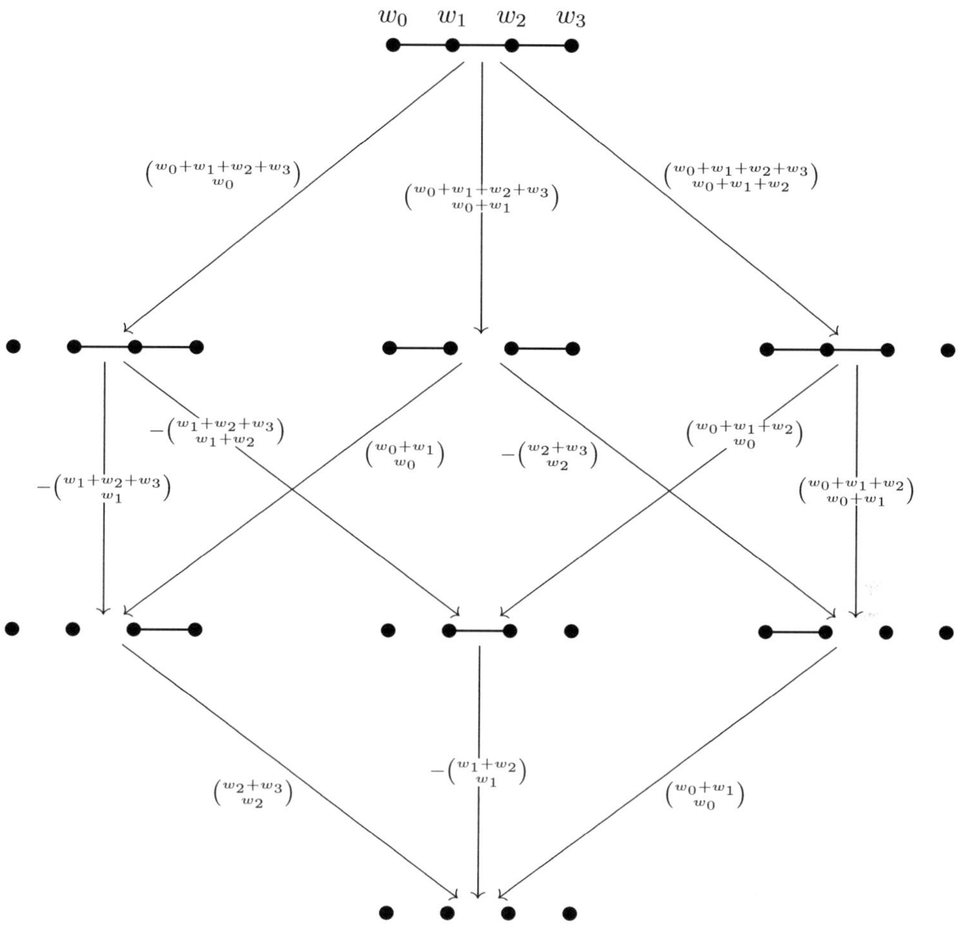

FIGURE 1. The complex $C_\bullet(w_0, w_1, w_2, w_3)$

general however, the calculation of homology and its dependence on $\mathbf{k}$ is an open problem.

PROBLEM 2.1. Describe the homology of the complex $C_\bullet(w_0, \cdots, w_d)$ over a field of characteristic p.

To indicate the complexity of Problem 2.1, we discuss the answer in the case when $w_0 = \cdots = w_d = 1$, which is proved in [**33**]. To do so, we first introduce some notation. For a prime $p > 0$, we enumerate non-negative integers $\equiv 0, 1 \pmod{p}$ as $0, 1, p, p+1, 2p, 2p+1, \cdots$. If m is in the list above then we write $|m|_p$ for its position, and call $|m|_p$ the p-**index** of m:

$$\text{if } m = pa + b, \text{ with } b \equiv 0, 1 \pmod{p} \text{ and } 0 \le b < p, \text{ then } |m|_p = 2a + b.$$

If $p = 2$ then $|m|_p = m$, but when $p = 3$ for instance, we get the following values of the p-index:

m	0	1	2	3	4	5	6	7	8	9	10	11	$\cdots$
$\|m\|_p$	0	1		2	3		4	5		6	7		$\cdots$

For a tuple $\alpha = (\alpha_0, \cdots, \alpha_k)$ of non-negative integers, with $\alpha_i \equiv 0, 1 \bmod p$, we write

$$|\alpha|_p = \sum_{i=0}^{k} |\alpha_i|_p,$$

and we define

$$A_{p,d} = \left\{ \alpha = (\alpha_0, \cdots, \alpha_k) : \sum_{i=0}^{k} \alpha_i \cdot p^i = d,\ \alpha_i \geq 0,\ \alpha_i \equiv 0, 1 \pmod{p} \right\}.$$

We have for instance $A_{p,d} = \emptyset$ when $d \not\equiv 0, 1 \pmod{p}$,

$$A_{2,4} = \{(0,0,1), (0,2), (2,1), (4)\},\ A_{3,4} = \{(1,1), (4)\}, \text{ and} \tag{2.3}$$

$$|(0,0,1)|_2 = 1, |(0,2)|_2 = 2, |(2,1)|_2 = 3, |(4)|_2 = 4, |(1,1)|_3 = 2, |(4)|_3 = 3. \tag{2.4}$$

THEOREM 2.2 ([**33**]). *Let $C_\bullet = C_\bullet(w_0, \cdots, w_d)$, where $w_0 = \cdots = w_d = 1$, and suppose that $p = \mathrm{char}(\mathbf{k}) > 0$. The following equality holds in $\mathbb{Z}[t]$:*

$$\sum_{i \geq 0} \dim_{\mathbf{k}} H_i(C_\bullet) \cdot t^i = \sum_{\alpha \in A_{p,d+1}} t^{d+1-|\alpha|_p}$$

We can then use (2.3), (2.4) and Theorem 2.2 to compute the homology of the complex $C_\bullet$ in (2.2) when $p = 2, 3$. We have $d = 3$ and therefore

$$\sum_{i \geq 0} \dim_{\mathbf{k}} H_i(C_\bullet) \cdot t^i = \begin{cases} 1 + t + t^2 + t^3 & \text{if } p = 2, \\ t + t^2 & \text{if } p = 3. \end{cases} \tag{2.5}$$

Since $4 \not\equiv 0, 1 \pmod{p}$ when $p \geq 5$, it follows that $C_\bullet$ is exact for such p.

It is difficult to imagine a general answer for arbitrary weights $w_0, \cdots, w_d$, but one could hope for a recursive description of the homology. This involves relating the homology of $C_\bullet(\underline{w})$ for different weight sequences $\underline{w}$. An easy consequence of Lucas' theorem shows for instance that

$$C_\bullet(w_0 + p^r, w_1, \cdots, w_d) \simeq C_\bullet(w_0, w_1, \cdots, w_d) \quad \text{if } p^r > w_1 + \cdots + w_d. \tag{2.6}$$

One can also check that for each i there is a short exact sequence of complexes

$$\begin{aligned} 0 \longrightarrow C_\bullet(w_0, \cdots, w_i) \otimes C_\bullet(w_{i+1}, \cdots, w_d) \longrightarrow C_\bullet(\underline{w}) \\ \longrightarrow C_{\bullet-1}(w_0, \cdots, w_i + w_{i+1}, \cdots, w_d) \longrightarrow 0 \end{aligned}$$

We end this section with a more subtle property of the complexes $C_\bullet(\underline{w})$. If we consider the ring of integer-valued polynomials

$$R = \bigoplus_{n \geq 0} \mathbb{Z} \cdot \binom{x}{n} \subset \mathbb{Q}[x], \quad \text{where} \quad \binom{x}{n} = \frac{x(x-1)\cdots(x-n+1)}{n!},$$

then we can define $C_\bullet(x, w_1, \cdots, w_n)$ as a complex of free R-modules for all non-negative integers $w_1, \cdots, w_n \geq 0$ (by choosing $w'(J, j_i)$ in (2.1) to be the weight of the interval not containing $w_0 = x$, so that the resulting binomial expression makes sense as a polynomial in x).

CONJECTURE 2.3. *If we take $w_1 = \cdots = w_d = 1$ then we have an isomorphism of complexes of R-modules*

$$C_\bullet(x, 1, \cdots, 1) \simeq C_\bullet(-x - 2d, 1, \cdots, 1).$$

Notice that we can specialize x to an arbitrary integer and obtain a complex of $\mathbb{Z}$-modules, and then base change to any field $\mathbf{k}$, thus obtaining complexes $C_\bullet(\underline{w})$ over an arbitrary field, where w_0 is allowed to be a negative number! Combining Conjecture 2.3 with (2.6), one gets that in characteristic $p > 0$

$$C_\bullet(w_0, 1, \cdots, 1) \simeq C_\bullet(p^r - w_0 - 2d, 1, \cdots, 1) \text{ for } p^r > w_0 + 2d,$$

which we can verify independently (see [**33**, Section 6.5]). We do not know if there is a way to generalize the construction of $C_\bullet(\underline{w})$ to allow multiple weights to be negative, but this could likely lead to more symmetries among the complexes $C_\bullet(\underline{w})$.

REMARK 2.4. Recently, Conjecture 2.3 was verified in [**21**].

3. Trivial group action and stabilization of cohomology

If we identify $Fl_n = \mathrm{GL}_n / B$ where B is the Borel subgroup of upper triangular matrices, then we can see that the GL_n-equivariant line bundles on Fl_n form a free abelian group of rank n. Algebraically, they are parametrized by one-dimensional representations of B (or equivalently, characters of the maximal torus). Geometrically, they are obtained as follows (see also [**8**, Section 9.2]): we write $\mathcal{Q}_i$ for the universal rank i quotient sheaf on Fl_n, let $\mathcal{L}_i = \ker(\mathcal{Q}_i \twoheadrightarrow \mathcal{Q}_{i-1})$ and define

$$\mathcal{O}_{Fl_n}(\lambda) = \mathcal{L}_1^{\lambda_1} \otimes \cdots \otimes \mathcal{L}_n^{\lambda_n} \quad \text{for } \lambda \in \mathbb{Z}^n.$$

The tautological surjection

$$\mathbf{k}^n \otimes \mathcal{O}_{Fl_n} \twoheadrightarrow \mathcal{Q}_{n-1}$$

induces an isomorphism on global sections, and we write x_i for the global section corresponding to the i-th standard unit vector in $\mathbf{k}^n$. We can then realize Fl_{n-1} as the zero locus of the section $x_n \in H^0(Fl_n, \mathcal{Q}_{n-1})$, and the inclusion $Fl_{n-1} \subset Fl_n$ is GL_{n-1}-equivariant. Moreover, for $\lambda \in \mathbb{Z}^n$ we have

$$\mathcal{O}_{Fl_n}(\lambda)_{|Fl_{n-1}} = \mathcal{O}_{Fl_{n-1}}(\overline{\lambda}) \text{ where } \overline{\lambda} = (\lambda_1, \cdots, \lambda_{n-1}) \in \mathbb{Z}^{n-1}. \tag{3.1}$$

Given a weight $\lambda \in \mathbb{Z}^r$ we define

$$\lambda^{[n]} = (\lambda_1, \cdots, \lambda_r, 0, \cdots, 0) \in \mathbb{Z}^n \quad \text{for } n \geq r,$$

and consider the maps

$$\psi_n^j : H^j\left(Fl_n, \mathcal{O}_{Fl_n}(\lambda^{[n]})\right) \longrightarrow H^j\left(Fl_{n-1}, \mathcal{O}_{Fl_{n-1}}(\lambda^{[n-1]})\right).$$

induced on cohomology by the inclusion $Fl_{n-1} \subset Fl_n$ and the identification (3.1).

THEOREM 3.1 ([**33**]). *Suppose that $\lambda \in \mathbb{Z}^r$ satisfies $|\lambda| = \lambda_1 + \cdots + \lambda_r \leq 0$.*

1. *The action of GL_n on $H^j\left(Fl_n, \mathcal{O}_{Fl_n}(\lambda^{[n]})\right)$ is trivial for $n \gg 0$.*
2. *The map ψ_n^j is an isomorphism for $n \gg 0$.*
3. *If $|\lambda| < 0$ then $H^j\left(Fl_n, \mathcal{O}_{Fl_n}(\lambda^{[n]})\right) = 0$ for $n \gg 0$.*

When $\mathrm{char}(\mathbf{k}) = 0$, the conclusions of Theorem 3.1 are easy applications of the Borel–Weil–Bott theorem, but in characteristic p they are more subtle. Theorem 3.1 allows us to define for $|\lambda| \leq 0$ the **stable cohomology groups**

$$H^j_{stab}(\lambda) = H^j\left(Fl_n, \mathcal{O}_{Fl_n}(\lambda^{[n]})\right) \text{ for } n \gg 0$$

and formulate the stable versions of Problems 1.1 and 1.2.

PROBLEM 3.2. Compute the stable cohomology groups $H^j_{stab}(\lambda)$ over a field of characteristic $p > 0$, and/or characterize their (non-)vanishing behavior.

The significance of Theorem 2.2 can now be explained by its relation to stable cohomology. We show in [**33**] that

$$(3.2)\quad H^j_{stab}(\lambda) = H_{d+1-j}(C_\bullet(\underline{w})) \text{ for } \lambda = (-d-1, d+1) \in \mathbb{Z}^2,\ w_0 = \cdots = w_d = 1.$$

In fact, we show more generally that if $w_0 > 0$, $w_1 = \cdots = w_d$, and if we let

$$(3.3)\qquad \lambda = (-w_0 - d, d+1, 1, \cdots, 1) \in \mathbb{Z}^{1+w_0}$$

then we have

$$H^j_{stab}(\lambda) = H_{d+w_0-j}(C_\bullet(\underline{w})).$$

It follows from (2.5) that for $\lambda = (-4, 4)$ the only non-zero stable cohomology is

$$H^1_{stab}(\lambda) = H^2_{stab}(\lambda) = H^3_{stab}(\lambda) = H^4_{stab}(\lambda) = \mathbf{k} \quad \text{when } \mathrm{char}(\mathbf{k}) = 2, \text{ and}$$
$$H^2_{stab}(\lambda) = H^3_{stab}(\lambda) = \mathbf{k} \quad \text{when } \mathrm{char}(\mathbf{k}) = 3.$$

Perhaps one familiar example is obtained by taking $d = 0$ in (3.3), where the cohomology is independent of the characteristic of $\mathbf{k}$ and is given by

$$H^{w_0}_{stab}(\lambda) = \mathbf{k} \quad \text{for } \lambda = (-w_0, 1, \cdots, 1) \in \mathbb{Z}^{w_0+1}.$$

There are many natural questions that one can pose now, regarding the effective bounds for stabilization as well as efficient algorithmic methods for computing the stable cohomology.

QUESTION 3.3. Is it possible to construct arithmetic complexes such as the ones in Section 2, which compute the stable cohomology $H^j_{stab}(\lambda)$ for all λ and all characteristics?

We note here that the homology of the complexes $C_\bullet(\underline{w})$ for a general $\underline{w}$ does have an interpretation in terms of sheaf cohomology, but it is the cohomology of certain higher rank vector bundles on projective space, as explained in the next section.

4. Polynomial functors of the cotangent sheaf on projective space

If we let $\mathbf{P}^{n-1} = \mathbb{P}(\mathbf{k}^n)$ denote the $(n-1)$-dimensional projective space, then there is a natural forgetful map

$$\pi : Fl_n \longrightarrow \mathbf{P}^{n-1}, \quad \pi(V_\bullet) = V_1,$$

and some of the cohomology groups from Section 3 can be realized as sheaf cohomology groups of vector bundles on $\mathbf{P}^{n-1}$, as follows. Recall that a partition $\mu = (\mu_1, \mu_2, \cdots)$ is a non-increasing sequence $\mu_1 \geq \mu_2 \geq \cdots \geq 0$. We write Ω for the **cotangent sheaf** on $\mathbf{P}^{n-1}$, and write $\mathbb{S}_\mu$ for the **Schur functor** associated to μ, noting that $\mathbb{S}_\mu\Omega = 0$ if $\mu_n > 0$ since $\mathrm{rank}(\Omega) = n-1$. If $\mu_n = 0$ then one can check using the projection formula (see [**8**, Theorem 9.8.5, and Section 9.3]) that

$$H^j\left(\mathbf{P}^{n-1}, \mathbb{S}_\mu\Omega(e)\right) = H^j\left(Fl_n, \mathcal{O}_{Fl_n}(\lambda)\right) \text{ where } \lambda = (e - |\mu|, \mu_1, \cdots, \mu_{n-1}) \in \mathbb{Z}^n.$$

In the special case when $e = 0$ and $\mu = (d, 0, \cdots)$ we obtain

$$H^j\left(\mathbf{P}^{n-1}, \mathrm{Sym}^d\,\Omega\right) = H^j\left(Fl_n, \mathcal{O}_{Fl_n}(\lambda)\right) \text{ where } \lambda = (-d, d, 0, \cdots),$$

whose (stable) cohomology can be computed based on (3.2) and Theorem 2.2.

In analogy with Theorem 3.1, we can prove a general stabilization result, as well as a statement regarding the trivial GL_n-action on cohomology.

THEOREM 4.1 ([**33**]). *Suppose that T is a polynomial functor of degree d.*

(1) *The action of GL_n on $H^j\left(\mathbf{P}^{n-1}, T(\Omega)\right)$ is trivial for all $j < n-1$.*
(2) *If $n-1 \geq d$ then the action of GL_n on $H^{n-1}\left(\mathbf{P}^{n-1}, T(\Omega)\right)$ is trivial.*
(3) *$H^j\left(\mathbf{P}^{n-1}, T(\Omega)\right)$ is independent of n as long as $n-1 \geq d$.*

It is easy to see that the condition $n-1 \geq d$ is necessary: for instance if $d = n = 2$ and $T = \mathrm{Sym}^2$, then $\Omega = \omega$ is the canonical line bundle on $\mathbf{P}^1$ and

$$H^1(\mathbf{P}^1, \omega^{\otimes 2}) = \mathrm{D}^2(\mathbf{k}^2) \otimes \bigwedge^2(\mathbf{k}^2)^\vee, \tag{4.1}$$

where D^d denotes the d-th divided power functor.

In light of Theorem 4.1, we can talk about **stable cohomology** for $T(\Omega)$ and ask the analogues of Problem 3.2 and Question 3.3 in this context. In the case when T is a Schur functor, we propose the following periodicity conjecture, which we can verify when $l = 1$ or $\mu_2 \leq 2$.

CONJECTURE 4.2. *Consider a partition $\mu = (\mu_1, \cdots, \mu_l)$ with $\mu_l > 0$, and write*

$$\mu[q] = (\mu_1, \cdots, \mu_l, 1, \cdots, 1)$$

for the partition obtained from μ by appending q parts of size 1. If $q = p^r > |\mu| - l$ and $\mathrm{char}(\mathbf{k}) = p$ then

$$H^j(\mathbf{P}^{n-1}, \mathbb{S}_\mu\Omega) \simeq H^{j+q}(\mathbf{P}^{n-1}, \mathbb{S}_{\mu[q]}\Omega).$$

Schur functors can be associated not only to partitions, but also to **skew shapes** λ/μ [**3**, Section II.1], where λ, μ are partitions with $\mu_i \leq \lambda_i$ for all i. We picture a shape λ/μ by subtracting the Young diagram of μ from that of λ, and say that λ/μ is a **ribbon** if it contains no 2×2 square. For instance if $\lambda = (7, 4, 3, 3, 3)$ and $\mu = (3, 2, 2, 2)$ then we get the ribbon skew-shape

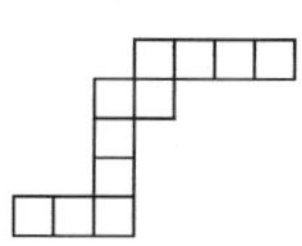

which we can also record using the sequence of column sizes $\underline{w} = (1, 1, 4, 2, 1, 1, 1)$. Note that if $w_0 \geq 1$ and $w_1 = \cdots = w_d = 1$ then the corresponding skew-shape is a hook shape, and hence an actual partition. The significance of the complexes $C_\bullet(\underline{w})$ from Section 2 is that they compute the stable cohomology of $\mathbb{S}_{\lambda/\mu}\Omega$, where λ/μ is the skew-shape corresponding to $\underline{w}$. The statement of Conjecture 4.2 is then natural in light of (2.6).

The relationship between the cohomology of $\mathbb{S}_{\lambda/\mu}\Omega$ and the homology of $C_\bullet(\underline{w})$ is based on the hypercohomology spectral sequence associated with the resolution of

$\mathbb{S}_{\lambda/\mu}$ by tensor product of exterior power functors. The question of resolving Schur functors by tensor products of exterior (or divided) powers has a long history, that played a significant part in David Buchsbaum's work [**1**, **2**, **11**–**14**]. Akin and Buchsbaum proved already in the 80s that such resolutions exist, but the quest to find an optimal one continues on: perhaps the best general results are due to Santana and Yudin [**36**], but their constructions give resolutions that are much longer than the global dimension of the relevant category of polynomial functors [**38**, Theorem 2].

PROBLEM 4.3. Find the (an) *optimal* resolution of a Schur functor $\mathbb{S}_\lambda$ (or $\mathbb{S}_{\lambda/\mu}$) by tensor products of exterior power functors.

The discussion above simplifies quite a bit if we consider Weyl functors $\mathbb{W}_\mu$ instead of Schur functors – a quick way to define them is by letting

$$\mathbb{W}_\mu(V) = (\mathbb{S}_\mu(V^\vee))^\vee$$

for a finite dimensional $\mathbf{k}$-vector space V. If $\mu_1 \geq 2$ and $n > |\mu|$ then one can show

$$H^j\left(\mathbf{P}^{n-1}, \mathbb{W}_\mu\Omega\right) = 0 \text{ for all } j.$$

If $\mu_1 = 1$ and $d = |\mu|$ then we have $\mathbb{W}_\mu\Omega = \bigwedge^d \Omega = \Omega^d$ is the sheaf of differential d-forms, in which case the only non-vanishing cohomology is $H^d\left(\mathbf{P}^{n-1}, \Omega^d\right) = \mathbf{k}$. To make the question of computing cohomology more interesting, one should then consider twists by a line bundle: we formulate the following general problem, whose difficulty is likely comparable to that of Problems 1.1 and 1.2.

PROBLEM 4.4. Given a polynomial functor T, describe (and characterize the vanishing behavior of) the cohomology of $T(\Omega) \otimes \mathcal{O}_{\mathbf{P}^{n-1}}(e)$ for all $e \in \mathbb{Z}$.

5. The incidence correspondence

Problem 4.4 is already open in the case when $T = \mathrm{D}^d$ is a **divided power functor** (the Weyl functor $\mathbb{W}_\mu$ for $\mu = (d, 0, \cdots)$). The vanishing behavior of the cohomology of $\mathrm{D}^d\,\Omega(e)$ is characterized in [**23**], but we do not have a general description of the cohomology groups. To provide some geometric context, we consider the variety X parametrizing partial flags

$$0 \subset V_1 \subset V_{n-1} \subset \mathbf{k}^n,$$

which in the case $n = 3$ agrees with Fl_3. If we write $\mathbf{P} = \mathbf{P}^{n-1}$, and let $\mathbf{P}^\vee$ denote the dual projective space, we can reinterpret X as the **incidence correspondence**

$$X = \{(p, H) : p \in H\} \subset \mathbf{P} \times \mathbf{P}^\vee,$$

which is a hypersurface cut out by the bilinear form

$$\omega = x_1y_1 + \cdots + x_ny_n. \tag{5.1}$$

The line bundles on X arise by restriction from $\mathbf{P} \times \mathbf{P}^\vee$, and we write

$$\mathcal{O}_X(a, b) = \mathcal{O}_{\mathbf{P}\times\mathbf{P}^\vee}(a, b)_{|X} \quad \text{ for } (a, b) \in \mathbb{Z}^2.$$

Along the first projection $\varphi : X \longrightarrow \mathbf{P}$, we can identify X with the projective bundle $\mathbb{P}_{\mathbf{P}}(\Omega) = \underline{\mathrm{Proj}}\,(\mathrm{Sym}(\Omega^\vee))$ which after some manipulations yields

$$H^j\left(\mathbf{P}, \mathrm{D}^d\,\Omega \otimes \mathcal{O}_{\mathbf{P}}(e)\right) \simeq H^{j+n-2}\left(X, \mathcal{O}_X(e-d+1, -d-n+1)\right) \otimes \left(\bigwedge^n \mathbf{k}^n\right)^\vee,$$

where the last tensor factor is there to make the isomorphism GL_n-equivariant. In terms of the forgetful map

$$\psi : Fl_n \longrightarrow X, \quad \psi(V_\bullet) = (V_1 \subset V_{n-1}),$$

we have that $\psi^*(\mathcal{O}_X(a,b)) = \mathcal{O}_{Fl_n}(a,0,\cdots,0,-b)$ and by the projection formula

$$H^j(X, \mathcal{O}_X(a,b)) = H^j(Fl_n, \mathcal{O}_{Fl_n}(a,0,\cdots,0,-b)),$$

showing that Problem 4.4 for $T = \mathrm{D}^d$ is just a special case of Problems 1.1, 1.2.

We find it more convenient to rephrase the cohomology calculations in terms of the **tautological sheaf** $\mathcal{R} = \Omega(1)$ (as in [**23**]), and note that we have

$$H^j\left(\mathbf{P}, \mathrm{D}^d\mathcal{R}\otimes\mathcal{O}_{\mathbf{P}}(e)\right) = H^j\left(\mathbf{P}, \mathrm{D}^d\Omega\otimes\mathcal{O}_{\mathbf{P}}(e+d)\right)$$

$$\simeq H^{j+n-2}(X, \mathcal{O}_X(e+1,-d-n+1)) \otimes \left(\bigwedge^n \mathbf{k}^n\right)^\vee.$$

For yet another equivalent interpretation of these cohomology groups, we let

$$R = \mathbf{k}[x_1,\cdots,x_n,y_1,\cdots,y_n], \tag{5.2}$$

and consider the (infinitely generated) local cohomology module

$$M = H^n_{(y_1,\cdots,y_n)}(R).$$

We define a bi-grading on M by letting

$$M_{d,e} = \bigoplus_{\substack{a_i,b_j\geq 0\\ |\underline{a}|=d,\ |\underline{b}|=e}} \mathbf{k}\cdot\frac{x_1^{b_1}\cdots x_n^{b_n}}{y_1^{1+a_1}\cdots y_n^{1+a_n}}.$$

For ω as in (5.1), $\mathcal{L} = \mathcal{O}_X(e+1,-d-n+1)$, $e \geq -1$, we have an exact sequence

$$0 \longrightarrow H^{n-2}(X,\mathcal{L}) \longrightarrow M_{d,e} \xrightarrow{\cdot\omega} M_{d-1,e+1} \longrightarrow H^{n-1}(X,\mathcal{L}) \longrightarrow 0$$

and $H^j(X,\mathcal{L}) = 0$ for $j \notin \{n-2, n-1\}$ (when $e \leq -2$ the cohomology can only be non-zero for $j = 2n-3$, and its dimension is the same in all characteristics). For a concrete example that illustrates the dependence on characteristic, take $n = 3$, $d = 2$, $e = 1$ and consider the element

$$f = \frac{x_1}{y_1y_2^2y_3^2} + \frac{x_2}{y_1^2y_2y_3^2} + \frac{x_3}{y_1^2y_2^2y_3} = \frac{\omega}{y_1^2y_2^2y_3^2} \in M_{2,1}. \tag{5.3}$$

We have that

$$f\cdot\omega = \frac{\omega^2}{y_1^2y_2^2y_3^2} = 2\cdot\left(\frac{x_1x_2y_1y_2 + x_1x_3y_1y_3 + x_2x_3y_2y_3}{y_1^2y_2^2y_3^2}\right)$$

which is zero if and only if $\mathrm{char}(\mathbf{k}) = 2$. In that case we have that $H^1(X,\mathcal{L}) = \mathbf{k}\cdot f$ is one dimensional.

As opposed to the examples in previous sections, the cohomology groups we consider here usually have a non-trivial GL_n-action, which should play a role in describing their structure. A first approximation of this comes from the action of the maximal torus $(\mathbf{k}^\times)^n$ of diagonal matrices. This is encoded by the notion of a **character**, or equivalently by a $\mathbb{Z}^n$-grading: in the preceding paragraphs, the natural $\mathbb{Z}^n$-grading comes from assigning $\deg(x_i) = \vec{e}_i$ (the i-th standard unit vector), and $\deg(y_i) = -\vec{e}_i$. We review the notion of characters next, followed by a discussion of some conjectural character formulas.

5.1. Characters. Representations of the algebraic torus $(\mathbf{k}^\times)^n$ are equivalent to $\mathbb{Z}^n$-graded vector spaces [**8**, Lemma 9.7.9], [**25**, Section 2.11], and for any such (finite dimensional) vector space W, its **character** is

$$[W] := \sum_{(i_1,\cdots,i_n)\in\mathbb{Z}^n} \dim(W_{(i_1,\cdots,i_n)})\cdot t_1^{i_1}\cdots t_n^{i_n} \in \mathbb{Z}[t_1^{\pm 1},\cdots,t_n^{\pm 1}].$$

If W is a GL_n-representation, then $[W]$ is invariant under the action of the symmetric group $\mathfrak{S}_n$ by permutations of $t_1,\cdots,t_n$. Examples include the **complete symmetric polynomials** h_d and **Schur polynomials** $s_{(a,b)}$, defined by

$$h_d = [\mathrm{Sym}^d(\mathbf{k}^n)] = \sum_{\substack{i_1+\cdots+i_n=d\\ i_j\geq 0}} t_1^{i_1}\cdots t_n^{i_n}, \quad s_{(a,b)} = [\mathbb{S}_{(a,b)}(\mathbf{k}^n)] = h_a\cdot h_b - h_{a+1}\cdot h_{b-1}.$$

Notice that although $\mathrm{Sym}^d(\mathbf{k}^n)$ and $\mathrm{D}^d(\mathbf{k}^n)$ are not isomorphic GL_n-representations in general, they are isomorphic as $(\mathbf{k}^\times)^n$-representations and therefore they have the same character. We have

$$M_{d,e} \simeq \mathrm{D}^d(\mathbf{k}^n)\otimes \mathrm{Sym}^e(\mathbf{k}^n)\otimes\left(\bigwedge^n \mathbf{k}^n\right), \quad [M_{d,e}] = h_d\cdot h_e\cdot(t_1\cdots t_n).$$

Since the element f in (5.3) has multidegree $(2,2,2)$, we get for $n=3$, $\mathrm{char}(\mathbf{k})=2$

$$\left[H^1(X,\mathcal{O}_X(2,-4))\right] = t_1^2t_2^2t_3^2 \quad\text{and}\quad \left[H^0\left(\mathbf{P},\mathrm{D}^2\mathcal{R}(1)\right)\right] = t_1t_2t_3.$$

For a non-polynomial example, if W is the representation in (4.1) then

$$[W] = t_1^2t_2^{-1} + 1 + t_1^{-1}t_2^2.$$

To describe further cohomology characters, it is useful to consider q-**truncated complete symmetric polynomials** $h_d^{(q)}$ and q-**truncated Schur polynomials** $s_{(a,b)}^{(q)}$, defined by (see also [**20**], [**39**])

$$h_d^{(q)} = \sum_{\substack{i_1+\cdots+i_n=d\\ 0\leq i_j<q}} t_1^{i_1}\cdots t_n^{i_n}, \qquad s_{(a,b)}^{(q)} = h_a^{(q)}\cdot h_b^{(q)} - h_{a+1}^{(q)}\cdot h_{b-1}^{(q)}.$$

as well as the q-**Frobenius endomorphisms** of the character ring

$$F^q : \mathbb{Z}[t_1^{\pm1},\cdots,t_n^{\pm1}] \longrightarrow \mathbb{Z}[t_1^{\pm1},\cdots,t_n^{\pm1}], \quad \text{defined by } t_i\mapsto t_i^q.$$

5.2. Weights close to the origin. It is a general phenomenon that the cohomology characters of $\mathcal{O}_{Fl_n}(\lambda)$ behave well (as in characteristic zero) when λ is close to the origin, and get increasingly more complicated the further away one gets from the origin. For the divided powers of $\mathcal{R}$ (or Ω) this translates into the fact that the cohomology of $\mathrm{D}^d\mathcal{R}(e)$ in characteristic p is the same as that in characteristic zero when $0\leq d<p$. Passing to the next order of magnitude ($p\leq d<p^2$) we have the following conjectural description of the cohomology characters (see [**23**] for the explanation of the reduction to the case $e\geq d-1$).

CONJECTURE 5.1. *Suppose that $tp\leq d<(t+1)p$, where $1\leq t<p$. We have for all $e\geq d-1$*

$$\left[H^1\left(\mathbf{P}^{n-1}, D^d\mathcal{R}(e)\right)\right] = \sum_{\substack{1\leq b\leq a\leq t\\ 0\leq j\leq a-b}} F^p(s_{(a-b,j)})\cdot s^{(p)}_{(e+(b-j)p,d-ap)}.$$

When $t = 1$, Conjecture 5.1 simplifies to the assertion that

$$\left[H^1\left(\mathbf{P}^{n-1}, \mathrm{D}^d\,\mathcal{R}(e)\right)\right] = s^{(p)}_{(e+p,d-p)}, \tag{5.4}$$

which is proved in [**23**, Theorem 1.6]. When $n = 3$, Conjecture 5.1 can be proved using [**23**, Theorem 1.7]. In general, we expect that a character formula for the cohomology of $\mathrm{D}^d\,\mathcal{R}(e)$ will rely in an essential way on truncated Schur polynomials, and we formulate a precise statement when $\mathrm{char}(\mathbf{k}) = 2$ below.

5.3. Nim polynomials. In addition to truncated Schur polynomials, the other ingredient in the description of cohomology characters when $\mathrm{char}(\mathbf{k}) = 2$ appears to come from the symmetric polynomials defining winning positions in the game of Nim (see [**10**] for the basic theory), which we introduce next.

We write $d = (d_k \cdots d_0)_2$ for the 2-adic expansion of a non-negative integer d:

$$d = \sum_{i=0}^{k} d_i \cdot 2^i, \text{ with } d_i \in \{0, 1\} \text{ for all } i.$$

The **Nim-sum** (or **bitwise xor**) $a \oplus b$ is defined by performing addition modulo 2 to each of the digits in the 2-adic expansion of non-negative integers a, b:

$$a \oplus b = c \text{ if and only if } a_i + b_i \equiv c_i \bmod 2 \text{ for all } i,$$

where a_i, b_i, c_i denote the digits of a, b, c in the 2-adic expansion. We define the n-variate **Nim symmetric polynomials** via

$$\mathcal{N}_m = \sum_{\substack{i_1+\cdots+i_n=2m \\ i_1\oplus i_2\oplus\cdots\oplus i_n=0}} t_1^{i_1} t_2^{i_2} \cdots t_n^{i_n}.$$

We have for instance that $\mathcal{N}_0 = 1$ and

$$\mathcal{N}_1 = \sum_{1\le i<j\le n} t_i t_j = s_{(1,1)}.$$

CONJECTURE 5.2. *If* $\mathrm{char}(\mathbf{k}) = 2$ *and* $e \geq d - 1$ *then*

$$\left[H^1\left(\mathbf{P}^{n-1}, D^d\mathcal{R}(e)\right)\right] = \sum_{(q,m)\in\Lambda_d} F^{2q}(\mathcal{N}_m) \cdot s^{(q)}_{(e-(2m-1)q,d-(2m+1)q)},$$

where $\Lambda_d = \{(q, m) | q = 2^r \text{ for some } r \geq 1,\ m \geq 0,\ \text{and } (2m + 1)q \leq d\}$.

When $d = 2, 3$ we have $\Lambda_d = \{(2, 0)\}$ and Conjecture 5.2 predicts

$$\left[H^1\left(\mathbf{P}^{n-1}, D^d\mathcal{R}(e)\right)\right] = s^{(2)}_{(e+2,d-2)}$$

which is a special case of (5.4). When $d = 6$ we have $\Lambda_d = \{(4, 0), (2, 0), (2, 1)\}$ and the conjecture asserts

$$\left[H^1\left(\mathbf{P}^{n-1}, D^6\mathcal{R}(e)\right)\right] = s^{(4)}_{(e+4,2)} + s^{(2)}_{(e+2,4)} + F^4(\mathcal{N}_1) \cdot s^{(2)}_{(e-2,0)} \quad \text{for } e \geq 5.$$

We have checked computationally that the dimensions of the corresponding representations (obtained by setting $t_1 = \cdots = t_n = 1$) agree for $n \leq 10$.

Conjecture 5.2 appears in [**22**, Chapter 4] where it is proved for $n = 3, 4$, and it is also discussed in an equivalent form for $n = 3$ in [**23**, Theorem 1.9]. Computational data suggest character formulas with a similar flavor in other characteristics, but we do not know how to construct an appropriate replacement for the Nim polynomials $\mathcal{N}_m$ in general.

6. Relation to determinantal ideals

Schur functors are intimately related to the study of determinantal rings and ideals, particularly through the standard monomial theory for the ring R of polynomial functions on the space of $m \times n$ matrices [**16**], [**8**, Chapter 3]. In characteristic zero it is now well-understood how to classify the ideals $I \subset R$ which are $\mathrm{GL}_m \times \mathrm{GL}_n$-invariant (for the action by row and column operations on the matrix entries) as well as how to compute basic homological invariants such as their Hilbert function. In characteristic $p > 0$, even the case $m = 1$ of the classification problem (GL_n-invariant ideals in $\mathbf{k}[x_1, \cdots, x_n]$) is non-trivial, and requires understanding the submodule structure for the GL_n-representations $\mathrm{Sym}^d(\mathbf{k}^n)$, which was worked out by Doty [**19**].

In what follows we focus on the case of $2 \times n$ matrices. We let R as in (5.2), and let I denote the ideal of 2×2 minors of the generic matrix

$$\begin{bmatrix} x_1 & x_2 & \cdots & x_n \\ y_1 & y_2 & \cdots & y_n \end{bmatrix}.$$

There is a bi-grading on R given by $\deg(x_i) = (1,0)$, $\deg(y_j) = (0,1)$ which makes the ideal I bihomogeneous. The GL_n-action (by column operations) makes R and I into GL_n-representations, so we can analyze the characters of their bigraded components. We have for $a \geq b$

$$\left[R_{(a,b)}\right] = \left[\mathrm{Sym}^a(\mathbf{k}^n) \otimes \mathrm{Sym}^b(\mathbf{k}^n)\right] = s_a \cdot s_b = \sum_{i=0}^{b} s_{(a+b-i,i)},$$

where the last equality comes from Pieri's rule. This identity at the level of characters can be explained using the I-adic filtration on R: we have

$$\left[(I^i/I^{i+1})_{(a,b)}\right] = s_{(a+b-i,i)} \text{ for } i = 0, \cdots, b, \tag{6.1}$$

and $(I^i/I^{i+1})_{(a,b)} = 0$ for $i > b$. We can refine this further using tableaux combinatorics, focusing on the case $i = b$ for simplicity. We recall that a Young tableau

$$T = \begin{array}{|c|c|c|c|c|c|}\hline u_1 & u_2 & \cdots & u_b & \cdots & u_a \\ \hline v_1 & v_2 & \cdots & v_b & \multicolumn{2}{c}{} \\ \cline{1-4}\end{array} \tag{6.2}$$

of shape (a,b) is **semi-standard** if

$$u_1 \leq u_2 \leq \cdots \leq u_a, \quad v_1 \leq v_2 \leq \cdots \leq v_b, \quad u_i < v_i \text{ for } 1 \leq i \leq b.$$

If we write $Tab_n(a,b)$ for the set of semi-standard tableaux of shape (a,b) with entries in $\{1, \cdots, n\}$, and let

$$t^T = t_{u_1} \cdots t_{u_a} \cdot t_{v_1} \cdots t_{v_b} \quad \text{for } T \text{ as in (6.2)},$$

then the Schur polynomial $s_{(a,b)}$ can be computed as

$$s_{(a,b)} = \sum_{T \in Tab_n(a,b)} t^T. \tag{6.3}$$

This suggests the existence of a basis of $(I^b)_{(a,b)} = (I^b/I^{b+1})_{(a,b)}$ indexed by $Tab_n(a,b)$, and indeed one such basis is constructed by letting

$$G_T = \left(\prod_{i=1}^{b} (x_{u_i} y_{v_i} - x_{v_i} y_{u_i})\right) \cdot x_{u_{b+1}} \cdots x_{u_a}.$$

The linear independence of the polynomials G_T follows for instance from the fact that the leading term of G_T with respect to the graded lexicographic order on R (where $x_1 > \cdots > x_n > y_1 > \cdots > y_n$) is given by

$$M_T = x_{u_1} \cdots x_{u_a} \cdot y_{v_1} \cdots y_{v_b}. \tag{6.4}$$

If one wants the analogues of this classical theory modulo Frobenius, then it is natural to consider (when $\mathrm{char}(\mathbf{k}) = p > 0$) the quotient ring

$$\overline{R} = R/\langle x_1^p, \cdots, x_n^p, y_1^p, \cdots, y_n^p \rangle,$$

and the corresponding ideal $\overline{I}$. The following analogue of (6.1) is already non-trivial.

CONJECTURE 6.1. *If $a - b \geq p - 1$ then*

$$\left[\left(\overline{I}^i / \overline{I}^{i+1} \right)_{(a,b)} \right] = s^{(p)}_{(a+b-i,i)} \text{ for } i = 0, \cdots, b.$$

We don't have a good explanation for the hypothesis $a - b \geq p - 1$ (see also Conjecture 6.4), but it is easy to see by example that it is necessary: for instance when $a = b = 1$, $i = 0$, and $p = 2$, we have $\left[(\overline{R}/\overline{I})_{(1,1)} \right] = h_2$ but $s^{(2)}_{(2,0)} = h_2^{(2)}$ is the second elementary symmetric polynomial.

As before, we can look for standard bases that match up with the (conjectural) character formulas. The following appears to be the appropriate replacement of semi-standard tableaux.

DEFINITION 6.2. We say that a tableau T is p-**semi-standard** if

(1) T is weakly increasing along rows and columns:

$$u_1 \leq \cdots \leq u_a, \quad v_1 \leq \cdots \leq v_b, \quad \text{and} \quad u_i \leq v_i \text{ for } 1 \leq i \leq b.$$

(2) Any constant sequence within each row has length at most $(p - 1)$:

$$u_i < u_{i+p} \quad \text{and} \quad v_j < v_{j+p} \quad \text{for all } i, j.$$

(3) If $u_j = v_j$, consider the unique indices $r \geq j$ and $s \leq j$ so that

$$u_j = \cdots = u_r < u_{r+1}, \quad v_{s-1} < v_s = \cdots = v_j.$$

We have $(r - j + 1) + (j - s + 1) \geq p$, that is, the total number of entries equal to $u_j = v_j$ to the right of (and including) u_j and to the left of (and including) v_j is at least p.

We write $Tab_n^{(p)}(a, b)$ for the set of p-semi-standard tableaux of shape (a, b) with entries in $\{1, \cdots, n\}$.

Note that T is 2-semi-standard if and only if the transposed tableau to T is semi-standard in the usual sense. When $p = 3$, it is not hard to check that

$$Tab_n^{(3)}(2, 1) = Tab_n(2, 1) \bigcup \left\{ \begin{array}{|c|c|} \hline i & i \\ \hline i & \\ \cline{1-1} \end{array} | i = 1, \cdots, n \right\} \tag{6.5}$$

One can show (without restrictions on a, b) the following analogue of (6.3):

$$s^{(p)}_{(a,b)} = \sum_{T \in Tab_n^{(p)}(a,b)} t^T,$$

so (6.5) translates into the identity

$$s^{(3)}_{(2,1)} = s_{(2,1)} + \sum_{i=1}^{n} t_i^3.$$

Note that truncated Schur polynomials may be *larger* then their classical analogues, which is perhaps counterintuitive. It is also the case that for a general partition, a truncated Schur polynomial does not have to be the character of a GL_n-representation (but will be a virtual character). Conjecture 6.1 asserts then that $s^{(p)}_{(a,b)}$ is indeed a character whenever $a-b \geq p-1$, and predicts a specific GL_n-representation realizing it. A natural approach to Conjecture 6.1 in the case $i=b$ (which is in fact the main case) comes from the following.

CONJECTURE 6.3. *If $a-b \geq p-1$ then for every tableau $T \in Tab^{(p)}_n(a,b)$, there exists an element $H_T \in \overline{I}^b$ whose leading monomial is M_T.*

If we write $F^p \subset \mathrm{Sym}^p$ for the Frobenius power functor, and consider the **truncated symmetric power** functors

$$T_p \,\mathrm{Sym}^d(\mathbf{k}^n) = \mathrm{coker}\left(F^p(\mathbf{k}^n) \otimes \mathrm{Sym}^{d-p}(\mathbf{k}^n) \longrightarrow \mathrm{Sym}^d(\mathbf{k}^n)\right)$$

then one can view the discussion above as an attempt to understand

$$\overline{R}_{(a,b)} = T_p \,\mathrm{Sym}^a(\mathbf{k}^n) \otimes T_p \,\mathrm{Sym}^b(\mathbf{k}^n),$$

the tensor product of two truncated powers. This is a representation of GL_n (with character $h^{(p)}_a \cdot h^{(p)}_b$) and as such it has a filtration with composition factors given by simple modules $L(\lambda)$ of highest weight λ, for certain dominant weights λ.

CONJECTURE 6.4. *If $a-b \geq p-1$ then every composition factor $L(\lambda)$ of $\overline{R}_{(a,b)}$ is p-**restricted**, that is, $\lambda_i - \lambda_{i+1} \leq p-1$ for all i.*

The hypothesis $a-b \geq p-1$ is again necessary: if $a=p-1$, $b=1$ then $\overline{R}_{(a,b)} = \mathrm{Sym}^{p-1}(\mathbf{k}^n) \otimes \mathbf{k}^n$ surjects onto $\mathrm{Sym}^p(\mathbf{k}^n)$ which contains $F^p(\mathbf{k}^n)$ as a submodule. Since $F^p(\mathbf{k}^n) = L(\lambda)$ for $\lambda = (p,0,\cdots,0)$, this is not p-restricted. The question of understanding composition factors for tensor products of (any number of) truncated symmetric powers was considered in [**18**], but the results there don't seem to be enough to settle Conjecture 6.4. There is of course no reason to restrict oneself to matrices with two rows, and we end with the hope that one day there will be a **standard monomial theory modulo Frobenius**.

REMARK 6.5. Recently, Conjectures 6.1 and 6.4 were proved in characteristic $p=2$ in [**15**].

Acknowledgments

We thank Alessio Sammartano for comments, corrections, and a careful reading of the manuscript.

References

[1] Kaan Akin and David A. Buchsbaum, *Characteristic-free representation theory of the general linear group*, Adv. in Math. **58** (1985), no. 2, 149–200, DOI 10.1016/0001-8708(85)90115-X. MR814749

[2] Kaan Akin and David A. Buchsbaum, *Characteristic-free representation theory of the general linear group. II. Homological considerations*, Adv. in Math. **72** (1988), no. 2, 171–210, DOI 10.1016/0001-8708(88)90027-8. MR972760

[3] Kaan Akin, David A. Buchsbaum, and Jerzy Weyman, *Schur functors and Schur complexes*, Adv. in Math. **44** (1982), no. 3, 207–278, DOI 10.1016/0001-8708(82)90039-1. MR658729

[4] Marian Aprodu, Gavril Farkas, Ştefan Papadima, Claudiu Raicu, and Jerzy Weyman, *Koszul modules and Green's conjecture*, Invent. Math. **218** (2019), no. 3, 657–720, DOI 10.1007/s00222-019-00894-1. MR4022070

[5] Marian Aprodu, Gavril Farkas, Ştefan Papadima, Claudiu Raicu, and Jerzy Weyman, *Topological invariants of groups and Koszul modules*, Duke Math. J. **171** (2022), no. 10, 2013–2046, DOI 10.1215/00127094-2022-0010. MR4484204

[6] Henning Haahr Andersen, *The first cohomology group of a line bundle on G/B*, Invent. Math. **51** (1979), no. 3, 287–296, DOI 10.1007/BF01389921. MR530635

[7] Henning Haahr Andersen, *Representation theory via cohomology of line bundles*, Transform. Groups **28** (2023), no. 3, 1033–1058, DOI 10.1007/s00031-022-09769-x. MR4633003

[8] Winfried Bruns, Aldo Conca, Claudiu Raicu, and Matteo Varbaro, *Determinants, Gröbner bases and cohomology*, Springer Monographs in Mathematics, Springer, Cham, 2022, DOI 10.1007/978-3-031-05480-8. MR4627943

[9] Raoul Bott, *Homogeneous vector bundles*, Ann. of Math. (2) **66** (1957), 203–248, DOI 10.2307/1969996. MR89473

[10] Charles L. Bouton, *Nim, a game with a complete mathematical theory*, Ann. of Math. (2) **3** (1901/02), no. 1-4, 35–39, DOI 10.2307/1967631. MR1502275

[11] D. A. Buchsbaum, *In search of lost resolutions*, Rend. Semin. Mat. Univ. Politec. Torino **64** (2006), no. 4, 373–379. MR2295445

[12] D. A. Buchsbaum and Gian-Carlo Rota, *Projective resolutions of Weyl modules*, Proc. Nat. Acad. Sci. U.S.A. **90** (1993), no. 6, 2448–2450, DOI 10.1073/pnas.90.6.2448. MR1208252

[13] D. A. Buchsbaum and Gian-Carlo Rota, *A new construction in homological algebra*, Proc. Nat. Acad. Sci. U.S.A. **91** (1994), no. 10, 4115–4119, DOI 10.1073/pnas.91.10.4115. MR1273361

[14] David A. Buchsbaum and Gian-Carlo Rota, *Approaches to resolution of Weyl modules*, Adv. in Appl. Math. **27** (2001), no. 1, 82–191, DOI 10.1006/aama.2001.0727. MR1835678

[15] Casabella, Laura; Yu, Teresa. *Standard monomial theory modulo Frobenius in characteristic two*, `arXiv:2311.05530`, 2023.

[16] C. de Concini, David Eisenbud, and C. Procesi, *Young diagrams and determinantal varieties*, Invent. Math. **56** (1980), no. 2, 129–165, DOI 10.1007/BF01392548. MR558865

[17] Stephen Donkin, *The cohomology of line bundles on the three-dimensional flag variety*, J. Algebra **307** (2007), no. 2, 570–613, DOI 10.1016/j.jalgebra.2006.04.039. MR2275364

[18] Stephen Donkin and Haralampos Geranios, *Composition factors of tensor products of truncated symmetric powers*, Adv. Math. **285** (2015), 394–433, DOI 10.1016/j.aim.2015.07.026. MR3406504

[19] Stephen R. Doty, *The submodule structure of certain Weyl modules for groups of type A_n*, J. Algebra **95** (1985), no. 2, 373–383, DOI 10.1016/0021-8693(85)90109-7. MR801273

[20] Stephen Doty and Grant Walker, *Truncated symmetric powers and modular representations of GL_n*, Math. Proc. Cambridge Philos. Soc. **119** (1996), no. 2, 231–242, DOI 10.1017/S0305004100074120. MR1357041

[21] Fiorindo, Luca; Reed, Ethan; Roshan-Zamir, Shahriyar; Yu, Hongmiao, *An isomorphism theorem for arithmetic complexes*, `arXiv:2309.17416`, 2023.

[22] Zhao Gao, *Cohomology of Line Bundles on the Incidence Correspondence*, ProQuest LLC, Ann Arbor, MI, 2022. Thesis (Ph.D.)–University of Notre Dame. MR4495082

[23] Zhao Gao and Claudiu Raicu, *Cohomology of line bundles on the incidence correspondence*, Trans. Amer. Math. Soc. Ser. B **11** (2024), 64–97, DOI 10.1090/btran/173. MR4685816

[24] Walter Lawrence Griffith Jr., *Cohomology of flag varieties in characteristic p*, Illinois J. Math. **24** (1980), no. 3, 452–461. MR573481

[25] Jens Carsten Jantzen, *Representations of algebraic groups*, 2nd ed., Mathematical Surveys and Monographs, vol. 107, American Mathematical Society, Providence, RI, 2003. MR2015057

[26] George R. Kempf, *Vanishing theorems for flag manifolds*, Amer. J. Math. **98** (1976), no. 2, 325–331, DOI 10.2307/2373887. MR409493

[27] George R. Kempf, *Linear systems on homogeneous spaces*, Ann. of Math. (2) **103** (1976), no. 3, 557–591, DOI 10.2307/1970952. MR409474

[28] Alain Lascoux, *Syzygies des variétés déterminantales* (French), Adv. in Math. **30** (1978), no. 3, 202–237, DOI 10.1016/0001-8708(78)90037-3. MR520233

[29] Linyuan Liu, *Cohomologie des fibrés en droites sur* SL_3/B *en caractéristique positive : deux filtrations et conséquences* (French, with English and French summaries), J. Eur. Math. Soc. (JEMS) **26** (2024), no. 4, 1365–1422, DOI 10.4171/jems/1310. MR4721034

[30] András C. Lőrincz and Claudiu Raicu, *Iterated local cohomology groups and Lyubeznik numbers for determinantal rings*, Algebra Number Theory **14** (2020), no. 9, 2533–2569, DOI 10.2140/ant.2020.14.2533. MR4172715

[31] Luke Oeding, Claudiu Raicu, and Steven V. Sam, *On the (non-)vanishing of syzygies of Segre embeddings*, Algebr. Geom. **6** (2019), no. 5, 571–591, DOI 10.14231/ag-2019-026. MR4009174

[32] Claudiu Raicu, *Regularity and cohomology of determinantal thickenings*, Proc. Lond. Math. Soc. (3) **116** (2018), no. 2, 248–280, DOI 10.1112/plms.12071. MR3764061

[33] Raicu, Claudiu; VandeBogert, Keller, *Stable sheaf cohomology on flag varieties*, arXiv:2306.14282, 2023.

[34] Claudiu Raicu and Jerzy Weyman, *Local cohomology with support in generic determinantal ideals*, Algebra Number Theory **8** (2014), no. 5, 1231–1257, DOI 10.2140/ant.2014.8.1231. MR3263142

[35] Claudiu Raicu, Jerzy Weyman, and Emily E. Witt, *Local cohomology with support in ideals of maximal minors and sub-maximal Pfaffians*, Adv. Math. **250** (2014), 596–610, DOI 10.1016/j.aim.2013.10.005. MR3122178

[36] Ana Paula Santana and Ivan Yudin, *Characteristic-free resolutions of Weyl and Specht modules*, Adv. Math. **229** (2012), no. 4, 2578–2601, DOI 10.1016/j.aim.2012.01.006. MR2880232

[37] Jean-Pierre Serre, *Représentations linéaires et espaces homogènes kählériens des groupes de Lie compacts (d'après Armand Borel et André Weil)* (French), Séminaire Bourbaki, Vol. 2, Soc. Math. France, Paris, 1995, pp. Exp. No. 100, 447–454. MR1609256

[38] Burt Totaro, *Projective resolutions of representations of* $\mathrm{GL}(n)$, J. Reine Angew. Math. **482** (1997), 1–13, DOI 10.1515/crll.1997.482.1. MR1427655

[39] Grant Walker, *Modular Schur functions*, Trans. Amer. Math. Soc. **346** (1994), no. 2, 569–604, DOI 10.2307/2154861. MR1273543

[40] Jerzy Weyman, *Cohomology of vector bundles and syzygies*, Cambridge Tracts in Mathematics, vol. 149, Cambridge University Press, Cambridge, 2003, DOI 10.1017/CBO9780511546556. MR1988690

Department of Mathematics, University of Notre Dame, 255 Hurley Notre Dame, Indiana 46556

Email address: zgao1@alumni.nd.edu

Department of Mathematics, University of Notre Dame, 255 Hurley Notre Dame, Indiana 46556

Email address: craicu@nd.edu

Department of Mathematics, University of Notre Dame, 255 Hurley Notre Dame, Indiana 46556

Email address: kvandebo@nd.edu

Proceedings of Symposia in Pure Mathematics
Volume **110**, 2024
https://doi.org/10.1090/pspum/110/02010

Castelnuovo-Mumford regularity and Schubert geometry

Alexander Yong

ABSTRACT. We study the Castelnuovo-Mumford regularity of tangent cones of Schubert varieties. Conjectures about this statistic are presented; these are proved for the *covexillary* case. This builds on work of L. Li and the author on these tangent cones, as well as that of J. Rajchgot-Y. Ren-C. Robichaux-A. St. Dizier-A. Weigandt and J. Rajchgot-C. Robichaux-A. Weigandt on tableau rules for computing regularity of some matrix Schubert varieties.

1. Introduction

Let GL_n/B be the *complete flag variety*; GL_n is the group of $n \times n$ invertible complex matrices and B is the Borel subgroup of invertible upper triangular matrices. B acts with finitely many orbits

$$X_w^\circ = BwB/B \cong \mathbb{C}^{\ell(w)};$$

$w \in \mathfrak{S}_n =$ the symmetric group on $[n] := \{1, 2, \ldots, n\}$ and $\ell(w)$ is the *Coxeter length* of w, that is,

$$\ell(w) = \#\{i < j : w(i) > w(j)\}.$$

Their closures

$$X_w := \overline{X_w^\circ} = \coprod_{v \leq w} X_v^\circ$$

are the *Schubert varieties*; here $v \leq w$ refers to *(strong) Bruhat order*. Let $T \subset GL_n$ be the maximal torus of invertible diagonal matrices. The T-fixed points are $e_v := vB/B$. To study the local structure of X_w, it suffices to study only the points e_v (for $v \leq w$), since B provides local isomorphisms to any other point of $X_v^\circ \subseteq X_w$. A book reference is [**7**].

Let $(\mathcal{O}_{p,Y}, \mathfrak{m}_p, \mathbb{k})$ be the local ring of a point p in a variety Y. The *associated graded ring* [**1**, Chapter 10] with respect to the $\mathfrak{m}_p$-adic filtration is

$$R_{p,Y} := \mathrm{gr}_{\mathfrak{m}_p} \mathcal{O}_{p,Y} = \bigoplus_{i=0}^{\infty} \mathfrak{m}_p^i/\mathfrak{m}_p^{i+1} \qquad (\mathfrak{m}_p^0 := \mathcal{O}_{p,Y}).$$

2020 *Mathematics Subject Classification.* Primary 14M15, 05E40.

Key words and phrases. Schubert varieties, tangent cones, Castelnuovo-Mumford regularity.

The author was partially supported by a Simons Collaboration Grant, an NSF RTG 1937241 in Combinatorics, and an appointment at the UIUC Center for Advanced Study.

$R_{p,Y}$ has a $\mathbb{Z}$-graded *Poincaré series*

$$\mathsf{PS}_{p,Y}(q) = \sum_{i=0}^{\infty} \dim(\mathfrak{m}_p^i/\mathfrak{m}_p^{i+1})q^i = \frac{H_{p,Y}(q)}{(1-q)^{\dim(Y)}}, \tag{1.1}$$

where $H_{p,Y}(q) \in \mathbb{Z}[q]$. $H_{p,Y}(1)$ is the *Hilbert-Samuel multiplicity*. In the case $p = e_v$ and $Y = X_w$, let

$$\mathsf{PS}_{v,w}(q) = P_{p,Y}(q), R_{v,w} = R_{p,Y}, \text{ and } H_{v,w}(q) = H_{p,Y}(q).$$

We study the *Castelnuovo-Mumford regularity* $\mathrm{Reg}(R_{v,w})$, viewed as a graded module over $\Bbbk[\mathfrak{m}_{e_v}/\mathfrak{m}_{e_v}^2]$. This statistic measures, in some sense, the "complexity" of $R_{v,w}$; see Section 3 for definitions. Outside of Schubert geometry, study of regularity of the associated graded ring appears in, *e.g.*, [**3, 25**] and the references therein.

CONJECTURE 1.1. $\mathrm{Reg}(R_{v,w}) = \deg H_{v,w}(q)$.

CONJECTURE 1.2 (Semicontinuity). *If $u \leq v \leq w$ in Bruhat order then*

$$\mathrm{Reg}(R_{u,w}) \geq \mathrm{Reg}(R_{v,w}).$$

CONJECTURE 1.3 (Upper bound). $\mathrm{Reg}(R_{v,w}) \leq \frac{\ell(w)-\ell(v)-1}{2}$.

Proposition 5.4 shows they follow from earlier conjectures with L. Li [**17, 18**]; see Section 5. Conjectures 1.1 and 1.2 imply that $\mathrm{Reg}(R_{u,v})$ is a singularity measure that falls into the framework of [**26**]. In particular, it would imply the locus of points $p \in X_w$ with "$\mathrm{Reg}(p) \geq k$" is described using *interval pattern avoidance.*

Speculatively, a strengthening of Conjecture 1.3 holds, namely,

$$\mathrm{Reg}(R_{v,w}) \leq \deg\, P_{v,w}(q)$$

where $P_{v,w}(q)$ is the *Kazhdan-Lusztig polynomial*; but, the evidence is not strong ($n \leq 6$).

The papers [**17, 18**] study the tangent cones in the case w is *covexillary*, *i.e.*, *w avoids* the pattern 3412 (there are not indices $i_1 < i_2 < i_3 < i_4$ such that $w(i_1), w(i_2), w(i_3), w(i_4)$ are in the same relative order as 3412). This defines a subfamily with a number of prior results. For example, *ibid.* gives formulas for $H_{v,w}(q)$ and related them to the *Kazhdan-Lusztig polynomials*; a combinatorial formula for the latter was already known due to work of A. Lascoux [**15**]. One also has a "diagonal Gröbner basis theorem" for *matrix Schubert varieties* [**14**].[1] These results play a role in our work. This is our main result:

THEOREM 1.4. *Conjectures* 1.1, 1.2, *and* 1.3 *hold if w is covexillary. In this case, there is a combinatorial rule for* $\mathrm{Reg}(R_{v,w})$ *(see Theorem* 4.4*), and* $\mathsf{Reg}(R_{v,w}) = \deg\, P_{v,w}$.

Our proof of the first part of Theorem 1.4 makes use of [**18**], which degenerates the tangent cone of the *Kazhdan-Lusztig ideal* $\mathcal{N}_{v,w}$ to the Gröbner limit [**14**] of the matrix Schubert variety $\overline{X}_{\kappa(v,w)}$ for a *different* covexillary permutation $\kappa(v,w)$. Thereby, $H_{v,w}(q)$ can be expressed in terms of *flagged Grothendieck polynomials* [**14, 16**]. We were inspired by the paper of J. Rajchgot–Y. Ren–C. Robichaux–A. St. Dizier–A. Weigandt [**23**], who determine the degree of a

[1] Some of these results are stated for *vexillary* rather than covexillary family; this is a matter of convention.

symmetric Grothendieck polynomial to find the regularity of $\overline{X}_w$ when w is *Grassmannian* (has at most one descent). Recent work of J. Rajchgot–C. Robichaux–A. Weigandt [**24**] extends that formula to vexillary permutations, which we apply (see also the generalization in O. Pechenik-D. Speyer-A. Weigandt's [**22**], and E. Hafner's [**9**]).

In Section 2, we recall the notion of Kazhdan-Lusztig ideals/varieties [**26**]. We also recapitulate necessary results about its tangent cone from [**17, 18**]. We summarize definitions and facts we need about regularity in Section 3. We then prove our main result in Section 4. Final remarks are collected in Section 5.

2. Kazhdan-Lusztig varieties

Let

$$\Omega_v^\circ = B_- vB/B$$

be the *opposite Schubert cell* where $B_- \subset GL_n$ consists of invertible lower triangular matrices. Ω_{id}° is the *opposite big cell*; it is an affine open neighborhood of $(id)B/B$. Hence $v\Omega_{id}^\circ \cap X_w$ is an affine open neighborhood of X_w centered at e_v. However, by [**12**, Lemma A.4],

$$X_w \cap v\Omega_{id}^\circ \cong (X_w \cap \Omega_v^\circ) \times \mathbb{A}^{\ell(w)}. \tag{2.1}$$

Hence it suffices to study the *Kazhdan-Lusztig variety*

$$\mathcal{N}_{v,w} := X_w \cap \Omega_v^\circ.$$

Explicit coordinates and equations for $\mathcal{N}_{v,w}$ were first studied in work with A. Woo [**26**]. Let $\mathsf{Mat}_{n\times n}$ be the set of all $n \times n$ complex matrices. The coordinate ring is $\mathbb{C}[\mathbf{z}]$ where $\mathbf{z} = \{z_{ij}\}_{i,j=1}^n$ are the functions on the entries of a generic matrix Z. Here z_{ij} corresponds to the entry in the i-th row from the *bottom*, and the j-th column to the right.

Let us realize Ω_v° as a affine subspace of $\mathsf{Mat}_{n\times n}$ consisting of matrices $Z^{(v)}$ where

$$z_{n-v(i)+1,i} = 1, z_{n-v(i)+1,s} = 0, z_{t,i} = 0, \text{for } s > i \text{ and } t > n - v(i) + 1.$$

Let $\mathbf{z}^{(v)} \subseteq \mathbf{z}$ be the unspecialized variables. Furthermore, let $Z_{st}^{(v)}$ be the southwest $s \times t$ submatrix of $Z^{(v)}$. The *rank matrix* is

$$r^w = (r_{ij}^w)_{i,j=1}^n$$

(which we index in the same manner), where

$$r_{ij}^w = \#\{h : w(h) \geq n - i + 1, h \leq j\}.$$

One combinatorial characterization of *Bruhat order* is that $v \leq w$ if and only if $r_{ij}^v \leq r_{ij}^w$ for all $1 \leq i, j \leq n$.

The *Kazhdan-Lusztig ideal* is $I_{v,w} \subset \mathbb{C}[z^{(v)}]$ generated by all $r_{st}^w + 1$ minors of $Z_{st}^{(v)}$ where $1 \leq s, t \leq n$. As explained in [**26**],

$$\mathcal{N}_{v,w} \cong \operatorname{Spec}\left(\mathbb{C}[z^{(v)}]/I_{v,w}\right);$$

this is reduced and irreducible.

EXAMPLE 2.1. Let $w = 7314562$, $v = 1423576$ (in one line notation). The rank matrix r^w and the matrix of variables $Z^{(v)}$ are, respectively,

$$r^w = \begin{pmatrix} 1 & 2 & 3 & 4 & 5 & 6 & 7 \\ 1 & 2 & 2 & 3 & 4 & 5 & 6 \\ 1 & 2 & 2 & 3 & 4 & 5 & 5 \\ 1 & 1 & 1 & 2 & 3 & 4 & 4 \\ 1 & 1 & 1 & 1 & 2 & 3 & 3 \\ 1 & 1 & 1 & 1 & 1 & 2 & 2 \\ 1 & 1 & 1 & 1 & 1 & 1 & 1 \end{pmatrix}, \quad Z^{(v)} = \begin{pmatrix} 1 & 0 & 0 & 0 & 0 & 0 & 0 \\ z_{61} & 0 & 1 & 0 & 0 & 0 & 0 \\ z_{51} & 0 & z_{53} & 1 & 0 & 0 & 0 \\ z_{41} & 1 & 0 & 0 & 0 & 0 & 0 \\ z_{31} & z_{32} & z_{33} & z_{34} & 1 & 0 & 0 \\ z_{21} & z_{22} & z_{23} & z_{24} & z_{25} & 0 & 1 \\ z_{11} & z_{12} & z_{13} & z_{14} & z_{15} & 1 & 0 \end{pmatrix}$$

The Kazhdan-Lusztig ideal $I_{1423576,7314562}$ contains among its generators, all 2×2 minors of $Z_{25}^{(v)}$ but also inhomogeneous elements such as

$$\begin{vmatrix} z_{51} & 0 & z_{53} \\ z_{41} & 1 & 0 \\ z_{31} & z_{32} & z_{33} \end{vmatrix} = z_{51}z_{33} + z_{53}z_{41}z_{32} - z_{53}z_{31}. \tag{2.2}$$

This generator, *per se*, does not imply $I_{1423576,7314562}$ is inhomogeneous; however one can confirm the ideal is in fact inhomogeneous with respect to the standard grading using Macaulay2's function `isHomogeneous`. These ideals (and their statistics) can be computed using `https://faculty.math.illinois.edu/~ayong/Schubsingular.v0.2.m2`. □

We also need the *Schubert determinantal ideal* I_w which is defined similarly as $I_{v,w}$ except that we repace $Z^{(v)}$ with the matrix $Z = (z_{ij})$. The zero-set is the *matrix Schubert variety.*

Given $f \in \mathbb{C}[z^{(v)}]$, let $\mathsf{LD}(f)$ denote the lowest degree homogeneous component of f. Now, define the *(Kazhdan-Lusztig) tangent cone ideal* to be

$$I'_{v,w} = \langle \mathsf{LD}(f) : f \in I_{v,w} \rangle.$$

E.g., if f is the polynomial in (2.2) then $\mathsf{LD}(f) = z_{51}z_{33} - z_{53}z_{31}$. The *tangent cone* of $\mathcal{N}_{v,w}$ is

$$\mathcal{N}'_{v,w} := \operatorname{Spec}\left(\mathbb{C}[z^{(v)}]/I'_{v,w}\right).$$

This can be computed using Macaulay2's `tangentCone` function.

3. Castelnuovo-Mumford regularity basics

The *Castelnuovo-Mumford regularity* of a finitely generated graded module

$$M = \bigoplus_{j \in \mathbb{Z}} M^{(j)}$$

over a standard $\mathbb{N}$-graded ring $S = \bigoplus_{j \geq 0} S^{(j)}$ is defined by

$$\operatorname{Reg}(M) = \max\{f_j(M) + j : j \geq 0\}$$

where

$$f_j(M) := \begin{cases} \sup\{n : H^j_{S_+}(M)_n \neq 0\} & \text{if } H^j_{S_+}(M) \neq 0, \\ -\infty & \text{otherwise.} \end{cases}$$

Here $S_+ = \bigoplus_{j>0} S^{(j)}$ is the irrelevant ideal of S and $H^i_{S_+}(M)$ is the i-th local cohomology module of M with respect to S_+ (and its endowed grading). We refer the reader to the book [**4**, Chapter 15] for further details. One has an expression for the Poincaré series

$$\mathsf{PS}_M(q) = \frac{\mathcal{K}_M(q)}{(1-q)^{\dim(M)}}, \tag{3.1}$$

where $\mathcal{K}_M(q) \in \mathbb{Z}[q]$; see, *e.g.*, [**5**, Corollary 4.1.8]. Let $h_M(q)$ be *Hilbert function* and $p_M(q)$ be the *Hilbert polynomial*. Hilbert's theorem states that

$$h_M(q) = p_M(q)$$

for all sufficiently large q. The *postulation number* is

$$\text{post}(M) = \max\{n : h_M(n) \neq p_M(n)\}.$$

By [**5**, Proposition 4.1.12],

$$\text{post}(M) = \deg \mathcal{K}_M(q) - \dim M.$$

It is known (and not hard) that when M is Cohen-Macaulay, $\text{Reg}(M) = \text{post}(M) + \dim M$. Hence

$$\text{Reg}(M) = \deg \mathcal{K}_M(q). \tag{3.2}$$

Now suppose $S = \mathbb{C}[x_1, \dots, x_N]$ and $M = S/J$ is the S module where $J \subseteq S$ is an ideal that is standard graded homogeneous. $M = S/J$ has a minimal free resolution

$$0 \to \bigoplus_j S(-j)^{\beta_{i,j}(S/J)} \to \bigoplus_j S(-j)^{\beta_{i-1,j}(S/J)} \to \cdots \to \bigoplus_j S(-j)^{\beta_{0,j}(S/J)} \to S/J \to 0.$$

Here $i \leq N$ and $S(-j)$ is the free S-module where degrees of S are shifted by j. Also,

$$\text{Reg}(M) := \max\{j - i : \beta_{i,j}(M) \neq 0\},$$

and

$$\mathsf{PS}_{S/J}(q) = \frac{\mathcal{K}_{S/J}(q)}{(1-q)^N},$$

where $\mathcal{K}(S/J, q) \in \mathbb{Z}[q]$. If S/J is Cohen-Macaulay, (3.2) says

$$\text{Reg}(S/J) = \deg \mathcal{K}(S/J, q) - \text{ht}_S(J), \tag{3.3}$$

where $\text{ht}_S(J)$ is the *height* of the ideal J in S. In our application, the algebraic set $V(J)$ is radical and equidimensional; $\text{ht}_S(J)$ is the codimension of the variety $V(J) \subseteq \mathbb{C}^N$.

EXAMPLE 3.1. Continuing Example 2.1, using Macaulay2's `resolution` and `betti` one computes that the Betti numbers for the minimal free resolution of $T_{1423576,7314562}$ as

	0	1	2	3	4	5	6	7	8	9	10
total:	1	12	61	176	322	392	322	176	61	12	1
0:	1	7	21	35	35	21	7	1	.	.	.
1:	.	5	40	140	280	350	280	140	40	5	.
2:	.	.	.	1	7	21	35	35	21	7	1

In Macaulay2 format, the entry in row j and column i is $\beta_{i,i+j}$. So

$$\text{Reg}(\mathbb{C}^{(v)}/T_{1423576,7314562}) = 2$$

is the largest row index of this table. Similarly one checks that

$$\mathrm{Reg}(\mathbb{C}^{(v)}/T_{1234567,7314562}) = 3,$$

in agreement with Conjecture 1.2. □

4. Proof of Theorem 1.4

4.1. Proof of Conjectures 1.1, 1.2, 1.3 in the covexillary case. Let

$$R'_{v,w} := \mathbb{C}[\mathbf{z}^{(v)}]/I'_{v,w}.$$

We claim

$$\mathrm{Reg}(R'_{v,w}) = \deg H_{v,w}. \tag{4.1}$$

By [**17**, Theorems 3.1 and 5.5], $\mathrm{Spec}\, R'_{v,w}$ Gröbner degenerates to $\mathrm{init}_{\prec} \overline{X}_{\kappa(v,w)}$ (up to a permutation of coordinates), the Gröbner limit in [**14**] of a matrix Schubert variety $\overline{X}_{\kappa(v,w)}$ of the covexillary permutation $\kappa(v,w)$. We will define $\kappa(v,w)$ in Section 4.2. At this moment, it suffices to know that $\mathrm{init}_{\prec} \overline{X}_{\kappa(v,w)}$ is a reduced union of coordinate subspaces, whose associated Stanley-Reisner simplicial complex is homeomorphic to a shellable ball or sphere [**14**, Theorem 4.4]. Shellable simplicial complexes are Cohen-Macaulay, which by definition, means the said union of coordinate subspaces is Cohen-Macaulay [**20**, Section 13.5.3]. Therefore $\overline{X}_{\kappa(v,w)}$ is Cohen-Macaulay, and hence $\mathrm{Spec}\, R'_{v,w}$ is also Cohen-Macaulay as it also Gröbner degenerates to it [**6**, Section 15.8].

In [**17**], one has

$$\mathcal{K}(R'_{v,w}, q) = \frac{H_{v,w}(q)(1-q)^{\ell(w_0 w)}}{(1-q)^{\ell(w_0 v)}}.$$

Thus by (3.3),

$$\mathrm{Reg}(R'_{v,w}) = \deg\, H_{w,v}(q) + \ell(w_0 w) - \ell(w_0 w),$$

since

$$\mathrm{ht}_{\mathbb{C}[\mathbf{z}^{(v)}]} I'_{v,w} = \ell(w_0 w)$$

(here we use the fact that the tangent cone of $\mathcal{N}_{v,w}$ has the same dimension as $\mathcal{N}_{v,w}$ itself, namely $\ell(w) - \ell(v)$, and that). Thus (4.1) holds.

Since the tangent cone of $\mathcal{N}_{v,w}$ is $\mathrm{Spec}\, R'_{v,w}$ it follows from (2.1) that

$$\text{tangent cone } (v\Omega^{\circ}_{id} \cap X_w) \cong \mathrm{Spec}\, R'_{v,w} \times \mathbb{A}^{\ell(v)}.$$

The tangent cone of any affine open neighborhood of $p \in Y$ is isomorphic to $R_{p,Y}$; see, *e.g.*, [**6**, Section 5.4] and [**21**, III.3]. Hence the Cohen-Macaulayness of $R'_{v,w}$ implies the same of $R_{v,w}$, since this property of an affine variety is preserved under cartesian product with affine space. Hence Conjecture 1.1 holds in this case by (3.1).

Conjecture 1.2 holds in our case since it is shown in [**18**] that $H_{v,w}(q)$ is semicontinuous. Also, in the covexillary case, one has from *ibid.* that

$$\deg H_{v,w}(q) = \deg P_{v,w}(q)$$

where $P_{v,w}(q)$ is the *Kazhdan-Lusztig polynomial.* By definition,

$$\deg P_{v,w}(q) \leq \frac{\ell(w) - \ell(v) - 1}{2};$$

this is Conjecture 1.3.

4.2. Permutation combinatorics and the formula. We recall some standard permutation combinatorics; our reference is [**19**] (although our conventions are upside down from theirs). The *graph* of $w \in \mathfrak{S}_n$ places a $\bullet$ in position $(w(i), i)$ (written in matrix notation). Cross out all boxes weakly right and weakly above a $\bullet$; the remaining boxes of $[n] \times [n]$ form the *Rothe diagram* of w, denoted $D(w)$. That is,

$$D(w) = \{(i,j) \in [n] \times [n] : i > w(j), j < w^{-1}(i)\}.$$

The vector $\mathsf{code}(w) = (c_n, c_{n-1}, \ldots, c_1)$ where c_i is the number boxes of $D(w)$ in row i. The *essential set* $E(w)$ of w consists of those maximally northeast boxes of any connected component of $D(w)$, *i.e.*,

$$E(w) = \{(i,j) \in D(w) : (i-1,j), (i,j+1) \notin D(w)\}.$$

EXAMPLE 4.1. Continuing our running example, where $w = 7314562$, diagram is graphically depicted in Figure 1. Hence

$$D(w) = \{(2,3), (4,2), (4,3), (5,2), (5,3), (5,4), (6,2), (6,3), (6,4)\}$$

and

$$E(w) = \{\mathfrak{e}_1 = (6,5), \mathfrak{e}_2 = (5,4), \mathfrak{e}_3 = (4,2), \mathfrak{e}_4 = (2,3)\}.$$

Moreover, $\mathsf{code}(w) = (0,4,3,2,0,1,0)$.

$D(w) =$ 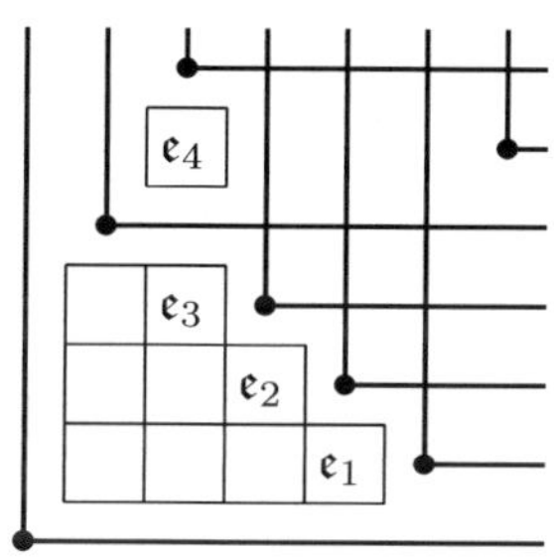

FIGURE 1. The diagram and essential set for $w = 7314562$.

A permutation in $\mathfrak{S}_n$ is uniquely identified by the values of the rank matrix (r_{ij}^w) when restricted to $D(w)$ or even merely $E(w)$.

Throughout the remainder of this subsection, we assume w is covexillary.

Let $\lambda(w)$ be the partition obtained by sorting $\mathsf{code}(w)$. It is useful to know the *graphical construction* of $\lambda(w)$: Since $(a,b), (c,d) \in E(w)$ then one is weakly northwest of the other [**19**], it follows there is a unique Young diagram (in French notation) obtained by pushing all boxes of $D(w)$ on a given antidiagonal to the southwest; that is the diagram of $\lambda(w)$.

EXAMPLE 4.2. Our running example $w = 7314562$ is covexillary with $\lambda(w) = (4,3,2,1)$. $\square$

Given $v \leq w$, [**17**] defines (and proves the existence of) a different covexillary permutation $\kappa(v,w)$. This is the unique permutation whose essential set is obtained by moving each $\mathfrak{e} = (i,j) \in E(w)$ southwest along its antidiagonal by r_{ij}^v squares to $\mathfrak{e}'$ and imposing that

$$r_{\mathfrak{e}'}^{\kappa(v,w)} = r_{ij}^w - r_{ij}^v.$$

By construction,

$$\lambda(w) = \lambda(\kappa(v, w)).$$

The graphical construction $\lambda(\kappa(v, w))$ induces a bijection of boxes:

$$\phi : \lambda(\kappa(v, w)) \to D(\kappa(v, w)).$$

Define a filling of each box $b \in \lambda(\kappa(v, w))$ with $r^w_{\phi(b)}$. We call this $\mathsf{RRW}(v, w)$, as its provenance is from [**24**].

EXAMPLE 4.3. One can check that $\kappa(1423576, 7314562) = 3472561$. □

The next result is the combinatorial rule of Theorem 1.4. It uses a similar result of J. Rajchgot-C. Robichaux-A. Weigandt [**24**, Theorem 1.3]:

THEOREM 4.4.

(4.2)
$$\mathrm{Reg}(R_{v,w}) = \mathrm{Reg}(R'_{v,w}) = \deg H_{v,w} = \sum_{k \geq 1} \sum_{\alpha \in \mathsf{Connected}(\lambda(\kappa(v,w))_{\geq k})} \mathsf{maxdiag}(\alpha),$$

where:

- $\lambda(\kappa(v, w))_{\geq k}$ *is the shape of the subtableau of* $\mathsf{RRW}(v, w)$ *that have entries* $\geq k$*;*
- $\mathsf{Connected}(\kappa(v, w))_{\geq k})$ *are the connected components of the aforementioned shape; and*
- $\mathsf{maxdiag}(\alpha)$ *is the largest northwest-southeast diagonal that appears in* α.

EXAMPLE 4.5. To complete our running example,

$$\mathsf{RRW}(1423576, 7314562) = \begin{array}{|c|c|c|c|} \hline 0 \\ \cline{1-1} 0 & 0 \\ \cline{1-2} 0 & 0 & 1 \\ \cline{1-3} 0 & 0 & 1 & 1 \\ \hline \end{array}$$

and hence Theorem 4.4 asserts Reg $= 2$ (the longest diagonal appearing in the unique 1's component), in agreement with Example 3.1. □

For any $u \in \mathfrak{S}_n$ let $\mathfrak{G}_w(x_1, \ldots, x_n)$ be the *Grothendieck polynomial* [**16**]. By definition, $\mathfrak{G}_{w_0} = x_1^{n-1} x_2^{n-2} \cdots x_{n-1}$ where w_0 is the longest element in $\mathfrak{S}_n$. If $\ell(us_i) > \ell(u)$ where $s_i = (i\ i+1)$ is a simple transposition, then $\mathfrak{G}_u = \pi_i(\mathfrak{G}_{us_i})$ where

$$\pi_i : \mathbb{Z}[x_1, x_2, \ldots, x_n] \to \mathbb{Z}[x_1, x_2, \ldots, x_n]$$

is the *isobaric divided difference operator* defined by

$$\pi(f) = \frac{(1 - x_{i+1})f(\cdots, x_i, x_{i+1}, \cdots) - (1 - x_i)f(\cdots, x_{i+1}, x_i, \cdots)}{x_i - x_{i+1}}.$$

4.3. Proof of Theorem 4.4. By [**17**, Theorem 6.6],

$$\mathsf{PS}_{v,w}(q) = \frac{G_\lambda(q)}{(1 - q)^{\binom{n}{2}}}, \tag{4.3}$$

where

$$G_\lambda(q) = \mathfrak{G}_{w_0 \kappa(v,w)}(1 - q, 1 - q, \ldots, 1 - q).$$

Comparing (4.3) with (1.1) and using the fact that $\dim(X_w) = \ell(w)$, we see that

$$\deg H_{v,w} = \deg \mathfrak{G}_{w_0 \kappa(v,w)} - \left(\binom{n}{2} - \ell(w)\right). \tag{4.4}$$

On the other hand, since $\lambda(\kappa(v,w)) = \lambda(w)$, one has $\ell(\kappa(v,w)) = \ell(w)$, and hence

$$\ell(w_0\kappa(v,w)) = \binom{n}{2} - \ell(w). \tag{4.5}$$

Moreover since $\kappa(v,w)$ is covexillary, $w_0\kappa(v,w)$ is vexillary (avoids 2143). The formula of J. Rajchgot-C. Robichaux-A. Weigandt [**24**, Theorem 1.3] shows (in our conventions) that for any vexillary $u \in \mathfrak{S}_n$ that

$$\deg \mathfrak{G}_u = \ell(u) + \sum_{k\geq 1} \sum_{\alpha \in \mathsf{Connected}(\lambda(w_0 u)_{\geq k})} \mathsf{maxdiag}(\alpha). \tag{4.6}$$

Hence the theorem follows by combining (4.4), (4.5) and (4.6) with $u = w_0\kappa(v,w)$. □

In general, there are no simple formulas to compute the degree of a Kazhdan-Lusztig polynomial $P_{v,w}(q)$ (we refer the reader to [**2**, Chapter 5]). This proves the final assertion of Theorem 1.4.

COROLLARY 4.6. *Let $w \in \mathfrak{S}_n$ be covexillary, then* $\deg P_{u,v}$ *is computed by the rule of Theorem* 4.4.

PROOF. [**18**, Theorem 1.2] shows $\deg H_{v,w}(q) = \deg P_{v,w}(q)$ when w is covexillary. Now apply Theorem 4.4. □

5. Further results and discussion

These conjectures were asserted in [**18**]:

CONJECTURE 5.1. *$R_{v,w}$ is Cohen-Macaulay. Consequently, $H_{v,w} \in \mathbb{N}[q]$.*

That X_w is Cohen-Macaulay does not imply Conjecture 5.1. In fact, C. Huneke [**11**] established $R_{p,Y}$ being Cohen-Macaulay implies the same for $(\mathcal{O}_{p,Y}, \mathfrak{m}_p, \Bbbk)$, and gave counterexamples for the converse. This is a strengthening of Conjecture 5.1:

CONJECTURE 5.2 (Semicontinuity). *If $u \leq v \leq w$ then $[q^t]H_{u,w} \geq [q^t]H_{v,w}$.*

CONJECTURE 5.3 ([**18**, Proposition 2.1]). $\deg H_{v,w} \leq \frac{\ell(w)-\ell(v)-1}{2}$.

PROPOSITION 5.4. *Conjectures* 5.1, 5.2, *and* 5.3 *imply Conjectures* 1.1, 1.2, *and* 1.3.

PROOF. The Cohen-Macaulay assertion of Conjecture 5.1 implies Conjecture 1.1 by the reasoning in our proof of Theorem 1.4. Combined with Conjecture 5.2 gives Conjecture 1.2. Separately, combined with Conjecture 5.3 one would obtain Conjecture 1.3. □

During the preparation of [**18**], Conjectures 5.1 and 5.3 were checked for $n \leq 7$. Conjecture 5.2 was checked for at least $n \leq 6$ and much of $n = 7$.

Let

$$\mathrm{maxReg}(n) = \max_{v \leq w \in \mathfrak{S}_n} \mathrm{Reg}(R_{v,w}).$$

CONJECTURE 5.5. $\mathrm{maxReg}(n) = \Theta(n^2)$.

Computational data was not directly useful to arrive at Conjecture 5.5. For $n = 4, 5, 6, 7$, $\mathrm{maxReg}(n) = 1, 2, 3, 5$, respectively. For example, when $n = 7$ the maximizer is the (non-covexillary) $w = 6734512$ at $v = id$. Here $I_{v,w}$ is inhomogeneous and

$$H_{id,6734512}(q) = 1 + 4q + 9q^2 + 9q^3 + 4q^4 + q^5.$$

Let

$$\overline{\text{maxReg}(n)} = \max_{v \leq w \in \mathfrak{S}_n, w \text{ covexillary}} \text{Reg}(R_{v,w}).$$

We apply Theorem 4.4 to prove the covexillary case of Conjecture 5.5.

PROPOSITION 5.6. $\overline{\text{maxReg}(n)} = \Theta(n^2)$.

PROOF. For the lower bound, first suppose $n = 3j - 1$ for $j \geq 1$. Let $v = id$ and $w \in \mathfrak{S}_n$ be the unique permutation with

$$\mathsf{code}(w) = (1, 2, 3, \ldots, j, 0, 0, \ldots, 0).$$

Then w is covexillary, with

$$\lambda(w) = (j, j-1, \ldots, 3, 2, 1).$$

For example, if $j = 4$ then $w = 7, 11, 6, 10, 5, 9, 4, 8, 3, 2, 1$. By our assumption, $\kappa(id, w) = w$. Hence $\mathsf{RRW}(\kappa(id, w))$ is the staircase $\lambda(w)$ where column c from the left is filled by $(c-1)$'s. In our example,

$$\mathsf{RRW}(\kappa(id, w)) = \begin{array}{|c|c|c|c|} \hline 0 \\ \cline{1-1} 0 & 1 \\ \cline{1-2} 0 & 1 & 2 \\ \cline{1-3} 0 & 1 & 2 & 3 \\ \hline \end{array} \; .$$

Hence, Theorem 4.4 asserts that

$$\text{Reg}(R_{id,w}) = (j-1) + (j-2) + \ldots + 2 + 1 = \binom{j}{2}.$$

Now, if $n = 3j$ or $n = 3j + 1$, use the same construction as for $n = 3j - 1$, except that $\mathsf{code}(w)$ will have an additional 0 or 0, 0 postpended, respectively. In those two cases, the same analysis implies

$$\text{Reg}(R_{id,w}) = \binom{j}{2}.$$

Hence $\overline{\text{maxReg}(n)} = \Omega(n^2)$ follows.

For the upper bound, since $w \in \mathfrak{S}_n$, $\lambda(\kappa(v, w)) \subseteq n \times n$ and $\mathsf{RRW}(\kappa(v, w))$ only uses labels $k \in [n]$. For each such k, the inner sum of (4.2) contributes $\leq n$. Hence $\text{Reg}(R_{v,w}) \leq n^2$. Therefore, $\overline{\text{maxReg}(n)} = O(n^2)$, as required. □

COROLLARY 5.7. *Conjecture* 1.3 *implies Conjecture* 5.5.

PROOF. The lower bound of Conjecture 5.5 is immediate from Proposition 5.6. If Conjecture 1.3 holds, then

$$\text{Reg}(R_{v,w}) \leq \frac{\ell(w) - \ell(v) - 1}{2} \leq \ell(w_0) = \binom{n}{2}. \qquad \square$$

Sometimes $I_{v,w}$ is homogeneous with respect to the standard grading; see [**27**] and the references therein. In those cases, trivially, $I'_{v,w} = I_{v,w}$ and Cohen-Macaulayness of $I_{v,w}$ and Conjecture 1.1 is automatic. As argued in [**18**], the covexillary case is interesting precisely because $I'_{v,w} = I_{v,w}$ need not hold in general (as in the case of our running example).

It is also natural to expect that our regularity conjectures are true for other Lie types. We remark that in the minuscule case studied by [**8**], it is again true that the Schubert varieties admit a dilation action of $\mathbb{C}^*$ and hence the analogue of Conjecture 1.1 holds for a similar reason as in the previous paragraph. This problem should be in reach:

PROBLEM 5.8. Determine the regularity of tangent cones of Schubert varieties for minuscule G/P.

We also mention that the *banner permutations* of Z. Hamaker-O. Pechenik-A. Weigandt [**10**] extend the vexillary permutations and have a description of the Gröbner basis (also, see a further extension by P. Klein [**13**]). It would therefore be interesting to see if the results of this paper (or of [**17, 18**]) extend to that setting.

With regards to Theorem 4.4, one can use any rule that computes $\deg(\mathfrak{G}_u)$. Another rule applicable to arbitrary $u \in \mathfrak{S}_n$ has been found by O. Pechenik-D. Speyer-A. Weigandt [**22**]. On the one hand, the tableau rule of [**24**] is fitting with the covexillary combinatorics we use. On the other hand, one wonders if that general rule can be adapted to compute $\mathsf{Reg}(R_{u,v})$? We also remark that both of these formulas can be regarded as solving a special case of our regularity problem; see [**27**, Corollary 2.6] and its proof.

Acknowledgments

We thank Daniel Erman, Shiliang Gao, Cao Huy Linh, Jenna Rajchgot, Hal Schenck, Colleen Robichaux, Anna Weigandt, and Alexander Woo for helpful communications. We are grateful to Allen Knutson, Li Li, Ezra Miller, and Alexander Woo for joint work on which this piece is based. We also thank the organizers of "Singularities of the Midwest (online edition)" and "Recent Developments in Gröbner Geometry"; these notes are based on the talks given at those Summer '21 events. We made use of Macaulay2 in our investigations.

References

[1] M. F. Atiyah and I. G. Macdonald, *Introduction to commutative algebra*, Addison-Wesley Publishing Co., Reading, Mass.-London-Don Mills, Ont., 1969. MR242802

[2] Anders Björner and Francesco Brenti, *Combinatorics of Coxeter groups*, Graduate Texts in Mathematics, vol. 231, Springer, New York, 2005. MR2133266

[3] Markus Brodmann and Cao Huy Linh, *Castelnuovo-Mumford regularity, postulation numbers and relation types*, J. Algebra **419** (2014), 124–140, DOI 10.1016/j.jalgebra.2014.07.014. MR3253282

[4] M. P. Brodmann and R. Y. Sharp, *Local cohomology: An algebraic introduction with geometric applications*, 2nd ed., Cambridge Studies in Advanced Mathematics, vol. 136, Cambridge University Press, Cambridge, 2013. MR3014449

[5] Winfried Bruns and Jürgen Herzog, *Cohen-Macaulay rings*, Cambridge Studies in Advanced Mathematics, vol. 39, Cambridge University Press, Cambridge, 1993. MR1251956

[6] David Eisenbud, *Commutative algebra: With a view toward algebraic geometry*, Graduate Texts in Mathematics, vol. 150, Springer-Verlag, New York, 1995, DOI 10.1007/978-1-4612-5350-1. MR1322960

[7] William Fulton, *Young tableaux: With applications to representation theory and geometry*, London Mathematical Society Student Texts, vol. 35, Cambridge University Press, Cambridge, 1997. MR1464693

[8] William Graham and Victor Kreiman, *Excited Young diagrams, equivariant K-theory, and Schubert varieties*, Trans. Amer. Math. Soc. **367** (2015), no. 9, 6597–6645, DOI 10.1090/S0002-9947-2015-06288-6. MR3356949

[9] E. Hafner, *Vexillary Grothendieck polynomials via bumpless pipe dreams*, Preprint, arXiv:2201.12432, 2022.

[10] Zachary Hamaker, Oliver Pechenik, and Anna Weigandt, *Gröbner geometry of Schubert polynomials through ice*, Adv. Math. **398** (2022), Paper No. 108228, 29, DOI 10.1016/j.aim.2022.108228. MR4379747

[11] Craig Huneke, *On the associated graded ring of an ideal*, Illinois J. Math. **26** (1982), no. 1, 121–137. MR638557

[12] David Kazhdan and George Lusztig, *Representations of Coxeter groups and Hecke algebras*, Invent. Math. **53** (1979), no. 2, 165–184, DOI 10.1007/BF01390031. MR560412

[13] Patricia Klein, *Diagonal degenerations of matrix Schubert varieties*, Algebr. Comb. **6** (2023), no. 4, 1073–1094. MR4635090

[14] Allen Knutson, Ezra Miller, and Alexander Yong, *Gröbner geometry of vertex decompositions and of flagged tableaux*, J. Reine Angew. Math. **630** (2009), 1–31, DOI 10.1515/CRELLE.2009.033. MR2526784

[15] Alain Lascoux, *Polynômes de Kazhdan-Lusztig pour les variétés de Schubert vexillaires* (French, with English and French summaries), C. R. Acad. Sci. Paris Sér. I Math. **321** (1995), no. 6, 667–670. MR1354702

[16] Alain Lascoux and Marcel-Paul Schützenberger, *Structure de Hopf de l'anneau de cohomologie et de l'anneau de Grothendieck d'une variété de drapeaux* (French, with English summary), C. R. Acad. Sci. Paris Sér. I Math. **295** (1982), no. 11, 629–633. MR686357

[17] Li Li and Alexander Yong, *Kazhdan-Lusztig polynomials and drift configurations*, Algebra Number Theory **5** (2011), no. 5, 595–626, DOI 10.2140/ant.2011.5.595. MR2889748

[18] Li Li and Alexander Yong, *Some degenerations of Kazhdan-Lusztig ideals and multiplicities of Schubert varieties*, Adv. Math. **229** (2012), no. 1, 633–667, DOI 10.1016/j.aim.2011.09.010. MR2854186

[19] Laurent Manivel, *Symmetric functions, Schubert polynomials and degeneracy loci*, SMF/AMS Texts and Monographs, vol. 6, American Mathematical Society, Providence, RI; Société Mathématique de France, Paris, 2001. Translated from the 1998 French original by John R. Swallow; Cours Spécialisés [Specialized Courses], 3. MR1852463

[20] Ezra Miller and Bernd Sturmfels, *Combinatorial commutative algebra*, Graduate Texts in Mathematics, vol. 227, Springer-Verlag, New York, 2005. MR2110098

[21] David Mumford, *The red book of varieties and schemes*, Second, expanded edition, Lecture Notes in Mathematics, vol. 1358, Springer-Verlag, Berlin, 1999. Includes the Michigan lectures (1974) on curves and their Jacobians; With contributions by Enrico Arbarello, DOI 10.1007/b62130. MR1748380

[22] O. Pechenik, D. Speyer, and A. Weigandt, *Castelnuovo-Mumford regularity of matrix Schubert varieties*, arXiv:2111.10681, 2021.

[23] Jenna Rajchgot, Yi Ren, Colleen Robichaux, Avery St. Dizier, and Anna Weigandt, *Degrees of symmetric Grothendieck polynomials and Castelnuovo-Mumford regularity*, Proc. Amer. Math. Soc. **149** (2021), no. 4, 1405–1416, DOI 10.1090/proc/15294. MR4242300

[24] Jenna Rajchgot, Colleen Robichaux, and Anna Weigandt, *Castelnuovo-Mumford regularity of ladder determinantal varieties and patches of Grassmannian Schubert varieties*, J. Algebra **617** (2023), 160–191, DOI 10.1016/j.jalgebra.2022.11.001. MR4513784

[25] Ngô Viêt Trung, *The Castelnuovo regularity of the Rees algebra and the associated graded ring*, Trans. Amer. Math. Soc. **350** (1998), no. 7, 2813–2832, DOI 10.1090/S0002-9947-98-02198-9. MR1473456

[26] Alexander Woo and Alexander Yong, *Governing singularities of Schubert varieties*, J. Algebra **320** (2008), no. 2, 495–520, DOI 10.1016/j.jalgebra.2007.12.016. MR2422304

[27] Alexander Woo and Alexander Yong, *A Gröbner basis for Kazhdan-Lusztig ideals*, Amer. J. Math. **134** (2012), no. 4, 1089–1137, DOI 10.1353/ajm.2012.0031. MR2956258

Department of Mathematics, University of Illinois at Urbana-Champaign, Urbana, Illinois 61801, USA

Email address: ayong@illinois.edu

Proceedings of Symposia in Pure Mathematics
Volume **110**, 2024
https://doi.org/10.1090/pspum/110/02009

Perfect matching problems in cluster algebras and number theory

Ralf Schiffler

Abstract. This paper is a slightly extended version of the talk I gave at the Open Problems in Algebraic Combinatorics conference at the University of Minnesota in May 2022. We introduce two strict order relations on lattice paths and formulate several open problems. The topic is related to Markov numbers, the Lagrange spectrum, snake graphs and the cluster algebra of the once punctured torus.

Our lattice paths are required to proceed by North and East steps and never go over the diagonal. To define the order relations, we first construct a snake graph $\mathcal{G}(\omega)$ and a band graph $\overline{\mathcal{G}(\omega)}$ for every such lattice path ω. The first order relation $<_M$ is given by the number of perfect matchings of the snake graphs. The second order relation $<_L$ is given by the Lagrange number of a quadratic irrational associated to the band graph.

Contents

1. Introduction

We introduce two strict order relations on lattice paths ω that proceed by North and East steps and never go over the diagonal. To define the order relations, we first construct a snake graph $\mathcal{G}(\omega)$ and a band graph $\overline{\mathcal{G}(\omega)}$ for every lattice path ω.

Snake graphs and band graphs arise in the theory of cluster algebras from surfaces, where they are used to compute the Laurent expansion of the cluster variables and to construct canonical bases of the cluster algebra. A survey of these results is given in [**S**].

2020 *Mathematics Subject Classification.* Primary 06A07, 11J06, 05C70, 13F60.
The author was supported by the NSF grant DMS-2054561.

The lattice path snake graphs in this paper form a special class of snake graphs. Since the lattice can be thought of the universal cover of the torus with one puncture, (a slight deformation of) the lattice path corresponds to an element of the cluster algebra of the once-punctured torus. In particular, the Christoffel paths correspond to the cluster variables.

The cluster variables are related to the Markov numbers. In our setting this relation can be expressed by saying that the number of perfect matchings of the snake graph of a Christoffel path is a Markov number and each Markov number arises this way.

A classical theorem of Markov gives a construction of the Lagrange spectrum below 3 in terms of the Markov numbers. Our order relation $<_M$ is related to Markov numbers and the relation $<_L$ is related to the Lagrange spectrum.

We review snake graphs and band graphs in section 2 and Markov numbers and the Lagrange spectrum in section 3. We define the two order relations in section 4. In section 5, we list known results about the order relations and section 6 contains the open problems.

I would like to thank the organizers of the OPAC conference for their invitation and encouragement to write up these notes.

2. Snake graphs and band graphs

Snake graphs and band graphs were used in [**MS, MSW, MSW2**] to construct canonical bases for cluster algebras from surfaces. The Laurent expansions of the basis elements are parametrized by the perfect matchings of these graphs. In [**MOZ**], some of these results were extended to the super cluster algebra setting by replacing the perfect matchings of snake graphs by double dimer covers.

2.1. Snake graphs. We now recall the definition of snake graphs following [**CS1**]. A *tile* is a planar graph with 4 vertices and 4 edges that has the shape of a square. All tiles will have the same side length. A snake graph $\mathcal{G}$ is a connected planar graph consisting of a finite sequence of tiles $G_1, \dots, G_d$ that is constructed recursively by the rule that the $i+1$-st tile G_{i+1} shares exactly one edge e_i with the previous tile G_i, and this edge is either the north edge of G_i and the south edge of G_{i+1} or the east edge of G_i and the west edge of G_{i+1}. An example is given in Figure 1.

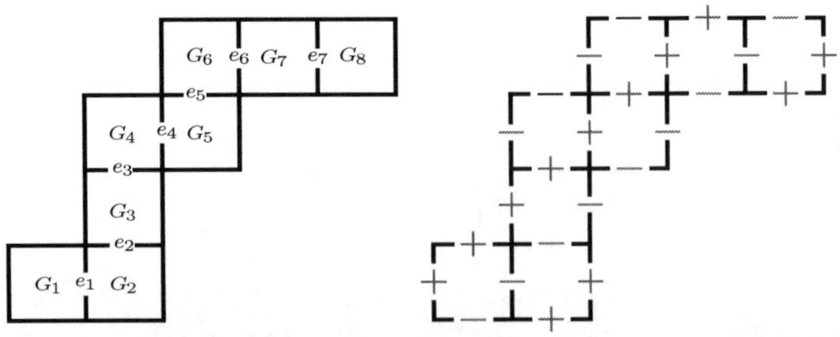

FIGURE 1. A snake graph with 8 tiles and 7 interior edges (left); a sign function on the same snake graph (right)

The $d-1$ edges $e_1, e_2, \ldots, e_{d-1}$ which are contained in two tiles are called *interior edges* of $\mathcal{G}$ and the other edges are called *boundary edges.* We denote by $\mathcal{G}^{NE}$ the 2 element set containing the north edge and the east edge of the last tile of $\mathcal{G}$.

A *sign function* f on a snake graph $\mathcal{G}$ is a map f from the set of edges of $\mathcal{G}$ to the set $\{+,-\}$ such that on every tile in $\mathcal{G}$ the north and the west edge have the same sign, the south and the east edge have the same sign and the sign on the north edge is opposite to the sign on the south edge. See Figure 1 for an example.

Note that on every snake graph there are exactly two sign functions. A snake graph is determined up to symmetry by its sequence of tiles together with a sign function on its interior edges.

2.2. Band graphs. Every snake graph $\mathcal{G}$ gives rise to two *band graphs* $\overline{\mathcal{G}}$ defined by identifying either

- the south edge of the first tile in $\mathcal{G}$ with the unique edge in $\mathcal{G}^{NE}$ that has the same sign;
- or the west edge of the first tile in $\mathcal{G}$ with the unique edge in $\mathcal{G}^{NE}$ that has the same sign.

2.3. Perfect matchings. A *perfect matching* of a graph $\mathcal{G}$ is a subset P of the edges of $\mathcal{G}$ such that each vertex of $\mathcal{G}$ is incident to exactly one edge of P. We denote by $\operatorname{Match}\mathcal{G}$ the set of perfect matchings of $\mathcal{G}$. If $\overline{\mathcal{G}}$ is a band graph, we define $\operatorname{Match}\overline{\mathcal{G}}$ to be the set of all perfect matchings P of the snake graph $\mathcal{G}$ such that P is a perfect matching of $\overline{\mathcal{G}}$.

2.4. Relation to continued fractions. A relation between snake graphs and continued fractions was discovered in [**CS1**].

A *finite continued fraction* is a function

$$[a_1, a_2, \ldots, a_n] = a_1 + \cfrac{1}{a_2 + \cfrac{1}{\ddots + \cfrac{1}{a_n}}}$$

of n positive integers $a_1, \ldots, a_n$. Notice that $[a_1, \ldots, a_{n-1}, 1] = [a_1, \ldots, a_{n-1} + 1]$. Similarly, an *infinite continued fraction* is a function

$$[a_1, a_2, \ldots] = a_1 + \cfrac{1}{a_2 + \cfrac{1}{a_3 + \cfrac{1}{\ddots}}}$$

of infinitely many positive integers $a_1, a_2, \ldots$. We use the notation $[\overline{a_1, \ldots, a_n}]$ for the periodic continued fraction $[\overline{a_1, \ldots, a_n}] = [a_1, \ldots, a_n, a_1, \ldots, a_n, \ldots]$.

Now let $\mathcal{G}$ be a snake graph with d tiles and sign function f. Let $e_1, \ldots, e_{d-1}$ be the sequence of interior edges, let e_0 be the south or west edge of the first tile and let e_d be the north or east edge of the last tile. The sequence of signs $(f(e_i))_{i=0}^{d}$ is called the sign sequence of $\mathcal{G}$. It determines the snake $\mathcal{G}$ completely.

Convention 2.1. In this paper, we shall always choose the edge e_0 to be the west edge of the first tile and the edge e_d to be such that the last two signs in the sign sequence are equal.

The sign sequence in the example in Figure 1 is $(+,-,-,+,+,+,+,-,-)$.

To the sign sequence of a snake graph, we also associate a sequence of positive integers $a_1,\dots,a_n$ that count the lengths of maximal constant subsequences in the sign sequence. In the example, this sequence $a_1,\dots,a_n$ is equal to $1,2,4,2$. The sequence $a_1,\dots,a_n$ uniquely determines the snake graph and we therefore denote the snake graph by $\mathcal{G}[a_1,\dots,a_n]$.

It is shown in [**CS1**, Theorems 3.4 and 4.1] that the construction above gives a bijection between snake graphs and continued fractions (whose last coefficient is greater than one), such that the continued fraction is computed by the following formula

$$[a_1,\dots,a_n] = \frac{|\operatorname{Match}(\mathcal{G}[a_1,\dots,a_n])|}{|\operatorname{Match}(\mathcal{G}[a_2,\dots,a_n])|}. \tag{2.1}$$

This is a combinatorial interpretation of continued fractions in terms of cardinalities of sets of perfect matchings.

The snake graph in Figure 1 has 29 perfect matchings, and we have $[a_1,\dots,a_n] = [1,2,4,2] = \frac{29}{20}$.

PROBLEM 2.2 (Solved). Find an analogue of (2.1) for band graphs.

This problem has been solved by PJ Apruzzese as follows. Given a band graph, associate a sign sequence to it considering only the interior edges, and denote the corresponding integer sequence by $a_0,a_1,\dots,a_n$ as usual. The corresponding periodic continued fraction $[\overline{a_0,a_1,\dots,a_n}] = [a_0,a_1,\dots,a_n,a_0,a_1,\dots,a_n,a_0,a_1,\dots]$ is equal to a quadratic irrational

$$[\overline{a_0,a_1,\dots,a_n}] = \frac{p-s+\sqrt{(p-s)^2+4rq}}{2q},$$

where $p/q = [a_0,a_1,\dots,a_n]$, and $r/s = [a_0,a_1,\dots,a_{n-1}]$. Let $D = (p-s)^2+4rq$ be the radical. Apruzzese obtains the following two formulas.

THEOREM 2.3 ([**Ap**]). *Let $\mathcal{G}[\overline{a_0,a_1,\dots,a_n}]$ be a band graph. Then*

$$\begin{aligned} |\operatorname{Match}\mathcal{G}[\overline{a_0,a_1,\dots,a_n}]| &= p+s \\ &= \sqrt{D+4}. \end{aligned}$$

REMARK 2.4. The sequence $a_0,a_1,\dots,a_n$ depends on the choice of the snake graph $\mathcal{G}$ that realizes the given band graph. Indeed, cutting the band graph an any interior edge will produce such a snake graph. However, the radical D and the sum $p+s$ doesn't depend on this choice.

EXAMPLE 2.5. Starting from the snake graph $\mathcal{G}[1,2,4,2]$ in Figure 1, we can form the two band graphs $\mathcal{G}[\overline{4,4}]$ and $\mathcal{G}[\overline{1,2,4,1}]$. They are obtained from the snake graph by identifying the south (respectively west) edge of the first tile with the north (respectively east) edge of the last tile.

For $\mathcal{G}[\overline{4,4}]$, we have $\frac{p}{q} = [4,4] = \frac{17}{4}$, $\frac{r}{s} = [4] = \frac{4}{1}$, thus $|\operatorname{Match}\mathcal{G}[\overline{4,4}]| = p+s = 18$. Using the radical formula instead, we have $D = (p-s)^2+4rq = 16^2 - 4^3 = 320$ and thus $\sqrt{D+4} = 18$, confirming the result.

For $\mathcal{G}[\overline{1,2,4,1}]$, we have $\frac{p}{q} = [1,2,4,1] = \frac{16}{11}$, $\frac{r}{s} = [1,2,4] = \frac{13}{9}$, and therefore $|\operatorname{Match}\mathcal{G}[\overline{1,2,4,1}]| = p+s = 25$. On the other hand, $D = (p-s)^2+4rq = 7^2-4\cdot13\cdot11 = 621$ and thus $\sqrt{D+4} = 25$, again confirming the result.

PROBLEM 2.6. Find an analogue of (2.1) for double dimer covers.

3. Markov numbers and the Lagrange spectrum

In 1879, Markov studied the equation

$$x^2 + y^2 + z^2 = 3xyz, \tag{3.1}$$

now known as the *Markov equation.* A positive integer solution (m_1, m_2, m_3) of (3.1) is called a *Markov triple* and the integers that appear in the Markov triples are called *Markov numbers.* For example $(1,1,1), (1,1,2), (1,2,5), (1,5,13), (2,5,29)$ are Markov triples and $1, 2, 5, 13, 29$ are Markov numbers.

In 1913, Frobenius conjectured that, for every Markov number m, there exists a unique Markov triple in which m is the largest number, see [**F**]. This uniqueness conjecture is still open today. It has inspired a considerable amount of research and the Markov numbers have important ramifications in number theory, hyperbolic geometry, combinatorics and algebraic geometry. For an overview, we refer to the recent textbooks [**Aig13**, **R**].

The Markov numbers can be represented in a binary tree called the *Markov tree.* This Markov tree is combinatorially equivalent to the Farey or Stern-Brocot tree of rational numbers. Thus there is a correspondence between $\mathbb{Q}_{[0,1)}$ and the Markov numbers by considering corresponding positions in these trees. We henceforth will write $m_0 = 1$ and refer to all other Markov numbers as $m_{\frac{b}{a}}$, where $b < a$ are relatively prime positive integers. For example, $m_{\frac{1}{1}} = 2, m_{\frac{1}{2}} = 5, m_{\frac{1}{3}} = 13, m_{\frac{2}{3}} = 29$.

The Markov numbers are related to the Lagrange numbers in approximation theory. Given a real number α, its *Lagrange number* $L(\alpha)$ is defined as the supremum of all real numbers L for which there exist infinitely many rational numbers $\frac{p}{q}$ such that $|\alpha - \frac{p}{q}| < \frac{1}{Lq^2}$. Thus the Lagrange number measures how well the real number α can be approximated by rational numbers.

The *Lagrange spectrum* is defined as the set of all Lagrange numbers $L(\alpha)$, where α ranges over all irrational real numbers. Considering it as a subset of the real line, it is known that the Lagrange spectrum is discrete below 3, it is fractal between 3 and a number $F \approx 4.5278$ called the *Freiman number*, and it is continuous above F, see Figure 2. We recommend [**CF**] as reference for the Lagrange spectrum.

Markov proved the following theorem in 1879.

THEOREM 3.1 ([**M**]). *The Lagrange spectrum below 3 is precisely the set of all* $\sqrt{9m^2 - 4}/m$, *where* m *ranges over all Markov numbers.*

Note that $m < m'$ implies $\sqrt{9m^2 - 4}/m < \sqrt{(9m')^2 - 4}/m'$. The first four Lagrange numbers are therefore $\sqrt{5}, \sqrt{8}, \frac{\sqrt{221}}{5} \approx 2.973, \frac{\sqrt{1517}}{13} \approx 2.996$.

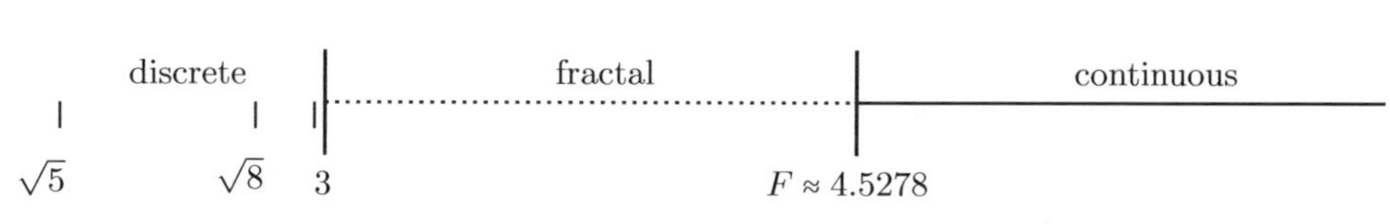

FIGURE 2. A schematic picture of the Lagrange spectrum

4. Lattice path snake graphs

Let a, b be relatively prime integers such that $0 < b < a$. We denote by $\mathcal{D}(a,b)$ the set of all lattice paths ω from $(0,0)$ to (a,b) that proceed by North (Up) and East (Right) steps and never go above the diagonal. Let $\mathcal{D} = \cup_{(a,b)}\mathcal{D}(a,b)$.

DEFINITION 4.1. For $\omega \in \mathcal{D}(a,b)$, we construct a snake graph $\mathcal{G}(\omega)$ and a band graph $\overline{\mathcal{G}(\omega)}$ as follows. An example is shown in Figure 3.

First place tiles of side length $1/2$ along ω. Then do one of the following.

- To construct $\mathcal{G}(\omega)$, remove the first and the last tile.
- To construct $\overline{\mathcal{G}(\omega)}$, add one tile on top of the last tile and glue the left vertical edge of the first tile to the right vertical edge of the last tile.

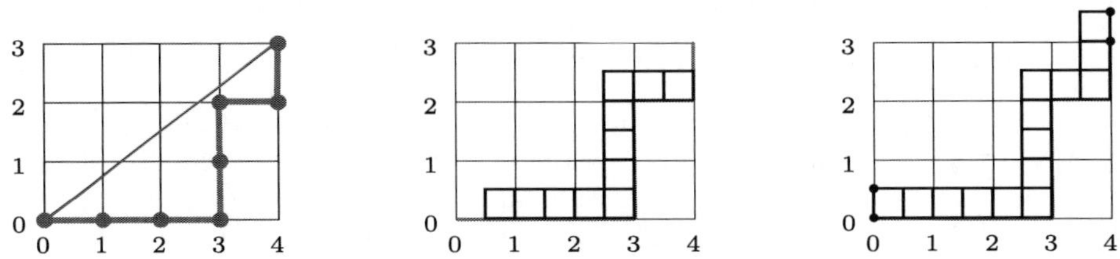

FIGURE 3. The lattice path ω = RRRUURU in $\mathcal{D}(4,3)$ is shown in the picture on the left. The center picture shows its snake graph $\mathcal{G}(\omega)$, and the picture on the right shows its band graph $\overline{\mathcal{G}(\omega)}$, where the gluing edge is indicated by bullet points.

In the example in Figure 3, we have ω = RRRUURU,

$$\mathcal{G}(\omega) = \mathcal{G}[1,1,1,1,2,1,1,2,2] \quad \text{and} \quad \overline{\mathcal{G}(\omega)} = \mathcal{G}\overline{[2,1,1,1,1,2,1,1,2,2]}.$$

For an arbitrary lattice path $\omega \in \mathcal{D}$, if $\mathcal{G}(\omega) = \mathcal{G}[a_1, a_2, \dots, a_n]$ then $\overline{\mathcal{G}(\omega)} = \mathcal{G}\overline{[2, a_1, a_2, \dots, a_n]}$.

Alternatively, we can construct the sequence of integers $a_1, \dots, a_n$ directly from the sequence of steps in ω by associating the subsequence $1,1$ to every subsequence RR and every subsequence DD in ω; and associating the subsequence 2 to every subsequence RU and every subsequence UR in ω.

REMARK 4.2. These graphs are significant in the cluster algebra of the once-punctured torus. Via the covering map from the lattice to the torus, a slight deformation of the path ω avoiding all intermediate lattice points corresponds to an arc γ on the torus that starts and ends at the puncture. This arc has selfcrossings unless it is the diagonal line from $(0,0)$ to (a,b) and ω is the path closest to the diagonal. The arcs without selfcrossings correspond to cluster variables and the selfcrossing arcs to so-called generalized cluster variables. In both cases their Laurent expansion is computed by the snake graph $\mathcal{G}(\omega)$. The band graph $\overline{\mathcal{G}(\omega)}$ computes the Laurent expansion a the closed curve $\overline{\gamma}$ obtained from γ by moving its endpoints infinitesimally away from the puncture.

4.1. Two strict partial orders on lattice paths. The following definition associates two numbers to every lattice path $\omega \in \mathcal{D}$: an integer $M(\omega)$ and a real number $L(\omega)$.

DEFINITION 4.3. Let $\omega \in \mathcal{D}$ be a lattice path and $\mathcal{G}(\omega) = \mathcal{G}[a_1, \dots, a_n]$ its snake graph. Define

$$M(\omega) = |\operatorname{Match}\mathcal{G}(\omega)| = \text{Numerator of } [a_1, \dots, a_n];$$

$$L(\omega) = \text{Lagrange number of } [\overline{2, a_1, \dots, a_n}].$$

REMARK 4.4. The number $L(\omega)$ can be computed by the formula $L(\omega) = \max(\alpha - \alpha')$, where α runs over all cyclic shifts of $[\overline{2, a_1, \dots, a_n}]$ and α' is the conjugate of α, see [**Aig13**, Proposition 1.29]. Recall that the conjugate of a quadratic irrational $\frac{a+\sqrt{D}}{b}$ is defined as $\frac{a-\sqrt{D}}{b}$.

This definition allows us to introduce two strict partial orders on lattice paths. Recall that a strict partial order is a relation $<$ that is irreflexive ($a \nless a$), asymmetric and transitive.

DEFINITION 4.5. Let $\omega, \omega' \in \mathcal{D}$ be two lattice paths.

(a) The *matching order* relation $<_M$ is defined by $\omega <_M \omega'$ if $M(\omega) < M(\omega')$.
(b) The *Lagrange order* relation $<_L$ is defined by $\omega <_L \omega'$ if $L(\omega) < L(\omega')$.

REMARK 4.6. The order relations are strict partial orders $<$ instead of non-strict partial orders $\leq$, because there can be two different lattice paths $\omega \neq \omega'$ such that $M(\omega) = M(\omega')$. Thus the non-strict partial order would not satisfy the antisymmetry condition.

REMARK 4.7. By definition, any two lattice paths ω, ω' are comparable under $<_M$, or $<_L$, unless $M(\omega) = M(\omega')$, or $L(\omega) = L(\omega')$, respectively.

PROBLEM 4.8. Study the posets $\mathcal{D}(a,b)$ and $\mathcal{D}$ with respect to $<_L$ and $<_M$.

5. Known results

The path in $\mathcal{D}(a,b)$ that is closest to the diagonal is called the *Christoffel path* of slope b/a. Here 'closest' means that there are no lattice points between the path and the diagonal.

THEOREM 5.1. *Let $\omega_0 \in \mathcal{D}(a,b)$ denote the Christoffel path. Then $M(\omega_0)$ is the Markov number $m_{\frac{b}{a}}$. Moreover ω_0 is the unique minimal element in the poset $(\mathcal{D}(a,b), <_M)$.*

PROOF. This follows from the work of several people. A correspondence between Markov numbers and the cluster algebra of a torus with one puncture was obtained in [**BBH, P**]. Therefore the equation $M(w_0) = m_{\frac{b}{a}}$ follows from the expansion formula for cluster variables in [**MSW**]. The description of the snake graph $M(\omega_0)$ we use here first appeared in [**CS2**]. The fact that w_0 is the unique minimal element follows from [**LLRS**, Theorem 3.5]. □

THEOREM 5.2. *Let $\omega \in \mathcal{D}$. Then $L(\omega) < 3$ if and only if ω is a Christoffel path. In that case, $L(\omega) = \sqrt{9m^2 - 4}/m$ as in Markov's theorem, where $m = M(\omega)$ is the Markov number of the Christoffel path. In particular, the Christoffel path is the unique minimal element in the poset $(\mathcal{D}(a,b), <_L)$.*

PROOF. This follows from a more general result by Markov which is restated (and proved) in terms of Christoffel words in [**BLRS**, Theorem 8.4]. The theorem states that for any bi-infinite sequence A of positive integers, a certain quantity $M(A) < 3$ if and only if A is the bi-infinite repetition of a Christoffel word ω'. Moreover, in that case, $M(A) = \sqrt{rc^2 - 4}/c$, where c is a Markov number associated to ω'.

Our setting is a special situation of this result, because we take the sequence A to be the bi-infinite sequence $\dots, 2, a_1, \dots, a_n, 2, a_1, \dots, a_n, \dots$, where the band graph $\overline{\mathcal{G}(\omega)}$ is $\mathcal{G}[\overline{2, a_1, \dots, a_n}]$. Then the quantity $M(A)$ becomes $\max(\alpha - \alpha')$, where α runs over all cyclic shifts of $[\overline{2, a_1, \dots, a_n}]$. Thus $M(A) = L(\omega)$ according to Remark 4.4.

We should point out that there is a slight difference in notation here. The ω' in [**BLRS**] is not the same as the ω in our setting. Instead, ω is the image of ω' under the morphism **G** in [**BLRS**, Lemma 2.2], which maps the Christoffel path in $\mathcal{D}(a, b)$ to the Christoffel path in $\mathcal{D}(a + b, b)$.

The statement that the Christoffel path is the unique minimal element in $(\mathcal{D}(a, b), <_L)$ follows because the Christoffel path is the only path $\omega \in \mathcal{D}(a, b)$ for which $L(\omega) < 3$. □

6. Open problems

In this section, we give a list of open problems and some of them are followed by comments. Throughout we denote by $\omega_0(a, b)$ the Christoffel path in $\mathcal{D}(a, b)$.

PROBLEM 6.1. Show that the unique maximal element in $\mathcal{D}(a, b)$ with respect to both partial orders is the $\lrcorner$-shape path $\omega = \mathrm{RR}\cdots\mathrm{RUU}\cdots\mathrm{U} = \mathrm{R}^a\mathrm{U}^b$.

This path ω first runs horizontally along the lower boundary of the (a, b)-rectangle and then runs vertically along the right boundary of the rectangle. This is the path furthest from the diagonal.

PROBLEM 6.2. Consider two lattice points $(a, b), (a', b')$ and their respective Christoffel paths $\omega_0(a, b)$ and $\omega_0(a', b')$. Is it true that

(a) $M(\omega_0(a, b)) = M(\omega_0(a', b'))$ if and only if $(a, b) = (a', b')$?
(b) $L(\omega_0(a, b)) = L(\omega_0(a', b'))$ if and only if $(a, b) = (a', b')$?

Part (a) is equivalent to the Uniqueness Conjecture, because of Theorem 5.1.

For part (b), let us point out that two irrational numbers α, β are said to be equivalent if their continued fractions eventually coincide. It is known that equivalent irrational numbers have the same Lagrange number. This is the case for the cyclic shifts of $[\overline{2, a_1, \dots, a_n}]$ that occur in the definition of the $L(w_0(a, b))$ when the point (a, b) is fixed. Note however, that in this problem we are dealing with two different points (a, b) and (a', b').

It is a conjecture that α and β are equivalent irrational numbers if and only if the Lagrange numbers $L(\alpha)$ and $L(\beta)$ are equal. This conjecture is known to be equivalent to the Uniqueness Conjecture, see for example [**Aig13**, page 39].

PROBLEM 6.3. If $\omega, \omega' \in \mathcal{D}(a, b)$ is it true that

$$L(\omega) = L(\omega') \quad \text{implies} \quad \overline{\mathcal{G}}(\omega) \cong \overline{\mathcal{G}}(\omega')?$$

Note that the previous Problem 6.2 is not a special case of this problem, since there we compared paths with different endpoints, whereas here the paths all have the same endpoint.

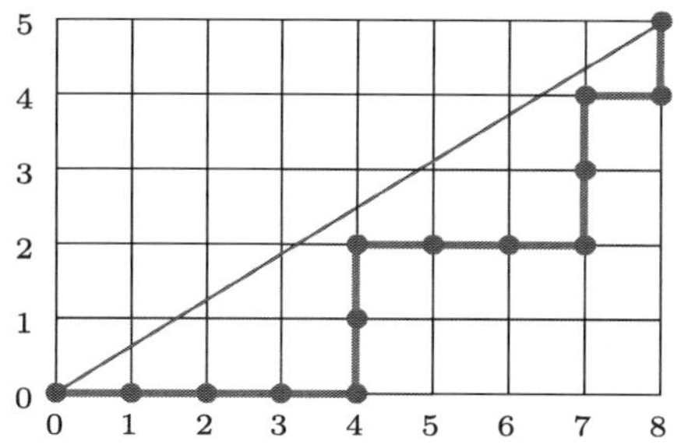

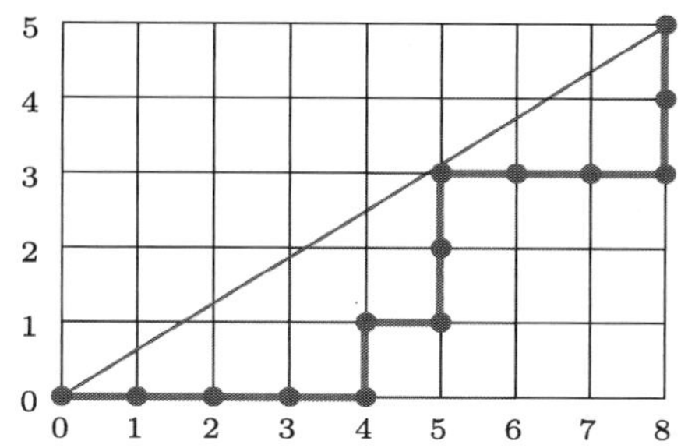

FIGURE 4. Two lattice paths ω_1, ω_2 in $\mathcal{D}(8,5)$ such that $L(\omega_1) = L(\omega_2)$. The snake graphs $\overline{\mathcal{G}(\omega_1)}$ and $\overline{\mathcal{G}(\omega_2)}$ are isomorphic.

In Figure 4, we give an example of two lattice paths ω_1 =RRRRUURRRUURU and ω_2 = RRRRURUURRRUU in $\mathcal{D}(8,6)$ such that $L(\omega_1) = L(\omega_2)$. In this case, the corresponding band graphs

$$\begin{aligned}\overline{\mathcal{G}(\omega_1)} &= \mathcal{G}[\overline{2,1,1,1,1,1,1,2,1,1,2,1,1,1,1,2,1,1,2,2}]\\ \overline{\mathcal{G}(\omega_2)} &= \mathcal{G}[\overline{2,1,1,1,1,1,1,2,2,2,1,1,2,1,1,1,1,2,1,1}]\end{aligned}$$

are isomorphic, which can be seen from the fact that one obtains the periodic integer sequence of the second graph by the one of the first graph by reversing the order and shifting. For completeness, we also note that $M(\omega_1) = 49396$ is not equal to $M(\omega_2) = 46900$.

On the other hand, let us also mention that the statement $(M(\omega) = M(\omega')) \Rightarrow (\mathcal{G}(\omega) \cong \mathcal{G}(\omega'))$ is false! For example the lattice paths ω = RRURURRURRUUU and ω' = RRRURURRUURUU shown in Figure 5 have the same value under the function M, namely 40199. The continued fractions are

$$[1,1,2,2,2,2,1,1,2,2,1,1,2,1,1,1,1] = \frac{40199}{23549}$$

and

$$[1,1,1,1,2,2,2,2,1,1,2,1,1,2,2,1,1] = \frac{40199}{24653}.$$

We further point out that, if we measure the distance from the diagonal by the number of lattice points between the lattice path and the diagonal, then the two paths do not have the same distance from the diagonal. It is 4 for ω and 6 for ω'. Moreover, the two paths have different Lagrange numbers

$$L(\omega) = \frac{\sqrt{16530502037}}{40189} \quad \text{and} \quad L(\omega') = \frac{\sqrt{16545934157}}{40547}.$$

PROBLEM 6.4. Let $\omega, \omega' \in \mathcal{D}$.

(a) Is it true that $\omega <_L \omega'$ implies $\omega <_M \omega'$ or $M(\omega) = M(\omega')$?
(b) Is it true that $\omega <_M \omega'$ implies $\omega <_L \omega'$ or $L(\omega) = L(\omega')$?

PROBLEM 6.5. Give a characterization of the cover relation in the poset $\mathcal{D}(a,b)$ for any of the two orders in terms of the lattice paths.

As a first step towards this problem, one can ask if ω is smaller than ω' for both orders if there exist lattice paths α, β such that $\omega = \alpha\text{RU}\beta$ and $\omega = \alpha\text{UR}\beta$.

PROBLEM 6.6. Prove that $\sup\{L(\omega) \mid \omega \in \mathcal{D}\} = 1 + \sqrt{5}$.

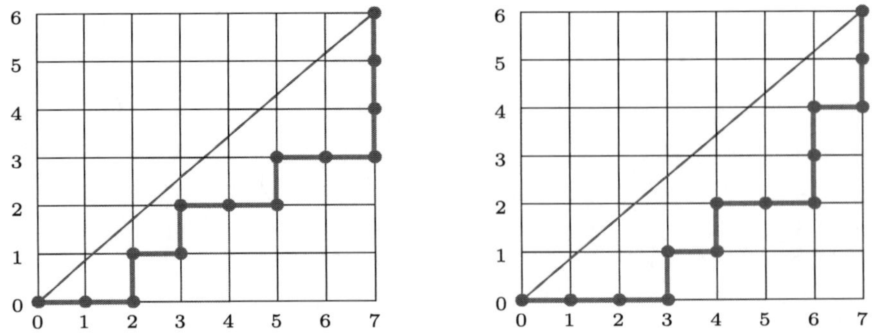

FIGURE 5. Two lattice paths ω, ω' in $\mathcal{D}(7,6)$ such that $M(\omega) = M(\omega')$.

There is some evidence for the above conjecture from computer calculations.

PROBLEM 6.7. By definition the set $\{L(\omega) \mid \omega \in \mathcal{D}\}$ is a subset of the Lagrange spectrum, and it contains the Lagrange spectrum below 3 by Theorem 5.2. What part of the Lagrange spectrum above 3 does it contain?

Note that upper bound conjectured in Problem 6.6 is equal to $1+\sqrt{5} \approx 3.23607$ which lies within the fractal part of the Lagrange spectrum. Assuming the conjecture is true, can every Lagrange number below $1+\sqrt{5}$ be realized as the limit of a sequence of Lagrange numbers $L(\omega)$ of lattice paths $\omega \in \mathcal{D}$?

As an example, we show that the number 3 can be realized in this way. Indeed, let α be given by the infinite continued fraction

$$\alpha = [(1,1)^1, 2, 2, (1,1)^2, 2, 2, (1,1)^3, 2, 2, (1,1)^4, 2, 2, \dots],$$

where the notation $(1,1)^a$ stands for the sequence $1, \dots, 1$ of $2a$ numbers 1. This continued fraction is associated to the infinite lattice path

$$\omega = \mathrm{R}^2\mathrm{U}\,\mathrm{R}^3\mathrm{U}\,\mathrm{R}^4\mathrm{U}\,\mathrm{R}^5\mathrm{U}\cdots.$$

The Lagrange number of α is equal to

$$L(\alpha) = \limsup_{n\to\infty} \left([a_n, a_{n+1}, \dots] + [0, a_{n-1}, a_{n-2}, \dots, a_1] \right).$$

To compute the supremum we may assume that n is such that $a_n = 2$. There are two cases. Consider first the case where a_n is the second number in a $2,2$ subsequence.

$$\left([2, (1,1)^b, 2, 2, (1,1)^{b+1}, \dots] + [0, 2, (1,1)^{b-1}, (1,1)^{b-2}, 2, 2, \dots, 2, 2, 1, 1] \right)$$

As b goes to infinity, the first continued fraction goes to $[2, \overline{1}] = 1 + [\overline{1}] = \frac{3+\sqrt{5}}{2}$, and the second continued fraction goes to $[0, 2, \overline{1}] = \frac{2}{3+\sqrt{5}}$. The sum of the two terms is $\frac{3+\sqrt{5}}{2} + \frac{2}{3+\sqrt{5}} = \frac{18+6\sqrt{5}}{6+2\sqrt{5}} = 3$. Now consider the case where a_n is the first number in a $2,2$ subsequence.

$$\left([2, 2, (1,1)^b, 2, 2, (1,1)^{b+1}, \dots] + [0, (1,1)^{b-1}, (1,1)^{b-2}, 2, 2, \dots, 2, 2, 1, 1] \right).$$

As b goes to infinity, the first continued fraction goes to $[2, 2, \overline{1}] = 2 + \frac{2}{3+\sqrt{5}} = \frac{8+2\sqrt{5}}{3+\sqrt{5}}$, and the second continued fraction goes to $[0, \overline{1}] = \frac{2}{1+\sqrt{5}}$. The sum of the two terms is $\frac{8+2\sqrt{5}}{3+\sqrt{5}} + \frac{2}{1+\sqrt{5}} = \frac{24+12\sqrt{5}}{8+4\sqrt{5}} = 3$. This shows $L(\alpha) = 3$.

PROBLEM 6.8. The set in Problem 6.7 is a union of sets $\cup_{(a,b)}\{L(\omega) \mid \omega \in \mathcal{D}(a,b)\}$ and each of these subsets contains exactly one number smaller than 3. Are these subsets pairwise disjoint? Do they provide an interesting additional structure for the Lagrange spectrum?

References

[Aig13] Martin Aigner, *Markov's theorem and 100 years of the uniqueness conjecture: A mathematical journey from irrational numbers to perfect matchings*, Springer, Cham, 2013, DOI 10.1007/978-3-319-00888-2. MR3098784

[Ap] P.J. Apruzzese, *Two formulas for the number of perfect matchings of band graphs*, in prepartion.

[BBH] Andre Beineke, Thomas Brüstle, and Lutz Hille, *Cluster-cyclic quivers with three vertices and the Markov equation*, Algebr. Represent. Theory **14** (2011), no. 1, 97–112, DOI 10.1007/s10468-009-9179-9. With an appendix by Otto Kerner. MR2763295

[BLRS] Jean Berstel, Aaron Lauve, Christophe Reutenauer, and Franco V. Saliola, *Combinatorics on words: Christoffel words and repetitions in words*, CRM Monograph Series, vol. 27, American Mathematical Society, Providence, RI, 2009, DOI 10.1090/crmm/027. MR2464862

[CS1] İlke Çanakçı and Ralf Schiffler, *Cluster algebras and continued fractions*, Compos. Math. **154** (2018), no. 3, 565–593, DOI 10.1112/S0010437X17007631. MR3778183

[CS2] İlke Çanakçı and Ralf Schiffler, *Snake graphs and continued fractions*, European J. Combin. **86** (2020), 103081, 19, DOI 10.1016/j.ejc.2020.103081. MR4058266

[CF] Thomas W. Cusick and Mary E. Flahive, *The Markoff and Lagrange spectra*, Mathematical Surveys and Monographs, vol. 30, American Mathematical Society, Providence, RI, 1989, DOI 10.1090/surv/030. MR1010419

[F] Ferdinand Georg Frobenius, *Gesammelte Abhandlungen. Bände I, II, III* (German), Springer-Verlag, Berlin-New York, 1968. Herausgegeben von J.-P. Serre. MR235974

[LLRS] Kyungyong Lee, Li Li, Michelle Rabideau, and Ralf Schiffler, *On the ordering of the Markov numbers*, Adv. in Appl. Math. **143** (2023), Paper No. 102453, 29, DOI 10.1016/j.aam.2022.102453. MR4505398

[M] A. Markoff, *Sur les formes quadratiques binaires indéfinies* (French), Math. Ann. **17** (1880), no. 3, 379–399, DOI 10.1007/BF01446234. MR1510073

[MOZ] Gregg Musiker, Nicholas Ovenhouse, and Sylvester W. Zhang, *Double dimer covers on snake graphs from super cluster expansions*, J. Algebra **608** (2022), 325–381, DOI 10.1016/j.jalgebra.2022.05.033. MR4444427

[MS] Gregg Musiker and Ralf Schiffler, *Cluster expansion formulas and perfect matchings*, J. Algebraic Combin. **32** (2010), no. 2, 187–209, DOI 10.1007/s10801-009-0210-3. MR2661414

[MSW] Gregg Musiker, Ralf Schiffler, and Lauren Williams, *Positivity for cluster algebras from surfaces*, Adv. Math. **227** (2011), no. 6, 2241–2308, DOI 10.1016/j.aim.2011.04.018. MR2807089

[MSW2] Gregg Musiker, Ralf Schiffler, and Lauren Williams, *Bases for cluster algebras from surfaces*, Compos. Math. **149** (2013), no. 2, 217–263, DOI 10.1112/S0010437X12000450. MR3020308

[P] James Propp, *The combinatorics of frieze patterns and Markoff numbers*, Integers **20** (2020), Paper No. A12, 38. MR4067101

[R] Christophe Reutenauer, *From Christoffel words to Markoff numbers*, Oxford University Press, Oxford, 2019. MR3887697

[RS] Michelle Rabideau and Ralf Schiffler, *Continued fractions and orderings on the Markov numbers*, Adv. Math. **370** (2020), 107231, 18, DOI 10.1016/j.aim.2020.107231. MR4103773

[S] Ralf Schiffler, *Cluster algebras from surfaces: lecture notes for the CIMPA School Mar del Plata, March 2016*, Homological methods, representation theory, and cluster algebras, CRM Short Courses, Springer, Cham, 2018, pp. 65–99. MR3823388

DEPARTMENT OF MATHEMATICS, UNIVERSITY OF CONNECTICUT, STORRS, CONNECTICUT 06269
Email address: `schiffler@math.uconn.edu`

SELECTED PUBLISHED TITLES IN THIS SERIES

110 **Christine Berkesch, Benjamin Brubaker, Gregg Musiker, Pavlo Pylyavskyy, and Victor Reiner, Editors,** Open Problems in Algebraic Combinatorics, 2024

109 **Paul M. N. Feehan, Lenhard L. Ng, and Peter S. Ozsváth, Editors,** Frontiers in Geometry and Topology, 2024

108 **Pramod N. Achar, Kailash C. Misra, and Daniel K. Nakano, Editors,** Categorical, Combinatorial and Geometric Representation Theory and Related Topics, 2024

107 **Ron Donagi, Adrian Langer, Piotr Sułkowski, and Katrin Wendland, Editors,** String-Math 2022, 2024

106 **Gilles Godefroy, Mohammad Sal Moslehian, and Juan Benigno Seoane-Sepúlveda, Editors,** Geometry of Banach Spaces and Related Fields, 2024

105 **A. Connes, C. Consani, B. I. Dundas, M. Khalkhali, and H. Moscovici, Editors,** Cyclic Cohomology at 40: Achievements and Future Prospects, 2023

104 **A. Kechris, N. Makarov, D. Ramakrishnan, and X. Zhu, Editors,** Nine Mathematical Challenges, 2021

103 **Sergey Novikov, Igor Krichever, Oleg Ogievetsky, and Senya Shlosman, Editors,** Integrability, Quantization, and Geometry, 2021

102 **David T. Gay and Weiwei Wu, Editors,** Breadth in Contemporary Topology, 2019

101 **Avraham Aizenbud, Dmitry Gourevitch, David Kazhdan, and Erez M. Lapid, Editors,** Representations of Reductive Groups, 2019

100 **Chiu-Chu Melissa Liu and Motohico Mulase, Editors,** Topological Recursion and its Influence in Analysis, Geometry, and Topology, 2018

99 **Vicente Muñoz, Ivan Smith, and Richard P. Thomas, Editors,** Modern Geometry, 2018

98 **Amir-Kian Kashani-Poor, Ruben Minasian, Nikita Nekrasov, and Boris Pioline, Editors,** String-Math 2016, 2018

97 **Tommaso de Fernex, Brendan Hassett, Mircea Mustaţă, Martin Olsson, Mihnea Popa, and Richard Thomas, Editors,** Algebraic Geometry: Salt Lake City 2015 (Parts 1 and 2), 2018

96 **Si Li, Bong H. Lian, Wei Song, and Shing-Tung Yau, Editors,** String-Math 2015, 2017

95 **Izzet Coskun, Tommaso de Fernex, and Angela Gibney, Editors,** Surveys on Recent Developments in Algebraic Geometry, 2017

94 **Mahir Bilen Can, Editor,** Algebraic Groups: Structure and Actions, 2017

93 **Vincent Bouchard, Charles Doran, Stefan Méndez-Diez, and Callum Quigley, Editors,** String-Math 2014, 2016

92 **Kailash C. Misra, Daniel K. Nakano, and Brian J. Parshall, Editors,** Lie Algebras, Lie Superalgebras, Vertex Algebras and Related Topics, 2016

91 **V. Sidoravicius and S. Smirnov, Editors,** Probability and Statistical Physics in St. Petersburg, 2016

90 **Ron Donagi, Sheldon Katz, Albrecht Klemm, and David R. Morrison, Editors,** String-Math 2012, 2015

89 **D. Dolgopyat, Y. Pesin, M. Pollicott, and L. Stoyanov, Editors,** Hyperbolic Dynamics, Fluctuations and Large Deviations, 2015

88 **Ron Donagi, Michael R. Douglas, Ljudmila Kamenova, and Martin Rocek, Editors,** String-Math 2013, 2014

87 **Helge Holden, Barry Simon, and Gerald Teschl, Editors,** Spectral Analysis, Differential Equations and Mathematical Physics: A Festschrift in Honor of Fritz Gesztesy's 60th Birthday, 2013

For a complete list of titles in this series, visit the
AMS Bookstore at **www.ams.org/bookstore/pspumseries/**.